CW01647148

Selections from *CRYPTOLOGIA*
History, People, and Technology

For a complete listing of the *Artech House Telecommunications Library*,
turn to the back of this book.

Selections from *CRYPTOLOGIA*
History, People, and Technology

Cipher A. Deavours
David Kahn
Louis Kruh
Greg Mellen
Brian J. Winkel
Editors

Artech House
Boston • London

Library of Congress Cataloging-in-Publication Data
Deavours, Cipher A.
Selections from Cryptologia : history, people, and technology / Cipher A. Deavours ...[et al.].
p. cm. — (Artech House telecommunications library)
"Third volume of selected papers from issues of Cryptologia"—Introd.
Includes bibliographical references and index.
ISBN 0-89006-862-3 (alk. paper)
1. Cryptography. 2. Ciphers. I. Deavours, Cipher A. II. Cryptologia.
III. Series.
Z103.S45 1998
952'.8—dc21 97-50519
CIP

British Library Cataloguing in Publication Data
Deavours, Cipher A.
Selections from Cryptologia : history, people, and technology
1. Cryptography 2. Cryptography - History
I. Deavours, Cipher A.
652.8'09

ISBN 0-89006-862-3

Cover design by Elaine Donnelly

© 1998 ARTECH HOUSE, INC.
685 Canton Street
Norwood, MA 02062

All rights reserved. Printed and bound in the United States of America. No part of this book may be reproduced or utilized in any form or by any means, electronic or mechanical, including photocopying, recording, or by any information storage and retrieval system, without permission in writing from the author

All terms mentioned in this book that are known to be trademarks or service marks have been appropriately capitalized. Artech House cannot attest to the accuracy of this information. Use of a term in this book should not be regarded as affecting the validity of any trademark or service mark.

International Standard Book Number: 0-89006-862-3
Library of Congress Catalog Card Number: 97-50519

10 9 8 7 6 5 4 3 2 1

Contents

Introduction to the History section of the third collection of *CRYPTOLOGIA* articles

Illumination from the Past
by David Kahn

This third volume of selected papers from issues of *CRYPTOLOGIA* again glows with the quality of its history. Some readers might have feared—as I did, both with the previous volume and with this one—that the quality would decline as the field was mined. In fact, the quality has grown. Mining has not depleted the history of cryptology; it has augmented it. It has raised new questions. It has repeatedly shown, in a variety of contexts, the importance of crytology in people's lives. It has demonstrated the richness of the sources available. The flow of papers to *CRYPTOLOGIA* continues to increase; the quality and variety from a growing number of authors are better than ever before.

Some of the superior examples of *CRYPTOLOGIA* articles are printed here. A few are works of the first order in the history of cryptology. For example, Ibrahim A. Al-Kindi's study of early Arab cryptology pushes the known origins of that field back some 300 years. Stephen Schlesinger's study of American solutions of foreign codes to help establish a United Nations the way the United States wanted it is one of the few illustrations of a major use of cryptanalysis in peacetime. David Alvarez's study of the cryptography of that almost legendary body, the *papal curia*, dispassionately analyzes the sad decline of an agency that once was supreme in its field and still is bathed in a nimbus of mystery messages.

Frederick Parker maintains that, had the Japanese messages intercepted before the Japanese attack on Pearl Harbor been solved at the time, instead of retrospectively after the War, as they were, the attack would have been foreseen and so perhaps prevented. I disagree with this, thinking that the conclusion is rooted in hindsight, but approved the article for publication because I thought the view was legitimate and worthy of dissemination. My own article holds that no cryptanalytic information pointed, or would have pointed, to an attack.

Also included is Susan Lujan's profile of a colorful and important, but too often forgotten, woman cryptanalyst. The two studies of Italian cryptology provide badly needed illumination of an area only sketched in elsewhere.

The other historical articles likewise add details to history or fill in chinks in the record. We therefore hope that the cryptologic community will concur with our confidence in the interest and worth of these articles.

THE CRYPTOGRAPHIC SERVICES OF THE ROYAL (BRITISH) AND ITALIAN NAVIES

A comparative analysis of their activities during World War II

Rear Admiral (disch.) Luigi Donini[1]
Translated by Augusto Buonafalce[2].

ADDRESS: (1) Via della Motonautica 9, 00194 Rome ITALY and (2) Via Rogoredo 113, 20138 Milano ITALY.

[This article is a translation of a paper which was published in the January 1983 issue of *Rivista Marittima*. *Rivista Marittima*, founded in 1868, is a monthly publication of the Italian Naval General Staff, Via Romeo Romei, n. 5 - 00136 Rome ITALY. The article is printed with permission of the author, the editors of *Rivista Marittima*, and the translator.]

1 INTRODUCTION

1.1

The interesting article by Captain (Engineering Duty Corps) Gino Jori appearing in the May 1982 issue of *Rivista Marittima* under the title *La crittologia nelle operazioni navali in Mediterraneo (1940-43)*, offers a clear and sufficiently enlightened picture of the influence that the secret services - especially the cryptological one - had on the naval operations in the Mediterranean in the Second World War. This article, however, conveys something less than the title seems to promise, because it deals extensively with the influence of these services in the direction Royal Navy - Italian Navy, but very little in the opposite direction. The author must not be blamed for this. In the first instance he had at his disposal the abundant documentation furnished with some reticence and adjustments, involuntary or deliberate, by the first two volumes of the official work *British Intelligence in the Second World War*[1]. In the second instance he had

[1] Hereinafter quoted as B.I.

nothing else, official or unofficial, than the precious material contained in the 21 volumes of the Ufficio Storico della Marina *La Marina italiana nella Seconda Guerra Mondiale.* This work says little or nothing of our secret services and particularly nothing about the cryptographic service in our Navy and even less about its influence on the conduct of the naval warfare. As of now an official, or at least unofficial, documentation is missing, and it is not foreseen that this gap may be filled in a totally credible manner, as all papers relating to our cryptological work performed during the war period, and even before, were destroyed by superior orders in the days immediately following the armistice.

Jori himself was conscious of this gap and has expressed in his article the wish that it may be filled by others more directly cognizant of this particular matter. This is what I am going to do.

The aforesaid abnormal situation, while it prevents at the present time a documented, equitable summing-up of the contribution furnished to the relevant war operations by the two rival cryptographic services, has caused even an authoritative historian to make some statements that should have been verified. For instance, dealing with the war in the Mediterranean, it was written that "our formidable decrypting ability" did not cause any damage to our adversaries because "almost always it was our ships that were involved, our men who died."

I regard this judgment not much considered and rather unfair because the British or Allied ships, whether military or merchant, sunk or damaged in the Mediterranean were more than one. In addition to that, it should be pointed out that the British naval ciphers [Transl. note: Includes *codes* everywhere in this article.] broken by our cryptographic service were in general use, i.e. not only in the Mediterranean, but in all the naval warfare theaters, and that the effects of those breaks weighed heavily on our opponents, especially in the Atlantic, both Northern and Southern, and in the Arctic Seas.

1.2

Although there are not sufficient elements to draw a final balance, I believe it conceivable to make a comparative analysis of the activities of both services regardless of whether their higher commands have utilized them or not in strategic and tactical fields. In other words I will make a purely technical comparison of the structure and quality of the cryptosystems, of the degree of difficulty of the relevant cryptographic work, of the capability/incapability shown of reaching a success or suffering a defeat. This analysis is in order because several authors show a tendency to minimize or criticize what was done by us in this field and they exalt what the British did. For example, some gave a general judgment on our enciphering systems defining them, *little sophisticated.* Our opponent has been defined as *skillful*, while it has been conceitedly written about our cryptanalysts that "after all they had obtained appreciable results." The fact that we had been unable to prevent the British from breaking our mechanical enciphering is described as "bordering guilt" while our opponent is not to be criticized for our cracking of their principal naval ciphers.

An analysis of this kind will help clarify the cause of the British success in breaking our mechanical encipherments, and must take into consideration both offensive cryptography or decryption, i.e., attacking the enemy signals, and defensive cryptography, i.e. the protection of one's encipherments.

It seems right to me to limit my analysis to the systems of general usage for the wireless messages of high degree of secrecy, i.e. the superenciphered naval ciphers (British and Italian) and the mechanical encipherments (Italian). About the latter I think I can disregard, in this article, the Enigma machine used by the SIS (Navy Intelligence). One reason is because it was not of general naval use, having been used by Supermarina (Navy Hqtrs.) in rare occasions (B.I. on page 210, vol 1 says: "only one or two signals per day" and only during the first 9 months of war.) Another reason is because its history has been absorbed by the history of the Enigma machine, used by the German armed forces, which was broken by the British in 1940.

1.3

In the following chart, whose banality will be forgiven by my readers, the signs + or - indicate respectively a success or failure in attacking enemy systems in case of offensive cryptography and failure in cracking (+) or actual cracking (-) of one's systems by the opponent in case of defensive cryptography.[2]

		Italy	Great Britain
Attack	Naval ciphers	+	–
	Cipher machine		+
Defense	Naval ciphers	+	–
	Cipher machine	–	

As the reader can see, while forgiving such oversimplification, we have two

[2]The Royal Navy, in contrast with the other Armed Forces, did not use mechanical encipherment during the conflict with Italy. They implemented it only in November 1943.

plus signs and a minus sign for Italy versus two minus signs and only one plus sign for Great Britain. Let's examine in detail what the systems were like and what happened to them in that secret war of means and brains.

2 THE SUPERENCIPHERED NAVAL CIPHERS

2.1

As is well known to the majority of readers, this cipher system is divided into two successive operations. The first operation consists of substituting the words or phrases of the *plaintext* with secret literal or numerical groups (4-digit groups for both Navies in question) found in the cipher book or code book proper. For instance (words and 4-digit groups imaginary):

Plaintext	Hush	most	secret	stop	Cincmed	from	Adm London	stop ...
First encipherment	1782	8130	6615	8469	0983	3561	1565	8469

The second operation, called second encipherment or superencipherment, consists of subtracting one digit at a time, without borrowing, i.e. modulo 10, the first - encipherment groups from the groups of the so called *superencipherment keystream* as taken from suitable tables. For instance:

Keystream	(V)	2916	3410	9226	0880	4534	8117	4036	1912	...
-1st enc.	(C)	-1782	8130	6615	8469	0983	3561	1565	8469	...
Superenciph.	(S)	1234	5380	3611	2421	4651	5656	3571	3553	

That is: $S = V - C$ to perform superencipherment and $C = V - S$ to eliminate superencipherment.

It is obvious that no cryptanalyst, even if in possession of the enemy cipher, will ever be able to decrypt a message superenciphered with a random or incoherent keystream, not known to him and used only for that message.

However, if in the mass of material under examination there are pairs, triples etc. of messages superenciphered with the same keystream, then a small gap is open for penetration on the basis of a very simple algebraic consideration.

In fact the difference between two second encipherment groups S_2 and S_1, superenciphered with the same keystream group V, is equivalent to the difference between the two first encipherment groups (C_1 and C_2) which have generated the relevant superenciphered groups:

$$S_1 = V - C_1$$
$$S_2 = V - C_2$$
$$S_1 - S_2 = C_2 - C_1$$

It follows that only the pairs, triples etc. of messages enciphered with the same cipher and superenciphered with the same keystream are the basis with which to start the penetration. To do this the cryptanalyst must:

- identify, if any, all pairs, triples etc. of messages superenciphered with the same keystream
- execute, for each identified pair, the difference between the corresponding groups, i.e. between the groups occupying the same position in both messages
- keep track of the differences which have appeared with frequency noticeably greater than that expected according to the laws of probability
- for each difference ($S_1 - S_2 = C_2 - C_1$) accepted as meaningful, assign arbitrary numerical values (C_2 and C_1) to the groups of code which have generated it. He should then try to put together such arbitrary values so as to reduce them all to the same relative basis. These *arbitrary* values should then differ from the corresponding *true* code groups (still unknown) by a common *constant* which in time may perhaps be identified, although this is not necessary.

The above work will have to be continued until a sufficient quantity of material is collected so as to assign meaning to some of the arbitrary code groups.

The reconstruction of the used code or cipher proper can then be started and developed together with the relevant superencipherment tables.

2.2

It is not necessary to show here how hard and laborious is the work I have briefly described and which the cryptographic service of our Navy managed to complete for the different British ciphers and codes which had been utilized from 1938 through 1942.

Similarly, "British attempts against our naval ciphers of the war period met with complete failure, honestly admitted by our adversaries and repeatedly regretted also in the two volumes of B.I.", as has been pointed out by Jori in his above mentioned article. For instance, at page 210 of B.I. vol. 1 one can read: "Italy's main naval book-cyphers, which were the cyphers used by her fleet for most of its communications, were never read again after July 1940, except for a few brief intervals as a result of captures after the middle of 1941." We do not know how much of this failure was due to their lesser ability or how much was due to the possibly smaller intensity of our radio traffic or how much was due to a greater efficiency of the expedients and countermeasures which we introduced in our *cipher/superencipherment* system in mid-July 1940. It was because we could use the knowledge of the faults found in the corresponding British system and the tricks we had learned to use in order to break it.

2.3

The weaknesses of the British system were varied and of differing weight. I will limit myself to citing a few:

a) *total* reuse of keystream, very frequently up to 3/4 times for the same keystream; *partial* re-use very frequent;

b) period of validity (even though for a few days) of the superencipherment tables was badly commensurated with the intensity of radio traffic;

c) imperfect control of the wearing out of the keystream tables. When rescission of overused keystreams was ordered by enciphered message, the damage had already occurred;

d) within the tables containing the superencipherment keystream whole lines existed here and there already existing in rescinded tables (perhaps for quicker compilation);

e) the *indicator groups* placed at the beginning of the cryptograms to indicate to the recipient which keystream was used were utilized for many years without *masking*, i.e. not disguised. The possible reuse of a keystream was then directly revealed by the identical indicators;

f) when, from December 1940 through the better part of 1941, the British introduced the *masking* of the indicators, this masking was so banal and moreover so badly used by the encipherers that it only caused us a two week crisis;

g) the ciphers did not contain *homophones*, i.e. did not assign two or more different secret groups to encipher a certain word of very frequent usage. For instance *fullstop* could be enciphered only in one manner (in our ciphers: a score of different ways), similarly for *from* etc.;

h) geographical names were coupled, as an alternate interpretation, to vocabulary entries beginning with the same two or three letters. e.g.: 7184 = Give/Gibraltar, 0921 = Last/La Spezia, 4650 = Make/Malta, 2935 = All Concerned/Alexandria, 7714 = Left/Leghorn. This was of great help in identifying the geographical term when we had already identified the vocabulary term coupled thereto and vice versa;

i) the signal almost invariably began with the complete address: (to)...from;

j) the cipher books were made of two sections. The first section, which was mostly used to encipher the address, contained the names of the Commands, of the Authorities, of the Naval Units, and, in addition to that, numbers, dates, ranks and minutes of latitude and longitude; the second section was the vocabulary proper. In both sections the secret enciphering groups (4 figures) were chosen at random from the natural series from 0001 to 9999, and almost all the 4 digit groups were found in both the first and second section.

If the address used up 1,2,3, ..., 9 groups, the first group of the message in first encipherment was 1111, 2222, 3333, ..., 9999 respectively to indicate to the addressee that for 1,2,3,..., 9 groups he was to use the first section of the cipher book. If the address used more than 9 groups, the first group was 9900 and at the end of the address the group 0099 was placed to indicate that the second section was to be used. 0099 was also placed as first group when the second section and not the first section was to be used from the start.

These peculiarities obviously had a mnemonic purpose, and were kept in use for several years. That meant that the differences pertaining to the first group of message pairs having identical indicator group, i.e. superenciphered with the same keystream[3], only had the following values 0000, 1111, 2222, ..., 9999, 1122, 2233, ..., 8899, 2211, 3322, ..., 9988, 1199, 9911. From the observation of this peculiarity to the formulation of the logical explanation was a short step for us and that marked the opening of the first crack for penetration.

Among other information, we were immediately able to know how many groups were used for the addresses, where the secret group for *from* (which had no homophones, see Subparagraph g) was located and where the actual text was beginning, with its stereotyped initial phrases (reference ..., answer ..., continuation, ..., etc.).

[3]See Paragraph 2.1. It should be remembered that the difference between two second encipherment groups superenciphered with the same keystream is equal to the difference between the relevant groups in first encipherment.

In my opinion, however, the most serious British carelessness or inadvertence was to keep in use for the whole first year of war with Germany the same principal naval cipher which was in force since the time of the Spanish civil war, broken by us (and by the Germans) in 1938.

When I had to leave the command of a submarine unit to assume the direction of the SIS Cryptographic Branch (where I had worked with few interruptions since 1934) I was surprised - but also elated - to notice that the main target was still the same I was confronted with from 1938 to the beginning of 1940.

This was the reason why our cryptographic section had decrypted, on July 4, 1940, the operation order of the Commander of the Mediterranean Fleet, which caused the passage of arms of Punta Stilo on the 9th of that month. This was the first occurrence in history of a clash between the two fleets; an encounter which was to be inconclusive[4].

Unfortunately only two weeks after my return to the cryptographic service, the British discontinued that cipher, substituting it with two different ciphers, and they introduced some devices which did not prevent us from re-assuming control of the situation for both ciphers within a few months.

To sum up the brief technical analysis I performed in reference to the *cipher warfare*, I believe I can state that our Navy's cryptographic service surpassed the homologous British service greatly.

3 MECHANICAL ENCIPHERMENT OF THE ITALIAN NAVY

I shall use the same designation C38m used by B.I. for our cipher machine, which was a modified version of almost no cryptographic importance of a Swedish machine commercially available.

3.1

The following describes the adoption of the C38m according to my personal recollection. It is well noted that a previous version of that machine, offered to our Navy by its inventor and producer in 1934-35, was refused by SIS after an accurate test performed by a board of 4 cryptographers, the present writer included. In 1935 that same machine, obviously considered efficient by the French specialists, was acquired in large quantities by their armed forces.

A few years later, perhaps in 1938, the Swedish maker presented to our General Staff a new model he had somewhat improved, increasing the number of

[4] As Jori mentioned in his article, it appears that the British had obtained information, albeit incomplete, on the intention of our fleet, by intercepting our plain language signals and also by decrypting some radio messages which were still enciphered with our pre-war cipher. This had already been broken by the British. It was substituted only in mid-July 1940, probably owing to organizational difficulties.

rotors from 5 to 6; but the original principle of the enciphering mechanism was still that of 1935.

I do not know whether the SIS was again consulted as I was absent and on sea duty. The fact is that later on, I believe during 1940, that machine was adopted, but limited to the secret communication of a few High Commands, on shore and on board. In the meantime it had also been acquired by the US Army, which utilized it as a medium level system from Division Command down to Battalion. I can only guess that the improvements introduced by the makers and the knowledge of the bulk purchases made by France and USA may have persuaded our technical administrative bodies to disregard the unfavorable opinion expressed by the SIS some years earlier. Moreover B.I. states in a note on page 22 of vol 2 that C38m was forced on us by the Germans, who considered our prewar system insecure, and perhaps with good reason (see Footnote 4). If this is true then one could assess that the Germans, involuntarily caused us a two-fold damage: with the Enigma and with the C38m. It is easy to be wise after the event and condemn the decision without an appeal. Some may even find it convenient. I maintain that considering the situation of cryptology at that time and the circumstances connected with it, this choice is more than justified[5]. Another discussion will have to be made hereafter on the manner in which the C38m was utilized, especially during the crucial period of our transport activities to North Africa.

3.2. Working on C38m - Encipherment and decipherment.

3.2.1

The C38m produced with its mechanism a polyalphabetic substitution encipherment, by using 26 inverse normal cipher alphabets obtained by shifting one place at a time the normal English alphabet in reverse order. These alphabets are partially shown in Table 1, which can be used to manually perform this type of polyalphabetic encipherment. Each letter of the plain text is substituted with the letter corresponding to it in the secret alphabet used in turn according to a succession prescribed by the *key* or enciphering *keystream.*

For instance, utilizing Table 1.

[5] It should not be forgotten that the C38 was used by the U.S. Army under the designation M-209 throughout the war.

		Cipher Alphabets													
plaintext		0	1	2	3	...	12	13	14	15	...	23	24	25	plaintext
0	a	A	B	C	D	...	M	N	O	P	...	X	Y	Z	a
1	b	Z	A	B	C	...	L	M	N	O	...	W	X	Y	b
2	c	Y	Z	A	B	...	K	L	M	N	...	V	W	X	c
3	d	X	Y	Z	A	...	J	K	L	M	...	U	V	W	d
4	e	W	X	Y	Z	...	I	J	K	L	...	T	U	V	e
5	f	V	W	X	Y	...	H	I	J	K	...	S	T	U	f
6	g	U	V	W	X	...	G	H	I	J	...	R	S	T	g
7	h	T	U	V	W	...	F	G	H	I	...	Q	R	S	h
8	i	S	T	U	V	...	E	F	G	H	...	P	Q	R	i
9	j	R	S	T	U	...	D	E	F	G	...	O	P	Q	j
10	k	Q	R	S	T	...	C	D	E	F	...	N	O	P	k
11	l	P	Q	R	S	...	B	C	D	E	...	M	N	O	l
12	m	O	P	Q	R	...	A	B	C	D	...	L	M	N	m
13	n	N	O	P	Q	...	Z	A	B	C	...	K	L	M	n
14	o	M	N	O	P	...	Y	Z	A	B	...	J	K	L	o
15	p	L	M	N	O	...	X	Y	Z	A	...	I	J	K	p
16	q	K	L	M	N	...	W	X	Y	Z	...	H	I	J	q
17	r	J	K	L	M	...	V	W	X	Y	...	G	H	I	r
18	s	I	J	K	L	...	U	V	W	X	...	F	G	H	s
19	t	H	I	J	K	...	T	U	V	W	...	E	F	G	t
20	u	G	H	I	J	...	S	T	U	V	...	D	E	F	u
21	v	F	G	H	I	...	R	S	T	U	...	C	D	E	v
22	w	E	F	G	H	...	Q	R	S	T	...	B	C	D	w
23	x	D	E	F	G	...	P	Q	R	S	...	A	B	C	x
24	y	C	D	E	F	...	O	P	Q	R	...	Z	A	B	y
25	z	B	C	D	E	...	N	O	P	Q	...	Y	Z	A	z
plaintext		0	1	2	3	...	12	13	14	15	...	23	24	25	plaintext
		Cipher Alphabets													

Table 1. Table for the manual polyalphabetic encipherment with normal reverse alphabets.

Keystream	23	14	00	12	25	02	13	01	15	12	00
Plaintext	r	i	f	e	r	i	m	e	n	t	o
Ciphertext	G	G	V	I	I	U	B	X	C	T	M

Note: Keystream refers to the number of secret alphabets used.

The same table can be used for decipherment:

Keystream	23	14	00	12	25	02	13	01	15	12	00
Ciphertext	G	G	V	I	I	U	B	X	C	T	M
Plaintext	r	i	f	e	r	i	m	e	n	t	o

3.2.2

The enciphering mechanism was made of:

a) a certain number of metallic elements both in the rotary and fixed parts of the machine: pins, guide-arms, lugs; each one of them could be set by the operator into an active or inactive position, thus forming the so called internal key. This was, in reality, made up of two components called W and X. The number of different internal keys that the operator could set was astronomical, nearly 10^{60} (a 60 digit number) but the setting was laborious and required much attention and time. The result was that the same internal key was kept in use for a long time.

b) 6 rotors having like diameters, mounted side by side on a horizontal axle, each one rotatable, manually or not, by means of gears having a different number of teeth for each rotor, precisely 26, 25, 23, 21, 19, 17.

On the outer rim of each rotor were carved, in normal order, the letters of the English alphabet, respectively A to Z, A to Z excluding W, A to X excluding W, A to U, A to S and A to Q.

Along the edge of the right face of each rotor were housed respectively 26, 25, 23, 21, 19, 17 pins corresponding to as many letters carved around the rim. As already mentioned each pin could be set by the operator into an active or inactive position.

The number and position of the active pins made up the X component of the internal key, as I said before. The number of different X components that the operator could form was extremely great, 2^{131} (a 40-digit number), even though many components were not advisable because they were too simple.

Note for Table 2: The external key EATJHP corresponds to position 30,000,000 in the keystream because dividing 30,000,000 by 26, 25, 23, 21, 19, and 17, yields respectively the following remainders:

$4(=\mathrm{E})$, $0(=\mathrm{A})$, $19(=\mathrm{T})$, $9(=\mathrm{J})$, $7(=\mathrm{H})$, and $15(=\mathrm{P})$.

	Rotor						Position within Keystream	Keystream (# of cipher Alphabet)		
	1	2	3	4	5	6				
# of teeth	26	25	23	21	19	17				
	A	A	A	A	A	A	0	8		
	B	B	B	B	B	B	1	11		
	⋮	⋮	⋮	⋮	⋮	⋮	⋮	⋮		
	Q	Q	Q	Q	Q	Q	16	17		
	R	R	R	R	R	A	17	2		
	S	S	S	S	S	B	18	11		
	T	T	T	T	A	C	19	12		
	U	U	U	U	B	D	20	1		
	V	V	V	A	C	E	21	18		
	W	X	X	B	D	F	22	2		
	X	Y	A	C	E	G	23	12		
	Y	Z	B	D	F	H	24	8		
	Z	A	C	E	G	I	25	13		
	A	B	D	F	H	J	26	5		
	⋮	⋮	⋮	⋮	⋮	⋮	⋮	⋮	See Table 1	
External Key	E	A	T	J	H	P	30 000 000	14	Plain text	Cipher text
	F	B	U	K	I	Q	30 000 001	23	r	G
	G	C	V	L	J	A	30 000 002	14	i	G
	H	D	X	M	K	B	30 000 003	0	f	V
	I	E	A	N	L	C	30 000 004	12	e	I
	J	F	B	O	M	D	30 000 005	25	r	I
	K	G	C	P	N	E	30 000 006	2	i	U
	L	H	D	Q	O	F	30 000 007	13	m	B
	M	I	E	R	P	G	30 000 008	1	e	X
	N	J	F	S	Q	H	30 000 009	15	n	C
	O	K	G	T	R	I	30 000 010	12	t	T
	P	L	H	U	S	J	30 000 011	0	o	M
	Q	M	I	A	A	K	30 000 012	3	v	I
	R	N	J	B	B	L	30 000 013	14	o	A
	S	O	K	C	C	M	30 000 014	24	s	G
	⋮	⋮	⋮	⋮	⋮	⋮	⋮	⋮		
	Y	Y	V	T	R	P	101 405 848	13		
	Z	Z	X	U	S	Q	101 405 849	15		
	A	A	A	A	A	A	0	8		
	B	B	B	B	B	B	1	11		

Table 2. Generation scheme of a complete keystream relevant to a given internal key. Encipherment of a message with a given external key.

	Rotors	(1)	(2)	(3)	(4)	(5)	(6)	(0)	
W component of internal key		w_1	w_2	w_3	w_4	w_5	w_6	w_0	Keystream (# of secret alphabet)
		5	1	8	2	6	3	2	
position	30 000 000	E	A	T	J	H	P		
		+5		+8	+2	6		+2	= 23
position	30 000 001	F	B	U	K	I	Q		
		·	+1	·	+2	+6	+3	+2	= 14
position	30 000 002								
		+5	·	+8	+2	+6	+3	+2	$= 26 \equiv 0$
position	30 000 003								
		·	·	+8	+2	·	·	+2	= 12
position	30 000 004								
		+5	+1	+8	·	+6	+3	+2	= 25
position	30 000 005								
		·	·	·	·	·	·	+2	= 2
position	30 000 006								
		+5	·	·	·	+6	·	+2	= 13
position	30 000 007								
		+5	+1	+8	+2	+6	+3	+2	$= 27 \equiv 1$
position	30 000 008								
		+5	·	+8	·	·	·	+2	= 15
position	30 000 009								
		·	+1	·	·	+6	+3	+2	= 12
position	30 000 010								
		+5	·	+8	+2	+6	+3	+2	$= 26 \equiv 0$
position	30 000 011								
		·	+1	·	·	·	·	+2	= 3
position	30 000 012								
		·	+1	+8	·	·	+3	+2	= 14
position	30 000 013	R	N	J	B	B	L		
		+5	+1	+8	+2	+6		+2	= 24
position	30 000 014								
		↑	↑	↑	↑	↑	↑	↑	Keystream

X component of internal key

(See Table 1 for secret substitution alphabets.)

+ pin in active position
· pin in inactive position

Table 3. Generation of encipherment keystream for message in Table 2.

3.2.3

Manually rotating each of the 6 rotors, their respective letters could be aligned along a reference line made up of 6 windows on the machine cover. This alignment executed by the operator was the so-called external key, which was to be formed before the beginning of the encipherment of the message. During the letter by letter enciphering operation the 6 rotors advanced simultaneously by one step at each stroke (see Table 2). Thus to encipher each struck letter one of the 26 secret substitution alphabets of the above cited Table 1 was used, according to a succession called *worm* (keystream), which was only apparently chaotic. It actually depended on the formation law generated by the two components W and X of the internal key as set by the operator.

To better demonstrate the effect of the combined W and X components of the internal key, it will be useful to make an example, even though somewhat simplified (see Table 3).

We have already described the X component; the 131 (26+25+23+21+19+17) pins partly extended and partly retracted. The W component[6] was in reality the sum of 7 components, one for each rotor ($w_1 w_2 w_3 w_4 w_5 w_6$) and one (w_0) independent of the rotors. When the operator had to set the internal key, he could assign to w_0 one of the values 1 or 2 and to each of the other w's the values from 1 to 9[7]. The sum however must always be 27. Let's consider the alignment which, in Table 2, corresponds to position 30,000,000 of the keystream generated by the internal key as set by the operator whose W component is shown below (see Table 3):

rotors	1	2	3	4	5	6	0	
W component	5	+ 1	+ 8	+ 2	+ 6	+ 3	+ 2	= 27
alignment (external key)	E	A	T	J	H	P		

According to the X component of the preset internal key, the 6 pins housed in the 6 rotors below the 6 letters of the given alignment (and the subsequent alignments caused by the enciphering operation) could be active or inactive.

The active pins interacted with their respective w while the inactive ones left their respective w ineffective.

[6] For brevity we will omit the description of the quite complicated mechanical structure of the W component. We will only say that it was a cylinder with a number of beveled lugs and guide arms which could be positioned in active or inactive position by the operator.

[7] What we have called for simplicity *values 1 or 2 and values from 1 to 9* were in reality the number of guide arms set in active position by the operator (see Footnote 6 above).

The w_0 was independent from the rotors, then always active. At each simultaneous advancement of the 6 rotors the internal mechanisms of the machine brought into play the secret substitution alphabet having the serial number (see Table 1) equal to the sum, modulo 26, of the active w's. This alphabet enciphered the struck letter. This stroke also caused the advancement of the rotors.

Now we can see, from Table 3, how that portion of the keystream which enciphered the text shown in Table 2 has been generated. As to the external key, we have already seen that it consisted in the alignment made by the operator before beginning the encipherment, during which the 6 rotors advanced simultaneously one step at a time. Since a complete turn of each of the 6 rotors involved a different number of steps: 26, 25, 23, 21, 19, 17, a given alignment set at the beginning presented itself identically only after $26 \times 25 \times 23 \times 21 \times 19 \times 17 = 101,405,850$ simultaneous steps of the 6 rotors (see Table 2). This was the total length of the keystream as generated by the internal key. Obviously each possible internal key generated a total keystream of identical length but having a different structure (i.e. succession of secret alphabets).

3.3. Decryption of the C38m

3.3.1

Let us resume the encipherment example shown in Subparagraph 3.2.1 and in Table 2:

Keystream	(V)	23	14	00	12	25	02	13	01	15	12	00	03	14	24
Plaintext	(C)	r	i	f	e	r	i	m	e	n	t	o	v	o	s
Ciphertext	(S)	G	G	V	I	I	U	B	X	C	T	M	I	A	G

Let us substitute for each letter the number indicating the position it occupies in the normal alphabet ($A = 0, Z = 25$ see Table 1) and let's transcribe everything (see Table 4, Section I). We can observe that each secret letter can be obtained by subtracting algebraically, modulo 26, the plaintext letter from the relevant keystream element ($V - C = S$):

$$23 - 17 = 6, 14 - 8 = 6, 0 - 5 = -5 + 26 = 21; \ \ldots; \ 2 - 8 = -6 + 26 = 20; \text{ etc.}$$

Now the enciphering process is identical with the superencipherment operation in the case of numerical ciphers described in Subparagraph 2.1, with the enormous advantage for the codebreaker that in this case there is no secret code. One may say that the code is so well known to him that he knows it by heart (if he masters the language).

Section I	V)	23	14	0	12	25	2	13
	C)	- [r] 17	[i] 8	[f] 5	[e] 4	[r] 17	[i] 8	[m] 12
	S)	[G] 6	[G] 6	[V] 21	[I] 8	[I] 8	[U] 20	[B] 1
Section II								
External key settings page/col/row								
S_1	18/3/20	G	G	V	I	I	U	B
S_2	18/3/20	I	X	M	U	V	A	T
	S_1	6	6	21	8	8	20	1
	$-S_2$	− 8	23	12	20	21	0	19
$S_1 - S_2 = C_2 - C_1$		24	9	9	14	13	20	8
$C_2 - C_1$		24	9	9	14	13	20	8
Hypothesis + C_1		+ [r] 17	[i] 8	[f] 5	[e] 4	[r] 17	[i] 8	[m] 12
		= 41	17	14	18	30	28	20
		-26				-26	-26	
Confirmation = C_2		= 15	17	14	18	4	2	20
		p	r	o	s	e	c	u
Section III								
	S_1	[I] 8	[X] 23	[M] 12	[U] 20	[V] 21	[A] 0	[T] 19
	$+C_2$	+ [p] 15	[r] 17	[o] 14	[s] 18	[e] 4	[c] 2	[u] 20
		= 23	40	26	38	25	2	39
			-26	-26	-26			-26
	V)	23	14	0	12	25	2	13

Table 4. Example of decryption of two cryptograms with same keystreams and recovery of relevant portion of keystream by the probable word method.

3.3.2

This being the situation, the decryption of two messages enciphered with the same keystream is no longer a serious problem. In fact, if one can assume the existence in one of the texts, of *probable* words or phrases (this often happens especially in the initial and final portions of the messages), one can directly verify the validity of the hypothesis on the other text (see Table 4, Section II). Since $S_1 - S_2 = C_2 - C_1$ then $S_1 - S_2 + C_1 = C_2$ and $S_2 - S_1 + C_2 = C_1$ as was said in Subparagraph 2.1. Usually a few trials are sufficient to start a decryption which can be continued step by step without great difficulty, thus furnishing at one time the elements of the keystream portion which has enciphered the two messages: $V - S_2 = C_2$ then $S_2 + C_2 = V$ (see Table 4, Section III).

3.3.3

After completing the decryption of the two messages (even though incomplete) the cryptanalyst has at his disposal a good portion of the keystream generated by the internal key set by the operator. At this point no assumptions would be possible as to the remaining keystream portions, in case of numerical ciphers superenciphered with random or incoherent keystreams (see Paragraph 2). In the case of C38m a skilled cryptanalyst was able, if he had a deep knowledge of the machine, to recover the whole keystream (101,405,850 elements) after reconstructing the W and the X components of the internal keys (see Table 3). He could do this by writing and solving equations with 1 to 6 unknowns or empirically by successive trials. Usually it is sufficient to have about 100 keystream elements to reconstruct the internal key. In luckier instances about 40 elements can suffice.

It should be noted that the keystream so obtained is identical with the keystream generated by the encipherer's machine but it is *shifted* with respect to it because the element which has arbitrarily been assigned the 0 position (external key AAAAAA) by the solver actually occupies an unknown position between 0 and 101,405,849 in the original keystream. The shifts will be initially different for different internal keys but it will be possible to gradually bring them to a common arbitrary shift.

3.3.4

Once in possession of the internal key and consequently of an entire keystream, the British cryptanalysts would have been able to decrypt all the messages enciphered with that internal key and with any external key without being disturbed by the above said shift. Of course, owing to the length of the keystream, the number of necessary trials was so great that it is impossible for me to hazard a statement as to the possible efficiency of this work, which depended on the quantity of people employed and, above all, on the substantial aid of special *ad-hoc* equipment which the joint British-American technology was in a position to produce in 1941[8]. Such decryptions were in general timely only on occasion and certainly the most productive work in terms of quantity and timeliness was that of *attacking* each message directly by assuming the existence of probable words or phrases. These, if confirmed, would furnish a portion of keystream which was not difficult to position within the total keystream, i.e. identify the

[8]B.I. repeatedly mentions the use of such equipment (the electro-mechanical *Bombes*; the *Colossus*, an electronic computer) in the penetration work of enciphered Enigma radio traffic. We also had developed a simple machine to accelerate the work against the British naval superenciphered codes in 1942. Using punched cards it would perform the many trials necessary to detect which portions of a long keystream, already recovered, had been used to superencipher other messages still to be decrypted.

(fictitious or arbitrary) external key set by the enemy operator and then decrypt the whole message. Table 5 shows an example of determination of the external key of a message (S), with known internal key X and W, starting from a portion of keystream (V) obtained by assumption on the beginning of plaintext (C):

(S)	G	H	R	V	O	L	J	Q	W	N	X	A REUHLDREYUY
	06	07	17	21	14	11	09	16	22	13	23	
+ (C)	s	u	p	e	r	m	a	r	i	n	a	
	18	20	15	04	17	12	00	17	08	13	00	
	24	27	32	25	31	23	09	33	30	26	23	
		-26	-26		-26			-26	-26	-26		
= (V)	24	01	06	25	05	23	09	07	04	00	23	

It may seem that such a method would depend on chance only, but when cryptanalysts have a deep experience and knowledge of the message subject matter, the encipherers' habits, the way the texts usually start and end, the stereotyped phrases etc., then the above method may become the normal routine method and yield timely solutions too.

3.3.5

Two or more messages enciphered with an identical keystream can be immediately identified among the intercepted traffic if the *indicator groups* of the external key used are not *masked*, as in the example of Table 4; otherwise like portions of ciphertext in different messages must be found at the beginning and at the end of the secret message.

3.3.6

If no cases exist of keystream *reuse*, (this is very possible and actually probable when the adversary has issued careful directions for the use of the machine), one can *attack* the messages one at a time by the *probable word or phrase* method until one has some portions of keystream sufficiently reliable and long enough to permit reconstruction of the internal keys and proceed as in Subparagraph 3.3.3. and 3.3.4.

3.3.7

To sum up, it is clearly shown from the above that the C38m enciphering machine lent itself to incomparably greater chances of decryption than the naval superenciphered code described in Paragraph 2.

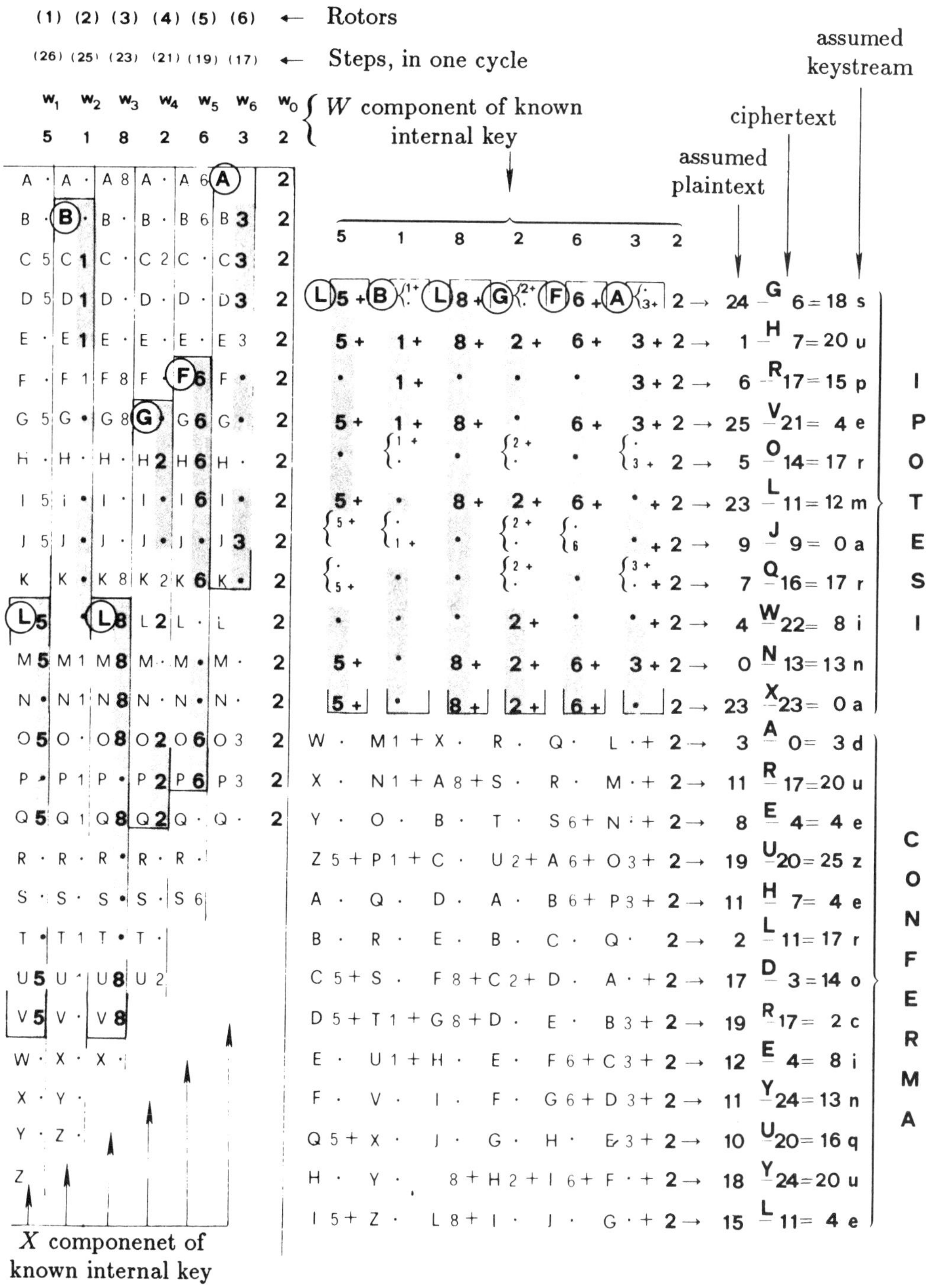

Table 5. Recovery of External Key (LBGFA pos. 30 082, 401) and consequent solution of a message by the "probable word" method.

3.4. Usage of the C38m by the Italian Navy during the war.

3.4.1

I will state in advance that the contents of this paragraph are based both on what the British say in B.I. and on what I could come up with by consulting a few documents of that time still available in the archives of the Naval Staff.

It should be kept in mind that in contrast with what has been repeatedly said and written, the choice, the production, the regulation, the distribution of the cryptosystems and the control over their utilization were not made by the SIS Cryptographic Branch. These were performed by the Telecommunications Branch (TLC). This was, in my opinion, a considerable fault of our secret communications organization. The Cryptographic Branch only furnished consulting when so required. I know with certainty that this was done and was decisive for the wartime enciphering-superenciphering system's security. I do not know, however, whether that was true for the C38m or to what extent it was done. I believe that direct or indirect co-responsibility for our Cryptographic Branch cannot be excluded.

3.4.2

The regulations of the time did not make a neat distinction between the material that could be enciphered with the superenciphered naval ciphers and that which could be enciphered with the C38m. The two systems were practically considered alternates and the guiding criterion for the commanders probably was, or in the long run became, that of greater practicality and speed, clearly favorable for the C38m. One can therefore think that in the crucial circumstances of more intense operations or radio traffic, those who had both systems, often preferred, perhaps to an excessive extent, the mechanical encipherment to the manual encipherment with cipher book and superencipherment tables, which was safer but more laborious and slow.

3.4.3

According to the *distribution lists* contained in the documents still available, the authorities who were issued the C38m were very few[9]: only High Commands on shore and on board (see Table 6 and Subparagraph 3.4.5), and no single surface or submarine unit.

I cannot exclude, however, that in some particular cases some single unit may

[9]They were few in comparison with the number of users of the general superenciphered naval cipher, which had been issued to almost every naval unit and minor Navy Command.

have been temporarily given the machine for use with assigned keys.

3.4.4

The astronomical number of possible *internal keys*, around 10^{60}, theoretically offered very good security, which could not be practically utilized because the setting of the internal keys was delicate, laborious and long. For the X component alone, one had to manipulate 131 pins. This could be not be done as frequently as desired; therefore it was decided to use internal keys common to all users and change them each month. This rule was in force at least throughout 1942, but probably until the armistice. It was certainly acceptable during the first war period, with C38m radio traffic not intense (B.I. says: "during the month of August 1941, 600 signals")[10]. In my opinion it was unsafe to keep it when the traffic was gradually growing (B.I. says: 4,000 messages during the most intense month, July 1942). In that situation the preference was to increase the number of *external keys* available, switching from 18,000 to about 40,000.

3.4.5

It was understood that the greatest danger was in the *reuse* of the keystream, that is two or more messages enciphered with the same external key, within the same month. For this reason this practice was positively forbidden and all authorities issued a C38m were assigned series of keys not to be used by other authorities. Of course I cannot exclude that violation of the rules may have existed but this must have been the exception, considering the large number of keys available.

In Table 6 a tabulation is given of the external keys assigned in the second and third edition of the so-called *Table of External Keys for Cipher Machine.*

I found no trace of the first edition. The second edition, enforced on March 1, 1942, contained 15,000 *normal* keys plus another 3,000 for the R.P. messages (Restricted-Personal). The third edition introduced in May or June of the same year contained the same number of *normal* keys and presumably another 3,000 for the R.P. On October 1, 1942 new editions were implemented with a total of 40,000 keys counting both *normal* and R.P. These quantities were, in my opinion, judiciously commensurate with the foreseeable radio traffic intensity for the C38m in a month, i.e. within the life span of the internal key.

3.4.6

All right so far; but let us see how the external keys were chosen, then we shall see how the utilized key was indicated. As we have seen in the preceding

[10] B.I. vol 2 page 283.

paragraph, the number of external keys theoretically possible were 101,405,850, but it was deemed necessary to avoid the possibility that the initial part of a message should be casually enciphered with the portion of keystream which had or would later encipher the *tail* of another message.

2nd edition in force March 1942		R.P.	3rd edition (April 1942) enforced in May/June 1942	
Supermarina e M.M.	3 500	300	Supermarina	1 200
Reparto Comunicazioni		200	Reparto Comunicazioni	1 800
Com. FF.NN. e Com 1 Sq.	500	100	Com. FF.NN. e Com 1 Sq.	300
Com. 2 Sq. e Com 5 Div	300	100	Com. 2 Sq. e Com 5 Div	200
Com. 3 Div.	200	100	Com. 3 Div.	100
Com. 7 Div.	200	100	Com. 7 Div.	100
Com. 8 Div.	200	100	Com. 8 Div.	100
Com. 9 Div.	200	100	Com. 9 Div.	100
Marina Napoli	200	100	Marina Napoli	200
" Taranto	200	100	" Taranto	200
" Messina	200	100	" Messina	200
" Brindisi	200	100	" Brindisi	00
" Albania	500	100	" Albania	500
" Morea	1 500	100	" Morea	1 200
" Sud est	800	100	" Sud est	800
" Egeo	800	100	" Egeo	1 200
" Lero	300	100	" Lero	600
" Libia	2 000	100	" Libia	1 500
" Tripoli	1 000	100	" Tripoli	800
" Bengasi	1 000	100	" Bengasi	800
" Tobruk	—	100		
Reserve	1 700	600	Reserve	2 900
Total	15 000	3 000	Total	15 000

Table 6. Quantities of External Keys Assigned to the Different Commands.

This was meant to avoid providing our opponent with a reusage of keystream, even partial. For that purpose, since a maximum length of 500 letters was prescribed for each message, the external keys, within the total keystream generated by the internal key, had to be distanced by a number of elements greater than 500. An average distance of about 1,000 elements would have been abundantly sufficient while not reducing the quantity of available keys. This would have reduced the number of possible keys to a still reasonable amount, i.e. about 100,000 (101,405,850/1,000).

A greater interval was chosen instead and this decreased the available external keys to about 40,000. Obviously, it was thought with good reason, this number was more than sufficient to cover the requirements of one month. The above reduction would have brought no appreciable advantage to the British cryptanalysts. After recovering, by lucky decipherments, the month's internal key, they would still have to *try* all the 100 million possible external keys to find the external key of any other message during that month. I have used the conditional because unfortunately a *fixed number of elements* (2497) was prescribed by our planning. This being the situation, our adversary, after identifying some external keys (arbitrary) by means of some decryption of intercepted messages within that month, could observe that they were all at the same distance of a multiple of 2497[11] in the total keystream. At this point he was able to identify, by means of a simple calculation, all the 40,611 (arbitrary) external keys prescribed by our plans, thus reducing the *trials* necessary to decrypt every other message in that month[12] from about one hundred million to only about 40,000 in the worst case[13]. I cannot believe that the British overlooked this unsecure feature of our planning; they must have noticed it fairly early. They certainly made great use of it to decrypt those messages believed to be important which had resisted the direct attack by the probable word or phrase (see Subparagraph 3.3.5) method.

3.4.7

Let us look at the *indicator groups*, or those agreed-upon groups that the encipherer placed as first and last groups of the cryptogram to communicate to the receiver which external key had been set on the machine before starting the encipherment of that message.

The books containing the *Tables of External Keys* had a certain number of pages, subdivided among the different users, on the basis of the quantity of keys assigned to each user. (See Table 6.)

Each page contained a number of keys chosen at random, not repeated, among the 40,611 of the general plan. They were listed in rows and columns. The encipherer would each time select within the relevant pages the key he was to use and checked it *off* in pencil to avoid reutilization before checking off all the keys at his disposal. After finishing the encipherment he would place an *indicator*

---

[11] Table 7 shows some *external keys* chosen at random from our *Tables of External Keys* in use in 1941/2 with their positions in the total keystream, the partial distance and their prime factors.

[12] Something similar happened during our work against the British superenciphered naval ciphers. We found out that the keystream extracted from like blocks of superencipherment tables were always at a distance of a multiple of 20 elements (5 tetragraphs). Thus we could reduce the necessary trials by 80%.

[13] Actually the cryptanalyst did not know the position in the original keystream where our breakdown of 40,611 keys with the interval of 2,497 was started. Since 101,405,850 is not a multiple of 2,497, the necessary trials were in the worst case 80,000 to start with. When the number of identified keys increased, the trials decreased to a maximum of 40,000 in the worst case.

group as first and last group of the cryptogram. By an appropriate index (not as explicitly as we have shown in Table 4, Section II for simplicity), he would furnish the receiver page, column and row numbers pertaining to the key.

Obviously with this procedure only those who had the above mentioned secret books were in a position to interpret the indicator groups, i.e. know the external key used in the messages[14]. With a certain imprudence, up to the fall of 1942 the indicator groups were used without *masking* in such a way that like indicator groups meant like external keys to the adversaries. This imprudence was similar to that committed by the British in superenciphering their naval ciphers (see Paragraph 2.3.e) which proved much more dangerous. It was probably caused by the confidence that the rules positively prohibiting reuse of the keys were strictly adhered to. Anyhow, let us see which consequences could come from *not masking* the indicator groups, to the advantage of the British cryptanalysts. We will list the different situations:

a) indicator groups used only once. No usefulness to the decryptors;

b) indicator groups used twice during a given period of validity of the external key (i.e. within the month). This was the typical case of two cryptograms enciphered with the same keystream: very serious consequences (see Subparagraph 3.3.2 through 3.3.5 and Table 4). However, I repeat, such a circumstance should not occur and if it rarely happened, this does not justify, in my opinion, the systematic and timely penetration alleged by B.I.;

c) indicator groups used twice during two different months, i.e. with different internal keys, i.e. not with the same keystream.

   This latter event, also forbidden by the regulations, could present itself in some months' time when the keys assigned to a specific user were exhausted (however reserve keys were provided, see Table 6).

   This fact could be to some advantage, if the two messages were decrypted, to match the total keystream relevant to the two recovered internal keys (see Subparagraph 3.3.3) and then utilize, at best, other possible reemployment of the indicator groups between the two relevant months.

With the new issue of the *Tables of External Keys* adopted on October 1, 1942 a good procedure for masking the indicator groups was finally implemented. This

[14]The above statement about the structure and function of the indicator groups has been derived by the examination of second and subsequent editions of *Tables of External Keys* in force after March 1, 1942. I cannot say anything about the structure and function of the indicators used before that date because I found no trace of the first edition of these *Tables* in the Navy Staff Archives.

series page/row	plan's serial # at step 2947	external key (26)	(25)	(23)	(21)	(19)	(17)	position in the total theoretical keystream	partial distances	prime factors
		(A	A	A	A	A	A)	(0)		
		0	0	0	0	0	0			
	(0)	(E	A	T	J	H	P)	(30 000 000)		
		4	0	19	9	7	15		23 906 278	2 · 11 · 227 · 4787
A 143/4	9574	K	D	F	N	K	J	53 906 278		
		10	3	5	13	10	9		15 988 291	11 · 19 · 227 · 337
E 7/1	15977	R	T	H	R	K	E	69 894 569		
		17	19	7	17	10	3		684 · 178	2 · 11 · 137 · 227
D 7/2	16251	F	X	E	P	R	A	70 578 747		
		5	22	4	15	17	0		2 · 651 · 778	2 · 3 · 11 · 179 · 227
C 5/4	17325	N	A	F	J	C	L	73 260 525		
		12	0	5	9	2	11		2 · 499 · 497	7 · 11 · 13 · 227
B 1/1	18326	A	X	A	C	L	P	75 760 022		
		0	22	0	2	11	15		32 · 161 360	2^4 · 5 · 7 · 11 · 23 · 227
B 21/2	31206	K	H	A	J	O	K	6 515 532		
		10	7	0	9	14	10	101 405 850 (1)		
								107 921' 382	1 922 690	2 · 5 · 7 · 11^2 · 227
B 14/2	31976	A	X	F	C	S	A	8 438 222		
		0	22	5	2	18	0		1 712 942	2 · 3 · 7^3 · 11 · 227
B 39/3	32662	K	O	X	Q	P	F	10 151 164		
		10	14	22	16	15	5		17 701 233	3 · 11 · 17 · 139 · 227
A 63/2	39751	B	X	S	N	M	F	27 852 397		
		1	22	18	13	12	5		2 147 420	2^2 · 5 · 11 · 43 · 227
	(40611)	(D	R	U	P	O	C)	(29 999 817) (2)		
		3	17	20	15	14	2	+101 405 860 (1)		
								131 405 667		11 · 227 = 2497.

(1) 101 405 850 = 26 · 25 · 23 · 21 · 19 · 17 = total length of keystream

(2) 29 999 817 divided by → 26, 25, 23, 21, 19, 17
gives as remainder → 3, 17, 20, 15, 14, 2
D R U P O C

Table 7. Determining the planning scheme for the external keys.

makes one think that during the summer of 1942, a period of maximum C38m radio traffic intensity, some violations of the rules may actually have been made and discovered by the control service performed by the Communication Branch of the Naval Staff. One can infer that serious breaches may not have previously happened since the *non-masking* of the indicator groups was maintained in the editions enforced in March and in May/June 1942.

3.4.8. Plaintext language of the secret messages.

In my opinion this was another weak point of our C38m radio traffic. It may have progressively helped - after an initially laborious and lucky penetration - the British cryptanalysts when they acquired experience and became familiar with the subject matters and the fashion in which they were discussed in the messages.

Stereotyped beginnings and endings, habits of the encipherers and other details may have allowed the decryptors to guess probable words and phrases (as mentioned in Subparagraph 3.3.2) with which to attack single messages, without waiting for a keystream to be reused, a very rare occurrence. During the period of maximum radio traffic intensity this may have been the more generalized system to widely penetrate our C38m communications.

3.4.9. Mistakes made by the encipherers.

In 1978, the noted British writer, Ronald Lewin, wrote in his work *Ultra Goes to War*: "Italian cipher-clerks were too often haphazard and negligent, whilst the cryptanalysts of Vincent's section – as elsewhere at BP – developed a rigorous and effective professionalism. And so, for example, it became almost customary to pass on advance information to British headquarters in the Middle East about the make-up, content, port of departure and port of arrival for the Axis convoys that transported supplies and reinforcements to Rommel's army in Africa."[15].

Nothing is surprising in Lewin's statement about pretended carelessness of our encipherers; this has always happened, in peace time and even more so in war time, to us and to others. I must stress that in the case of C38m, such negligence as the reuse of keystreams or other typical errors, which I will not enumerate, help the cryptanalysts in an often decisive way. This help cannot be compared with similar circumstances that may occur with more complex systems such as,

[15]The fact that the British had largely penetrated our secret communications relevant to the ship and convoy traffic to North Africa was given out by some daily press as an unheard of discovery or revelation in 1981. It had been known for at least three years.

for instance, the superenciphered ciphers.

3.5. British penetration of the Italian Navy's C38m.

3.5.1

The C38m, like other Hagelin machines which had preceded it in the market for two decades utilizing for the encipherment the same original concept devised by the inventor and constructors, was not unknown to the British. They did not make the confusion found in a book published last year where the Italian author has written that the Hagelin machine was a German ciphering machine firm which had made a variant of the Enigma machine for the Germans. This is an error: the firm was Swedish with a plant in Stockholm and its proprietor, a businessman, also Swedish, was not even pro-German. Just after the Nazi occupation of Norway he had adventurously crossed Germany and Italy with the dismantled parts of the last version of this machine (almost identical with the C38m) hidden in his luggage. From Genoa he reached the United States which he had visited a few years before and he immediately placed about 50 units with the United States armed forces. Back in 1935, as mentioned in Paragraph 3.1., he had supplied the French armed forces. The machine must have been well known to the British who keenly studied it from the cryptographic security angle. It is not unlikely that the diffidence which was shown up to November 1943 - as stated in B.I. - by the Royal Navy, in contrast with the other armed forces, toward the ciphering machines in general may have originated or been strengthened precisely after the study of the Hagelin (and Enigma) machines. One may suppose, with grounds, that owing to the alliance with France and the tight bonds with the Americans, some specimens very similar to C38m may have been in their hands at the beginning of the conflict.

Nothwithstanding this markedly advantageous situation, the British, after observing the use of the C38m in the secret radio communications of the Italian Navy, did not obtain any results for a good six months, from December 1940 to June 1941. I have heard someone's opinion that the Royal Navy had not concerned itself with this problem during those months.

I cannot wrong our adversaries of that time. In fact - according to B.I.[16] - "after the end of July 1940 the black-out of SIGINT from Italian naval communications was broken only by fitful gleams from the systems of the lower grades and by the first success of GC and CS (Government Code and Cypher School) against an Italian naval machine cypher ENIGMA, broken in September 1940. Unfortunately, it carried only one or two messages a day up to the summer of

[16]B.I. vol 1, page 210.

1941, when it was withdrawn from naval use."

It is unlikely that in the darkness only broken by those few Enigma messages, and with the persistent failure of any attempts to penetrate our superenciphered naval ciphers, the British Intelligence service should have neglected for so many months trying to break those mechanical ciphers of ours that must have been regarded by their specialists as vulnerable, thanks to their knowledge, albeit theoretical, of the Hagelin machine.

3.5.2

Let's see what the British say. In vol. 2 of B.I. page 22, one can read: "The Italian C38m, which had first been intercepted in December 1940, was broken by GC and CS since the summer of 1941," and on pages 283-284: "By the end of June 1941 GC and CS had broken the C38m settings for May and June and was reading a few messages currently. The first intelligence from the source for the Middle East was sent out on June 23 .... On July 10 GC and CS broke the July settings. Thereafter, with routine and progressively shorter interruptions at the end of each month, when the Italians introduced new settings, it was going on reading the traffic with little or no delay until the Italian armistice. By the end of 1941 the monthly interruption had been reduced to less than three days; by the end of 1942 it was less than 24 hours."

Thus between *December 1940* and *June 1941*;

a) there were no cases of keystream reutilization, which is reasonable owing to the positive prohibition by our rules, especially in relation to the still sluggish C38m cipher traffic; or

b) if some reutilization occurred, the British did not manage to avail themselves of it, which is not likely;

c) having failed to decrypt our messages, the British cryptanalysts acquired neither such experience as to apply the *probable word or phrase* method (see Subparagraphs 3.3.4 and 3.3.6) nor the knowledge of the weak point in our external keys planning (see Subparagraph 3.4.3).

Nothwithstanding all this, at the end of June 1941 the British *cracked* the C38m so as to recover the internal keys for the previous month, reconstruct the internal keys for July in ten days and from that moment onwards began a systematic and timely decryption "...to give from July 1942 the operational authorities advance notice of virtually every convoy and independent ship that sailed with troops or supplies across the Mediterranean, and usually to identify the ships and escorting forces involved."[17]

[17]B.I. vol 2, page 284.

What sudden flash had enlightened the minds of the British cryptanalysts toward the end of spring 1941? Frankly such a rapid and substantial jump in quality and quantity puzzles me. Even more so if I consider what had happened at the beginning of March 1942, i.e. the serious interruption of the decryption exactly when the situation for the British cryptanalysts was supposed to be better with respect to June 1941, owing to the much greater intensity of our C38m traffic (B.I. reads: "600 messages in August 1941, 4000 in July 1942") and the experience acquired in the preceding 10 months of intensive decryptions. In fact B.I. Vol 2, page 348 reads: "At the beginning of March 1942 the Italians introduced a new indicator for the C38m signals and GC and CS temporarily read only a small portion of the traffic and with considerable delay. At the end of March GC and CS regained its ability to read C38m currently." What the British thought was an *introduction of new indicators*, was the implementation of our second edition of the *Tables of External Keys* (see Subparagraph 3.4.5.) where a good number of keys were also contained in the first edition, since they were all taken from the series of 40,611 keys envisaged by our planning.

If the appearance of new indicators had caused such a serious interruption in the decryptions, one must conclude that previously the indicators (those of the first edition) had been precious for the penetration work. We have seen (Subparagraph 3.4.7 c) that the indicators could be of some limited use for the cryptanalysts only if they had previously appeared in decrypted messages. This was not the case in March 1942 because the indicators were new. One can understand the crisis which followed to some extent. It was not the case in June/July 1941 either, because there were no messages previously decrypted. Still the British penetration of C38m was begun and suddenly developed with great success at that time. These considerations make one think that upon the discontinuance of the first edition of the *Tables of External Keys* on March 1, 1942, the British must have missed *some detail* that had to do with the indicators. This detail had permitted a particularly profitable type of work which had facilitated the initial cracking and the subsequent systematic and timely utilization of it from 1941 through February 1942.

The sudden lack of this detail must have forced the British to devise some different working method, which required almost a month's running in. This method must have been independent from the consideration of the indicators. This seems to be proved by the fact that the next change of the indicators accompanying the third edition of our *Tables of External Keys*, implemented in May or June 1942, does not seem to have caused any interruption in the decipherments. B.I. only mentions one or two serious interruptions, that of March 1942 and that considerably less serious of October 1942, when we introduced the masking of the indicators with our fourth edition (see last section of Paragraph 3.4.7). I am not in a position to say with certainty what this *detail* connected with the indicators of the first edition of our *Tables of External Keys* consisted of. Perhaps before March 1942, with the first edition of said Tables, the indicators, instead of furnishing the key *indirectly*, through an index as described in Subparagraph 3.4.7., furnished it *explicitly*, or with some simple disguise that the British ended up discovering after some months' attempts.

As I have already said, I have found no trace of the first edition of the *Tables of External Keys* in the Archives of the Navy Staff, therefore I cannot verify this hypothesis.

CONCLUSION

From what I have explained, it is confirmed, according to my conviction, that during WW II the Italian Navy's Cryptographic service had nothing to envy the British counterpart from the strictly technical viewpoint, and allowing for the different degrees of difficulty and vulnerability of the opposed systems. The contrary is true. Had I to reexamine the banal chart shown at the beginning of this article I would revise it as below.

		Italy	Great Britain
Attack	Naval ciphers	+++	–
	Cipher machine		+
Defense	Naval ciphers	+++	– – –
	Cipher machine	– – –	

I wish to make it clear that the above oversimplified evaluation has no absolute value for a comparison with our opponents of that time. I cannot overlook the vast and efficient British penetration of the radio communications enciphered by the German Enigma. That problem comprised a greater degree of difficulty than the C38m. I lack, for the time being, sufficient information to make a critical treatment of that interesting subject.

BIOGRAPHICAL SKETCHES

Luigi Donini, a retired Admiral, served the Italian Navy on surface ships and submarines. For a long time he served as executive in the Cryptographic Branch of Naval Intelligence, largely decrypting, above all, before World War II, French naval and diplomatic codes (CST, BDG, and TBM) and British naval codes (Fleet code, Aux. code, and Nav. cipher 1). During the War he worked on the principal British naval ciphers (Nav. ciphers 1-2-3, Nav. codes 1-2). These performances earned him the Second Class Gold Medal for works of interest to the Navy.

Augusto Buonafalce was born in Livorno (Leghorn) in 1930 and attended high school there. He studied Economics at Pisa University and worked in electronic data processing during the 1960's. At that time he became interested in information theory and efficient codification in general. He now heads the Patents and Licensing office with an Italian State oil concern. He was "hooked" into cryptography by David Kahn's book, *The Codebreakers.* He is a member of the American Cryptogram Association and regularly submits solutions of ciphers to the ACA publication *The Cryptogram.* Doing research on cryptology is one of his many hobbies, which also include archaeology and ancient music.

PEARL HARBOR AND THE INADEQUACY OF CRYPTANALYSIS

David Kahn

ADDRESS: 120 Wooleys Lane, Great Neck NY 11023 USA.

ABSTRACT: American codebreakers achieved remarkable success in solving Japanese codes, but could not prevent the surprise attack on Pearl Harbor because the Japanese never transmitted anything about the attack on the airwaves.

KEYWORDS: Pearl Harbor, PURPLE, JN25, Rowlett, Friedman, Grotjan, J19.

Fifty years later, Pearl Harbor still haunts Americans. The nation remains mystified about how Japan's surprise attack on the proud Pacific Fleet could have succeeded. The joint congressional committee that in 1945 and 1946 investigated the attack put the question sharply:

> Why, with some of the finest intelligence available in our history, with the almost certain knowledge that war was at hand, with plans that contemplated the precise type of attack that was executed by the Japanese on the morning of December 7 — Why was it possible for a Pearl Harbor to occur?[1]

The "finest intelligence" came from codebreaking. This is nearly always the best form of intelligence. It is faster and more trustworthy than spies, who have to write up and transmit their reports and who are always suspected of setting up a deception. It sees further into the future than aerial reconnaissance, which detects only what is present. It is broader in scope than the interrogations of prisoners, who know little more than what they have experienced. And it is usually cheaper and less obtrusive, hence more secret, than all of these. But it has a serious double-barreled failing. It cannot provide information that a nation has not put onto the airwaves, while its apparent omniscience and its immediacy seduce its recipients into thinking they are getting all the other nation's secrets.

[1] U.S., Congress, Joint Committee on the Investigation of the Pearl Harbor Attack, *Investigation of the Pearl Harbor Attack*, Report, 79th Congress, 2d Session (Document No. 244) (Washington: Government Printing Office, 1946), 253.

This is one of the lessons of Pearl Harbor. American codebreakers performed prodigies, giving remarkable insight into Japanese thinking. But that insight was not total, and so even the extraordinary U.S. cryptanalysis could not warn policymakers of Japan's secret intentions.

This article attempts, therefore, two things: to provide information about the codebreaking leading up to Pearl Harbor that has become available since the 1967 publication of *The Codebreakers*, and to explain why even the voluminous information that cryptanalysis provided did not prevent surprise.

After World War I, which taught or retaught the great powers the value of cryptanalysis, they extended or refounded their codebreaking agencies to provide in peace the benefactions gained in war.[2] The United States was one of these. In 1919, the head of the wartime Section 8 of the Military Intelligence Division of the Army General Staff, or MI-8, persuaded officials to establish a joint State Department-War Department codebreaking agency, formally called the Cipher Bureau but later widely known as the American Black Chamber. He was Herbert O. Yardley, 30, short, balding, and charismatic; he set up shop in New York. The Navy, which after a short-lived effort by naval intelligence had let MI-8 do its codebreaking during the war,[3] talked about organizing its own agency[4] but in fact did nothing in cryptanalysis until 1923. In that year, naval intelligence rifled the steamer trunk of a Japanese naval office in New York, yielding a 1918 Japanese naval codebook.[5] This trove inspired the Navy to establish a codebreaking agency[6] under the director of naval communications, calling it the Research Desk for security.

Its first head was Lieutenant Laurance F. Safford, who thus became the father

---

[2] "Codebreaking in World Wars I and II: The Major Success and Failures, Their Causes and Their Effects," in David Kahn, *Kahn on Codes: Secrets of the New Cryptology* (New York: Macmillan, 1983), 102-103; David Kahn, *Seizing the Enigma: The Race to Break the German U-Boat Codes, 1939-1943* (New York: Houghton Mifflin, 1991), 82-85.

[3] J. S. Holtwick, "Naval Security Group History to World Was ii," Appendix I, p. 003, SHR-355, Record Group 457, National Archives (since all SRH documents are from this record group, that designation is omitted in other references); William F. Friedman, "A Brief History of the Signal Intelligence Service," SRH-029, p. 01.

[4] SRH-355, p. 026.

[5] Jeffery M. Dorwart, *Conflict of Duty: The U. S. Navy's Intelligence Dilemma, 1919-1945* (Annapolis: Naval Institute Press, 1983), 39-40, giving March 1923 as the terminus ad quem; I have been unable to find the report in the source given by Dorwart, apparently through some error; Laurance F. Safford, "A Brief History of Communications Intelligence in the United States," pp. 3-12 in Ronald H. Spector, ed., *Listening to the Enemy: Key Documents on the Role of Communications Intelligence in the War with Japan* (Wilmington, Delaware: Scholarly Resources, 1988), 6, giving date as "about 1922"; Ladislas Farago, *The Broken Seal: The Story of "Operation Magic" and the Pearl Harbor Disaster* (New York: Random House, 1967), 36-37, giving 1920 as the date. Farago's source is the memoir of Rear Admiral Ellis M. Zacharias, *Secret Missions* (New York: Putnam, 1946), which Farago co-authored.

[6] L. F. Safford, "The Undeclared War: 'History of R[adio]. I[ntelligence].' 15 November 1943," SRH-305, p. 03.

of U. S. naval cryptography. Fifteenth in the Annapolis class of 1916, "Saffo" – rumpled, with wavy blond hair that seemed permanently on end ("as if he had been scratching his head in perplexity," a staff member said), whose speech emerged in disconnected bursts – had a flair for mechanics and mathematics and a love for cryptology,[7] though his abilities lay more in making ciphers and cipher machines than in breaking them. He began work in January 1924[8] with four civilians in Room 1621[9] of Main Navy, a temporary building on what is now Constitution Avenue, near the Lincoln Memorial. One of the first things he did was to start setting up radio intercept stations in the Pacific to furnish more material for codebreaking than was provided by the haphazard monitoring by ships and the naval radio station in Shanghai.[10]

Safford concentrated on Japan because it was generally felt that she constituted the greatest danger to the United States. She had, not many years before, defeated China and then Russia to become mistress of the western Pacific. Now she was building a fleet to match America's and had occupied, under a League of Nations mandate, islands that menaced the ocean routes to the Philippines. Japan was a main concern of America's foreign policy, a main focus of America's naval war planning,[11] a main target of America's codebreaking. Yardley, for example, sent Japanese cables that he had cryptanalyzed to U. S. diplomats at the Washington disarmament conference of 1921-22. This gave them knowledge of Japan's negotiating position that they used to drive her to a lower capital-ship ratio with the United States than she wanted, costing her the equivalent of three battleships.

Soon thereafter, Safford took one of his most important strides in building up his agency when, in August 1924, he hired Mrs. Agnes Meyer Driscoll, 32, as a cryptanalyst. Tall, slender, secretive, a graduate of Ohio State University, she had taught mathematics and music in Texas, had worked in cryptography for the Navy, had studied cryptanalysis at the Riverbank Laboratories, had worked for Yardley's Cipher Bureau for five months and then for cipher-machine inventor Edward H. Hebern in California before accepting Safford's offer and rejoining the Navy, this time as a cryptanalyst.[12] She soon proved outstanding and became

[7]Edwin T. Layton with Roger Pineau and John Costello, *"And I Was There": Pearl Harbor and Midway – Breaking the Secrets* (New York: Morrow, 1985), 32.

[8]Russell Willson, "The Birthday of the Naval Security Group," in *A History of Communications Intelligence in the United States*, ed. Graydon A. Lewis (Denver: Naval Cryptologic Veterans Association, 1982), 41.

[9]Layton, 33

[10]Safford, "A Brief History," 5; Willson, 41.

[11]Edward S. Miller *War Plan Orange: The U. S. Strategy to Defeat Japan, 1897-1945* (Annapolis: Naval Institute Press, 1991).

[12]Susan M. Lujan, "Agnes Meyer Driscoll," *Cryptolog* (Naval Cryptologic Veterans Association), 9 (August

the Navy's principal civilian cryptanalyst (most of the Navy's cryptanalysts were officers). Among her first assignments was to work on the photographed code. The Research Desk had found that it was not used plain; rather, its codegroups were themselves enciphered by a columnar transposition. The first job of Miss Aggie, as she was called, was to solve the transposition. At first, she found, the key remained the same for several weeks; later it changed more frequently. Incessantly turning the pages of the photographed book with the rubber tip of her eraser to check codegroups, she took a couple of years to complete the job,[13] enabling the Research Desk to turn the coded messages into Japanese. A husband-and-wife pair translated the Japanese into English.[14]

By then, in 1926, Safford had gone to sea and had been succeeded by one of the first American naval officers to have studied Japanese in Japan: Lieutenant Joseph J. Rochefort. He was a "mustang"–a former enlisted man who had won a commission. This had made him tough and independent in a world dominated by Annapolis graduates; he neutralized his caustic speech with a conciliatory smile. Rochefort became one of the very few Americans capable both in Japanese and in codebreaking.[15]

One of his subordinates described work at the unit:

> Hours went by without any of us saying a word, just sitting in front of piles of indexed sheets on which a mumbo jumbo of figures or letters was displayed in chaotic disorder, trying to solve the puzzle bit by bit like fitting together the pieces of a jigsaw puzzle. We were just a few then ..., young people who gave ourselves to cryptography with the same ascetic devotion with which young men enter a monastery.[16]

The hardest part of breaking a code is the beginning. Rochefort explained it in colorful terms:

> It first off involved what I call the staring process. You look at all of these messages that you have, you line them up in various ways, you write them one below the other, and you'd write them in various forms and you'd stare at them. Pretty soon you'd notice a pattern; you'd

Special 1988), 4-6.

[13]"Reminiscences of Captain Joseph J. Rochefort, U. S. N., Ret." Transcript of oral history interview (U. S. Naval Institute, 1969), 30; Layton, 31.

[14]SRH-305, p. 01; Layton, 31-32.

[15]David kahn, *The Codebreakers: The Story of Secret Writing* (New York: Macmillan, 1967), 7 (henceforth as Kahn); Layton, 33, 34; Jasper Holmes, *Double-Edged Secrets: U. S. Naval Intelligence Operations in the Pacific During World War II* (Annapolis: Naval Institute Press, 1979), 3.

[16]Zacharias 89.

> notice a definite pattern between these messages. This was the first clue.... You notice a pattern that when you follow through, you say this means so-and-so; you'd run that through, and it doesn't work out. Then you'd proceed on some other effort and eventually, if you're lucky and the other fellow makes mistakes, which he invariably will, then you come up with a solution that will stand up under test, and this gives your first lead-in.[17]

Rochefort said he felt good while doing this work "because you have defied these people who have attempted to use a system they thought was secure, that is, it was unreadable. It was always somewhat of a pleasure to defeat them or challenge them." But the work took its toll. While engaged in the actual cryptanalysis, he said, he generally felt frustrated. The tension was so great that after work he had to lie down for two or three hours before he could eat anything; he developed ulcers anyway, and this, together with the fact that duty in communications intelligence hurt a man's career, drove him to get out of the work when his tour in the Research Desk ended in 1927.[18] At least for a while.

Safford returned in June 1929. He had the photographed code, called Code No. 1, whose translation had originally been put together in 10 "volumes" with metal-strip Acco office binders, retyped in four copies on huge 12-by-18-inch forms and bound in two volumes in red buckram McBee binders, far more convenient to use. This gave the code its more common name, the Red Code.[19] On 1 December 1930, however, the Japanese replaced it with a new code. But Miss Aggie, who had by then learned the ships' names, the communications patterns, and the frequently used phrases of the Japanese fleet, solved its transposition encipherment and then reconstructed the 85,000-group, two-part code, later called, from its binding, the Blue Code. It was a remarkable feat of cryptanalysis, and it gave the U. S. Navy insight into Japanese forces and tactics for years.[20]

Two events in 1929 led the Army to expand its codebreaking. In May, after giving the new secretary of state a little time to understand the realities of the job, Yardley passed him some solved messages. Henry L. Stimson was shocked at what he regarded as a dishonorable and counterproductive activity–"Gentlemen do not read each other's mail," he said later, maintaining that "the way to make men trustworthy is to trust them." He withdrew State Department support from

[17]Rochefort, 17.

[18]Rochefort, 16, 17, 45-46.

[19]SRH-305, p. 03.

[20]Safford, 8-9; Layton, 46; SRH-305, pp. 11-12.

the Cipher Bureau.[21] In the meantime, the Army had decided that Yardley was not doing what it needed most: training cryptanalysts for immediate use in case of war.[22] These events doomed the unit, which was dissolved as of October 31, 1929[23]–two days after the great stock market crash.

Its people were released; its papers went to the Army Signal Corps, which had charge of Army codes and ciphers.[24] The Signal Corps had in 1921 hired a 29-year-old who was on his way to becoming the world's greatest cryptologist. William F. Friedman, natty, uptight, brilliant, had written some theoretical treatises of landmark importance and had solved German codes in France during World War I. His new job was nominally to improve the Army's own cryptosystems. But doing this properly required him to test cryptosystems offered to it. This gave him experience in cryptanalysis and expanded the Army's knowledge of it.[25] With the closing of Yardley's agency, the Signal Corps added codebreaking to its responsibilities, and Friedman became the head of a new Signal Intelligence Service. He hired as junior cryptanalysts at $2,000 a year three young men who knew languages and mathematics. The first to report was Frank B. Rowlett, 21, a Virginian, a former teacher, fair-haired, with an all-American look to him, who at 8 a.m. April 1, 1930, found himself entering Room 3406 of the Munitions Building on B Street (now Constitution Avenue) near the Lincoln Memorial in Washington.[26]

Two months later, he and his colleagues were excitedly combing through the secret files of Yardley's organization. This most clandestine and most valuable form of intelligence thrilled them. They went on to study basic cryptography and the solution of machine ciphers, clearly the wave of the future. In 1932, their training completed, they attacked Japanese diplomatic cryptosystems, working on messages provided by the Army's new intercept service. They first cracked a simple code, the LA, so called from the indicator group that preceded its messages. It did little more than put kata kana, a Japanese syllabary, into Latin letters for telegraphic transmission and to secure some abbreviations for cable economy. It was rather regular in construction. For example, all kata kana syllables that ended in *e* had as code equivalents groups beginning with A,

[21]Louis Kruh, "Stimson, The Black Chamber, and the 'Gentlemen's Mail' Quote," *Cryptologia*, 12 (April 1988), 65-89.

[22]SRH-029, Paragraph 6; Frank B. Rowlett, "Comments on [an early draft of this] David Kahn Article," 22 May 1991.

[23]Kahn, 360.

[24]SRH-029, Paragraph 8.

[25]Kahn, 384-385.

[26]Theodore M. Hannah, "Frank B. Rowlett: A Personal Profile," *Cryptologic Spectrum*, 11 (Spring 1981), 4-21 at 5.

Frank D. Rowlett on 2 May 1930, his 22nd birthday, a month after joining the Signal Intelligence Service.

so that *ke*=AC and *se*=AD. This arrangement facilitated solution. When the young cryptanalysts discovered that LA encrypted only insignificant messages, such as those dealing with expenses or vacations, and when they had gained experience and knowledge of Japanese diplomatic language and communications practices, they turned their attention to the more important messages, apparently protected by a machine system. It served on two main Japanese diplomatic communication networks — one covering the Far East, one linking Tokyo with major world capitals.[27]

The cryptanalysts recognized messages in it because they consisted of 5-letter groups like RWOBA preceded by a 5-digit group like 77181. Study showed that vowels had a relatively higher frequency than consonants. It appeared that the machine divided the 26 letters of the Latin alphabet into two subsets, the 6 vowels and the 20 consonants. They perhaps got some help from Navy Lieutenant Jack S. Holtwick's solution of a Japanese cipher machine that enciphered kana. Eventually, the Army cryptanalysts discovered that the machine they were working on used the same cryptographic principles as the Holtwick machine but enciphered not 50 syllables but the 26 letters of the Latin-alphabet form of Japanese, called Romaji, with a mechanism appropriately adapted.[28] Rowlett and one of the other original junior cryptanalysts, Solomon Kullback, one day struck gold. Among their tentative recoveries of partial plaintext were three letters, an unknown, and another letter: *oyo?i*. They knew then they had broken the system, because *oyobi* was Japanese for *and*. They named it RED, the first of the primary colors.[29] By 1937, for the first time in American hist(solutions began going to the White House,[30] probably to President Franklin D. Roosevelt. The intercepts

[27] Ibid., passim.

[28] Rowlett, "Answers to David Kahn's Questions," 22 March and 31 May 1991. The U. S. Navy solved the following Japanese cipher machines: (1) The IKA machine, perhaps so called because messages encrypted it used the Japanese character *ika*, written as a double slash, as a separator. (2) Its successor, called the M1, a kana machine used by the Japanese navy for about four years starting in 1933 whose analogue was designed by Holtwick, who presumably had solved the machine. (3) M2, contemporaneous with M1 but used by naval attaches and producing text in Latin letters rather than kana. It passed through several models, each more complicated than its predecessor. (4) M3, used by the Japanese Foreign Office. The Navy gave this an additional, alphabetical designation and called it the A machine. The Army called it RED. (5) M4. No other information available. (6) M5, which the Navy called the B machine and the Army PURPLE. (SRH-355, Appendix VIII.) M2 would appear to be the Romaji machine utilizing the same cryptographic principles as the kana machine on which it was based, namely the machine solved by the Army, but M3 is stated to be RED, or that machine. I cannot clear up the discrepancy.

[29] Hannah, 14-15. I have written (Kahn, 20) that the machine solved by the Army was probably codenamed RED as a successor to the likely codename ORANGE — the Navy's codeword for Japan — for the machine Holtwick solved, and that RED's successor, PURPLE, was so named because the colors deepened as did the strength of the machine. Alas for this elegant theory, both Rowlett, in his Answers, and Holtwick, in SRH-355, Appendix VIII, pp. 090E-090F, deny that this was the basis for the naming. Holtwick even specifies that "These colors had no connection with color covernames for foreign nations in war plans, such as ORANGE for Japan."

[30] Frank Rowlett, telephone interview, 22 April 1986.

Women work with the U. S. Army analogue of the Japanese PURPLE cipher machine.
U. S. Government photo.

revealed, for instance, advance information about Italy's possible adherence to the German-Japanese Anti-Comintern Pact — providing this in March 1937, six months before American diplomats began reporting on it. Later they provided part of the text of the treaty.[31]

The next year, messages began to appear suggesting that a new machine would supplement and probably eventually replace the older one, which was physically wearing out. On March 20, 1939, three messages in the new system were intercepted, and over the next three months messages in RED gradually disappeared.

[31]Solved Japanese messages declassified by the National Security Agency; United States, State Department, *Foreign Relations of the United States: Diplomatic Papers, 1937*, 1 (Washington: Government Printing Office, 1954), 608.

Japan's major diplomatic messages had become unreadable. Faced with the loss of the nation's paramount intelligence, the Signal Intelligence Service mounted a concentrated attack to solve the new machine. Friedman put Rowlett in charge, with he himself exercising overall supervision. The Americans called the new machine PURPLE.[32]

In the absence of any American spies almost anywhere in the world,[33] these half-dozen cryptanalysts would be able to provide the United States with its best secret intelligence on Japan as relations with that nation, which persisted in its aggression against China, deteriorated. They plunged into their work in Rooms 3416 and 3418 in the Munitions Building. Room 3418, about 25 feet square with a steel door and barred windows, was known as the vault. As additional cryptanalysts were assigned to the PURPLE problem, the group moved into larger quarters, finally occupying about eight rooms. Rowlett worked in 3416. He arrived at 7, an hour early, and left at 5, an hour late. His desk was usually neat, for he spread out his worksheets on a nearby table. Rowlett was extremely focused on the work. He never hummed or chewed his pencil or muttered to himself; he looked out the window only when something attracted his attention; he never drank coffee at work, but did puff on a pipe. His mind didn't dwell on the cryptanalytic problems during the 15-minute drive to work from Arlington County, but he would awaken each night after a few hours of sleep and, lying there in the dark, would review the day's work and think of ways of improving it.[34] In the morning, at work he would exchange ideas with the other cryptanalysts — Assistant Cryptanalyst Robert O. Ferner, Rowlett's principal helper, and Junior Cryptanalysts Albert W. Small, Genevieve Grotjan, and Samuel S. Snyder, occasionally assisted by Cryptographic Specialists Cyrus C. Sturgis, Jr., Kenneth D. Miller and Glenn S. Landig[35] (their names deserve to be remembered!). After the conference, all would return to their desks. Quiet reigned as they pored over the intercepts, most of which had been teletyped in from the monitoring stations; sometimes they puzzled over statistical and alphabetical tables compiled by hand from intercepts. To Rowlett, the atmosphere was like that of a library. Only the rustling of papers and the scratching of

[32]Hannah, 15-16.

[33]United States, Congress, Joint Committee on the Investigation of the Pearl Harbor Attack, *Pearl Harbor Attack*, Hearings (79th Congress, 1st and 2nd Sessions) (Washington: Government Printing Office, 1946), 2:785 (henceforth *PHA*, following by volume and page number); Anthony Cave Brown, *Wild Bill Donovan: The Last Hero* (New York: Times Books, 1982), 176-177; Thomas F. Troy, *Donovan and the CIA: A History of the Establishment of the Central Intelligence Agency* (Frederick, Maryland: University Publications of America, 1981), 105-107.

[34]Rowlett, Answers.

[35]William F. Friedman, "Preliminary Historical Report on the Solution of the 'B' Machine," October 14, 1940, SRH-159, p. 8.

pencils and an occasional discussion broke the brooding silence.

PURPLE had carried over from RED the 6-and-20 division of letters. But this time the "sixes" were not exclusively vowels. Nevertheless, within a few weeks, the cryptanalysts ascertained how the sixes were enciphered. This enabled them to recover the plaintext for those letters. The process was slow and painstaking. A new recruit, Leo Rosen, was assigned to devise a way to mechanize this pencil-and-paper process. He hit upon the idea of using standard 6-level, 25-point telephone selector switches, used in dialing. They worked like a dream, and the solution process was considerably speeded.[36]

Despite Rosen's remarkable advance, the totality of PURPLE still resisted the Americans. Friedman, who had been supervising the work rather loosely, was asked by his bosses — all extremely supportive, financially as well as psychologically[37] — to participate personally.[38] But he was busy with administrative and communication security and, aside from the fundamental and important job of assembling the team, did not contribute directly to solving PURPLE.[39] The Navy, though its main effort continued to be against Japan's naval codes, lent a hand. It and the Army organized their files the same way to facilitate cooperation. After about four months, the Navy dropped its effort[40] and returned to Japanese naval cryptosystems. The S. I. S. pushed ahead. Within Rowlett's group, teamwork was extremely close. No one complained that a task was too menial. Determination pervaded the team. Rowlett was confident from the start that they would break PURPLE the way he and others had broken RED. He never grew depressed, even as one month after another went by without a solution.[41]

As the breakthrough was being sought, the cryptanalysts spent much of their time matching proposed plaintext — a guess as to the text of the original message — against the text of the cryptograms, the ciphertext. The proposed plaintext came from various sources. During the gradual replacement of RED with PURPLE, circular messages were sent to embassies and legations around the world in both systems. RED's solution gave the plaintext to the circulars in PURPLE. In a few cases, the State Department gave them the texts of notes to or from the Japanese ambassadors, which the codebreakers used as cribs. And the cryptanalysts' guessed at plaintexts. They knew, for example, that many diplomatic dispatches began "I have the honor to inform Your Excellency that ..." and they

[36] Rowlett, Answers.

[37] Ibid.

[38] *PHA*, 34:84-85.

[39] Rowlett, telephone interview.

[40] Ibid.

[41] Ibid.

often tried that as the plaintext start.

Among the things they were looking for, apparently, were letter intervals that would reveal the advancement of PURPLE's mechanism. Exactly what these intervals were has never been revealed, but one possibility is that two occurrences of a probable plaintext letter, such as *a*, stood at a certain distance from one another, say two places, as in *Japan*, and that the ciphertext letters that represented them were separated in the alphabet by the same interval, as X and Z. This would show that the machine had moved forward one space each time a letter was enciphered. C. A. Deavours thinks that the search was for identical ciphertext letters, so that in a situation like

```
1 2 3 4 5 6 7
a d v a n c e
X G A R X G V
```

the cryptanalysts looked for an alphabet where the *a* in alphabet 1 was represented by the same letter as plaintext *n* in alphabet 5, and *d* in alphabet 2 was represented by the same ciphertext letter as *c* in alphabet 6. A plugboard, which monoalphabetically enciphered the cipher alphabet table, complicated things. Either way, two or more intervals were needed for the hypothesis to be confirmed.

More than a year of tedious work passed.[42] Then, about 2 p. m. on Friday, 20 September 1940, a sunny day on which the temperature reached 88 degrees, in the middle of Roosevelt's running for an unprecedented third term, as Britain anxiously awaited a German invasion from occupied France, Albert Small noticed that Grotjan, a dignified, 26-year-old statistician, seemed to be concentrating extremely intently. He spoke to her about it, and learned that she had just discovered a couple of the needed intervals and was looking hard for others. He took her in to see Rowlett. He was conferring with Ferner. She said she had something to show him.[43] All moved to her desk. She pointed to her instances, then a third leaped out at the codebreakers. At once, they grasped the significance of what she was showing them. The ebullient Small dashed around the room, hands clasped above his head. Ferner, normally phlegmatic, shouted "Hooray!" Rowlett jumped up and down, crying "That's it!" Everybody crowded around. Friedman came in. "What's all the noise about?" he asked. Rowlett showed him Grotjan's findings. He understood immediately. Grotjan's discovery verified the team's theory of how the PURPLE machine worked. It

[42] Rowlett, Answers, says that nothing in the romantic and hypothetical paragraph on pages 21-22 of Kahn is accurate.

[43] Grotjan, interview, 12 May 1991.

marked the climax of one of the greatest cryptanalyses of all time. And how did the egghead codebreakers celebrate this great moment in the history of American codebreaking? They sent out for Cokes![44]

When the euphoria and the effects of the Coca-Colas had worn off, the cryptanalysts drifted back to work. Grotjan, who seems to have gotten excited about the breakthrough mainly because everybody else was, regarded it soberly as just one step in a series of steps. A week later, the day after Japan started to occupy French Indo-China and the very day the Tripartite Pact establishing the Rome-Berlin-Tokyo axis was signed, the Signal Intelligence Service handed in its first two solutions of PURPLE messages.[45] This did not mean its work was done. The settings for the machine changed each day, and the cryptanalysts had to recover these. But this work was facilitated by Rosen's construction of two American analogues of the Japanese PURPLE machine. The cost: $684.65.[46] Later, other copies of the machine were built — several at the Washington Navy Yard; some were given to the Navy, which had rejoined the PURPLE work to help with the heavy volume of solutions, and some to the British, so they could read the messages without having to wait for American solutions to be forwarded to them. Soon Navy Lieutenant Francis A. Raven discovered a pattern to the daily machine setting changes. With this knowledge, the United States was able to read reports from and instructions to Japan's ambassadors on average within a day or so, sometimes within hours. It had gained access to the most secret diplomatic dispatches of Japan as U. S. relations with that empire continued to worsen.

In cryptology, the beginning of wisdom is to know that there is no such thing as "the" Japanese code. PURPLE was not the only cryptosystem of the Foreign Ministry, much less of the empire. The Foreign Ministry employed a hierarchy of systems, of which PURPLE was the apex. Under it lay several codes that — unlike PURPLE, which served embassies exclusively — were used at both embassies and consulates. LA, the simplest, lay at the bottom.[47] Above it rested a two- part system, PA-K2. The PA was a code, like LA, though more extensive and with its codegroups disarranged in relation to their kata kana plaintexts. The K2 part shuffled the letters of the PA codegroups. American cryptanalysts could restore them to their original positions and then solve the PA code message in three days on the average.[48] More complex still was the J series of codes. These codes were

[44] Hannah, 5.

[45] Ibid., 18

[46] Ibid.

[47] Kahn, 14-15.

[48] Ibid., 15-16. This does seem long for this system, but this was in fact the time given.

more extensive and more thoroughly disarranged than PA, and their codegroups were transposed in a far more complicated manner than the K2: an irregular columnar transposition with a pattern of blanks that broke up the columns. (See Appendix.) Moreover, the code was changed every several months, and the transposition blank pattern — called K plus a number by the Americans — was, by November 1941, changed every 10 days; the transposition's columnar transcription key changed daily, so the codebreakers had to make a fresh analysis for each day's messages.[49] Thus J17-K6 was replaced on 1 March 1941 by J18-K8, and that in turn by J19-K9, whose K9 transposition was itself replaced by K10, which served from 11 to 20 November.[50] Some 10 to 15 percent of J-code messages were not solved at all, and those that were took an average of a week from interception through translation to distribution. This situation contrasts with that of PURPLE, in which most messages were solved within hours and all but 2 to 3 percent of keys were recovered. Did the Japanese err in assessing the security of their cryptosystems? Yes and no. PURPLE was a much more difficult system to solve in the first place, but once solved it was easier to keep up with.

While the Army was concentrating on Japan's diplomatic systems (though the Navy later joined in recovering daily keys and solving and translating messages), the Navy's codebreaking agency — except for some help to the Army — focused on Japan's naval systems. This agency, again under Safford, now a commander, was now called OP-20-G: the OP for the Office of the Chief of Naval Operations, the 20 for Office of Naval Communications, the G for its Communication Security Section, which included codebreaking. During the 1930s it read messages in the Blue code, gaining considerable knowledge about Japan's naval maneuvers.[51] The Japanese replaced this with a new code on 1 November 1938. But the paucity of intercepts in it meant that almost no progress was made in reading it.[52]

On 1 June 1939, the Japanese introduced yet another code. Called JN25 by the Americans, for the 25th Japanese naval code attacked, it soon became the main fleet code, enshrouding messages dealing with naval operations. It was a two-part code of 33,333 five-digit groups, each of which had the peculiarity of being divisible by 3: for example, 52884; this checked against garbles. It was superenciphered with an additive — a string of numbers, printed in a book, that the cipher clerk added to the codegroups, disguising them. Miss Aggie attacked

[49] Ibid., 15-18.

[50] Some J19-K10 solved intercepts and transposition worksheets are in Box 13, Records of the Pearl Harbor Liaison Office, Record Group 80, National Archives.

[51] Safford in *Listening to the Enemy*, 8-9, 12; SRH-305, 12-19; Layton, 46.

[52] Safford in *Listening to the Enemy*, 8; Layton, 77.

it , greatly helped by reserve Lieutenant Prescott Currier. She stripped off the additive, exposing the original codegroups, then recovered their meanings. A year and a half later, in almost the very week that the Army was producing its first PURPLE solutions, the first JN25 solutions emerged.[53] The Navy's satisfaction did not last long. On 1 December 1940, the Imperial Navy replaced it with a new edition, called by the Americans JN25b. But Japan's cryptographers foolishly kept the first edition's last additive book in force for the first two months of the second edition's service. Since this additive book had been reconstructed, OP-20-G gained easy access to the basic codegroups of JN25b and quickly determined the meaning of 2,000 of them. When new additive tables went into effect, its I. B. M. tabulators generated the difference tables necessary for stripping off the additives.[54] As Safford remarked, "We could always trust the Japanese, themselves, to do something that would assist us in the solution of their systems: they have never yet failed us."[55] Nevertheless, progress continued slow.

So in March 1941, OP-20-G in Washington ordered the radio intelligence unit on Corregidor Island in the Philippines to help with the solution of JN25b. A British codebreaking unit in Singapore exchanged recoveries of JN25b codegroups with Corregidor.[56] But despite all these efforts, the code remained readable to but a very small degree. By December 1941, only 10 to 15 percent of each message could be understood.[57]

This, then, was the cryptanalytic situation early in December 1941: the main Japanese diplomatic system could be read rapidly and completely; Japanese consular systems could be read with a few days' delay; the main naval system could be read only spottily. The diplomatic solutions provided Washington with the instructions to and reports from Japan's ambassadors to the United States

---

[53]Safford in *Listening to the Enemy*, 9; Layton, 77-78.

[54]Layton, 78.

[55]SRH-305, 12.

[56]Kahn, 10. James Rusbridger and Eric Nave, in *Betrayal at Pearl Harbor: How Churchill Lured Roosevelt into World War II* (New York: Summit Books, 1991), claim that the British unit was reading JN25 pretty fully and was passing its knowledge of Japanese fleet movements, and hence its foreknowledge of the Pearl Harbor attack, to Britain, where Prime Minister Winston Churchill suppressed it to make sure the attack would bring the United States into the war on the side of Britain. Aside from the fact that Churchill wanted the United States to fight Germany, not Japan, the claim is not only not substantiated by any documents (it is based chiefly on hypotheses and "must have beens"), but is vitiated by technical errors. The chief one is the belief that JN25b was merely a new additive book to the unchanged basic code. In fact JN25b was a new, improved, and enlarged codebook, which is why the Americans had made such little progress into it. It is improbable that the British would have made much more, and also unlikely that they would have limited exchanging codegroup recoveries with the Americans, when they would have benefited as much if not more than the Americans from learning as much as they could about the Japanese.

[57]Layton, 231-232, 547. Captain Thomas H. Dyer, U. S. N., states, in his U. S. Naval Institute oral history, p. 231, that JN25 produced no intelligence about the Pearl Harbor attack.

```
Kancho fugo atsukai   Kokusai jigyo no hip pa ku no kekka
        XEIC         NC   ST    WYNY  KY  ES NI CU KY MTAN

itsu saiaku no jitai ni tachi itaru kamo hakararezaru tokoro
WEUF DB   TH  ZW JXHZ US GKIY  IO    WV      MTGS        WU

kakaru baai waga ho to aitekoku to no tsushin wa tadachi ni
  YK   UQEQ   XF   UX KZRS LJ   SC   FW  AO  AD   CE     CY

teishi serarubeki wo waga ho no gaiko kankei kiken ni hinsuru
SI LW    BS  BN   FK    XF   ZW LUGS   XEYM  LZFF  US  TR GD

baai ni wa waga kaigai hoso no kakuchi muke nihoho news no
UQEQ  XG    FH  EK  FG XJKC KY  PE IY  ZTVE  FJ FX VAIX ZW

chukan oyobi saigo ni oite tenki yoho to shite    1  .  Nichibei
MS  ZP  KR    DB    NX  AE  NE LZ LJXJ   CT      NC LA BO   TMWD

kankei no baai ni wa "  higashi no kaze  ame "     2  .  Nichiso
 XEYM  KY UQEQ  XG   HL CUOM LW KY FEXD DWLD VB NC JG BO  EF  PB

kankei no baai ni wa "  kita no kaze kumori "     3  .  Nichiei
 XEYM  ZW UQEQ  XG   HL  VM  KY FEXD CUCNJM VB NC TK BO  OCXP

kankei no baai (  "  tai "  shinchu mare  "
 XEYM  KY UQEQ WM HL HZ  VB  AO MS  MD QC VB

Netherlands East Indies koryoku oboe fukumu "  "  nishi no kaze
         MTOV           ER CX    NV  FAXPZT VB HL ZA LW KY FEXD

hare "     o  2  do   zutsu kurikaeshi hoso seshimeru koto to
KPQV VB NC NV JG UP QOAF UF    XEEQ    XJKC    GG      BV  UX

seru wo motte migi ni yori ango  .  Shorui to tekito shobun
 NJ    FX      DM    MG    RA VD BO AD BE  RV  TMKD  AD  JB

aritashi    Nao migi wa gen ni gokuhi atsukai to seraretashi
   GQ     NC  NU  DM  AG VP  US  BDZK    -B    UX     ZH
```

Figure 1. The "East wind rain" message encoded in J19.

and so with considerable insight into Japan's foreign plans, and it did so with great speed, accuracy, and volume. By November 1941, 50 to 75 messages a day poured out of the U. S. codebreaking agencies.

Why did not all this information enable the United States to foresee the Pearly Harbor attack? The most basic answer is this: Japan never sent any message to anybody saying anything like "We shall attack Pearl Harbor." The ambassadors in Washington were never told of the plan. Nor were any other Japanese diplomats or consular officials. The ships of the strike force were never radioed any message mentioning Pearl Harbor. It was therefore impossible for the cryptanalysts to solve one.

Some historians have contended that, if only the intelligence officers of the Army and the Navy, and perhaps State Department officials as well, had found the time to analyze all the intercepts as a group, they would have discerned a pattern that pointed to Pearl Harbor. This argument resembles that made by Roberta Wohlstetter in her book *Pearl Harbor: Warning and Decision*,[58] in which she holds that the noise of the false evidence drowned out the indications of the true signals. "We failed to anticipate Pearl Harbor not for want of the relevant materials, but because of a plethora of irrelevant ones."[59] This is wrong. There were no true signals, no clear indications of the attack.

What then is the answer to the joint congressional committee's question? What about that "finest intelligence"? The simple answer is that fine though it was, it was not good enough. A nation can't cryptanalyze what is not there. In that elementary fact lies the inadequacy of cryptanalysis at Pearl Harbor.

---

[58](Stanford: Stanford University Press, 1962).

[59]Ibid., 387.

APPENDIX: A JAPANESE CRYPTOSYSTEM

The most complicated of Japan's pencil-and-paper cryptosystems was the J19-K10 enciphered code. J19 was a two-part code. A Japanese plaintext was first encoded using the J19 code, then this codetext was transposed using the K10 transposition.

An example of this system is one of the messages from Tokyo to Washington setting up the so-called "winds" code. If it had been drafted in English, Circular 2352 would have read:

> Regarding the broadcast of a special message in an emergency.
>
> In case of emergency (danger of cutting off our diplomatic relations), and the cutting off of international communications, the following warning will be added in the middle of the daily Japanese language short wave news broadcast.
>
> (1) In case of Japan-U. S. relations in danger: *HIGASHI NO KAZE AME.*
>
> (2) Japan-U. S. S. R. relations: *KITA NO KAZE KUMORI.*
>
> (3) Japan-British relations (including Thailand, Malaya, and Netherlands East Indies): *NISHI NO KAZE HARE.*
>
> This signal will be given in the middle and at the end as a weather forecast and each sentence will be repeated twice. When this is heard please destroy all code papers, etc. This is as yet to be a completely secret arrangement.
>
> Forward as urgent intelligence.

The phrases relate to the position of the various countries from Japan and mean, respectively, "East wind rain," "North wind cloudy," and "West wind clear."

The text in romaji (Japanese syllabary in Latin letters) version began: *Kancho fugo atsukai. Kokusai jigyo no hippaku no kekka itsu saiaku no jitai ni tachi itaru kamo....* This was first encoded in J19 (see Figure 1).

Most of the codewords are two-letter groups that replace syllables in kata kana, punctuation, numbers, and whole words. But some are four-letter groups that replace words. An M was added at the end, perhaps as a dummy to fill out the last five-letter group. The dash before the B in the third to last group indicates that that letter is missing from the intercepted texts.

3	17	12	4	5	18	2	10	19	7	11	9	14	1	6	16	13	15	8
X	■	■	■	■	■	E	I	C	■	■	■	■	N	■	■	■	C	S
T	■	■	■	■	■	W	Y	N	■	Y	K	Y	E	■	S	N	I	C
U	■	■	■	■	■	K	Y	M	■	T	A	N	W	E	U	F	D	B
T	■	H	Z	■	■	W	■	■	■	J	X	H	Z	U	S	G	K	I
Y	■	I	O	■	■	W	V	M	T	G	S	W	U	Y	K	U	Q	E
Q	X	F	U	X	■	K	Z	R	S	K	H	S	C	F	W	A	O	A
D	C	E	C	Y	■	S	I	L	W	B	S	B	N	F	K	X	F	Z
W	L	U	G	S	■	X	E	Y	M	L	Z	F	F	U	S	T	R	G
D	U	Q	E	Q	X	G	F	H	E	K	F	G	X	J	K	C	K	Y
P	E	I	Y	Z	T	V	E	F	J	N	X	V	A	I	X	Z	W	M
S	Z	P	K	R	D	B	N	X	A	E	N	E	L	Z	L	J	X	J
C	T	N	C	L	A	B	O	T	M	W	D	X	E	Y	M	K	Y	U
Q	E	Q	X	G	H	L	G	U	O	M	L	W	K	Y	F	E	X	U
D	W	L	D	V	B	N	C	J	G	B	O	E	F	P	B	X	E	Y
M	Z	W	U	Q	E	Q	X	G	H	L	V	M	K	Y	F	E	X	D
C	U	C	N	J	N	N	B	N	C	T	K	B	O	O	C	X	P	X
E	Y	M	K	Y	U	Q	E	Q	W	M	H	L	H	Z	V	B	A	O
M	S	B	L	M	D	Q	V	M	O	V	B	M	T	O	V	E	R	C
X	N	V	F	A	X	P	Z	T	V	B	H	L	Z	A	L	W	K	Y
F	E	X	D	K	P	Q	V	V	B	N	C	N	V	J	G	N	P	Q
O	A	F	U	F	X	E	E	Q	X	J	K	C	G	G	B	V	U	X
N	J	F	X	D	M	M	G	R	A	V	D	B	O	A	D	B	E	R
V	T	M	K	D	A	D	J	B	G	Q	N	C	N	U	D	M	A	G
V	P	U	S	B	D	Z	K	-	B	U	X	Z	H	M				

Figure 2. Code clerk's sheet of the day's tranposition block.

The code clerk next took out a sheet of paper with that day's transposition block printed on it (see Figure 2). This consisted of a grid of lines and columns in which several of the cells were blacked out (■) topped by a row of numbers from 1 to 19 in scrambled order. He wrote the coded message — XEICNCSTWYNYKYES... — horizontally into the grid, skipping the blacked out squares (■).

Then he transcribed the letters vertically by columns beginning with the column numbered 1. The result

NEWZUCNFXALEKFKOHTZVGONHEWKWWK...

was the cryptogram that would be sent. He divided it into five-letter groups for ease of transmission and to detect any dropped letters, placed the system and key indicators in two five-letter groups at the top, some other indicators at the end, typed it up, and handed it to a communications officer for transmission (probably by a commercial radio company) to the embassy of the United States.

When a cryptogram like this was intercepted by U. S. Army or Navy radio monitors and teleprinted to Washington, cryptanalysts there faced two steps.

They had, first, to determine the pattern of blanks in the grid and the order of the keynumbers at the top. They did this by trying various arrangements until repetitions of pairs of quadruplets or letters indicated that that arrangement had produced codewords. Then the codebreakers had to ascertain the original Japanese words or syllables for which the codewords stood. These two steps could not be completed on the basis of a single message; dozens were needed. When a message was finally solved, translators would render it into English. Then it would be distributed to the very limited circle of those high officials entitled to see the intercepts.

```
SF DE JAB                           8 NOV 41
361 SCDE TOKIO 90 19 9 40S JG P1/50                    3805
RIYOJI SANFRANCISCO
MWZHU BUWTJ NEWZU CNFXA LEKFK OHTZV GONHE WKWWK SXGVB BLNQV
QQPQE MDZXT UTYQD WDPSC QDMCE MXFON VVZOU CGEYK CXDUN KLFDU
XKSXY SQZRL GVQJY MAKFD DBEUY FFUJI ZYYPY OZOAJ GAUMT SWMEJ
AMOGH CWOVB XAGBS CBIEA ZGYMJ UDYDX OCYQX RGKAX SHSZF XNDLO
VKHBH CKDNX IYYVZ IEFEN OGCXB EVZVE GJKYT JGKBL
                                       1320  S RJ
                                       9160

SF DE JAB                           8 NOV 41
P2/360 41W KOS WASHN                                   3804

KNEWM BLTMV BNJVQ UHIFE UQIPN QLWCM BVXFF MUNFG UAXTC ZJKEX
EXBEW UVBMY NHWSB FGVEX WEMBL MLNCB CZCID KQOFR KWXYX EXPAR
KPUEA SUSKW KSKXL MFBFC VVLGB DDXCL UEZTE WZUYS NEAJT PXTDA
HBENU DXPXM ADCNM MRLYH FXTUJ GNQMT VQRB 84444 BDSHD TAKAO
TOGO
     ( VQRB ) AS SENT                         1317 S RJ
                                              9160
```

Figure 3. The Japanese cryptogram of Circular 2353, as intercepted in two part by American monitors.

BIOGRAPHICAL SKETCH

David Kahn, an op-ed editor and member of the editorial board at *Newsday*, the Long Island daily, is the author of *The Codebreakers* and, most recently, of *Seizing the Enigma: The Race to Break the German U-Boat Codes, 1939-1943* (Houghton Mifflin).

Genevieve Grotjan in 1935.

THE WOMAN WHO MADE THE KEY DISCOVERY

In a small brick ranch house exactly like the others lined along the wooded suburban street near Washington, D. C., lives an important piece of America's secret history — and of its feminist. The young neighbors on either side have no inkling that the structure between them houses a woman who detected data that led to perhaps the nation's most important cryptanalytic breakthrough. To them, she and her husband are just an old couple with a less than manicured lawn.

But the woman, the former Genevieve Grotjan, discovered, on 20 September 1940, significant characteristics in intercepted Japanese code messages. A week later, the team of U. S. Army civilian codebreakers, of which she was a part, had solved Japan's main diplomatic cipher machine, which the Americans called the PURPLE machine. Solutions of the messages enciphered in it gave President Franklin D. Roosevelt and other high-ranking members of the U. S. Government extraordinary insight into the thoughts and plans of the Foreign Ministry and its envoys in Washington through all of 1941.

Grotjan was born in Buffalo on April 30, 1913. She attended the University of Buffalo, hoping to become a mathematics teacher. But no such work was available at the time, so she accepted a position in Washington as a statistical clerk with the Railroad Retirement Board. Later she took an examination for a professional grade of mathematician and was offered a job by the Signal Intelligence Service, which naturally never told her that she would be breaking codes but asked only whether she would be interested in working with codes. She was; she was hired; she proved good enough to be assigned to the premier task of the unit — and she became the one who made the critical find.

Why her? "Maybe I was just lucky in getting the right series of papers," she said. "I perhaps had a little more patience than [Al] Small. [Bob] Ferner might have done it but he was working on something else." When she spotted the intervals, she aside, "I was elated up to a point but I didn't think 'This is it!' because there was so much more to do. I was looking forward to working on the mechanism. I regarded it more as just one step in a series of steps."

So she makes no great claims for herself. Interviewed in her cluttered living room, she thought long before offering a carefully worded answer to each question. She was evidently trying both to reconstruct transient events of more than half a century ago and to be as accurate as she could. She did not exaggerate her importance, did not suggest that without her discovery PURPLE would not have been cracked or the war lost. She recognized that only chance gave her the breakthrough. It was a commendable display of modesty.

Though the solution of PURPLE did not prevent Pearl Harbor, it had enormous effects on the war. Not the war in the Pacific, but that in Europe. "Our main basis of information regarding Hitler's intentions in Europe is obtained from [Japanese ambassador] Baron Oshima's messages from Berlin," wrote Army Chief of Staff General George C. Marshall. These and other solutions, he noted, "contribute greatly to the victory and tremendously to the saving in American lives." It was a remarkable tribute to America's cryptanalysts: including Grotjan, who lives obscure, forgotten, unsung in a low house with a scruffy lawn, but who helped, more than most, to shorten America's biggest war.

THE UNSOLVED MESSAGES OF PEARL HARBOR

Frederick D. Parker

ADDRESS: 721 Holly Dr. N., Annapolis MD 21401

ABSTRACT: In 1941, the United States Navy intercepted enough Japanese naval messages to predict the attack on Pearl Harbor if the code which protected them had been solved. The messages would have disclosed that, while the Japanese government cynically conducted diplomatic negotiations with the United States, the Japanese Combined Fleet under Admiral Isoroku Yamamoto, beginning as early as September 1941, systematically prepared ships, crews, weapons, tactics, and a complicated fuel supply system for the most ambitious operation ever undertaken by the Japanese navy. Details of Japan's intentions were hidden from Navy cryptanalysts because of limited manpower resources and because all Japanese navy codes were assigned a lower priority than Japanese diplomatic codes and German submarine threat.

KEYWORDS: Communications intelligence, Pearl Harbor, World War II, cryptanalysis, Japanese codes, JN25, Purple, OP-2o-G, traffic analysis, 14th Naval District, 16th Naval District, radio silence, tactical warning, Pacific Fleet, Japanese navy, Imperial Fleet, Combined Fleet, First Air Force, STrike Force, Second Fleet, Third Fleet, intentions, capabilities, Southern Operation, Northern Operation, strategic warning, Mount Niitaka.

A study of cryptologic archives in Crane, Indiana, and Washington, D. C., reveals that, in 1941, the United States Navy intercepted enough encoded and enciphered Japanese naval messages to predict the attack on Pearl Harbor. If they had been solved, the intercepts would have disclosed Japanese preparations for the surprise attack, including most significantly the positioning of a tanker where its only purpose could have been to refuel a long-range strike force. But the code that protected these messages could not then be penetrated by U. S. Navy codebreakers working in an obscure but important element of the Navy Communications Directorate, OP-20-G.

Beginning in September 1941, the Japanese Combined Fleet under Admiral Isoroku Yamamoto systematically prepared ships, crews, weapons, tactics and a complicated fuel supply system for the most ambitious operation ever undertaken by the Japanese navy. Their goal was to duplicate the successful British carrier assault on the Italian fleet at Taranto in November 1940 by attacking the U. S. Pacific fleet at Pearl Harbor.

Details of Japan's intentions were hidden from OP-20-G cryptanalysts by the Japanese navy's General Purpose code, designated by OP-20-G as JN25. The first, or Able, version of this code was introduced in 1939 and gradually became the primary cryptographic system used by the Japanese navy. The failure of OP-20-G to develop code group meanings in this five-digit enciphered code, particularly in 1941, can be traced, however, to causes other than superior Japanese cryptography. Simply stated, in Washington in 1941, Japanese navy codes were assigned a lower priority than Japanese diplomatic codes and the German submarine threat.[1]

Since the inception of OP-20-G in the 1920s, the U. S. Navy entrusted cryptography (the development of codes and ciphers) and cryptanalysis (the exploitation of codes and ciphers) almost exclusively to commissioned officers.[2] The goal of the officers assigned to OP-20-G in Washington was to improve the security of U. S. Navy cryptography. Japanese navy codes and ciphers were studied only for their cryptographic characteristics, not for the intelligence value of the underlying messages. Accordingly, the meaning of a code group was only incidental to their work, and, unless an additive cipher period lasted for an unusual length of time, they seldom if ever worked with current material. Conversely, the officers in the 14th and 16th Naval Districts, who supported the Pacific and Asiatic Fleets respectively, were vitally concerned with developing codegroup meanings and solving the current additive cipher. This knowledge would be critical to their attempts to read in real time any intercepted Japanese navy message sent in an enciphered code such as JN25.[3]

By late 1941, 80 officers were assigned to OP-20-G, and the 14th and 16th Naval Districts, 53 of them to OP-20-G. Of these only eight studied JN25. this small number represented a significant increase from 1939, 1940, and 1941, when usually two and never more than five analysts were assigned to analyze Japanese navy codes including JN25; JN20, a minor system for construction activity; JN22, JN39, and JN40, merchant vessel-navy liaison systems; JN61, a merchant vessel-navy five-letter cipher; and JNA20, a naval attaché cipher. The other 45 officer-analysts at OP-20-G, in addition to administrative duties, developed secure codes and ciphers for the navy, operated the communications and processing centers

[1]History of OP-20-GYP, Classified Naval Archives, Crane, Indiana. Hereafter History of OP-20-GYP.

[2]History of OP-20-GYP.

[3]Military Study [of] Communications Intelligence Research Activities, Author, Admiral Joseph Wenger USN (Ret) SRH-151 Record Group 457, National Archives. This study was done in 1937, in conjunction with an update of the Navy's Orange (Japan) War Plan. Page 003 of the study contains Article 102, Mission, of the Naval Communications Service, about 1929, which, in subparagraphs 5 and 6, specifies the two distinct efforts of research and exploitation. On pages 007 and 012 the study specifically notes that Washington and Pearl Harbor and the Philippines are charged with research and exploitation. A similar study was done in 1935, also in conjunction with War Plan Orange. It too assigned research activity to OP-20-G in Washington. Since all SRH studies are from Record Group 457 that designation will be omitted henceforth.

for a 24-hour Atlantic direction-finding (D/F) net tracking German submarines, and solved Japanese diplomatic messages in cooperation with the Army.[4]

Throughout 1941, intelligence staffs in Washington and London increased pressure on OP-20-G for more rapid processing of D/F bearings and dissemination of fixes to convoy managers. At the same time pressure increased on Army and Navy cryptanalysts and linguists working on the diplomatic messages to supply the president and the State, War and Navy departments with the Japanese government's latest instructions to its ambassadors.[5] The officers at OP-20-G excelled in both these endeavors, and the unrelenting pressures to produce results surely led OP-20-G to concentrate on diplomatic solutions and to deflect attention from the unreadable Japanese naval codes and ciphers.

Another cause of the Navy's failure to read any of the Japanese navy's codes before the war was the primitive communications facilities which existed at the time. Movement of intercepted radio traffic and the exchange of ideas between OP-20-G and the 14th and 16th Naval Districts was, even by those days' standards, incredibly slow. Some Japanese navy radio messages, probably those which contained a coding or enciphering error, were airmailed to the West Coast and thence sent to Washington by rail. This process probably consumed more than a week since the rail journey alone in 1940 was about six days. The bulk of Japanese navy radio traffic and all technical exchanges except warnings, which late in 1941 were given special treatment, were sent in weekly shipments by a combination of ship and rail, which at best took weeks and in prolonged bad weather might take months. Under these circumstances it is easy to see why little or no headway against current material was possible either in Washington or in the 14th and 16th Naval Districts.[6]

Nevertheless, success came early to the OP-20-G cryptanalysts, if only briefly. By 30 November 1940, the original Japanese code, JN25 Able, was fully diagnosed and analysts were certain that they were able to produce an accurate and original five-digit text, though lacking most code group meanings. But on 1 December all progress stopped. Japanese cryptographers introduced a greatly enlarged five-digit enciphered code, designated by OP-20-G JN25 Baker. It proved to be a much larger code with a new and larger additive cipher book. It also

[4]History of OP-20-GYP. See also SRH-355 pp. 158, 159, where Holtwick describes how a lack of trained manpower in sufficient number meant that no current system could be read, and how difficult it was to find competent trainees.

[5]History of OP-20-GYP.

[6]History of OP-20-GYP. See also SRH-012, *The Role of Radio Intelligence in the American-Japanese Naval War, August 1941-June 1942*, Vol. I. Prior to November 1941, daily comint reports written in Hawaii were customarily received in Washington 10 days to two weeks later! The first electrically transmitted reports from the Pacific were sent on 26 and 27 November 1941! See Com 14 26011ZNov41 and Com 16 261331ZNov41 in Com 14TI Summaries, SRMN012, RG457, NA.

employed several new and unfamiliar cipher techniques. Thanks to Japanese communications errors each successive JN25 Baker cipher, numbered Baker 1, Baker 2, etc., was successfully penetrated by the analysts in Washington and Corregidor until the additive book designated Baker 8 was introduced on 4 December 1941. This cipher, ironically, proved to be only a minor variation on its predecessor though it completely frustrated analysis for several months. The 1 December 1940 change in the code itself also resisted all efforts at a solution until after the Japanese attack on Pearl Harbor. Thus, JN25 messages could not be solved in time to play a part in U. S. policy or military decisions during this critical period. Thousands of Japanese navy messages intercepted in the last six months of 1941, particularly those in JN25, were not read because, due to manpower shortages and different priorities, the underlying code values remained unrecovered.[7]

To appreciate fully the value of these unread radio messages in 1941 it is necessary to review briefly what the United States knew about Japanese intentions from communications intelligence not derived from the text of naval messages, i. e. , from traffic analysis, direction finding, and from Japanese diplomatic messages.

By recovering the daily keys for the B Japanese diplomatic cipher machine, called PURPLE, the U. S. could read all messages between Tokyo and the two Japanese ambassadors in Washington. Often they were being read as soon as the Japanese recipients in Washington read them and sooner than the addressees in Tokyo did. The messages gave the U. S. government a complete picture of the diplomatic negotiations. All those reading the diplomatic messages were aware of Japan's warlike intentions and, beginning in November 1941, knew that the Japanese diplomats were waiting for a war between the United States and Japan to begin.[8]

After monitoring Japanese navy communications for many years, U. S. Navy communications analysts were completely familiar with the Japanese fleet organization both afloat and ashore. This included the Combined Fleet and its communications infrastructure down to individual fleets, divisions, home ports and navy yards. They were also completely familiar with communications serving Guard Districts and Base Forces, as well as the Combined Air Force, including the 11th Air Fleet, and its airfields with their resident squadrons.[9]

At the very top of the structure the analysts were familiar with the radio networks serving Japanese radio intelligence, the imperial and navy general staff

[7]History of OP-20-GYP.

[8]SRH-012.

[9]SRMN-012.

organizations, and the navy minister. They also routinely identified battleships, cruisers, carriers, submarines and many of the Navy's cargo vessels, particularly while they were at sea.[10]

Augmented by direction finding, message headings, and the knowledge gained from following the maneuvers and imperial fleet operations since the 1930s, communications analysts in Hawaii and Corregidor throughout 1941 frequently issued reports which today would be called strategic and tactical warnings. They regularly recognized and reported the formation of task forces, the creation of new fleets, the movements of large and small air and fleet units, build-ups of ground-based air units, and deployments of key fleet units such as carrier divisions. It was clear in the daily reports filed after 1 July 1941 that Admiral Thomas Hart, Commander in Chief, Asiatic Fleet, and Admiral Husband Kimmel, Commander in Chief, Pacific Fleet, wanted the analysts to follow the imperial fleet wherever it went and report on any threat to the Asiatic and Pacific Fleets. Primary emphasis was obviously placed on carriers, battleships, cruisers, and submarines. Until December 1941, even without the substance of the Japanese navy's messages, simply from the form of Japanese navy communications, the analysts more than fulfilled this mission.[11]

Beginning as early as July 1941, but becoming more emphatic by October, communications intelligence reports from the 14th and 16th Naval Districts warned of a "Southern Operation." The warnings stemmed from ground-based air unit deployments to Taiwan, Hainan, Viet Nam, and to the Japanese mandate islands of the Marianas, Carolines and Palaus, and on the movements in late November and on 2 December 1941 of two new task forces formed around the Second and Third Fleets.[12] Warnings of a threat to "Southern Objectives" dominated comint reports until 27 November when Admiral Harold R. Stark, chief of naval operations, issued the first of his "war warning" messages to the Pacific commands.[13] The alarming news concerning Japanese fleet movements provided by the comint reports from Hawaii and Corregidor was probably responsible for Stark's warnings.[14] Figure 1, depicting the deployment locations for Japanese gound-based air units, shows an unmistakable threat to the Philippines and the Netherlands East Indies and reflects what the chronology of comint reports and Stark warnings suggests: that comint played a major role in the decision-making

---

[10]Ibid.

[11]Ibid.

[12]Ibid.

[13]SRH-012.

[14]Edwin T. Layton, with Roger Pineau and John Costello, *And I Was There* (New York: Morrow, 1985) 187,212,234.

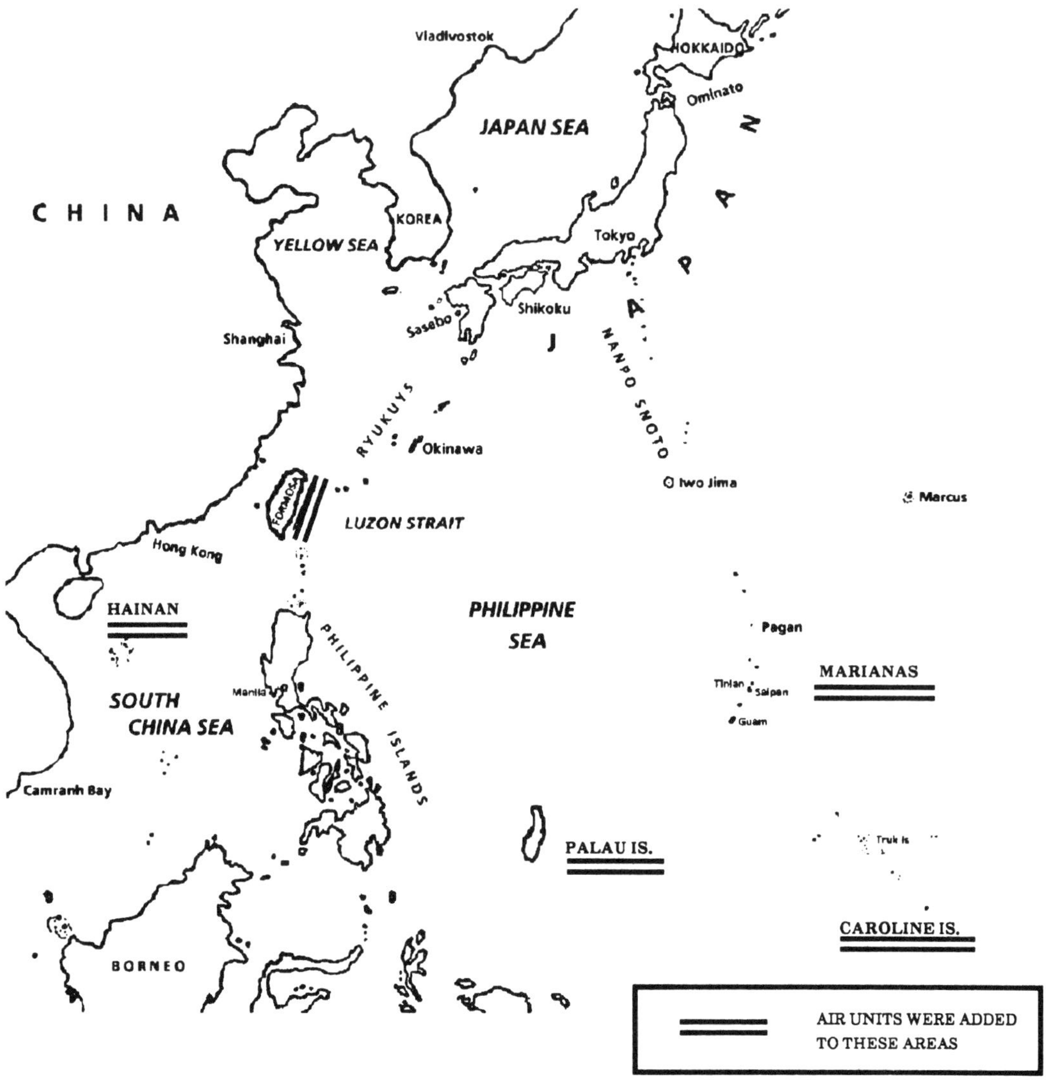

Figure 1. Japanese ground-based air buildups in the Pacific, 1941.

processes in Washington.[15] But there were no warnings from any quarter of a threat to Pearl Harbor.

As startling and ominous as the reports from Hawaii and Corregidor were, particularly after 1 September, they pale when compared with the texts of unread messages which the analysts in the 14th and 16th Naval Districts dutifully forwarded to OP-20-G for study. After the war, Navy analysts in Washington discovered that, during the period from July to December 1941, Navy monitors had intercepted more than 7000 original messages per month–and more than twice this volume if duplicates were counted. After studying this enormous backlog of unexploited prewar material, analysts in 1945 and 1946 decrypted 26,581 Japanese navy messages in seven different cryptographic systems intercepted between 5 September and 4 December 1941.[16]

Between 15 March 1946 and 20 August 1947, OP-20-G analysts and linguists from ONI undertook to study this huge body of solved messages to discover what Japanese secrets might be hidden in their contents. Of the 26,581 decryptions, only 2413 were considered important enough for full translation. Of these, 188 were isolated as pertaining specifically to the events of 7 December 1941 (8 December Japanese time). Had they been available at the time, decision makers in 1941 would have recognized these messages for what they were; pieces missing from the picture of Japanese fleet intentions already partially revealed by the comint reports written by the analysts in the 14th and 16th Naval Districts between September and December 1941.

As early as 5 September, long before the diplomatic messages began to reflect the possibility, the naval messages contained an indirect reference to an approaching war when the chief of staff, Second Fleet, directed this dispatch to the ships of the entire Second Fleet:[17]

A STATE OF COMPLETE READINESS FOR BATTLE OPERATIONS MUST BE ACHIEVED BY THE FIRST OF NOVEMBER ... EXPECT EVEN GREATER EFFORTS TOWARD ACHIEVING MAXIMUM FIGHTING STRENGTH...

Similar sentiments were reflected four days later in a dispatch from the chief of staff, Combined Fleet, to all fleet chiefs of staff and all fleet commanders in

[15]SRMN-012.

[16]Unpublished study of Pre-Pearl Harbor Japanese Naval Dispatches by OP-20-G, Classified Naval Archives, Crane, Indiana.

[17]This message, and others that follow, have been lightly edited for clarity. Two messages, 5 September and 4 November, do not have SRN numbers. All SRNs are individual translations of Japanese naval messages found in Record Group 457. Where no SRN number is given this message was found only in the unpublished OP-20-G study of Japanese naval dispatches.

chief that referred to impending war operations.[18]

AS CONDITIONS BECOME MORE AND MORE CRITICAL, EACH AND EVERY SHIP UNIT WILL AIM AT BEING FULLY PREPARED FOR COMMENCING WAR OPERATIONS BY THE FIRST PART OF NOVEMBER ... COMPLETING ALL PERSONNEL CHANGES ORDERED 3 AUGUST AS SOON AS POSSIBLE.

Beginning in late September and early October 1941, specific details concerning Japanese plans for war began to appear in the messages. The Combined Fleet and particularly the First Air Fleet were observed exchanging messages almost daily, organizing the forces and conducting drills on various aspects of forthcoming operations. In addition to providing these vital details, the message headings revealed the structure of a First Air Fleet carrier Strike Force and parts of its support train.

References to torpedo and dive-bombing drills against battleships occurred in messages from the First Air Fleet staff between 6 October and 4 November 1941. The following is typical:[19]

FROM: FIRST AIR FLEET STAFF 6 OCTOBER 1941

ACTION: COMMANDER CARRIER DIVISION 2

KAGOSHIMA AIR GROUP COMMANDER

COMMANDER CARRIER DIVISION 1

COMMANDING OFFICER YUUZUKI

COMBINED FLEET STAFF

IN FIRST AIR FLEET AERIAL TORPEDO ATTACK DRILL #13, WHICH IS SCHEDULED TO BE CONDUCTED ON 21 OCTOBER AGAINST BATTLESHIP DIVISION 1, AKAGI AND KAGA ARE EACH ALLOTED 9 TORPEDOES AND SOORYUU AND HIRYUU ARE EACH ALLOTED 6 TORPEDOES.

Several important details about the torpedoes being tested were revealed in a group of messages concerning torpedoes, of which this is typical:[20]

---

[18]SRN 115533.

[19]SRN 117453. Akagi, Kaga, Sooryuu, and Hiryuu are all carriers. All names are given as they appear in the translation.

[20]SRN 117301. Parentheses added by translator.

FROM: CHIEF OF STAFF FIRST AIR FLEET 28 OCTOBER 1941

TO: CHIEF GENERAL AFFAIRS SECTION NAVAL TECHNICAL BUREAU

GENERAL AFFAIRS SECTION AERONAUTICS BUREAU

CHIEF FIRST SECTION NAVAL SECTION IMPERIAL HEADQUARTERS

CHIEF SECOND SECTION NAVAL SECTION IMPERIAL HEADQUARTERS

INFO: CHIEF OF STAFF COMBINED FLEET

KURE NAVY YARD

ON 30 OCTOBER, THIS FLEET WILL PICK UP FROM 5 TO 10 NEAR SURFACE(?) TORPEDOES AT SASEBO MILITARY STORES DEPARTMENT(.) CLASSES ON THIS TORPEDO WILL BE HELD AT KANOYA FOR ABOUT FIVE DAYS FROM THE 31ST AND THEN WILL BE SHIFTED TO FIRING PRACTICE .... BY WORKING NIGHT AND DAY, IT SHOULD BE POSSIBLE TO COMPLETE 10 (A SPECIAL ATTACHMENT FOR TORPEDOES, PROBABLY BOW OR STERN PLANES) BY 5 NOVEMBER.

The most significant point was that modifications being made on a crash basis were probably intended to produce shallow-running torpedoes. This would have been of particular interest in the 14th Naval District and Washington, where a current assessment stated that Pearl Harbor was too shallow for a conventional torpedo attack.[21]

One message revealed that three carrier divisions totaling six carriers were to be equipped with the new torpedoes[22] and another disclosed that practice torpedo drills were to be conducted against anchored capital ships.[23]

Of equal significance, given the Japanese reputation for surprise attacks, was the message on 3 November from the staff of the First Air Fleet to the commander of Saeki Air Base:[24]

IN THE 3RD SPECIAL DRILL IN AMBUSHING, 54 SHIPBOARD BOMBERS WILL CARRY OUT A BOMBING AND STRAFING ATTACK IN SIGHT OF SAEKI BASE FROM 0815 ON THE 4TH, 0715 ON THE 5TH AND 0815 ON THE 6TH, AND ABOUT AN HOUR OR AN HOUR AND A HALF AFTERWARDS 54 SHIPBOARD ATTACK PLANES WILL CARRY OUT A SIMILAR BOMBING ATTACK.

[21] Layton, 74.

[22] SRN 116323.

[23] SRN 117665.

[24] SRN 117665.

The evidence of these translations is overwhelming: the Combined Fleet was planning another Taranto. The First Air Fleet was feverishly training its carrier pilots in dive bombing and torpedo attack techniques. And shipyards were hurriedly modifying large numbers of conventional torpedoes to run in shallow water. These preparations were for a surprise carrier ambush of an unsuspecting fleet at anchor, just as the British had done to the Italian fleet anchored at Taranto on 12 November 1940. The question was, What was the objective?

Other messages revealed that the objective of the Japanese First Air Fleet was in the north Pacific, and extraordinary distance from Japan and far beyond the Combined Fleet's normal capabilities. One group of messages, of which this intercept is typical, concerned the preparation of tankers by the First Air Fleet and the almost daily drills between 4 October and 15 November in refueling carriers and other vessels while under way:[25]

```
FROM: CHIEF OF STAFF FIRST AIR FLEET        30 OCTOBER 1941

ACTION: KUROSHIO MARU OR SHINKOKU MARU

INFO: KURE NAVY YARD
      KURE SANDA SECTION
      MAIZURU NAVY YARD CHIEF
      COMMANDER CARRIER DIVISION 2

WHEN INSTALLATION OF GEAR FOR REFUELING UNDER TOW AND PREPARATIONS FOR
ACTION HAVE BEEN COMPLETED, KUROSHIO (KOKU/CHOO) MARU AND SHINKOKU
(KAMI/KUNI) MARU WILL DEPART SASEBO AND KURE RESPECTIVELY ON THE 13TH AND
PROCEED TO KAGOSHIMA BAY, CONDUCTING EXERCISES WITH CARRIERS ENROUTE.

REQUEST THEY LOAD FUEL OIL FOR REFUELING PURPOSES BEFORE THEY DEPART.
```

Another group of messages concerned the preparation of large quantities of fuel in drums and tins. For example:[26]

```
FROM: YOKOSUKA MILITARY STORES DEPARTMENT CHIEF    3 NOVEMBER 1941

TO: CHIEF OF STAFF FIRST AIR FLEET

INFO:  TOKYO SUPPLY DEPARTMENT CHIEF

...THE AMOUNT WHICH WE CAN PREPARE BY THE 7TH IS 150 TONS (750 DRUMS) AND
```

[25]SRN 116588. Parentheses added by translator.

[26]SRN 117180.

140 TONS (8000 KEROSENE TINS) REQUIRED OF US BY MILITARY AFFAIRS BUREAU SERIAL ______,READY BY THE 7TH. WE HAVE ASKED THE BUREAU TO MAKE ARRRANGEMENTS FOR OBTAINING THIS FROM SOME OTHER MILITARY STORES DEPARTMENT. HOWEVER, WE CAN PUT THIS UP IN DRUMS, IF SATISFACTORY. PLEASE ADVISE.

By the end of October 1941 it was clear to the First Air Fleet staff that they could not come up with enough tanker capacity to supply the carriers and their escorts. This conclusion is based on a message originating with the First Air Fleet on 1 November.[27] A second message not only revealed how much fuel in tins and drums was to be carried but also described how the additional weight was to be distributed on board the carriers.[28]

FROM: CHIEF BUREAU MILITARY AFFAIRS SECTION 5 NOVEMBER 1941

TO: CHIEF OF STAFF FIRST AIR FLEET

INFO: CHIEF OF STAFF COMBINED FLEET

1. REGARDING THE LOAD OF DRUMS OF FUEL OIL ON SHIPS OF YOUR FLEET, ONLY THAT WHICH IS ABSOLUTELY NECESSARY FOR OPERATIONAL REASONS [should be loaded], AND THEN ONLY AFTER CAREFULLY CONSIDERING THE FOLLOWING POINTS SHOULD THIS BE DONE, AS IT WILL EFFECT [sic] THE STRENGTH OF THE HULL AND THE SHIPS PERFORMANCE.

A. AMOUNT TO BE LOADED: AKAGI, UNDER 600 TONS; SOORYUU AND HIRYUU UNDER 400 TONS: AND IN EITHER CASE THE MINIMUM REQUIRED AMOUNT, AND WHENEVER POSSIBLE, AN EQUIVALENT WEIGHT WILL BE REMOVED.

B. IN THE CASE OF AKAGI AND HIRYUU, LOAD AMIDSHIPS AND AVOID BOW AND STERN.

C. IN THE CASE OF SOORYUU, LOAD EVENLY, DISTRIBUTING THE WEIGHT OVER THE LENGTH OF THE SHIP.

2. WE HAVE PREPARED 1400 TONS OF FUEL OIL IN DRUMS.

One of the most significant of the intercepted messages was sent on 1 December 1941. It offered a tantalizing clue as to the whereabouts of the objective. One of the tankers assigned to the Strike Force was *Shiriya*. It was at Yokosuka delayed by repairs from joining the operation. Destined to support Destroyer Division 7,

[27]SRN 117150.

[28]SRN 116566.

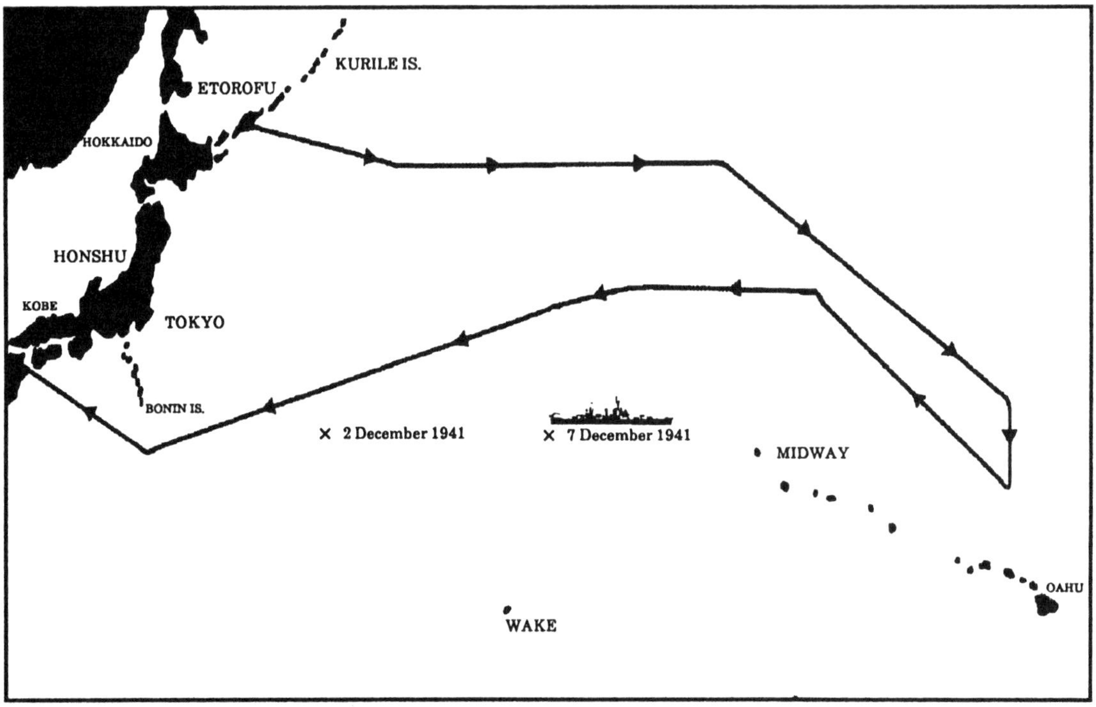

Figure 2. Aproximate position of Japanese *Shiriya* on 2 and 7 December 1941 shown against the route of the Japanese strike force.[30]

the escort of Carrier Division 1, one of the last anxious messages concerning this vessel pointed directly toward the future whereabouts of the First Air Fleet:[29]

```
FROM: SHIRIYA                                   1 DECEMBER 1941

ACTION: COMMANDER DESTROYER DIVISION 7

INFO: CHIEF OF STAFF COMBINED FLEET

THIS SHIP IS PROCEEDING DIRECT TO POSITION 30.00 N, 154.20E.  EXPECT TO
ARRIVE AT THAT POINT ON 3 DECEMBER.  THEREAFTER WILL PROCEED EASTWARD ALONG
30 DEGREE NORTH LATITUDE LINE AT SPEED 7 KNOTS.
```

Starting at 30 degrees north latitude 154.20 degrees east longitude, this course would have placed *Shiriya* at approximately 30 north, 170 east on the 8th of

[29]SRN 115398.

December (7 December in Hawaii), northeast of Wake and northwest of Midway. If solved before 7 December, this message would have shown clearly that the carriers of the First Air Fleet were not involved in a southern operation (see Figure 2).

In mid-November 1941, the messages intercepted since September showed the Japanese fleet planned to go on a war footing and planned a torpedo and dive bombing attack on anchored capital ships in the northern Pacific with a force of six carriers. But where was the Japanese Strike Force? Seven messages intercepted between 17 and 20 November would have provided the answer: Hitokappu Wan or Tankan Bay on a tiny island in the remote Kurile Islands.

The first three messages concerned a person named Suzuki. He was to be transported from Tokyo to the headquarters of the First Air Fleet, temporarily situated aboard the battleship *Hiei* located at Hitokappu Wan. This person was probably Lieutenant Commander Shugru Suzuki, who had been a Japanese spy at Pearl Harbor until early November 1941.[31,32,33,34]

FROM: CHIEF FIRST SECTION NAVY GENERAL STAFF 17 NOVEMBER 1941

TO: CHIEF OF STAFF FIRST AIR FLEET

INFO: COMMANDER BATTLESHIP DIVISION 3

SUZUKI (1776) IS BEING SENT TO YOUR HEADQUARTERS ON BOARD HIEI TO REPORT INSPECTION RESULTS.

FROM: CHIEF FIRST SECTION NAVY GENERAL STAFF 18 NOVEMBER 1941

TO: CHIEF OF STAFF, OMINATO GUARD DISTRICT

INFO: CHIEF OF STAFF, FIRST AIR FLEET

PLEASE ARRANGE TO HAVE SUZUKI (1776) WHO WAS SENT TO 1ST AIR FLEET ON BUSINESS PICKED UP AT ABOUT 23 OR 24 NOVEMBER AT HITTOKAPU WAN BY...OF YOUR COMMAND.

---

[31] Gordon Prange, *At Dawn We Slept*, (New York: McGraw-Hill, 1984), 315-319, 365-372.

[32] SRN 116436.

[33] SRN 116643.

[34] SRN 116920. Kunajiri, a heavy cruiser, was in the Kurile Islands Base Force.

Figure 3. Japanese submarine route.

FROM: CHIEF OF STAFF OMINATO GUARD DISTRICT 19 NOVEMBER 1941

TO: CHIEF FIRST SECTION NAVY GENERAL STAFF

HE (SUZUKI) WILL BE TAKEN ABOARD THE KUNAJIRI.

These messages show this spy being taken to the "standby location" mentioned in a a fourth message on 19 November from Admiral Isoroku Yamamoto, Commander in Chief of the Combined Fleet:[35]

FROM: COMMANDER IN CHIEF COMBINED FLEET 19 NOVEMBER 1941

TO: ALL FLAGSHIPS COMBINED FLEET

1. THE FOURTH SERIES OF COMMUNICATIONS TESTS FOR THE COMBINED FLEET WILL BE

[35]SRN 115678. Parentheses added by translator.

HELD AS FOLLOWS: (ON NOVEMBER 22 AND 23)

2. PARTICIPATING FORCES: (HOWEVER, THE FORCES PRESENTLY ENROUTE TO THE STANDBY LOCATION WILL NOT RECEIVE TESTS.)

3. PRINCIPAL TOPIC FOR CONSIDERATION IN THEIR DRILL: INVESTIGATION AND STUDY OF THE COMMUNICATIONZ REQUIRED TO EFFECTIVELY HANDLE THE SITUATION UPON OPENING OF HOSTILITIES.

The last three of the seven messages also point toward the Tankan Bay location by suggesting that the headquarters of the First Air Fleet was beyond Ominato. As this one does, each concerned a submarine departing Yokosuka on the 20th and 21st. The messages traced the route of each submarine from Yokosuka to Ominato and, from there to the First Air Fleet Flagship communications Zone (see Figure 3).[36]

FROM: STAFF SUBMARINE DIVISION TWO 20 NOVEMBER 1941

TO: COMMANDER OMINATO COMMUNICATIONS UNIT

YOKOSUKA COMMUNICATIONS UNIT

STAFF SIXTH FLEET

STAFF FIRST AIR FLEET

I-19 WILL LEAVE YOKOSUKA COMM ZONE ON NOVEMBER 21, AND ENTER OMINATO COMM ZONE. AT 1600 NOVEMBER 22 WILL ENTER 1ST AIR FLEET FLAGSHIP COMM ZONE.

But what was the Strike Force's objective? Four other messages intercepted in late November would, if solved, have helped unravel this mystery. Navy weather analysts, who in 1946-47 reviewed three weather messages intercepted on 27, 28, and 30 November 1941, were positive in their evaluation that these messages could easily have been solved at the time with the aid of the weather information available from the Pan American Airways weather service. Addressed to the "Striking Force," these messages, of which the following is typical, probably reflected weather conditions along the route to be followed in the northern Pacific by the carriers and their escorts.[37]

---

[36]SRN 116329/116990. See also SRN 117673, 117666 and 117674.

[37]SRN 116668. See also SRN 115460 and 115690.

FROM: TOKYO COMMUNICATIONS UNIT 27 NOVEMBER 1941

TO: ALL FLEETS

FROM: FIRST SECRETARY NAVY SECTION CHIEF, IMPERIAL HEADQUARTERS

TO: STRIKING FORCE

CHIEF OF STAFF COMBINED FLEET

1. WEATHER REPORT. THE LOW PRESSURE CENTER OF 740MM WHICH WAS "NERI O NA" TODAY AT NOON IS ADVANCING AT A SPEED OF 45KM. WIND SPEED IS OVER 15M [i.e., 15 meters per second] WITHIN 1000 KM TO SW OF CENTER AND ABOUT 26M NEAR THE CENTER. THE HIGH PRESSURE CENTER IN "RU" AREA CONTINUES TO PROCEED EASTWARD AT A SPEED OF 45KM. (Quotes as in text)

2. PRINCE HIROYASU FUSHIMI SENDS THE FOLLOWING MESSAGE TO COMMANDER IN CHIEF NAGUMO: I PRAY FOR YOUR LONG AND LASTING BATTLE FORTUNES.

Another message concerned Russian ships in the northern Pacific. It registered definite Strike Force interest in the area as emphatically as the reported intentions of the *Shiriya* to follow the 30 degree north line eastward.[38]

FROM: FIRST SECRETARY NAVY SECTION IMPERIAL HEADQUARTERS
27 NOVEMBER 1941

TO: STRIKING FORCE
CHIEF OF STAFF, COMBINED FLEET

ALTHOUGH THERE ARE INDICATIONS OF SEVERAL SHIPS OPERATING IN THE ALEUTIANS AREA, THE SHIPS IN THE NORTHERN PACIFIC APPEAR CHIEFLY TO BE RUSSIAN SHIPS...THEY ARE UZBEKISTAN (ABOUT 3000 TONS...12 KNOTS) AND AZERBAIDJAN (6114 TONS LESS THAN 10 KNOTS). BOTH ARE WEST BOUND (FROM SAN FRANCISCO).

The messages cited so far, if solved, would have been viewed by U. S. Navy communications analysts and decision-makers as conclusive evidence that the Japanese navy was planning a northern offensive in addition to the southern one. In the messages associated with a northern operation, those concerning ambushing of anchored capital ships by carrier planes, torpedoes modified to run near the surface, a six-carrier strike force, fueling at sea, fuel as deck cargo and weather surveys of the northern Pacific would have been viewed as unambiguous indications that the attack was aimed at Pearl Harbor. Finally, the messages

[38]SRN 116667.

contained evidence which suggested an answer to the final question: When will the attack take place?

Radio silence was recognized by Navy communications analysts in Hawaii and Corregidor as an indication of either impending operational activity or total inactivity when a ship or fleet was in home port. The three messages which follow, intercepted between 10 and 25 November 1941, would probably have inspired warnings to Admirals Hart and Kimmel. The most critical message by far was the last of this group, that from Admiral Yamamoto on the 25th. In view of the other messages intercepted during this period, there is little doubt that any analyst would have misunderstood their implications.[39,40]

FROM: COMMANDER IN CHIEF, SIXTH FLEET 10 NOVEMBER 1941

TO: SIXTH FLEET SUBMARINES

INFO: ALL FLEET COMMANDERS IN CHIEF

MAINTAIN WARTIME RADIO SILENCE ON SHORTWAVE COMMENCING 0000 NOVEMBER 11.

FROM: COMMANDER IN CHIEF ELEVENTH AIR FLEET 20 NOVEMBER 1941

TO: ELEVENTH AIR FLEET

FROM 26 NOVEMBER WHEN PLANES (OR PLANE UNITS) ARE SHIFTED, MAINTAIN PRECAUTIONARY SHORTWAVE SILENCE.

FROM: COMMANDER IN CHIEF COMBINED FLEET 25 NOVEMBER 1941

TO: ALL FLAGSHIPS

FROM 26 NOVEMBER SHIPS OF COMBINED FLEET WILL OBSERVE COMMUNICATIONS PROCEDURE AS FOLLOWS:

1. EXCEPT IN EXTREME EMERGENCY THE MAIN FORCE AND ITS ATTACHED FORCE WILL CEASE COMMUNICATING.

The last message is familiar to all students of the Pearl Harbor attack.[41]

---

[39] SRN 117687.

[40] SRN 116866.

[41] SRN 115376.

FROM: COMMANDER IN CHIEF COMBINED FLEET 2 DECEMBER 1941

TO: COMBINED FLEET

THIS DISPATCH IS TOP SECRET. THIS ORDER IS EFFECTIVE AT 1730 ON 2 DECEMBER. CLIMB NIITAKAYAMA 1208, REPEAT 1208. (CLIMB MOUNT NIITAKA DECEMBER 8).

Intercepted on 2 December 1941, rather than the 6th as recorded in the congressional hearings, this message was the signal to open hostilities. There was no doubt in 1945 that the underlying meaning of this message–"Attack on 8 December"–would have been understood by communications analysts in December 1941.[42]

In 1941, Japan had one the newest and strongest fleets in the world. However, alarming gaps were discovered in the Japanese navy's capability to actually conduct a successful attack against the U. S. Pacific Fleet at Pearl Harbor. By mid-1941, the Japanese government was covering these gaps with a cynical continuation of diplomatic negotiations designed not to avert war but to buy time. Many of the hurried Japanese preparations for war were detailed in secret naval radio messages intercepted by the U. S. Navy's communications intelligence installations in Hawaii and the Philippines. Ironically the Navy's principal capability to solve these secret radio messages was diverted to higher priority tasks in the last six months of 1941 with the result that the Japanese navy, in spite of its shortcomings, attained an almost total surprise on 7 December 1941.

The accuracy of the prewar reports from Hawaii and the Philippines based only on traffic analysis (the analysis of communications activity, message headings, and operational characteristics) show without a doubt that the analysts who wrote about task forces aimed at southern objectives were intimately familiar with the operational habits of the Japanese navy. If the Japanese navy messages had enjoyed a higher priority and were assigned more analytic resources, could the U. S. Navy have predicted the Japanese attack on Pearl Harbor from Japanese navy radio communications? Most emphatically yes! If the secret messages were read at the time, the analysts would have quickly deduced Japanese intentions to conduct a major northern campaign with a six-carrier strike force. Only for one place did all the northern Pacific pieces and clues fit precisely together. An objective extraordinarily far from Japan, the ambush of anchored capital ships, shallow-running torpedoes, six major carriers in a strike force, carrier fuel stored on deck, and a demonstrated interest in the waters of the northern Pacific–all these pointed inexorably to Pearl Harbor.

[42] U. S., Congress Joint Committee on the Investigation of the Pearl Harbor Attack, *Pearl Harbor Attack*, Hearings, 79th Congress, 2nd Session (Washington: Government Printing Office, 1946), 1:185.

BIOGRAPHICAL SKETCH

Frederick D. Parker is a senior historian in the Center for Cryptologic History at the National Security Agency. His field of concentration is the role of communications intelligence in the U. S. Navy's decision-making processes during the American-Japanese war in the Pacific. He lives in Annapolis, Maryland. This article is based on his still classified book, "A New View to Pearl Harbor," published in 1988, which traced the development of communications intelligence in the U. S. Navy from 1924 to 7 December 1941. His latest book, *A Priceless Advantage: The Role of Communications Intelligence in the Battles of Coral Sea, Midway, and the Aleutians*, will soon be published in the Cryptologic History Series by the National Security Agency. He is currently working on a book concerning the role of communications intelligence in the Battle for Guadalcanal.

Mr. Parker, a former Marine Corps Reserve Officer who served on active duty from 1943-1945 and 1950-1952, is a 1950 graduate of Georgetown University, School of Foreign Service, where he majored in history and economics. Retiring from NSA in 1984, he began a second career as a cleared part-time employee producing classified histories based on still classified material. The views expressed in this unclassified article are his own and not those of the National Security Agency.

THE FALL AND RISE OF CRYPTANALYSIS IN CANADA

Bill Robinson

ADDRESS: Conrad Grebel College, Waterloo, Ontario N2L 3G6 CANADA.

ABSTRACT: In the late 1970s, Canada's Communications Security Establishment (CSE) was an organization in decline. By 1979 only one person was still doing real cryptanalysis. When Ed Cheramy died late in 1980, Canadian cryptanalysis almost died with him. Earlier in 1980, however, CSE had sent Tom Johnston to NSA to investigate what it would take to re-enter the field. This effort culminated in the purchase of a Cray X-MP supercomputer, installed in 1985, and the hiring of a new staff of cryptanalysts to make use of it. Only with these developments did CSE enter the modern age of cryptanalysis.

KEYWORDS: Canada, CBNRC, Cray, cryptanalysis, CSE, NSA.

The Communications Security Establishment (CSE) is Canada's "ultra secret" signals intelligence agency.[26] Assisted by the radio intercept stations of the Canadian Forces Supplementary Radio System, it collects SIGINT "to provide the government with foreign intelligence on the diplomatic, military, economic, security and commercial activities, intentions, and capabilities of foreign governments, individuals, and corporations."[8] The primary target of Canada's SIGINT efforts is the Soviet Union, which CSE monitors in close co-operation with the American and British SIGINT agencies, the National Security Agency (NSA) and Government Communications Headquarters (GCHQ). In comparison to these SIGINT giants, CSE is a tiny intelligence agency. But Canada's arctic geography has made CSE an important player in the SIGINT game. According to Seymour Hersh, the protocol that divides monitoring responsibilities between the SIGINT allies, the SIGINT Combined Operating List, assigns CSE primary responsibility "for collection in the polar regions," including the northern Soviet Union.[11]

SIGINT satellites have reduced the role of Canadian geography in SIGINT collection, but CSE's contribution remains an important one. It was CSE, for example, that wrote the SIGINT community's "final after-action report" on the shoot-down of Korean Air Lines flight 902 over the Kola peninsula in 1978.[11]

It is CSE's ability to contribute useful data and analyses to the broader SIGINT community that ensures it significant access to the SIGINT data and analy-

CSE headquarters, the Sir Leonard Tilley building, June 1990. The virtually-windowless concrete structure is the new addition to the Tilley building under construction. Photo credit: Bill Robinson

ses produced by NSA and GCHQ. Pàtrick Spearey, who served as the Canadian Special Liaison Officer to NSA (CANSLO/W) from 1978 to 1980,[1] became director-general production late in 1980,[33] and is now serving as the Canadian Special Liaison Officer to GCHQ (CANSLO/L),[1] has stated that CSE's reputation as a useful contributor is "extremely important" in maintaining the flow of information from its allies: "CSE must maintain its ability to inform; it must continue to gather and obtain information from other countries."[33] To do otherwise risks being reduced to a mere collection arm of NSA/GCHQ, receiving in return only the scraps those agencies consider it useful to throw.

During the late 1970s CSE was running exactly this risk. Canada's codebreaking abilities were literally on the point of extinction, and CSE as a whole was understaffed and ill-equipped. It was to take a concerted and expensive renewal effort lasting most of the 1980s, involving both extensive hiring and the purchase of a multi-million dollar supercomputer, to revive Canadian cryptanalysis and the Canadian SIGINT effort in general. This article tells the story of that revival. Emphasis is placed on the details of the Canadian cryptanalysis effort.

Canada first entered the codebreaking business during the Second World War. A permanent postwar SIGINT agency, the Communications Branch of the National Research Council (CBNRC), was created in 1946, the product of the amalgamation of Canada's wartime civilian and military SIGINT efforts.[15] Starting with a staff of 62,[15] CBNRC was fully grown by 1962, at which time it had a staff of about 600.[29] It had about the same number in 1975, when it was transferred to the Department of National Defence and became the Communications Security Establishment.[14,28] CSE operates out of the Sir Leonard Tilley building in Ottawa, which has been its headquarters since 1962.

By the late 1970s Canadian cryptanalysis was virtually extinct. CSE's targets "had become very sophisticated and difficult to analyze" and its cryptanalysis unit, the O1 section of O Group, "had a poor reputation as a dead end, being unproductive." The section's computer resources were limited and much of its work was still being done by hand. Only one O1 staffer, O1A member Ed Cheramy, was doing full-fledged cryptanalysis, and he worked only on manual cryptosystems. The other members of O1A had "moved away from cryptanalysis" and were engaged instead in "linguistic analysis," the recovery of garbled plain text, and "depth reading," which this author believes to be a form of superimposition, a method for recovering the keys of known cryptosystems (see Frank Lewis, "The Day of the Dodo," *Cryptologia*, January 1990, p. 19). O1B, the other subunit of O1, recovered high-frequency radio call signs through traffic analysis.

It was widely recognized within CSE that it was "no longer its commitment to serve Canada and the other allied countries." O1 "had become obsolete and unreliable" and CSE felt that it "had to bring this area up to what was expected."[33] The decision to modernize the section, and the rest of CSE, appears to have been made early in 1980. The beginning of the 1980s was a time of growing tension between East and West, and allied (particularly American) pressure is the most probable explanation for the timing of the modernization decision. However, it may also have had something to do with the 1979 appointment of Peter Hunt to the position of director-general production, replacing Jack Dornan, who had held the job since 1957.[15,31] Hunt was a man on his way to the top. Serving in all the right jobs, analyst/reporter, directorate assistant (a key policy and planning position),[31] CANSLO/W,[2] L group head, and director-general administration, he had probably already been tapped to become CSE chief by the time he returned from a year of French language training to take up the position of director-general production. By the beginning of 1980, Hunt may already

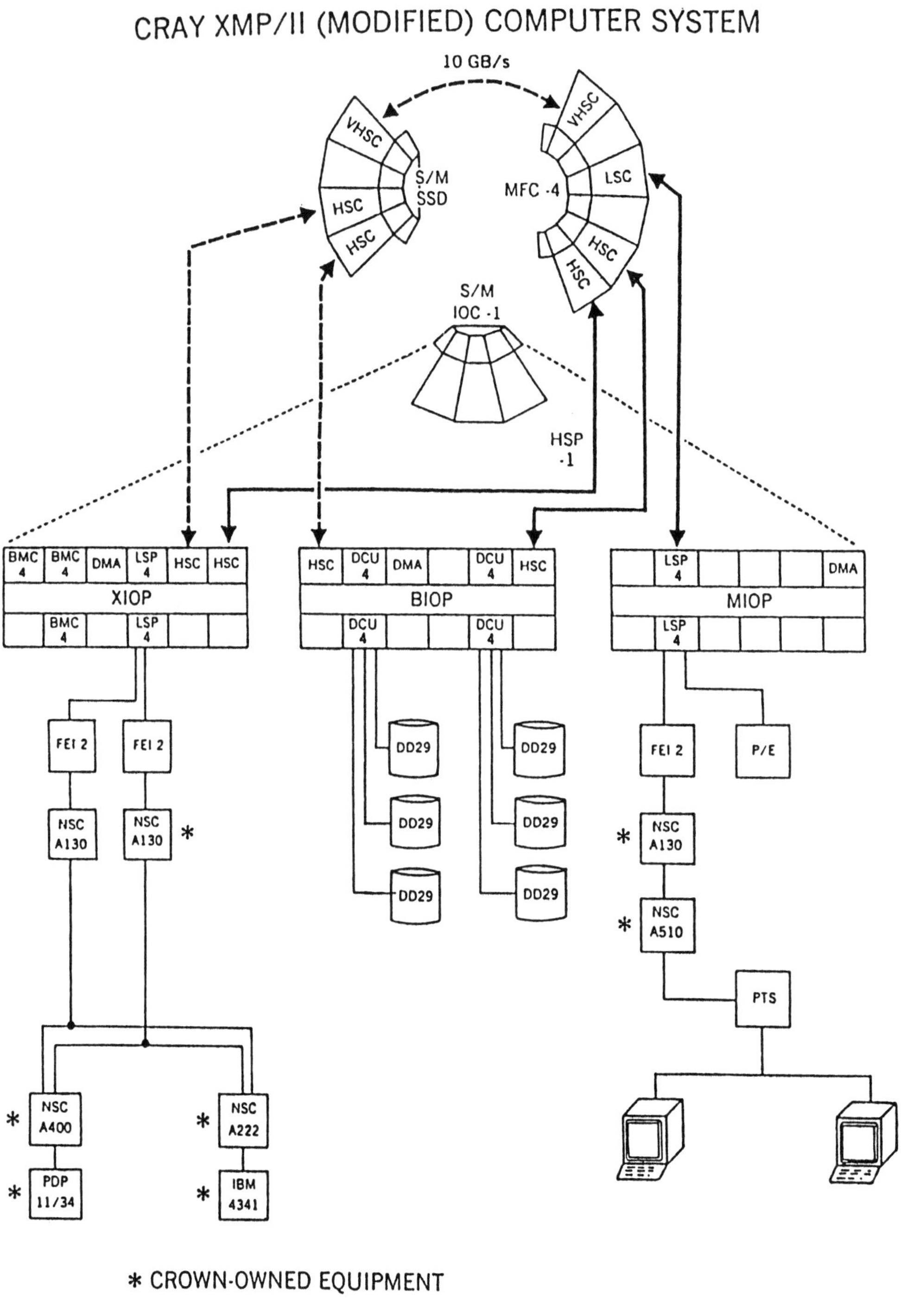

Figure 1. The proposed system.[4]

have been reshaping the organization he was soon to direct (he became chief in September of 1980).[31]

Whatever the cause, by 1980 change was on its way. In December 1979 O1 head John Orobko and two of his Second World War SIGINT vets, Frank Cumming and Camille Fiset, retired[22,30,33] and in the spring of 1980 Tom Briscoe, the new head of O1, sent O1A staffer Thomas Johnston south to NSA for a week of consultations. Johnston's mission was "to assess and study the cryptoanalysis [*sic*] performed there and recommend to CSE how it could re-enter the area of cryptanalysis." A physics PhD and registered engineer who was at home in the world of advanced mathematics, Johnston had joined CSE only in 1977. Now he was to become the key figure in the organization's cryptanalytic renaissance.[33]

Johnston's report made it clear that the only way for CSE to re-enter the cryptanalysis business was to hire a new cryptanalysis staff with advanced mathematical training and purchase a powerful supercomputer to support their work. However, although CSE's management committee accepted this conclusion, in 1980 the agency didn't have the money for a supercomputer.[33] CSE's largest computers at the time were its IBM 370 mainframes;[3] O1 was accessing the SIGINT databases kept on the mainframes, but using obsolescent PDP-8 and PDP-11 computers to run its linguistic analysis and call sign recovery software. The section's budget for extra purchases was only $8,000, and it would take time to get approval for a multi-million dollar computer purchase.[33]

While they were waiting, therefore, Johnston recommended that O1 try to use one of the IBMs for cryptanalysis. NSA doubted that any useful results could be achieved with this system, but it nonetheless assisted Johnston's efforts, supplying daily key information on the system he was trying to break.[33]

Meanwhile, O1A went ahead with hiring new mathematicians to replace the people retiring or transferring out of the unit. The first of this crop, John Mulholland (who was already a CSE employee) and Luc Laplante, joined O1A later that year, and immediately set to work helping Johnston (who had since become head of O1A) on the "machine cryptanalysis" project.[33]

It was none too soon. Late in 1980, while Johnston's team was still labouring over its software, Canada unexpectedly found itself out of the codebreaking business. Ed Cheramy, CSE's sole working cryptanalyst, had died.[32] During the period between Cheramy's death and January 1981, when Michel Jetté was transferred from M Group and took over Cheramy's work, nobody at CSE was actually "breaking codes." (Jetté later did "outstanding" work on manual systems, but in 1982 he was reassigned to machine cryptanalysis and CSE's manual work was effectively dropped. Former cryptanalyst Celia McInnis has stated that she saw only one "manually generated code" during the entire period – August 1985 to March 1988 – that she worked in O1.)[33]

In the spring of 1981, Johnston and co. finally achieved a minimal cryptanalysis capability on the IBM. Nevertheless, all was not well on the machine side, either. Estimating that the IBM "would cease to be effective [for cryptanalysis] within one to two years," Johnston (by then head of O1) recommended again in 1981 that CSE purchase a "suitable scientific computer" for cryptanalysis. At the time NSA was considering buying the Control Data Corporation's Cyber 740, so Johnston recommended this computer to his bosses. The Cyber 740 would have cost only $3-5 million, not as much as some supercomputers, but CSE was still unconvinced it could afford such an expenditure, so the purchase was again put off.[33]

Fortunately for Johnston, CSE's money problems were not to last forever. In fact, it was that same year that CSE's budget constraints began to ease and the agency entered its first significant period of growth in nearly 20 years.[9]

Late in 1983 or early in 1984, the purchase of O1's cryptanalysis computer was finally approved (the computer approved may have been the Cyber 740 or a Cray supercomputer; the evidence is unclear on this point). At the same time, 14 new positions were authorized in O1A: 8 cryptanalyst- mathematicians, 3 computer operators, 2 engineers, and 1 assistant. By the end of 1984, most of these positions had been filled,[33] and CSE had grown to a total staff of about 670.[10]

In April 1984, with the computer project approved and under way, Johnston again visited NSA, where he learned to his dismay that the Americans had decided to use a different model of computer. CSE had "a 40-year catch-up" to accomplish in machine cryptanalysis, and Johnston knew it would be necessary to use NSA software and expertise in the effort, so he returned to Ottawa and recommended that CSE purchase a more expensive Cray X-MP. The X-MP is Cray's second generation of supercomputer; developed in 1982, it superseded the original Cray-1 model developed in 1976. The purchase was formally approved in June 1984, under the name Project PM-40 (or Project Elevator), and a purchase contract was signed with Cray Research Inc. the following December.[4,33]

CSE's Cray X-MP/11 (modified) supercomputer (see Figure 1) was delivered to the Sir Leonard Tilley building in March 1985. Purchased for a mere $8,536,000 (Canadian), CSE's Cray was the budget version of the X-MP, containing just 1 central processing unit and 1 megaword of central memory (hence the designation X-MP/11).[4] The top-of-the-line X-MP/416 contains 4 CPUs and 16 megawords of central memory.

Nonetheless, it was an impressive machine. At the time of its purchase, it was the most powerful computer in Canada (Environment Canada and the University of Toronto have since purchased more powerful X-MPs). It is still more powerful than anything possessed by the Soviet Union.

Six 600-megabyte disk drives were needed just to keep the data-hungry computer fed. "Front-end interfaces" and a magnetic tape processing system gave the Cray access to additional data stored on CSE's other computer systems. Sixteen terminals connected the cryptanalysts and operators to their number-cruncher.[4]

Even the maintenance of the machine was a multi-million dollar job. Cray Research's Canadian subsidiary, Cray Canada Inc., was contracted to conduct all major maintenance under a five-year deal worth $3,545,000 (later increased to more than $4 million to allow for inflation), making the complete purchase and maintenance cost of the computer $12,082,000.[4]

The maintenance of the Cray is no simple job. The machine is required to run 24 hours a day, 7 days a week, 90 percent of the time. Cray Canada's maintenance personnel work on-site in the Tilley building and must be Canadian nationals cleared for Top Secret-Special Access, the same general clearance that CSE's cryptanalysts hold[4,33] (of course, the cryptanalysts also hold additional, specific SIGINT clearances).

The computer has undergone continuous upgrades since its arrival in 1985. The first element to be upgraded was the disk subsystem, which was traded in for the DD-49 system, with twice the capacity, in April 1986. The trade cost CSE an extra $849,848.[5] In 1987-1988, a much more ambitious upgrade was undertaken. Three megawords of central memory were added to the X-MP/11, making it an X-MP/14. This additional central memory enables more complex problems to be tackled without slowing the system down by resorting to external storage systems. At the same time, a special 64-megaword Solid-state Storage Device (SSD) was added to the system. According to Cray Research, an SSD "allows the development of algorithms to support [input/output]-intensive applications such as out-of-memory solution techniques. It is used for large prestaged or intermediate files generated and manipulated repeatedly by user programs." This upgrade – apparently planned from the beginning – cost CSE a total of $4,340,880,[6] more than half the cost of the original X-MP/11, for a total expenditure on the Cray of about $13,726,938 (more than $19 million including maintenance fees to date).

The NSA contribution to CSE's Cray was indispensable. NSA not only supplied the basic Cray SIGINT software, the Folklore operating system,[4,33] it also trained a number of O1's cryptanalysts in the use of the machine. "Operations required that O1 have in residence the expertise to continuously modify the operating system. Thus, Dr. Johnston arranged for the American counterparts

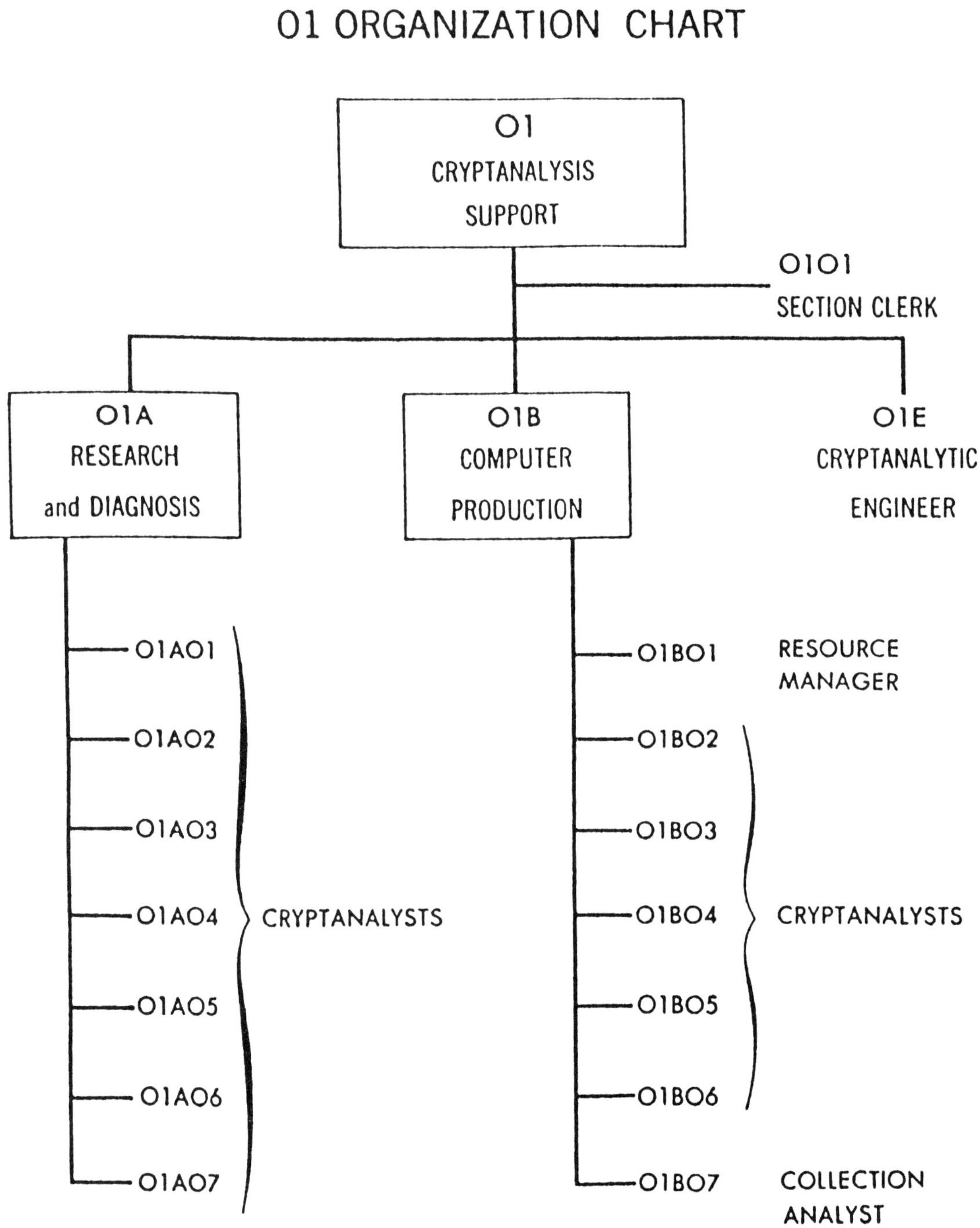

Figure 2. Organization Chart of O1, 1985.

to train the computer operators and the O1 staff for CSE."[33]

The first O1 staff member sent for training at NSA was Christina Sattler, a CO-3 (trainee) cryptanalyst, who had joined O1B in 1982 to run the call sign software on the PDP-11. Early in March 1984, Sattler started a 14-month stint at Fort Meade, returning to Ottawa late in April 1985, just in time to help get the Cray up and running. Sattler was promoted to CO-4 (a fully-trained cryptanalyst) upon her return to CSE. Ultimately, however, she proved to be something of a disappointment, spending 6 months analyzing "particular sign targets" and having "difficulties breaking the system." Johnston "found her more competent in support work to the computer than as a cryptanalyst" and in 1986 Sattler transferred to another section of O Group where she helped run the Cray for O1.[33]

John Mulholland, then head of O1A, also went to NSA for training in 1984, starting a 12-month stint at Fort Meade in April and returning, like Sattler, just in time to help get the Cray running.[33]

Finally, upon the return of Mulholland and Sattler, Luc Laplante was sent to NSA for 3 years of training, returning to CSE in the summer of 1988.[33]

In addition to NSA's training assistance, CSE also conducted its own training to prepare for the Cray's arrival. One such course, run in February 1985 "as a result of direct operational requirements," was shift-register cryptology, which required 3 weeks of full-time study. Among the 16 students who took the course were 5 members of O1, nearly half of O1's working cryptanalysts. (Most of the other students were from S Group in CSE's COMSEC arm.)[23]

Modern cryptanalysts need an extensive background knowledge of both mathematics and computer programming. As Johnston wrote in 1986, "Although it is fair to say that one need not necessarily be a graduate Mathematician to be a good cryptanalyst, modern cryptologic developments make it extremely unlikely that one can be a good cryptanalyst without having a good basis in Mathematics. Our recruitment processes have therefore been aimed at the selection of personnel with good analytic abilities who also happen to have the requisite formal mathematical education."[16]

Even with this background, new members of O1 undergo up to 3 years of classroom and on-the-job training before they become fully-qualified cryptanalysts.[24] Required internal courses include:

- CA-105 - Introduction to Cryptography and Exploitation of Manual Cryptosystems
- CA-107 - Exploitation of Manual Cryptosystems
- CA-123 - Shift-Register Cryptology

- CA-223 - Advanced Shift-Register Cryptology
- CA-247 - Vocoders for Cryptanalysts
- MA-213 - PTAH
- MA-246 - Cryptomathematics for Mathematicians
- MA-256 - Signal Processing Mathematics[16]

The cryptanalysts also require extensive computer programming knowledge and specific training on using the Cray, starting with Introduction to IMP/Folklore.[24] (See Appendix for more details on qualifications for cryptanalysts.)

In addition to the acquisition of the Cray, a number of organizational changes were also undertaken as O1 was transformed during the 1980s. In 1982 depth reading was transferred out of O1A.[33]

In June 1984 O1B was disbanded entirely, with its call sign recovery work, more formally known as the "[deleted] Identification Systems Task," transferred to M Group, which previously had been the recipient of O1B's product.[21] The 4 staff positions in O1B were transferred to M4B,[33] leaving only the head of O1, the section clerk, a cryptanalytic engineer, and O1A in O1 section.

By the summer of 1985 O1's Cray was finally running and O1A had grown to a staff of about 16.[20] Administering the unit had become too much work for one unit head (Mulholland) to handle and so, effective 2 July 1985, O1A was broken into 2 parts: a smaller O1A (Cryptanalysis Research and Diagnosis) with 8 members, and a new O1B (Computer Production) also with 8 members. O1E (the Cryptanalytic Engineer) remained unchanged with its single member, leaving the new O1 with a total staff of 19 (see Figure 2).[19]

Finally, in 1986 O Group itself was reorganized. O1B's computer support personnel were transferred out of the section and combined with O2's computer support personnel (O2 was a traffic analysis section) to create a new O3 section, with 6 or 7 members, which became responsible for the "pure computer sciences support" for O Group. Production cryptanalysis runs remained with the 6 remaining members of O1B, and O1A was unchanged.

These final changes, although minor, effectively completed O1's six-year transformation into a real cryptanalysis section. About 17 personnel were left in the O1 section as a whole: 15 of these had extensive and up-to-date cryptanalytic training and 12 were working cryptanalysts.

The renewal of Canadian cryptanalysis was expensive, but, according to Patrick Spearey, it was worth it: "since the acquisition of the Cray computer and the hiring of mathematical and computer sciences graduates, CSE's reputation has gone up immeasurably and it has enhanced its capability in the business."[33]

According to one source, the overall modernization of CSE during the 1980s cost Canadians about $100 million.[27] In addition to a continuing increase in staff (now more than 900)[7] and the purchase of the Cray, several other initiatives were undertaken to bring CSE into the modern age of SIGINT. Since 1988, a number of CSE's analysis-reporting sections, apparently including M4, have become 24 hour/7 day operations and have established quick reaction reporting teams, enabling them to provide more timely SIGINT reporting and work more closely with CSE's Watch Office, the CANSOC (CANSOC probably stands for Canadian SIGINT Operations Centre).[17] The Supplementary Radio System detachment at the Tilley building has been enlarged and upgraded to the status of an independent unit, the 771 Communications Research Squadron.[13] Finally, construction has begun on a $35.1 million addition to the Tilley building that will nearly double the size of the facility.[25] The addition, scheduled for completion in 1992, uses "certain innovative construction techniques... to prevent the emission of potentially compromising electromagnetic radiation." Its construction will enable CSE to consolidate its newly-enlarged staff in a single, modern, secure site.[34]

Ironically, CSE has managed to accomplish all this just as the "Soviet Threat" it was largely directed toward crumbled away. There is still much to be monitored in the Soviet Union (in some respects, the usefulness of such monitoring is greater now than ever), but many future targets will be elsewhere. What, and who, will be the targets in the post-Cold War world? CSE's long-range plan, the CSE 2000 document, reportedly underwent a rapid rewrite in 1990 to try to answer just that question. But the results of the rewrite will remain highly secret. Will CSE continue to grow? Will it remain the same size? Will its targets change? Where will CSE's eavesdropping powers be directed?

As the parliamentary committee reviewing the Canadian Security Intelligence Service recently reported, Canadians are right to be concerned about the potential for the abuse of CSE's powers.[12] The Canadian government currently shows few signs of permitting any public "oversight" of the operations of CSE, refusing even to tell members of Parliament how many employees CSE has (905 as of December 1990) or how much money the agency spends ($36.2 million in salaries alone in fiscal year 1989-1990).[7]

Nevertheless, the day of oversight is probably coming. CSE is said to be already preparing internally for the advent of some form of oversight committee, parliamentary or otherwise. The concern that such a committee would spill too many of CSE's secrets is unjustified. Indeed, if Canada's experience with the Security Intelligence Review Committee is any guide, the public will learn less about CSE's "secrets" from the committee than it will from this article, which was collected entirely from public sources.

APPENDIX: CSE REQUIREMENTS FOR CRYPTANALYSTS, EARLY 1980S [18] (FORMERLY CLASSIFIED TOP SECRET UMBRA)

This position requires a very good knowledge of Mathematics, acquired through University graduation in a relevant discipline, and a good understanding of its application to cryptanalysis. The incumbent would normally have up to eight years experience in Mathematics and cryptanalysis. Rated experience would include:

- relevant undergraduate and graduate studies in related Scientific/Engineering disciplines,
- training as a cryptanalyst as a result of participation in [NSA?] integration/training programs, and
- in-house training and experience as a cryptanalyst.

Knowledge of the following areas of Mathematics is required:

- Probability and Statistics including Stochastic and Markov Processes,
- Linear Algebra and Abstract Algebra including Finite Field Theory, Shift Register and Polynomial Theory,
- Communications, Information and Coding Theory,
- Sampling Theory,
- Digital Filtering,
- Numerical Methods, and
- Fast Transformation Algorithms.

Extensive knowledge of computer programming and operating systems is required. Operating system proficiency is necessary for the VAX, IBM 370 and the CRAY XMP and entails familiarity with:

- VMS and UNIX,
- VM/SP, CMS, and VS1SYS,
- FOLKLORE.

Programming languages required are:

- DEC Assembler,
- IBM Assembler,
- CRAY Assembler,
- Register IMP Assembler,
- FORTRAN,
- PL1,
- IMP.

The incumbent must be familiar with networking and protocol hardware, systems and architectures employed throughout the cryptanalytic community such as:

- Network Systems hardware,
- Netex protocol system,
- Transf bulk transfer system.

Adoption of other computer languages, operating systems, networking and forwarding systems may be required as the task demands. The incumbent must be flexible and quite adaptable to change as technology and practices improve.

The incumbent must develop and maintain familiarity with standard data formats and retrieval procedures used throughout the [cryptanalytic community?] to determine their impact upon our processing systems.

The incumbent must be fully aware of the various applications software packages available in the cryptanalytic community and their particular relevance and usefulness to current projects.

A broad knowledge of various modulation methods and multiplexing schemes is required such as:

- Amplitude Modulation,
- Frequency Modulation,

- Phase Modulation,
- Pulse Modulation,
- Frequency Division Multiplexing (FDM),
- Time Division Multiplexing (TDM),
- Time Division Multiple Access (TDMA).

Knowledge of various signal types, employed by potential targets, and their accessibility is necessary such as:

- teletypewriter,
- analog and digital voice and vocoder,
- facsimile,
- data systems.

REFERENCES

1. *Canadian Representatives Abroad.* 1978. Ottawa, Ont: Department of External Affairs. June. Also November 1980, November 1988, November 1990.

2. *Canadian Representatives Abroad and Representatives of Other Countries in Canada.* 1967. Ottawa, Ont: Department of External Affairs. August.

3. Computer Programmer (careers ad). 1981. *Ottawa Citizen.* January 10. 40.

4. Contract 12213-4-9120. 1984. Ottawa, Ont: Department of Supply and Services. Also Amendments 001-004. Released in severed form under Access to Information Act.

5. Contract 12213-5-0250. 1985. Ottawa, Ont: Department of Supply and Services. Released in severed form under Access to Information Act.

6. Contract W2213-7-4161. 1987. Ottawa, Ont: Department of Supply and Services. Also amendment 001. Released in severed form under Access to Information Act.

7. Dominguez, Christine, analyst, Statistics Canada. 1991. Letter to author. May 8.

8. Elcock, Ward, Deputy Clerk, Security and Intelligence and Counsel, Privy Council Office. 1990. *Minutes of Proceedings and Evidence of the Special Committee on the Review of the CSIS Act and the Security Offences Act.* Ottawa, Ont: House of Commons. Number 27, April 24. 5-6.

9. *Federal Government Employment, Catalogue 72-004 Quarterly*. 1981. Ottawa, Ont: Statistics Canada. January-March. Also April-June, July-Sept, Oct-Dec.

10. *Federal Government Employment*. 1984. Ottawa, Ont: Statistics Canada. October-December.

11. Hersh, Seymour. 1987. *"The Target is Destroyed"*. New York, NY: Vintage. 67n, 347n.

12. *In Flux But Not in Crisis, Report of the House of Commons Special Committee on the Review of the CSIS Act and the Security Offences Act*. 1990. Ottawa, Ont: House of Commons. September. 152-153.

13. *National Defence Telephone Directory, National Capital Region*. 1989. Ottawa, Ont: Department of National Defence. Spring. E-161.

14. *Ninth Annual Report, 1975-1976*. 1976. Ottawa, Ont: Public Service Staff Relations Board (PSSRB). 114.

15. O'Neill, Kevin, ed. 1987. *History of CBNRC*. Ottawa, Ont: CSE. Released in severed form under Access to Information Act.

16. PSSRB file 166-13-17850, exhibit 31 (*Requirements for a Career in Cryptanalysis*, CSE file O.109-86, 3 March 1986). 1990. Ottawa, Ont: PSSRB.

17. PSSRB file 166-13-17850, exhibit 63 (*Edwina Slattery: CSE competition no. M-144*, [1987]). 1990. Ottawa, Ont: PSSRB.

18. PSSRB file 166-13-17850, exhibit 107 (untitled, undated). 1990. Ottawa, Ont: PSSRB.

19. PSSRB file 166-13-17850, exhibit 110-A (*Reorganization of O1*, CSE file O.350-85, 17 June 1985). 1990. Ottawa, Ont: PSSRB.

20. PSSRB file 166-13-17850, exhibit 111 (*O1 Position Code Numbers*, CSE file O.359-85, 17 June 1985). 1990. Ottawa, Ont: PSSRB.

21. PSSRB file 166-13-17850, exhibit 112 (*Resubordination of the [deleted] Identification Systems Task*, CSE file ORG/6-4/ORG/6-2, 3 May 1984). 1990. Ottawa, Ont: PSSRB.

22. PSSRB file 166-13-17850, exhibit 113 (*O1 - 1979-1986*, undated). 1990. Ottawa, Ont: PSSRB.

23. PSSRB file 166-13-17850, exhibit 132 (*Staff training and development: course notice*, CSE file U2/012-85, 7 January 1985). 1990. Ottawa, Ont: PSSRB.

24. PSSRB file 166-13-17850, exhibit 133 (*O1 training requirements*, CSE file O.19-86, 9 January 1986). 1990. Ottawa, Ont: PSSRB.

25. *Public Works Canada Annual Report 1989-90*. 1991. Ottawa, Ont: Department of Public Works. 42.

26. Seaborn, Blair, former Intelligence and Security Co-ordinator, Privy Council Office. 1990. *Minutes of Proceedings and Evidence of the Special Committee*

on the Review of the CSIS Act and the Security Offences Act. Ottawa, Ont: House of Commons. Number 16, February 20. 10.

27. Spy agency was given millions, CBC reports. 1989. *Vancouver Sun.* March 21. A8.

28. *Tenth Annual Report, 1976-1977.* 1977. Ottawa, Ont: PSSRB. 118.

29. *Tillian (CBNRC/CSE newsletter).* 1964. Ottawa, Ont: CSE. May. 4. Released in severed form under Access to Information Act. Data for other years collected by the author from a variety of sources.

30. *Tillian.* 1980. Ottawa, Ont: CSE. Spring. 16, 38. Released in severed form under Access to Information Act.

31. *Tillian.* 1980. Ottawa, Ont: CSE. Autumn. 2. Released in severed form under Access to Information Act.

32. *Tillian.* 1981. Ottawa, Ont: CSE. Number 1. 17. Released in severed form under Access to Information Act.

33. Wexler, Muriel Korngold. 1990. *Record of decision, file 166-13- 17850.* Ottawa, Ont: PSSRB. March 12. All statements attributed to CSE personnel from this source are paraphrases by Wexler of oral testimony presented during the adjudication proceedings.

34. Yonson, Doug. 1989. The secret's out: Spy headquarters to get $30-million addition. *Ottawa Citizen.* January 22. A1-A2.

BIOGRAPHICAL SKETCH

Bill Robinson is a researcher for Project Ploughshares, the Canadian Council of Churches project on militarism and disarmament. In his spare time he is working on a book about Canadian signals intelligence from the Second World War to the present. He would appreciate contact with anyone who has information on this subject.

ORIGINS OF CRYPTOLOGY: THE ARAB CONTRIBUTIONS*

Ibrahim A. Al-Kadi†

ADDRESS: Advanced Electronics Company, Ltd., P. O. Box 90916, Riyadh 11623, SAUDI ARABIA.

ABSTRACT: Recently discovered old manuscripts show that the origin of cryptology, and the Arab contributions to it, are older and more extensive than previously thought. The word 'cipher' in European languages comes from the Arabic word *ṣifr*. The 9th-century Arab scientist al-Kindī is the author of the oldest known book on cryptology, antedating any other by more than 300 years. This paper highlights the specific contributions of some Arab cryptologists based on newly discovered documents that include books of al-Kindī, ibn Adlān and ibn ad-Duraihim. Factors behind the emergence and advancement of Arab cryptology are discussed. The discoveries reported in this paper push the frontiers of the history of cryptology back by about 500 years.

KEYWORDS: History of cryptology, Arab cryptology, al-Khalīl, al-Kindī, ibn Dunainīr, ibn Adlān, ibn ad-Duraihim, combinatorics, statistics, statistical cryptanalysis.

1. INTRODUCTION

In his celebrated history of cryptology, *The Codebreakers*, David Kahn asserts that "Cryptology was born among the Arabs" [9, p. 93]. Kahn attributes most of his account of Arab cryptology to the section on cryptology in the 1412 encyclopedia *Ṣubḥ al-A'sha*, written by the Arab scholar al-Qalqashandī [2,9,13]. Nearly all of al-Qalqashandī's writings on cryptology was taken from an earlier and more extensive book of ibn ad-Duraihim (1312-1361) entitled *Miftāḥ al-Kunūz fī Iḍāḥ al-Marmūz* (Treasured Key for Clarifying Ciphers), which was considered among the lost books of cryptology. This book was recently discovered and published [12].

*This paper is based on two presentations given by the author at the Swedish Royal Institute of Technology in Stockholm and at the University of Linköping in Linköping SWEDEN during September 1990.

†The author is on temporary leave from his permanent position as an Associate Professor, Electrical Engineering Department, College of Engineering, King Saud University, P. O. Box 800, Riyadh 11421 SAUDI ARABIA.

Actually, Arab cryptology is much older than Kahn's account suggests, and Arab contributions to it are much more extensive than previously thought. Recently discovered old manuscripts show that the origin of cryptology is attributed to the 9th-century Arab scientist al-Kindī. His study of cryptology is the oldest available book on the subject, antedating any other by about three centuries, although his predecessor al-Khalīl (718-786) is reported to have written *Kitāb al-Mu'ammā* (The Book of Cryptographic Messages) about a century earlier. Unfortunately this book has not been found.

The aim of this paper is to report and discuss not-widely-known Arab contributions to cryptology. The study is largely based on the recent discovery of three old Arab manuscripts of cryptology including al-Kindī's book, a book by ibn Adlān, in addition to the long-reported but never-seen book by ibn ad-Duraihim. These books were edited and published (in Arabic) by a group of researchers at the Arab Academy of Damascus [12].

The next section briefly discusses the factors that led to the emergence and advancement of cryptology among the Arabs. The origin of the word 'cipher' is touched upon in Section 3. Section 4 introduces some known Arab cryptologists and their works in general. The following five sections (5 to 9) highlight the contributions of al-Khalīl, al-Kindī, ibn Adlān, ibn Dunainīr and ibn ad-Duraihim, respectively. The original contributions of al-Kindī, in particular, are discussed in some detail in Section 6.

2. FACTORS BEHIND THE ARAB ADVANCEMENT IN CRYPTOLOGY

Many factors contributed to the early emergence and advancement of cryptology in the Arab civilization. These can be grouped into five main related factors, each of which is discussed in the following.

2.1 Translation

Like many other rising civilizations, the Arabs considered knowledge as a shared human wealth. They started, shortly after their emergence from Arabia, to absorb the learning of older neighboring civilizations. One important means of their acquiring knowledge was through translation. In particular, the first three centuries of the Islamic civilization (700-1000 A.D.) witnessed intensive translation activities into Arabic of major works written in Greek, Indian, Farsi, Syriac, Armenian, Hebrew, Roman and other languages. The Caliph al-Ma'mūn (reigned 813-833) established the famous Bait al-Hikmah (The House of Wisdom) in Baghdad as a national translation academy, state library and research center [1,3]. So much emphasis did the Arabs place on acquiring other cultures' knowledge that Greek manuscripts were obtained from the Byzantine Empire through peace treaties in return for maintaining the status quo. The main subjects of translation included philosophy, astronomy, mathematics, physics, chemistry and medicine. Some source books were written in dead languages forcing the Arab scholars to study them as ciphertexts and to classify their letters and symbols. They developed cryptanalytic techniques to try to understand those writings. In addition, some writings, were enciphered, especially in the fields of philosophy, theology[1], witchcraft and chemistry[2].

A number of Arab scholars have written on other languages and alphabets. Thobān al-Maṣry (? - 859) wrote a book entitled *Ḥall ar-Rumūz wa Brā'a al-Asqām fī Oṣūl al-Lughāt wal-Aqlām* (Solution of Symbols and Answers of Doubts on Principles of Languages and Scripts) [12]. The book of Aḥmad ibn Waḥshiyyah (? - 919), *Shawq al-Mustahām fī Ma'rifat Rumūz al-Aqlām* (Seekers' Joy in Learning about Other Languages' Written Symbols) contained 93 alphabets of different languages [12]. Nine centuries later, this book was translated from Arabic into English by the orientalist Joseph von Hammer and was published in both languages in London in 1806 under the title *Ancient Alphabets and Hieroglyphic Characters Explained* [2,9,7]. The French scholar Sylvestre de Sacy published a study about the book in Paris in 1810 that might have helped the famous archeologist J. F. Champollion in deciphering hieroglyphics [12].

2.2 Linguistic Studies

Being the language of the *Holy Qur'ān*, Arabic spread swiftly across vast areas of the world. It also served in different parts of the Islamic empire as a unifying official and scientific language. The first four centuries of the Arab civilization, in particular, witnessed intensive activities in studies of different linguistic aspects of Arabic, including phonetics, morphology, syntax, semantics, lexicography, grammar, prosody and computational linguistics. Such studies were prerequisites for the advancement of cryptography and cryptanalysis. Consequently many of the Arab cryptologists were also prominent linguists (see Sections 4-9).

[1] Some extremist sects of Islam cultivated cryptography to conceal their writings from the orthodox [9, p. 93].

[2] Chemistry was considered a precious secret because it was believed that through chemistry cheap metals could be converted to gold!

2.3 Administrative Studies

The need of the emerging Islamic state for administrative organization and services led the Arabs early on to concentrate on developing secretarial and management skills especially in the fields of composition and correspondence. The practical need to protect sensitive state affairs, particularly in the mails, was one of the main driving forces behind using, and therefore writing about, cryptography and cryptanalysis. The hands-on experience of practicing message concealment and codebreaking is evident in the available biographies of many Arab writers and secretaries who were considered among the most important personnel in government offices, especially in the Caliph's court [1,12]. Use of special codes by tax officials and accountants is well documented [9,10].

Many Arab administrative scientists wrote on cryptology either within their books of general administration or separately. Among those were Abū Bakr aṣ-Ṣūlī (? - 946) in his book *Adab al-Kuttāb* (The Secretaries' Manual), *Is-ḥāq al-Kātib* (~10 Cent.) in his book *al-Burhān fī Wōjōh al-Bayān* (The Authoritative Guide to the Art of Articulation), the 'father of sociology' Abdur-Raḥmān ibn Khaldūn (1322 - 1406) in his book *al-Muqaddimah* (The Introduction [to his history Encyclopedia]) which was called by the British historian Arnold Toynbee "undoubtedly the greatest work of its kind that has ever yet been created by any mind in any time or place", and al-Qalqashandī in Ṣubḥ al-A'shā [2,9,10,12,13].

2.4 Public Literacy

The Arab-Islamic civilization witnessed an expansion of reading and writing skills never approached by any previous civilization. This unprecedented level of literacy advanced cryptology in two ways:

- It increased the need to protect sensitive written messages. (In illiterate societies a mere writing of messages guaranteed their secrecy).
- The larger pool of educated people bred many qualified scientists in fields related to cryptology, such as translation, linguistics, mathematics, and administrative sciences.

2.5 Advanced Mathematics

The Arabs acquired the mathematical knowledge of their neighbors, especially the Greeks and Indians. During the first three centuries of Islam, in particular, they embarked on an extensive campaign of translation and absorption of major mathematical works of the time including the books of Euclid (*Elements*),

Ptolemy (*Tetrabiblos* and *Al-magest*), Archimedes, Aristotle, Diophantus (*Arithmetica*) and Brahmagupta (*Siddhānta* which the Arabs called *Sindhind*). By the early ninth century, Arab mathematicians started to produce major contributions of their own in the established fields of mathematics (arithmetic, trigonometry and geometry), but, more importantly, they pioneered in establishing two new fields, namely algebra and statistics, both of which are extremely important in modern cryptology. The glorious history of Arab mathematics cannot possibly be given a fair treatment here, but excellent references abound (see for example [1-6, 11, 14-16]). Here only aspects related to cryptology will be discussed.

Al-Khwārizmī (780 - 847?) is the most famous Arab mathematician. Algorismus, as he was known in Latin, was probably the best mathematician of his time [11]. He wrote two books on arithmetic and algebra, which played important roles in the history of mathematics. The original Arabic version of his first book (written around 820) on arithmetic is lost, but its contents have survived in different Latin translations. It was through those Latin versions of al-Khwārizmī's book that the (Hindu-)Arabic numerals were introduced into Europe (see Section 3). Al-Khwārizmī's second book, *al-Jabr wal-Muqābalah*, established the new field of algebra. It was translated in Spain into Latin by the Englishman Robert of Chester in the 12th century under the title *Algebra et Almucabala* [3, 11, 9]. From al-Khwārizmī and his works stem a number of common technical terms such as 'zero', 'cipher', 'algorithm', 'algebra' and 'Arabic numerals'. As Boyer wrote in his history of mathematics [3, p. 252], to al-Khwārizmī belongs the title "the father of algebra."

The Arabic foundation of algebra is widely acknowledged, but this is hardly the case with statistics. Historians of mathematics (see Boyer [3, p. 397] for example) attribute the first writings on probability and statistics to correspondence between Pascal and Fermat in 1654. In his recently discovered manuscript, al-Kindī gave the first description of statistical methods in cryptanalysis. He even explicitly required texts to be long enough to allow letter statistics to be meaningful (see Section 6). This is the world's first known writings in statistics, antedating those of Pascal and Fermat by about 800 years[3]. Statistical techniques were routinely used by Arab cryptologists after al-Kindī (see Sections 6 to 9).

[3]Historians of science and mathematics are urged to investigate the evidences of the new findings reported here.

3. ORIGIN OF "CIPHER"

Up to the thirteenth century (A.D.), the West was using Roman numerals (I, V, X, L, C, D, M), which were very cumbersome and made the simplest arithmetic operation (e.g. addition) extremely complicated. The reason is that the Roman number system is not a decimal system and has no symbol for an empty space (the zero). The lack of the concept of zero was the great stumbling block for the medieval mathematicians in Europe [11].

The decimal number system and the zero were originally developed in India. The Arabs became aware of the new numerals,in the early ninth century, through their translation of Brahmagupta's *Siddhānta* from Sanskrit into Arabic. The new numerals together with the zero were quickly adopted and put into widespread use throughout the Islamic empire from China in the east to Spain in the west. The Hindu-Arabic numerals were later introduced to Europeans through their contacts with the Arab civilization in Spain, where these numerals became known simply as the Arabic numerals. The spread of the new numerals throughout Europe started in the twelfth century with the Latin translations of al-Khwārizmī's book on arithmetic first by Robert of Chester and then by John of Seville [11]. In the thirteenth century, many authors helped to popularize the new Arabic numerals. Among those were the French Alexander de Villa Die (~1225), the English John of Halifax (~1200-1256), also known as Sacrobosco, through his book *Algorismus vulgaris*, and the Italian Leonardo of Pisa (~1180-1250), best known as Fibonacci, who studied under a Muslim teacher and traveled in Egypt, Syria and Greece. Fibonacci later wrote his book *Libre abaci*, in which he strongly advocated the use of Arabic numerals [3].

The new concept of zero confused people in the Middle Ages Europe[11]. In Sanskrit, the zero was called sunya or "empty". The Arabs translated the Indian name into its Arabic equivalent *ṣifr*. When the Europeans became aware of the new digit, they no longer translated the name, but adopted the concept and the symbol as well as its Arabic name *ṣifr* which was transformed into the Latin words *cifra* and *cephirum* (as written by Fibonacci). The two Latin words worked their way into different European languages. In Italian *cephirum* was changed to *zefiro*, *zefro*, and *zevero* which was later shortened to *zero*.

The French formed the word *chiffre*, and also took over the Italian *zero*. The English used *zero* and *cipher* from which the word ciphering was some times used in the sense of 'computation'. The German used *ziffer* and *chiffer* [11].

The concept of the *zero* or *ṣifr* or *cipher* was so confusing and ambiguous to common Europeans at first that a person used to say in arguments or discussions that he is "talking about something clear and comprehensible, and not about some ambiguous and far-fetched thing like the cipher." From such common confusion developed the use of the word "*cipher*" to mean concealment of clear meaning of messages or simply encryption [12]. In conclusion, the Arabic word *ṣifr* for the digit "*zero*" (0) developed into European technical terms that mean encryption.

	Scientist Name	Period	Comments	References
1	Al-Khalīl	718 - 786	see Section 5	[9,12]
2	Jābir ibn Ḥayyān (The Chemist) [known in Europe as Geber]	? - 815	*Ḥall ar-Rumūz wa Miftāḥ al-Kunūz* (Code Solving and Key to Treasures)	[1,4,12]
3	Thobān al-Maṣry	? - 859	See Subsection 2.1	[12]
4	Al-Kindī	801 - 973	See Section 6	[1,4,12]
5	Ibn Waḥshiyyah	? - >919	See Subsection 2.1	[2,9,12,15]
6	Moḥammad ibn Aḥmad ibn Kaisān	9th Cent.	Wrote a book on cryptography and prosody	[12]
7	Moḥammad ibn Aḥmad ibn Tabātabā	? - 934	Wrote a book on cryptanalysis preserved in the Ottoman Archive	[12]
8	Is-ḥāq al-Kātib	~10th Cent.	See Subsection 2.3	[12]
9	As'ad ībn Muhadhdhab ibn Mammāti	1149 - 1209	*Khaṣā'iṣ al-Ma'rifah fī al-Mu'ammiyāt* (Guides of Knowledge on Cryptograms)	[12]
10	Ibn Adlān	1187 - 1268	See Section 7	[12]
11	Ibn Dunainīr	1187 - 1229	See Section 8	[12]
12	Ibn ad-Duraihim	1312 - 1361	See Section 9	[2,9,12,13]
13	Ali ibn Moḥammad ibn Aidamur al-Jaldakī	? - >1341	*Kanz al-Ikhtiṣāṣ fī Ma'rifat Asrār Ilm al-Khawāṣ* (Treasures of Specialization in Learning Secrets of Elites' Science)	[12]
14	Al-Qalqashandī	1355 - 1418	See Sections 1 and 9	[2,9,13]

Table 1. Some known Arab Cryptologists.

4. EARLY ARAB CRYPTOLOGISTS

Cryptography has been practiced to conceal messages since antiquity by different civilizations, including the ancient Egyptian, Chinese, Indian, Mesopotamian, Greek and Roman. But in none of them was there any cryptanalysis; only cryptography existed [9]. Cryptology, the science of both making ciphers (cryptography) and breaking them (cryptanalysis), was born among the Arabs shortly after the rise of the Arab-Islamic empire. Many Arab scholars wrote on, and excelled in practicing, both branches of cryptology. Table 1 lists, in a chronological order, Arab cryptologists known to us who wrote on the subject. The space limitations, and the meager information available on some of them, preclude writing in detail on each of the individuals in the table, but we will highlight the contributions of some of them in the following five sections.

5. AL-KHALĪL

Abū Abdur-Raḥmān al-Khalīl ibn Aḥmad ibn ‘amr ibn Tammām al-Farāhīdī (718-786), who was among the greatest Arab linguists of all times, was also the world's first great philologist. Among other achievements, his record of ‘firsts’ includes the following:

1. He was the first to conceive the idea of, and produce, a comprehensive dictionary. *Al-Ain*, a dictionary of Arabic vocabulary, is the oldest known Arabic book of its kind.
2. He was the first to study and classify the prosody of Arabic poetry.
3. He wrote the first known book on cryptology, *Kitāb al-Mu‘ammā* (The Book of Cryptographic Messages). This book is apparently lost [9,12].
4. He used combinatorics to count all possible Arabic words by considering all different combinations of letters with and without vowels. Appendix B presents a sample of al-Khalīl's writings on this subject. His writings on the subject are eloquent and easy to understand.

Although considered the greatest linguist of his time, al-Khalīl was not known to be a mathematician. Yet his spontaneous use and apparent command of combinatorics (which may as well be the first such use in history) without much fanfare or elaboration raise a number of interesting questions that are worth consideration by historians of mathematics. Did the Arabs develop combinatorics and use it to such an extent that a nonmathematician linguist found it natural to employ it implicitly within a dictionary? Or was it the modesty of this great scholar that made him unaware of the importance of his ideas that could have contributed greatly to mathematics early in the 8th century?

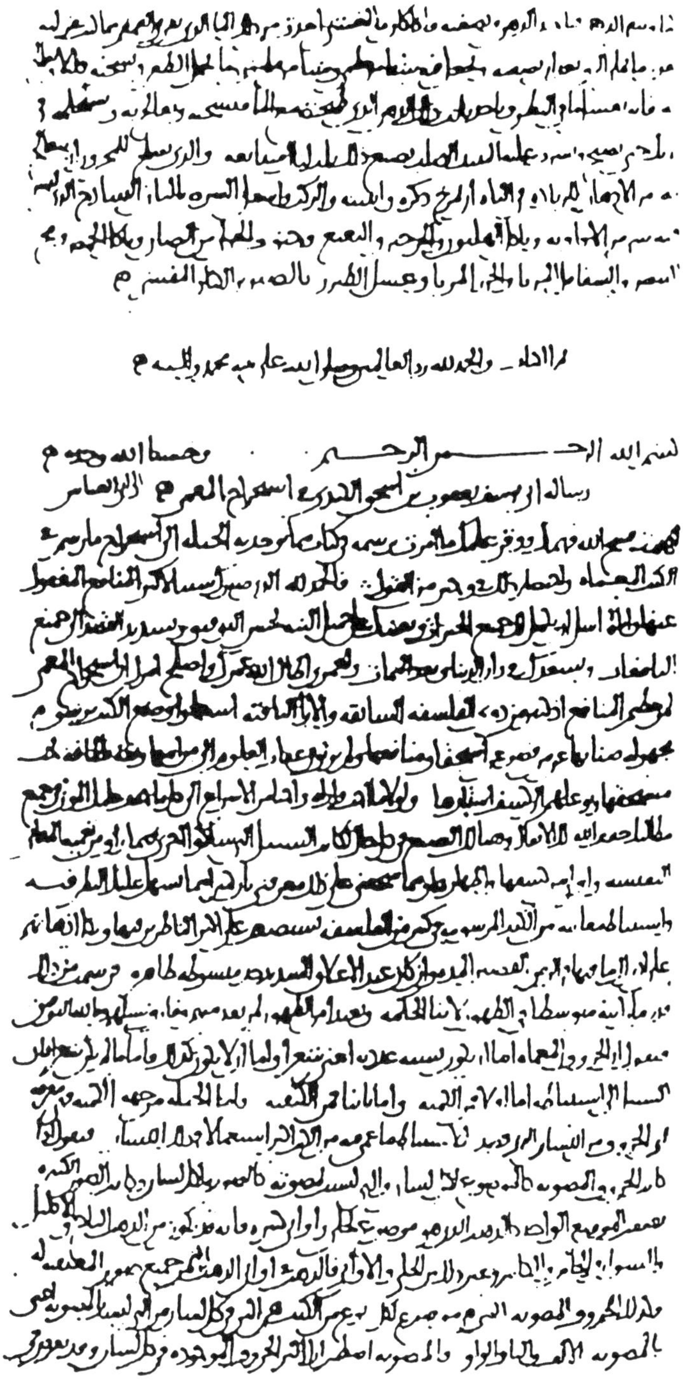

Figure 1. The first page of al-Kindī's manuscript [12].

6. AL-KINDĪ

Abū Yūsūf Ya'qūb ibn Is-ḥāq ibn aṣ-Ṣabbāh ibn 'omrān ibn Ismaīl al-Kindī, the author of the oldest known book on cryptology, was born around the year 801 A.D. in al-Kufah, where his father was a governor. He grew up in Baghdad, where he got his education. He excelled in many fields including philosophy, medicine, astronomy, mathematics, linguistics and music. Al-Kindī, the first philosopher of Islam, was known as "The Philosopher of the Arabs"[4]. He became the caliph's court physician, and was later named the first director of Bait al-Hikmah Library and Translation Academy, where he led the ongoing translation campaigns (see Subsection 2.1). Three of the caliphs al-Kindī served supported his research pursuits, but the unsympathetic attitude of the fourth caliph towards his philosophical thoughts brought about loss of prestige and personal fortune. His large famous personal library, the al-Kindiyyah, was confiscated and he was publicly beaten. He returned home sad and with a broken spirit; he later died in Baghdad in 873 at the age of seventy-two.

Al-Kindī left a wealth of knowledge in diversified fields. His writings were authoritative and encyclopedic, covering about 290 titles in different subjects. One of his books, *Risālah fī Istikhrāj al-Mu'ammā* (A Manuscript on Deciphering Cryptographic Messages), is the oldest available book on cryptology, antedating any other by more than 300 years. A manuscript of it was recently discovered in the Sulaimaniyyah Ottoman Archive in Istanbul. It was edited and published by the Arab Academy of Damascus in 1987 [12]. Figure 1 shows the first page of al-Kindī's manuscript.

The main contributions to cryptology of al-Kindī are briefly discussed in the following:

1. **Cryptanalysis Techniques**: Al-Kindī started by discussing principles of cryptanalysis. He divided cryptograms into two types–those of normal texts and those of poetry. For normal texts he specified four cryptanalysis methods: quantitative techniques (which he called 'quantitative tricks'), qualitative techniques ('tricks'), probable words (which he called 'opening statements and praises'), and vowel-consonant combinations. For poetry he called for using prosody in addition to all the above methods of normal texts.

2. **Cipher Types**: Al-Kindī gave a tree-diagram classification of the major types of cipher systems, described each, and explained how to cryptanalyze

[4] The Arabs reserved the more prestigious title, "The Philospher" for Aristotle, for whom they held much respect and admiration.

them. Figure 2a shows his tree diagram as it appeared in the manuscript, while a translation of it is given in Figure 2b. Al-Kindī also introduced super encipherment, but he did not discuss it in much detail. All he wrote about it is the following:

> ... [Super-encipherment] is composed of a number of simple [ones], by using two or more possible combinations of simple encipherments ... Solving it is more difficult [12, pp. 224 and 234].

3. **Arabic Phonetics**: He classified Arabic phonetics into consonants (he used the term 'silent'), long vowels (which he called "al-muṣṣawiytāt al-'iẓām" = major louds) and short vowels ("al-muṣṣawiytāt aṣ-ṣighār" = minor louds).
4. **Arabic Syntax**: Al-Kindī gave a detailed study of Arabic syntax, in which he explained possible and impossible letter combinations. Figure 3 shows his study in the manuscript.
5. **Statistical Cryptanalysis**: The above contributions are remarkable, but the most important and original of al-Kindī's contributions comprises the world's first description and use of statistical techniques in cryptanalysis (or in any other application, for that matter). In his discussion of the quantitative techniques in cryptanalysis, al-Kindī clearly described how to use letter frequency statistics of the cryptogram to solve it. He also explained how to find these letter frequencies (by using a sample of the same language). Equally important and original is the condition he set on the length of the ciphertext under consideration; he required texts to be long enough to allow letter statistics to be meaningful. This concept, introduced by al-Kindī more than 1100 years ago, is extremely important in statistics today. The ideas of al-Kindī are remarkably clear in his manuscript. He wrote:

> One way to solve an encrypted message, if we know its [original] language, is to find a [different clear] text of the same language long enough to fill one sheet or so and then we count [the occurrences of] each letter of it. We call the most frequently occurring letter the "first", the next most occurring the "second", the following most occurring the "third" and so on, until we finish all different letters in the cleartext [sample]. Then we look at the cryptogram we want to solve and we also classify its symbols. We find the most occurring symbol and change it to the form of the "first" letter [of the cleartext sample], the next most common symbol is changed to the form of the "second" letter, and the following most common

Figure 2a. Al-Kindī's tree diagram classification of cipher types as it appears in his manuscript [12].

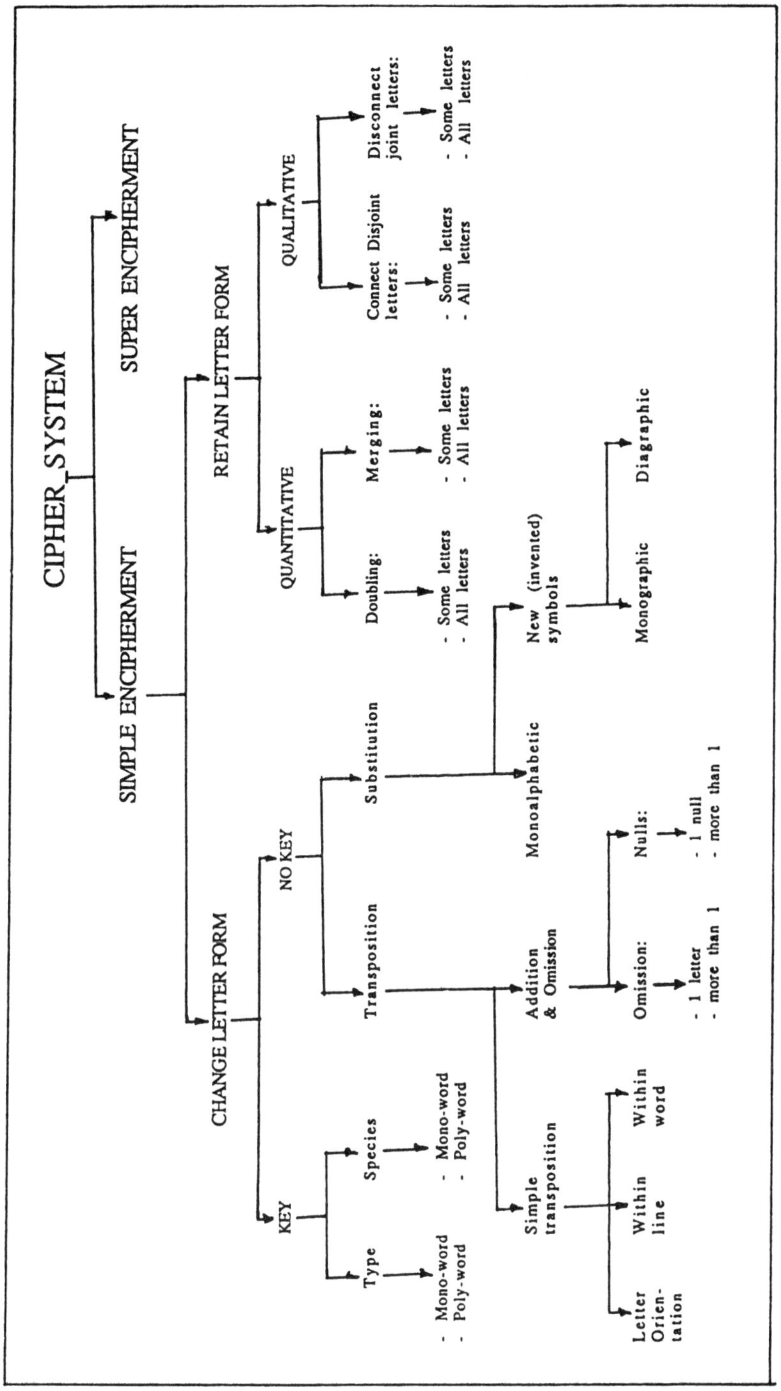

Figure 2b. A translation of al-Kindī's tree-diagram classification of cipher types.

symbol is changed to the form of the "third" letter and so on, until we account for all symbols of the cryptogram we want to solve.

It could happen sometimes that the cryptogram is too short to have all different letters. The high and low [frequency] counts will not be correct, for high and low counts are only correct in long enough messages to correspond to all places of frequent and rare occurrences so that if some letters are [too] few in one segment of the message, they will be [too] many in others. But if the cryptogram is too short, equivalence does not apply, letter ranks are not correct and [consequently] a second trick should be used to recover letters. Such a trick is qualitative which is ... [here al-Kindī wrote in detail on possible letter combinations in a language like 'al' in Arabic... etc.] [12, p. 216].

6. **Arabic Letter Frequency**: Later in his book, al-Kindī set out the first tally of Arabic letter frequencies. His sample consisted of "seven pages of Arabic" totaling 3667 letters. Al-Kindī's results are shown in Figure 4, together with results of a much more recent study [8]. Not only did al-Kindī pioneer in computing letter frequencies, but he also analyzed his results and explained what might appear as a partial contradiction between the assertion at the beginning of his book, that vowels are the most common letters on all languages, and the fact that his results show that although the Arabic vowel 'Alif' is the most common letter, the other two vowels 'Wāw' and 'Yā' come only fifth and sixth after the consonants 'Lām' (l), 'Mīm' (m) and 'Hā' (h). He explained that vowels are underrepresented in written Arabic texts because short vowels are not usually written explicitly in Arabic scripts (see Appendix A). Al-Kindī wrote:

> We said before that vowels are naturally the most common letters in all languages... Our results show that [the consonant] Lām (l) is more frequent in the Arabic language than [the vowels] Wāw and Yā. So are [the consonants Mīm (M) and] Hā (h). This is not a contradiction to what we previously said, because vowels appear in Arabic writing [only] if they are major [or long]. Minor [i.e., short] vowels do not appear in Arabic scripts... We have explained this in our [other] book [entitled] *fī Ṣinā'at ash-Sh'ir* [On the Art of Poetry]... [12, p. 236].

Al-Khalīl's combinatorics and al-Kindī's statistics are, most probably, the world's first writings in the field of computational linguistics.

Figure 3. A page of al-Kindī's manuscript, in which he discussed Arabic Syntax and impossible letter combinations [12].

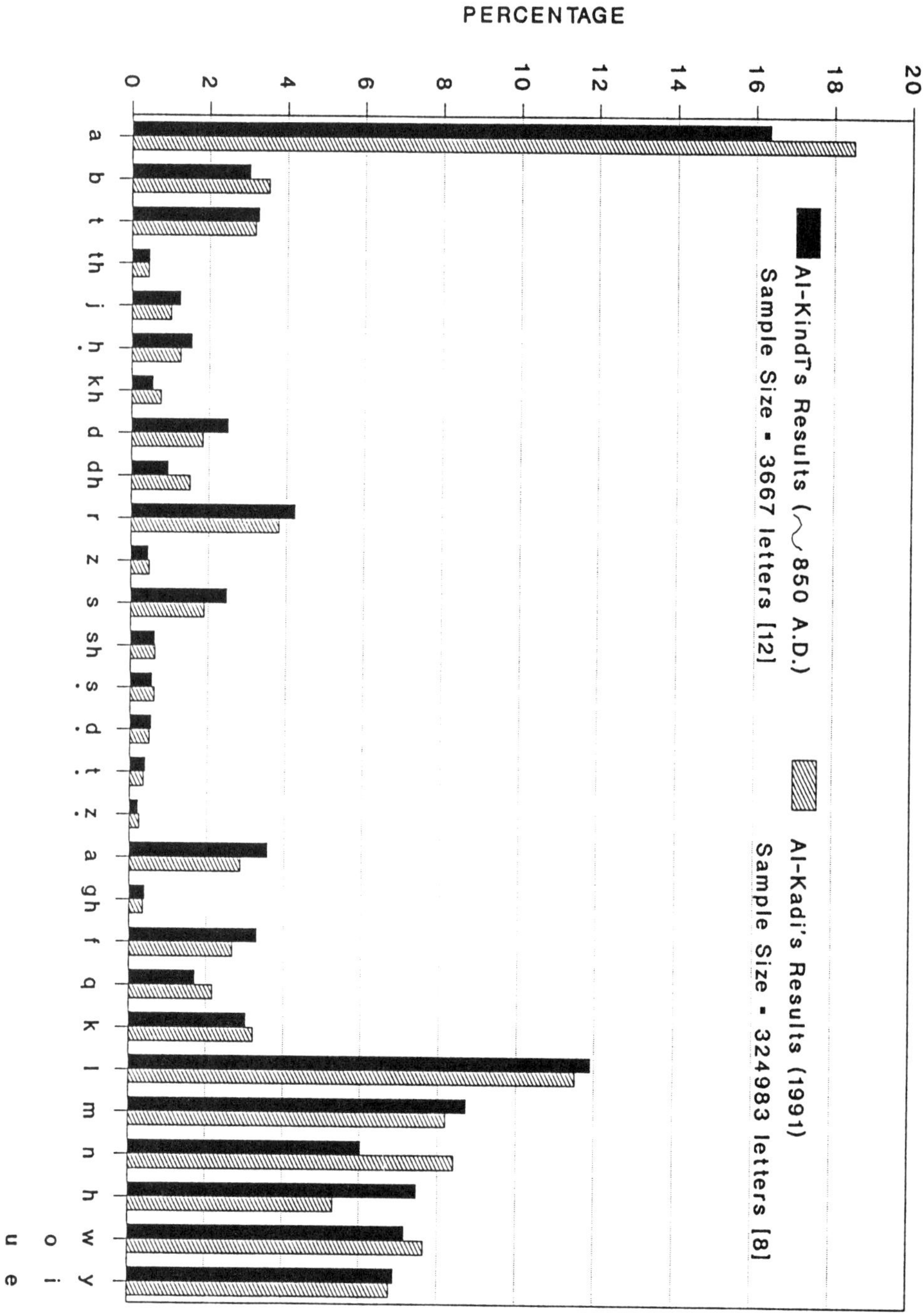

Figure 4. Al-Kindī's computed letter frequencies in Arabic, compared with recent results.

Thus it appears that al-Kindī is the father not only of cryptology but of statistics as well (see Subsection 2.5).

7. IBN ADLĀN

Afīf ad-Dīn Alī ibn Adlān ibn Ḥammād ibn Alī al-Mouṣilī an-Naḥwi al-Mutarjim was born in Mouṣil (1187), educated in Baghdad, lived in Damascus for some time, and taught in Cairo, where he died in 1268 at the age of eighty-one. He was a poet who also excelled in grammar and linguistics, but he did not write many books. He left four, two of which were on cryptology. The first, *al-Mu'lam* (The Told [Book]), is lost, while the second, *al-Mu'allaf lil-Malik al-Ashraf* (The [Book] Written for King al-Ashraf), was recently discovered and published [12]. Figure 5 shows the title page and the first page of the document. The book is a manual or user-guide for cryptanalysis; unlike al-Kindī, ibn Adlān does not go into cryptography much. At the beginning, he discusses qualifications necessary for cryptanalysts, examples of simple substitution ciphers and Arabic letter combinations. The main body of the book is devoted to techniques of cryptanalysis where Ibn-Adlān gives 20 rules in 9 topics. In the conclusions, he works out a detailed example of solving a cipher as a training exercise for the reader. Ibn Adlān's main contributions include the following:

- The concept of variable key for simple substitution. (He called it the 'controller', and used different verses of Arabic poetry as a variable key.)
- A detailed study of word spaces, identification of word starting points in a cryptogram, probable words and doubled letters.
- An explicit lower bound on the required ciphertext length. He wrote:

> The text to be solved should be [at least] 90 letters long approximately because letters will have circulated about three times [$3 \times 28 = 84$]. [12, p. 276]

- A classification of Arabic alphabet characters according to their frequencies into three groups–common (7), medium (11) and rare (10).
- An eloquent articulation of cryptanalysis methodology. In Rule 18, ibn Adlān clearly stated the consecutive steps of solving a cryptogram–moving from ciphertext to plaintext systematically from the unknown (ciphertext) through the set of all possible solutions, to suspected solutions, to probable ones and finally to the confirmed solution (the correctly solved plaintext).

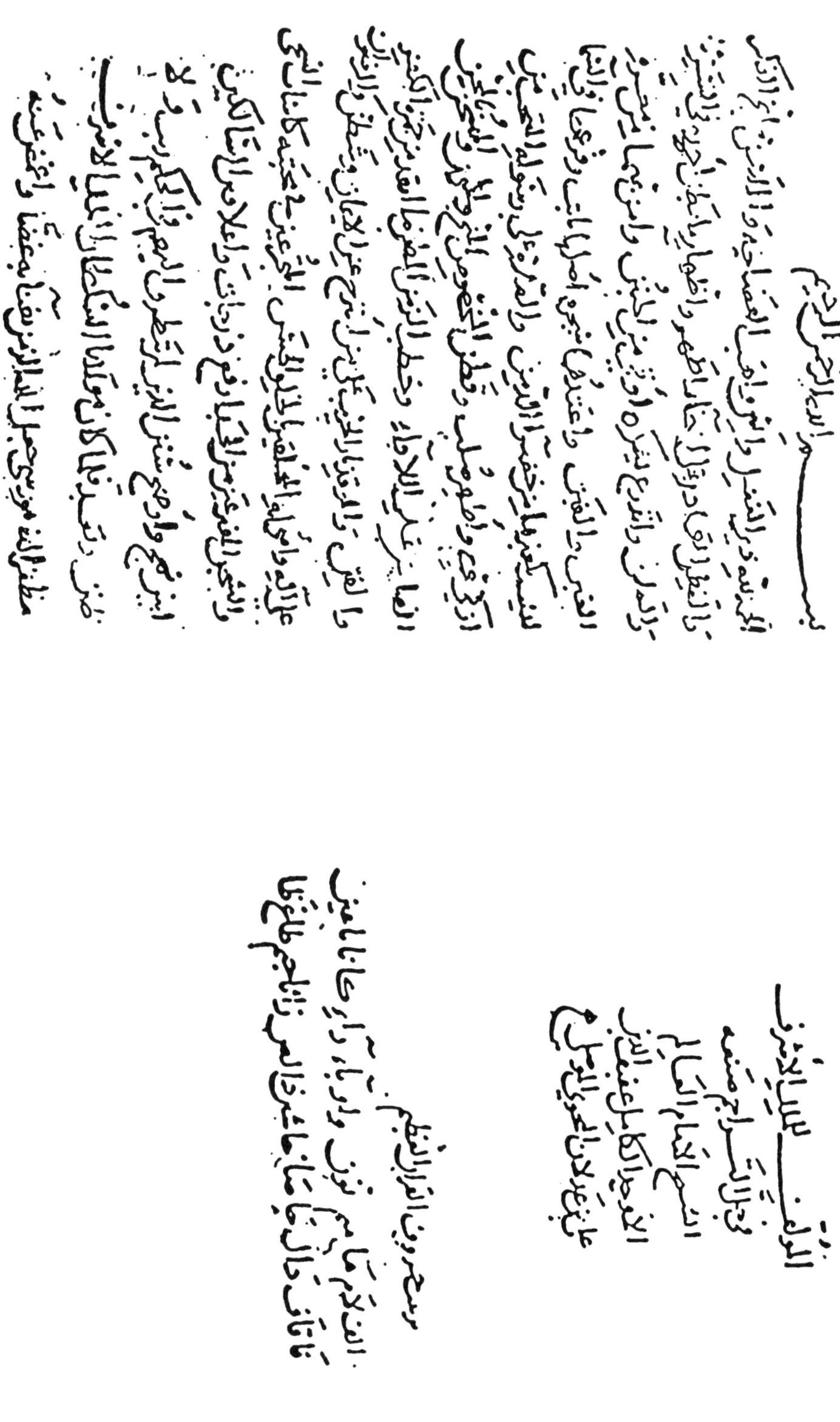

Figure 5. The title page and the first page of ibn Adlān's manuscript [12].

He briefly but lucidly explained that one should consider all possible solutions, fix the sure parts of them, neglect the impossible combinations, and try all suspected solutions using elimination to arrive at the final solution. This is probably his most impressive contribution.

8. IBN DUNAINĪR

Ibrāhīm ibn Moḥammad ibn Dunainīr was born in Mouṣil (1187), lived in Cairo and Damascus, and died in Baniyas (1229) at the age of forty-two. He wrote at least one book on cryptology, entitled *Maqāṣid al-Fuṣūl al-Mutarjamah an Ḥall at-Tarjamah* (Clear Chapters' Goals on Solving Ciphers). The historical manuscript was recently discovered but has not yet been published [12]. In addition to his use of statistical technique in cryptanalysis pioneered earlier by al-Kindī (see Section 6), ibn Dunainīr was the first to describe an arithmetical cipher, or what he called "sentence reckoning", in which cleartext characters are converted into numbers and then some simple arithmetic operations are performed on them (see Figure 7 for some examples). He was also the first to describe some elementary cipher mechanisms, such as the chessboard [12].

9. IBN AD-DURAIHIM

Tāj ad-Dīn Alī ibn Moḥmmad ibn Abdul'azīz ibn ad-Duraihim was born in Mouṣil in 1312. A wealthy merchant who traveled between Damascus and Cairo, he held various teaching and official positions in both cities. Ibn ad-Duraihim died in Qaus on his way to Eritrea in 1361 at the age of forty-nine. He excelled in theology, arithmetic, linguistics and cryptology. He wrote about 80 books on different subjects. Ibn ad-Duraihim's most extensive work on cryptology is contained in his book *Miftāḥ al-Kunūz fī Iḍāḥ al-Marmūz* (Treasured Key for Clarifying Ciphers) to which most of al-Qalqashandī' writings on the subject in *Ṣubḥ al-A'shā* are attributed. The book of ibn ad-Duraihim was considered lost [9, p. 95] until fairly recently when it was found, in Istanbul's Sulaimaniyyah Ottoman Archives, and published in a book by Mrayati et.al. [12]. Figure 6 shows two pages of the manuscript. Our discussion of ibn ad-Duraihim's contributions will be brief and will concentrate on aspects not sufficiently covered by other authors [2,9,13].

Ibn ad-Duraihim starts his book by briefly outlining the prerequisites of successful cryptanalysis. He then, also briefly, discusses the alphabets of 15 different languages including Greek, Farsi, Hindi and French. (Al-Qalqashandī included less than a third of this discussion in his book [13]). Like his predecessors, ibn

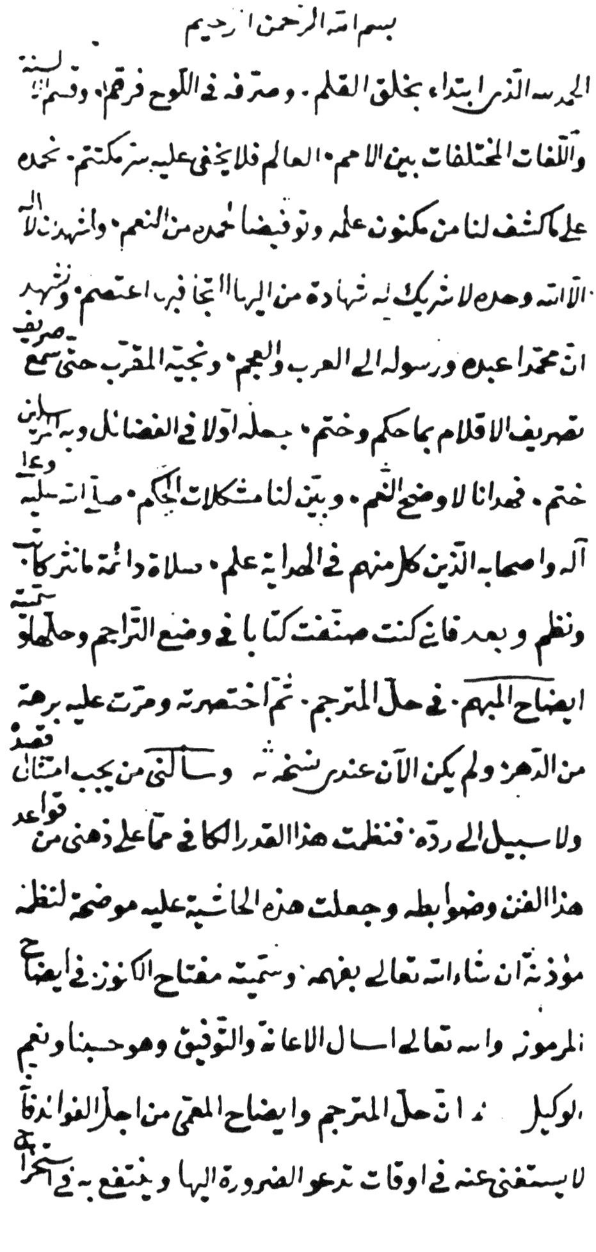

بسم الله الرحمن الرحيم

الحمد لله الذي ابتدأ بخلق القلم . وصرفه في اللوح فرقم . وقسم الألسنة
واللغات المختلفات بين الأمم . العالم فلا يخفى عليه سر مكتتم . نحمده
على ما كشف لنا من مكنون علمه وتوفيقنا لحمده من النعم . وأشهد أن لا إله
إلا الله وحده لا شريك له شهادة من إليها الالتجا في اعتصم . ونشهد
أن محمدا عبده ورسوله إلى العرب والعجم . ونجيبه المقرب حتى سمع صريف
تصريف الأقلام بما حكم وختم . جعله أولا في الفضائل وبه المرسلين
ختم . فهدانا لأوضح النعم . وبين لنا مشكلات الحكم . صلى الله عليه وعلى
آله وأصحابه الذين كل منهم في الهداية علم . صلاة دائمة ما نثر كلام
ونظم وبعد فإني كنت صنفت كتابا في وضع التراجم وحلها سميته
إيضاح المبهم في حل المترجم . ثم اختصرته ومرت عليه برهة
من الدهر ولم يكن الآن عندي نسخة منه وسألني من يجب امتثال قصده
ولا سبيل إلى رده . فنظمت هذا القدر الكافي مما على ذهني من قواعد
هذا الفن وضوابطه وجعلت هذه الحاشية عليه موضحة لنظمه
مؤذنة إن شاء الله تعالى بفهمه . وسميته مفتاح الكنوز في إيضاح
المرموز والله تعالى أسأل الإعانة والتوفيق وهو حسبنا ونعم
الوكيل . هذا حل المترجم وإيضاح المعمى من أجل الفوائد التي
لا يستغنى عنه في أوقات تدعو الضرورة إليها وينتفع به في أخراه

Figure 6a. The first page of ibn ad-Duraihim's manuscript [12].

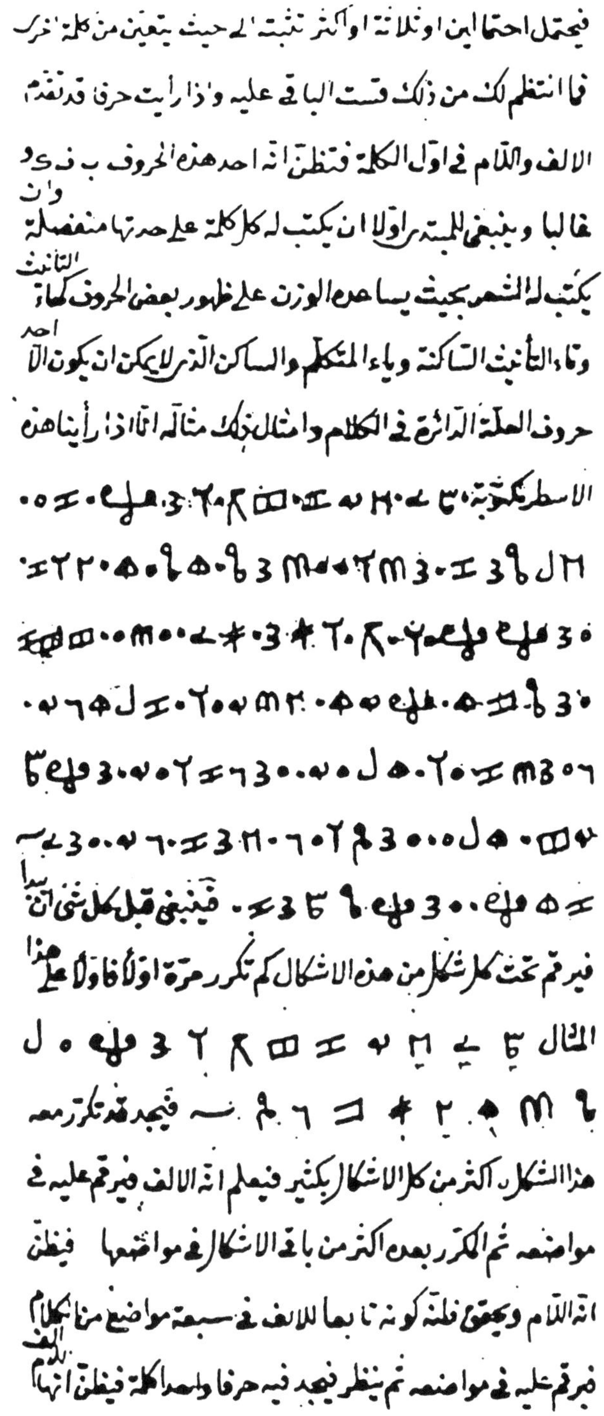

Figure 6b. An example of cryptanalysis as it appears in ibn ad-Duraihim's manuscript [12].

ad-Duraihim gives an extensive study of the linguistic characteristics of Arabic, which was preserved nearly completely in al- Qalqashandī's writings [2,13].

One of the most important contributions of ibn ad-Duraihim is his detailed and exquisite description of 8 systems of cipher with many variations and examples. His systems include the following:

1. Transposition with 24 variations and 31 examples including column transposition.
2. Substitution with 11 variations and 24 examples including the variable key substitution pioneered by ibn Adlān.
3. Letter addition (such as "ssifr" in "sifr") and omission (like "ifr" in "sifr").
4. Simple cipher mechanisms discussed earlier by ibn Dunainīr. An example of such mechanisms is a board with holes corresponding to letters of the alphabet and a string passed through holes of the intended message–a device described a millennium earlier by the Greek writer Aeneas the Tactician.
5. Arithmetical cipher. Ibn ad-Duraihim expanded on this method, pioneered by ibn Dunainīr, and gave 8 variations and 8 examples. Some of these examples are illustrated in Figure 7.
6. So-called letter-word substitution of four main types. One of these hides plaintext in the first (or second or last ...etc.) letter of words in innocent-looking but meaningful sentences. An example of which is hiding the plaintext Ali in the ciphertext sentence *pAint aLl pInk.*
7. Substituting for plaintext letters, under the control of a key, names from some species or families (e.g. animals, birds, cities, plants, etc.).
8. Use of some invented symbols for letters.

In his discussion of substitution cipher (No. 2 above), ibn ad-Duraihim analyzed the various possibilities and introduced the idea of what is now known as a 'Vigenère' table two centuries before Vigenère. Ibn ad-Duraihim's extensive and lucid study of cipher types was condensed beyond recognition (to about 10 percent) in al-Qalqashandī's book. Even though the presentation in *Ṣubḥ al-A'shā* was deformed by contraction, ibn ad-Duraihim's study of cipher types was commended by Kahn as remarkable and important [9, p. 98].

At the end, ibn ad-Duraihim gives a systematic methodology for cryptanalysis including the statistical techniques of al-Kindī and ibn Adlān. He falls short of them, however, in not giving actual frequency counts of Arabic letters but only

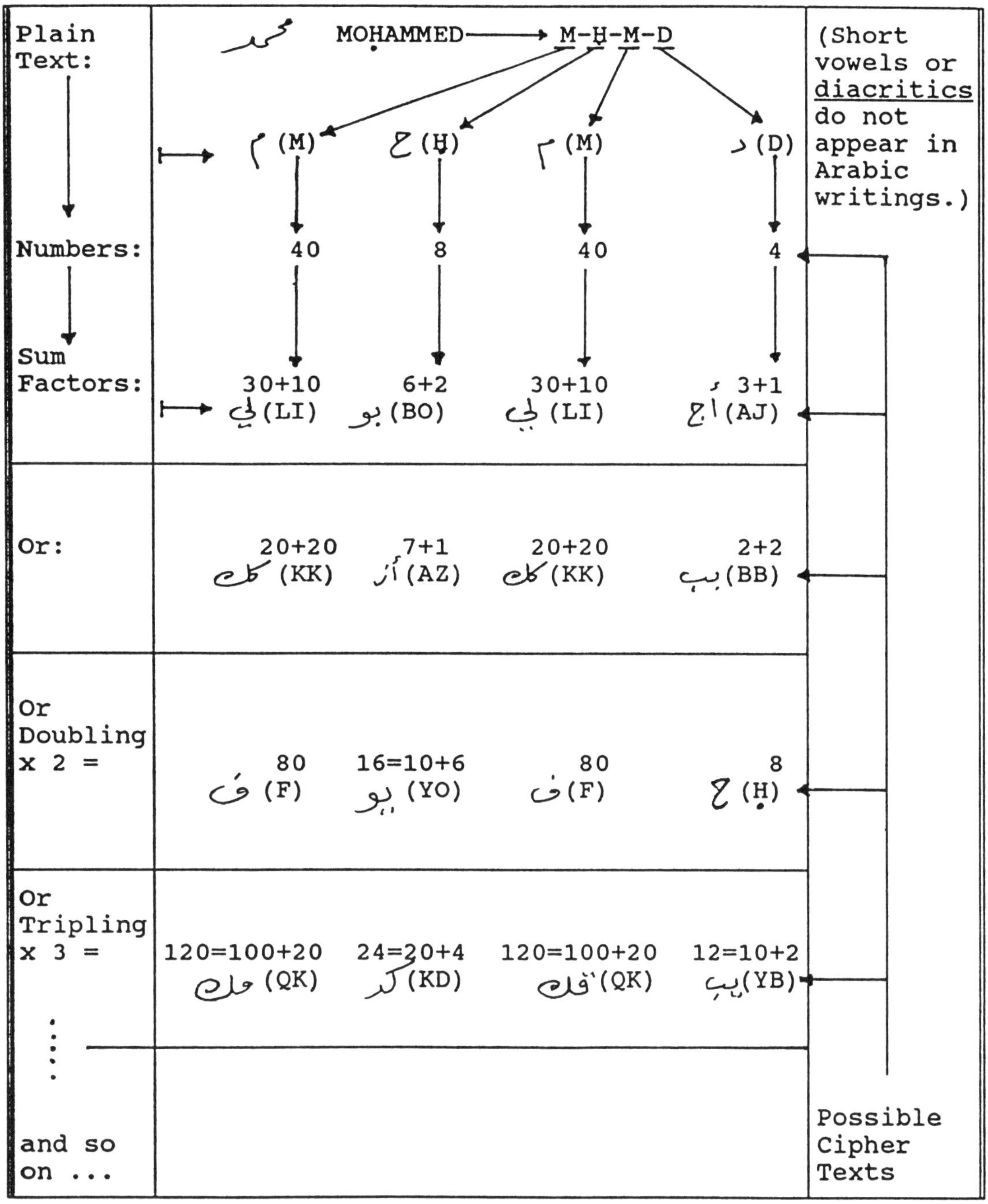

Figure 7. Some variations and examples of arithmetic cipher as given by ibn ad-Duraihim. (See Appendix A.)

their order of usage (rank). He also differs in that his ranking is based on the *Holy Qur'ān*. Next, ibn ad-Duraihim works out two detailed examples of solving sample cryptograms (see Figure 6b). Both the methodology and the two examples of ibn ad-Duraihim appear in their entirety in al-Qalqashandī's book, and were later described by Kahn as "the first exposition on cryptanalysis in history"[5] [9].

10. CONCLUSIONS

Cryptology originated, and developed to an advanced state, in the Arab-Islamic civilization (650 - 1400 A.D.). Among the factors that led to the early emergence and advancement of Arab cryptology were translation, linguistic studies, administrative sciences, public literacy and advanced mathematics. The origin of cryptology is much older, and Arab contributions to it are much more extensive, than previously reported. The new findings, reported in this paper, push back the frontiers of the known history of cryptology by more than 500 years. These findings are largely based on 8th century and later manuscripts written by the prominent Arab cryptologists al-Kindī, ibn Adlān and ibn ad-Duraihim. The contributions of these scientists, in addition to ibn Dunainīr, to cryptology are remarkable and important. Al-Kindī's contributions, in particular, are original and pioneering. He is the father of cryptology.

Among the new findings, reported in this paper, is the fact that the first writings in statistics are due to al-Kindī about 800 years before Pascal and Fermat. In addition, the great Arab linguist, al-Khalīl, used combinatorics more than 1200 years ago.

There are, however, some unanswered questions on Arab cryptology that cannot be cleared up yet. Arab cryptologists do not mention unbreakable ciphers, unlike many later inventors of cipher systems who claimed, and genuinely believed, that their systems were unbreakable.

It should be kept in mind that the sudden and quick decline of Arab civilization may have led to the loss of many books on the subject, and may have set back cryptology [9, pp. 98-99]. That decline seriously delayed the progress of human knowledge in many other fields as well.

[5] Al-Kindī's exposition was 500 years earlier!

APPENDIX A
THE ARABIC SCRIPT

The basic Arabic alphabet consists of a set of 28 letters. The Arabic script is extended to some 90 elements by additional shapes, marks and short vowels formally recognized in Arabic morphology. Arabic, written from right to left in elegant cursive form, does not differentiate between upper and lower case. Table A-1 shows Arabic letters, together with their transliteration as used in the etymologies and the numerical values of the different letters as adopted by early Arab linguists and cryptologists. In standardizing these values, early Arab scholars adopted the "abjadi" order of letters, which is different from the alphabetical order shown in the table. They gave the first nine "abjadi" letters consecutive numbers 1 to 9; the following nine letters were assigned numbers 10, 20, 30, ..., 90; letters 19 through 27 were given numerical values 100,200,300, ...,900; the last letter in the "abjadi" order, which is ghayn, is assigned the numerical value of 1000.

Arabic letters are all primarily consonants; three of which (alif, wāw and yā) are also used to represent long vowels. Full indication of short vowels or phonetic effects are not usually provided in normal Arabic scripts except in important texts or where confusion may arise. When they are provided, it is by means of a number of strokes and diacritical signs which are similar to accent marks in some European languages. Those signs are written above or below characters to mark short vowels and emphasize or loosen a letter's sound. They can also be mixed to produce composite phonetic effects.

APPENDIX B
AL-KHALĪL'S COMBINATORICS

The great Arab linguist al-Khalīl, the author of the oldest dictionary of Arabic words, used combinatorics in the eighth century to count all possible Arabic words by considering all different combinations of letters with or without vowels. In his dictionary, *Al-Ain*, al- Khalīl wrote:

> If you want to exhaustively know all of the Arabic language double-letter words, either meaningful or not, which the Arabs either used or rejected, such as *qd*, *km*, *an*...etc., take the [Arabic] alphabet letters which are 28, then multiply them with each other to get 784 [$= 28^2$]. A single letter is not a word. If you take two letters [without reversal], you get 392 [= 784/2] such as *dm* and the like. If you reverse [the two letter positions] it comes back to 784, 28 of which have identical letters

Letter Name[1]	Translit-eration	Arabic Form: Letter...... stand alone[2]	Arabic Form: Joined........ from right[3]	Arabic Form: Joined....... from left[4]	Arabic Form: Joined...... from left and right[5]	Numerical Value
Alif	ʻ,a[6]	ا	ـا			1
Bā	b	ب	ـب	بـ	ـبـ	2
Tā	t	ت	ـت	تـ	ـتـ	400
Thā	th	ث	ـث	ثـ	ـثـ	500
Jim	j	ج	ـج	جـ	ـجـ	3
Ḥā	ḥ	ح	ـح	حـ	ـحـ	8
Khā	kh	خ	ـخ	خـ	ـخـ	600
Dāl	d	د	ـد			4
Dhāl	dh	ذ	ـذ			700
Rā	r	ر	ـر			200
Zā	z	ز	ـز			7
Sin	s	س	ـس	سـ	ـسـ	60
Shin	sh	ش	ـش	شـ	ـشـ	300
Ṣād	ṣ	ص	ـص	صـ	ـصـ	90
Ḍād	ḍ	ض	ـض	ضـ	ـضـ	800
Ṭā	ṭ	ط	ـط	طـ	ـطـ	9
Ẓā	ẓ	ظ	ـظ	ظـ	ـظـ	900
ʻayn	ʻ	ع	ـع	عـ	ـعـ	70
Ghayn	gh	غ	ـغ	غـ	ـغـ	1000
Fā	f	ف	ـف	فـ	ـفـ	80
Qāf	q	ق	ـق	قـ	ـقـ	100
Kāf	k	ك	ـك	كـ	ـكـ	20
Lām	l	ل	ـل	لـ	ـلـ	30
Mim	m	م	ـم	مـ	ـمـ	40
Nūn	n	ن	ـن	نـ	ـنـ	50
Hā	h[7]	ه	ـه	هـ	ـهـ	5
Wāw	w	و	ـو			6
Yā	y	ي	ـي	يـ	ـيـ	10

Table A-1. (Opposite) Arabic letters, their forms, transliteration, and "abjadi" values.

Notes:

1. ā, ī, and ū are pronounced like *a* in *father*, *i* in *machine* and *u* in *rude*, respectively.
2. The form of the letter when it stands alone.
3. The form of the letter when it is joined to the preceding letter only.
4. The form of the letter when it is joined to the following letter only.
5. The form of the letter when it is joined to both the preceding and following letters.
6. Alif represents no sound in itself, but is used principally as an indicator of the presence of a glottal stop (transliterated ' medially and finally; not transliterated when initial) and as the sign of a long *a* (transliterated ā; see Note (1) above).
7. When 'o' has two dots above it (ö), it is called *ta marbūtah* and, if it precedes a vowel, is transliterated *t* instead of *h*.

like *hh* which do not change when reversed. 600 of these [$784 - 28 = 28 \times 27 = 756$ words] are perfect words [i.e., consonants only] with no *Wāw*, *Yā* or *Hamzah* [these are the three basic vowels in Arabic], which come to 300 before reversal [$(28-3)(27-3)/2 = 300$]. 150 words [of the 756] contain one of these [vowels]: *Yā*, *Wāw* and *Hamzah*, with 75 before reversal [25×3]. 6 words [of the 756] contain two [different] vowels [3×2], with three before reversal. 3 double-letter words [of the 784] contain the same vowel, 25 [double-letter words], contain identical consonants. You should understand what I just explained to you of the double-letter word counts which the Arabs spoke or rejected.

If you want to count all triplets [three-letter words], multiply the 3 vowels by the 9 double-letter words, they become 27 structures with three letters all of which are vowels [$3^3 = 27$]. Then multiply the 3 vowels by the 150 double-letter words where one letter is a consonant and the other is a vowel; you get 450 forms of three letters, two of which are vowels and one is a consonant. Also multiply the 3 vowels by the 600 double-consonant words; you get 1800 forms of three letters-2 consonants and 1 vowel. Multiply the 25 [consonant] letters by the 600 double-consonant words; they become 15625 three-consonant words [al-Khalīl might have made a computational error here (it should be 15000), but we believe this to be a slip; see below]. This is all that can result of three letters.

> If you want to count the quaternary [four-letter] words, use the same method. Multiply the 3 vowels by the 27 all-vowel triplets, then multiply by 450, then by 1800. Then multiply the 25 consonants by the 15000 three-consonant words [al-Khalīl is right this time, which proves that the 15625 above was indeed a slip]. These are all the quaternary structures. The same method is used for five-letter words. Six-letter [root] words are not possible except by using superfluous letters [a property of Arabic morphology].

In another part of al-Ain, al-Khalīl wrote:

> A double letter word has two variations like *qd*, *dq*, and *ṣd*, *dṣ* [= 2!]. A triple-letter word has 6 variations [= 3!]; the collection of the six variations is called a hexa-word, an example of which is: *ḍrb*, *ḍbr*, *brḍ*, *bḍr*, *rḍb*, *rbḍ*. A four-letter word has 24 variations because its letters, which are 4 are multiplied by the three-letter word variations which are 6 and the result is 24 [= 4! = 4 × 3!]. A five-letter word has 120 variations because its letters which are 5, are multiplied by the four-letter word variations which are 24, and the result is 120 [= 5!]...[6]

ACKNOWLEDGEMENTS

The author spent part of his sabbatical leave as a Visiting Researcher at the Data Transmission Division of the Electrical Engineering Department (Institutonen för Systemteknik) of Linköping University, Linköping SWEDEN, during which time this paper was written. I would like to express my gratitude to Prof. Thomas Ericson for hosting my stay in Sweden, and to Linköping University for using its services and facilities. I am grateful to Dr. Shakir Abdul-Jabbar for the helpful discussions and critical comments on the manuscript. I would also like to acknowledge the critical review and expert editing of Dr. David Kahn, which helped make the paper more focused and, hopefully, easier to read.

REFERENCES

1. Badeau, J. S. *et al.* 1983. *The Genius of Arab Civilization: Source of Renaissance*. Second Edition. Cambridge: MIT Press.

[6]The exact Arabic text of these two quotations is given by Mrayati et al [12]. The translation given here is the attempt of the author [texts between square brackets are added to clarify the original writings]. Apology is extended to the great al-Kahlīl for the inevitable weakness and loss of eloquence of this translation in conveying the actual meaning and beauty of the original words.

2. Bosworth, C. E. 1963. The Section on Codes and Their Decipherment in Qalqashandī's Ṣubḥ Al-A'shā. *Journal of Semitic Studies.* VIII-i (Spring): 17-33.

3. Boyer, Carl B. 1985. *A History of Mathematics.* Princeton: Princeton University Press.

4. Bynum, W. F., E. J. Browne and R. Porter, eds. 1983. *Dictionary of History of Science.* London: Macmillan Press Ltd.

5. ad-Dafffa', Ali. A 1978. *Nawabigh Ulama'i al-Arab wal-Muslimeen fi ar-riyadiyyat* (The Genius of Arab and Muslim Mathematicians). in Arabic. New York: John Wiley & Sons.

6. Fauvel, John and Jermy Gray, eds. 1987. *The History of Mathematics: A Reader.* London: Macmillan Press Ltd.

7. Hammer, Joseph von. 1806. *Ancient and Hieroglyphic Characters Explained.* London: W. Bulmer & Co.

8. al-Kadi, Ibrahim A. 1991. Cryptography and Data Security: Cryptographic Properties of Arabic. In Arabic. *Proceedings of the Third Saudi Engineering Conference.* Riyadh, Saudi Arabia: Nov 24-27. Vol. 2: 910-921.

9. Kahn, David. 1967. *The Codebreakers: The Story of Secret Writing.* New York: Macmillan.

10. Khaldūn, Abdur-Raḥmān ibn. 1958. *The Muqaddimah : An Introduction to History.* Trans. Franz Rosenthal. New York: Pantheon Books.

11. Menninger, Karl. 1969. *Number Words and Number Systems: A Cultural History.* Trans. Paul Broneer. Cambridge: MIT Press.

12. Mrayati, Mohammad, Yahya Meer Alam and Hassan al-Tayyan. 1987. *Ilm at-Ta'miyah wa Istikhrāj al-Mu'ammā Ind al-Arab.* (Origins of Arab Cryptography and Cryptanalysis), Volume I : *Analysis and Editing of Three Arabic Manuscripts of Al-Kindī Ibn-Adlān and Ibn ad-Duraihim.* In Arabic. Damascus: The Arab Academy of Damascus.

13. al-Qalqashandī, Aḥmad ibn Ali. 1412. *Ṣubḥ al-A'shāfī Ṣina'at al-Inshā.* (The Light of the Blind in the Profession of Writing). in Arabic. Section 8 of Chapter 2 in Part 4. Reprinted in Cairo (1957): Ministry of Culture and National Guidance.

14. Sarton, George. 1927-1948. *Introduction to the History of Science. 3 vol.* Baltimore: Carnegie Institution.

15. Waerden, B. L. van der. 1985. *A History of Algebra: From al-Khwārizmī to Emmy Noether.* Berlin: Springer-Verlag.

16. Youshkevitch, A. P. 1964. *Geschichte der Mathematik im Mittelatter.* Liepzig, Germany: Teubner.

BIOGRAPHICAL SKETCH

Ibrahim Al-Kadi was born in Onaizah, SAUDI ARABIA on 19 January 1954. He received the BS from Riyadh University (currently King Saud University), Saudi Arabia in 1978, the MS from the University of Michigan, Ann Arbor MI, in 1980 and the PhD from Stanford University, Stanford CA, in 1984 all in electrical engineering. He worked as a graduate assistant at Riyadh University (1978-79), as a research assistant at the Communication Satellite Planning Center of Stanford University (1981-84), and as an assistant professor at King Saud University (1984-88) where he is now an associate professor. He teaches undergraduate and graduate courses in systems, electromagnetics and communications, and performs research and consultation. He is a senior member of the IEEE. His areas of interest include communication systems, coding and cryptography, spectrum management, radio wave propagation, remote sensing, technology transfer, satellite positioning, and education.

Starting September 1991, he is on a temporary leave from the University working as Vice President for Engineering at the Advanced Electronics Company (AEC).

ROOSEVELT, MAGIC, AND ULTRA*

David Kahn

ADDRESS: 120 Wooleys Lane, Great Neck NY 11023 USA.

ABSTRACT: Exhaustive research finds little about what solved German and Japanese intercepts went to President Franklin D. Roosevelt during World War II and nothing about how he used them. However, a picture is given of Military Intelligence's Special Branch, which selected and edited the intercepts for presentation to the president and other high officials, and of the intercepts' presentation to and reception by the president.

KEYWORDS: Roosevelt, MAGIC, ULTRA, PURPLE, Special Branch, McCormack, Clarke, and Corderman.

I

What MAGIC and ULTRA information went to President Franklin D. Roosevelt during World War II, and how did he use it?

It may as well be said at once that exhaustive research provides only a sketchy answer to the first question and merely unproved hypotheses for the second.

Extant records specify only a few of the solved German and Japanese messages that went to the president. Moreover, no documents or recollections have come to light suggesting any action that he took based on, or even influenced by, these intercepts.

Despite these disappointing results, it may be worthwhile to set out what has been learned about the production of MAGIC (Japanese solutions) and ULTRA (German) and their selection for and presentation to Roosevelt. Doing so will at least outline how one intelligence organization operated in preparing information for the highest authority in the land. And it will provide a framework in case any papers are discovered that answer, or at least shed more light on, the fundamental questions about Roosevelt, MAGIC, and ULTRA.

Foreign intercepts began going to the White House for the first time in 1937.[1]

*Reprinted with slight changes from *Historians and Archivists*, ed. George O. Kent (Fairfax, Va.: George Mason University Press, 1991).

[1]Rowlett. I know of no evidence that earlier American signals intelligence results—during the Civil War, World War I, or the Washington Disarmament Conference of 1922, to name the three cryptanalytically most likely events—went to the White House.

They came from the Army's Signal Intelligence Service, the Signal Corps' codemaking and codebreaking agency. Founded in 1929[2] under the great cryptanalyst William F. Friedman, SIS directed its first nontraining codebreaking at Japan, starting in 1932.[3] At first it solved Japanese diplomatic codes.[4] But these turned out to encrypt only insignificant messages. High-level, important traffic was apparently protected by a machine cipher.[5] In 1935, when SIS realized this and had obtained more experience and greater knowledge of Japanese communication habits from its code solutions, it attacked the machine.[6] By 1936, cryptanalysts Frank B. Rowlett and Solomon Kullback,[7] perhaps helped by information about a simpler Japanese Navy machine cipher that the U. S. Navy had solved,[8] cracked what the Japanese Foreign Office called its "A" machine and what the Americans called RED. With this, the United States gained access to more and better intelligence than ever before.[9]

This included advance information starting in March 1937 about Italy's possible adherence to the German-Japanese anti-comintern pact of the previous November. The intercepts revealed the early indications that Italy might join, the start of negotiations, the discussion over whether Italy should accede to the existing treaty or sign a separate agreement with Japan, a statement that Hitler

---

[2] *The Origin and Development of the Army Security Agency 1917–1929* (Laguna Hills, Calif.: Aegean Park Press, 1978), 8–9.

[3] Hannah, 11–14.

[4] Rowlett; SRH-001, 303- 04.

[5] *Ibid* ., 305.

[6] *Ibid.*, 301; Hannah, 14–15.

[7] Hannah, 14.

[8] Lewis, ed., 12; SRH-305, 020; SRH-149, 8; SRH-355, 161- 62; Layton, 79. Rowlett disputes this (Hannah, 15), but the several naval officers' statements suggest that the Navy contributed something, though perhaps more in a general than a specific way.

[9] Hannah, 15; Rowlett. On RED and PURPLE, see Cipher A. Deavours and Louis Kruh, *Machine Cryptography and Modern Cryptanalysis* (Dedham, Mass.: Artech House, 1985), 211–245. Rowlett says, on p. 2 of a memorandum of 1 March 1980 prepared for author John Toland, that "G2 showed no more than a passing interest in the S. I. S. output [of Japanese diplomatic solutions] until a partially decoded circular message from Tokyo indicated that a secret codicil was being formulated to the tripartite treaty which Japan was negotiating with Germany and Italy." Actually, this intercept could only have concerned not the "tripartite treaty," which was not signed until 17 September 1940, but negotiations for the Anti-Comintern Pact between Germany and Japan. This was signed 25 November 1936, the year in which the A machine was solved; negotiations for Italy's adherence did not begin until 1937. The intercepts reflect this. Italy adhered to the pact on 6 November 1937 but was never told of its secret codicil between Germany and Japan. Parrish, 43, likewise errs in ascribing the army interest to negotiations on a "tripartite treaty." Ironically, the Japanese message of 14 November 1936 that provided a synopsis of the secret codicil was not solved until 13 November 1937 (from solved Japanese diplomatic messages dealing with the Anti-Comintern Pact kindly declassified for me by the National Security Agency). No "partially decoded circular message" was included in the messages supplied to me. (The agreement between Germany and Italy that in effect created the Axis, the so-called Pact of Steel, was signed 25 October 1936.)

wanted Italy to participate in the existing agreement, Italy's apparent acceptance, the emperor's approval, and part of the text of the treaty, which was signed 6 November 1937.[10]

U. S. diplomats did not start reporting until October on this rapprochement between three aggressive nations, to the east and to the west of the United States—and even then their information was far less specific than the intercepts.[11] So these seized the attention of officials in Army and Navy intelligence, in the State Department, and in the White House. Information from solved foreign cryptosystems, which had never before been of sufficient value to reach the topmost level of government, began going to the White House,[12] probably to the president.

On 20 February 1939, the Japanese introduced a new and more complicated cipher machine to supersede RED, whose mechanisms were wearing out. The Japanese called it their "B" machine; the Americans, PURPLE.[13] There followed nineteen months of concentrated cryptanalytic attack by a team under the general supervision of Friedman and the immediate direction of Rowlett, with some help from the Navy (which resumed diplomatic cryptanalysis to assist with PURPLE after having abandoned that field around 1938 or 1939 to concentrate on Japanese naval solutions). Their efforts were crowned with success in September 1940.[14] Thus, even after Japan discontinued use of the RED machine on 21 August 1941,[15] the PURPLE solution assured American officials of being able to read Japan's most secret diplomatic messages.

---

[10] Solved Japanese diplomatic messages declassified by N. S. A. In the following citations, the date of the telegram is given first, followed, in parentheses, by the date the codebreakers submitted the translation to the Army G-2. Note the increasing speed of solution. All dates are 1937. Berlin-Tokyo 19 January (8 March) and Berlin-Tokyo 9 March (5 April) for early indications; Tokyo-Rome 15 September (1 October) for start of negotiations; Rome-Tokyo 2 August (23 August) for existing treaty or separate; Berlin-Tokyo 7 October (15 October) for Hitler; Tokyo-Rome 26 October (3 November) for acceptance; Tokyo-Rome 4 November (8 November) for emperor's approval; Tokyo-Rome 31 October, 31 October and 1 November (all 5 November) for part of text. Some of the memoranda to G-2 bear the handwritten initials of Friedman, Rowlett, and John B. Hurt, a translator.

[11] *FRUS, 1937*, 1:608. British cryptanalysts likewise seemed to have obtained information about these negotiations only in September-October 1937 (John D. Ferris, "From Broadway House to Bletchley Park: The Diary of Captain Malcolm D. Kennedy, 1934–1946," *Intelligence and National Security*, 4 [July 1989], 421–450 at 447–448).

[12] Rowlett.

[13] SRH-145, 001, 008–09.

[14] Hannah, 5, 15–19; Kahn, 20–22. Lewis, ed., 14–15 for Navy diplomatic solutions.

[15] SRH-001, 308.

II

Growing cryptanalytic production—from an average of about one diplomatic solution a day in the first half of 1937[16] to fifty a day in the last half of 1941[17]—compelled the Army and the Navy to share the burden. In August 1940, SIS and the Navy's OP-20-G arranged that diplomatic messages originating in Tokyo on odd-numbered days would be handled by the Navy, those on even-numbered days, by the Army.[18] The two agencies exchanged the solutions.

Then, on 23 January 1941, the Army and the Navy agreed that their code-breaking results, which had been given the covername MAGIC, would be delivered to the White House by the Army in odd months, by the Navy in even.[19] But after a few months the Army stopped taking MAGIC to the White House.[20] In part this was because it felt that diplomatic intelligence should go to the president through the State Department, in part because the president's military aide, Brigadier General Edwin M. (Pa) Watson, had breached security by forgetfully throwing a MAGIC summary into his wastebasket some time after the Army's turn either in May or in July.[21] The final delivery in one or the other of those months seems to have been the last time an Army officer brought intercepts to Roosevelt. Henceforth, the naval aide was the person by whose hand MAGIC went to him.[22] This apparently came about because of the following events.

Near the end of September, an Army month, during which no intercepts were delivered to the White House, the president said he wanted to see them. In October, the Navy continued to send him memoranda based on MAGIC. Early in November, an Army month, during which again nothing was delivered to him, Roosevelt again told his naval aide, Captain John R. Beardall, that he wanted to see MAGIC. The aide told him it was an Army month. The president replied that he knew that and that he was either seeing MAGIC or being told what it said by Secretary of State Cordell Hull but that he still wanted to see the original intercepts. Beardall thought the reason was that the president grasped things more quickly by eye than by ear. On Monday, 10 November 1941, a

[16] S. I. S. serial 320 was solved on 12 February, serial 461 on 26 June—141 messages in 135 days.

[17] Kahn, 29.

[18] *PHA*, 8:3923, 18:3313-14; Lewis, ed., 15; SRH-270, 024.

[19] *PHA*, 2:288, 4:1734.

[20] Kahn, 30.

[21] *Ibid* ., 26. *PHA*, 11:5475, implies May. But Watson was receiving occasional deliveries of MAGIC in July.

[22] Clarke, in reply to my question of 2 December 1985 as to who brought solutions to Roosevelt, wrote: "Navy—always. They never let their poor relations near the old con man." *PHA*, 2:789, which states that from 1 November 1941 intercepts to the president were delivered through the naval aide, may refer only to the time up to Pearl Harbor.

conference changed the distribution setup. Henceforth, the Army would bring MAGIC to the State Department and the Navy to the White House. At 4:15 p. m. Wednesday, 12 November, the first intercepts were delivered to the White House under this system.[23]

Neither war nor changes in the production of MAGIC affected this arrangement. On 25 January 1942, a month and a half after hostilities began, the cryptological officers of the Army and the Navy agreed orally that the Army would intercept and cryptanalyze all diplomatic traffic as well as enemy army and air force messages. This was formalized in a written agreement of 30 June 1942, which specified that the Navy would solve enemy naval messages and that the Army would continue to give the Navy the diplomatic solutions. The Joint Chiefs of Staff approved this on 6 July.[24] Another agreement, on distribution matters, made 25 May 1942 between representatives of the Army, the Navy, and the Federal Bureau of Investigation, likewise did not alter Navy responsibility for White House delivery. It merely stated that, while the president should see diplomatic intelligence, he should not be given enemy naval, enemy military, western hemisphere clandestine (spy messages), or international clandestine intercepts.[25] (The reason was certainly not to conceal information from him but rather not to burden him with that which he could barely use.) Thus, although the Navy no longer generated the cryptanalytic intelligence that went to the White House, the naval aide continued to take it in to Roosevelt, who had been an assistant secretary of the Navy in World War I. By then, it was coming from a new Army agency.

III

After recovering from the shock of Pearl Harbor, the secretary of war, Henry L. Stimson, decided that a centralized analysis of the intercepted Japanese messages might have prevented the surprise.[26] He felt that an organization should be set up to study the intelligence from cryptanalysis more fully and carefully than before. Stimson, a lawyer, consulted with his assistant secretary, John McCloy, a lawyer, and they decided that the job should be handled by a lawyer, particularly one having experience in large cases involving many and complicated facts. They settled upon a former partner of McCloy in one of New York's most

[23] *PHA*, 11:5475-76.

[24] SRH-200, 184, 198, 042.

[25] Howe, 6; SRH-200, 038.

[26] SRH-116, 006-07; Henry L. Stimson and McGeorge Bundy, *On Active Service in Peace and War* (New York: Harper & Row, 1947), 454–55.

prestigious Wall Street firms who had represented Edwin Armstrong in his suits to protect his invention of the regenerative circuit. This was Alfred McCormack, a big, brilliant, demanding Brooklyn-born graduate of Princeton University and Columbia Law School.[27] He was appointed special assistant to Stimson on 19 January 1942, six days after his 41st birthday.[28]

Because the Signal Corps had for years produced the raw material—the intercepted and solved messages—that went to Army intelligence, one of the two slots allotted to the Signal Corps in the General Staff was in G-2, or intelligence. It was filled, at that time, by Colonel Carter W. Clarke, a tall, slim, energetic career officer with strong likes and dislikes, salty modes of expression, and very good common sense.[29] Clarke had no signal intelligence background—he had replaced an officer who had signal intelligence experience but had been shifted to prepare for censorship as war approached[30]—but was extremely capable. On 15 May 1942, he was placed in charge of the new organization; McCormack, who was commissioned in June, became his deputy.[31] They worked very well together, each recognizing that the other had strengths that he lacked. McCormack was ideally qualified to run an intelligence unit while Clarke was just right to fit it into the Army.

The new unit, named Special Branch, was a highly secret element of the Military Intelligence Division;[32] the coordinate branches were administration, intelligence, counterintelligence, operations, and plans. All depended directly from the Army's assistant chief of staff for intelligence, the G-2.[33] A reorganization of the Military Intelligence Division in June 1944 changed some titles—McCormack became director of intelligence within the Military Intelligence Service, of which Clarke was deputy director[34]—but had little effect on the work and output of Special Branch.

The intercepts being distributed early in 1942 still consisted merely of the

[27] Kai Bird, *The Chairman: John J. McCloy: The Making of the American Establishment* (New York: Simon & Schuster, 1992), 142; SRH-185, 77–83; *Who Was Who in America*, 3:572, s. v. McCormack, Alfred; "Alfred McCormack, Lawyer, Dies; Former Director of Intelligence," *The New York Times* (12 July 1956), 23:4–5; Shute; Nelson, Fontana.

[28] SRH-116, 007.

[29] Shute; Ervin; Nelson; Clarke (1963), "Carter W. Clarke Dies at 90, An Army Intelligence Officer," *The New York Times* (7 September 1987), A24:5.

[30] Corderman.

[31] Howe, 6; SRH-185, 002.

[32] SRH-146, 001.

[33] SRH-185, 015.

[34] SRH-062, 001, 080; SRH-146, 002. For the 1944 reorganization, see Otto L. Nelson, *National Security and the General Staff* (Washington: Infantry Journal Press, 1946), 526–535.

translations of Japanese messages with some footnotes. These gave the dates and intercept numbers of earlier messages referred to in those texts or said that those earlier messages were "not available."[35] This was raw intelligence. McCormack felt that the persons who got this exceptionally valuable information were entitled to have each detail in it illuminated by other information and the intercept itself placed into a larger context.[36] This would provide the recipients with finished—and therefore much more useful—intelligence. It would be presented to them in the form of a daily report. Moreover, surveying the intercepts over days or weeks would develop data that could be processed into long-range and specialized intelligence.[37]

For these jobs, McCormack sought men whose training "teaches them to deal with evidence, to be inquisitive and skeptical, to pursue an investigation through to a conclusion, to meet unfamiliar situations of fact and, most important, to do very detailed work without losing their sense of values."[38] He found these abilities not among historians, not among journalists, not among scientists nor detectives nor even parents of quarreling siblings, but exclusively among lawyers, and he began recruiting among members and former members of his law firm, Cravath, de Gersdorff, Swaine and Wood.

Clarke, meanwhile, handled administrative, security, and external matters. He prevailed upon the Signal Corps to expand its intercept facilities—the radio posts that monitored the foreign transmissions and so supplied the codebreakers with cryptograms to solve. He instructed the Signal Corps on assignment priorities—which cryptosystems intelligence wanted broken, which enemy circuits should be monitored. Perhaps most important of all, he alleviated the residual friction between SIS, a Signal Corps unit, and Special Branch, the glamorous organ of intelligence.[39]

IV

When Special Branch started, its offices were on the second floor of the front wing of Washington's Munitions Building, an ugly World War I temporary structure on Constitution Avenue between the Washington Monument and the Lincoln Memorial. In the rear wing of the same floor was SIS. Both moved in the fall of 1942: Special Branch to the newly built Pentagon, SIS to Arlington Hall, a former

[35] See intercepts in *PHA*, 12:1-248.

[36] SHR-116, 008-9.

[37] *Ibid.*, 009.

[38] SRH-185, 004.

[39] SRH-185, 00203; SRH-116, 010.

girls' school in Arlington, Virginia, half a dozen miles from the Pentagon.[40] As Special Branch grew, it expanded from its original single room into neighboring offices. The walls were temporary partitions that were constantly being shifted for new people. From time to time the branch moved;[41] for a while, at least, it was located in Rooms 2D715, 719, and 732.[42]

Though Clarke, McCormack, and some of the specialists had individual offices or offices shared with one or two people, the men preparing the daily intelligence reports—the heart of the operation— worked in a largish open area with desks facing each other in two rows. This enabled them to talk easily back and forth, to pass papers to one another, to go quickly and informally to another's desk to look at something to discuss it.[43]

The branch grew rapidly. From the dozen officers and civilians at the start, it swelled to 400 officers, enlisted men, and civilians in Washington and 80 abroad.[44]

Officers of Special Branch could not at first enter the supersensitive codebreaking agency,[45] even though the two organizations adjoined one another. But many questions arose in their work:[46] Was a message translated correctly? Did the codebreakers have other messages on the same subject? Could the intercept facilities pick up wanted traffic? In the first months, these matters were resolved by Clarke's consulting personally with the head of the Signal Intelligence Service, a fellow Signal Corps officer, at first Lieutenant Colonel Rex Minckler, and after April 1942, Lieutenant Colonel Frank W. Bullock.[47] Clarke carried written questions from his staff members to the codebreakers. Later a single officer of each agency handled liaison. But it was still a bottleneck.

Then, in February of 1943, Lieutenant Colonel W. Preston (Red) Corderman, who had studied in the SIS School in the 1930s, took charge of the codebreaking agency, now called the Signal Security Agency (SSA). It was Corderman who had been moved to handle censorship and whom Clarke had replaced in intelligence in the General Staff; they had known each other since 1927, when Corderman reported at Fort Monmouth, New Jersey, to the 51st Signal Battalion's B Com-

---

[40] SRH-117, 007, 008.

[41] Ervin; Shute; Fontana.

[42] SRH-062, 084-85.

[43] *Ibid.*

[44] SRH-145, 009.

[45] SRH-117, 008.

[46] SRH-132, 6.

[47] Kahn, 575.

pany, whose commander was Lieutenant Clarke.[48]

At once the liaison situation eased. Special Branch added several officers to handle the work. Secret teletype and secure telephone communications were established between Arlington Hall and the Pentagon.[49] The chief liaison officer with the codebreakers, Captain Thomas Ervin, a lawyer who had a flair for telling stories, would spend half his day with Special Branch and half with the codebreakers. He brought them questions Special Branch's analysts wanted answered and got from them both question and requests for information that would help in their solving, such as which one of five possible solutions of a name was the right one.[50] Finally, toward the end of the war, the ultimate step was taken and many Special Branch evaluators began working in the same room as the codebreakers,[51] greatly facilitating the work of both sides.

Once a day, and perhaps later in the war several times a day, a courier driven in a staff car brought the hundreds of daily translated intercepts to Special Branch from Arlington Hall.[52] Urgent ones were teletyped.[53] Most were Japanese army or diplomatic—the Japanese army material comprised 150 intercepts a day in the middle of the war, 700 at the end—a few were low-level German diplomatic, the higher-level German diplomatic codes not having been solved.[54] Italian messages did not figure prominently. It seems likely that every message solved and translated was sent to Special Branch: if Arlington Hall thought it worth translating, it was sent over.[55] In addition, an American unit under Colonel Telford Taylor at the British codebreaking agency at Bletchley Park, the Government Code and Cypher School, selected and then cabled to Special Branch, as it did to high military commands in the European theater, German military, air, and naval solutions, called ULTRA.[56] This began 27 August 1943.[57]

All messages, when they arrived at Special Branch, went to McCormack. He skimmed them to keep track of what was going on,[58] then, from 1943, divided

---

[48] *Ibid.*; Corderman.

[49] SRH-117, 009.

[50] Ervin.

[51] SRH-113, 010.

[52] SRH-132, 6; Ervin.

[53] SRH-062, 43.

[54] Ervin.

[55] *Ibid.*

[56] Shute. For the origins of this unit: Howe, 10–12, 90, 125– 131.

[57] Parrish, 181–2.

[58] Fontana.

Brigadier General Carter Clarke, head of the Special Branch, which processed German and Japanese intercepts for delivery to President Franklin D. Roosevelt and other high officials, congratulates his deputy, Alfred McCormack, a lawyer in civilian life, whom he has just awarded the Distinguished Service Medal for his work during World War II.

All photographs are from the National Archives.

They carried MAGIC and ULTRA to Roosevelt.

Captain John R. Beardall, naval aide

Captain John McCrea, naval aide

Rear Admiral Ross T. McIntire, physician and acting naval aide

Lieutenant Commander William C. Mott, assistant naval aide

Brigadier General W. Preston Corderman, head of the Signal Security Agency

Marine Corps Colonel Ronald A. Boone, deliverer of MAGIC and ULTRA to the president, arrives on assignment in Alaska.

Roosevelt smiles as he pins a Distinguished Service Medal on his departing naval aide, Vice Admiral Wilson Brown, 28 April 1942.

them into diplomatic and military and distributed them respectively to Henry Rigby and Benjamin Shute. These two lawyers, both in their early thirties (Rigby formerly of Cravath, and Shute on leave from it),[59] read the intercepts more closely. They kept those that dealt with immediate matters and gave those that dealt with longer-range or background matters to the specialists.[60]

Many of the specialists were academics with backgrounds in particular areas. They related the intercepts to other sources of information that flowed to Special Branch,[61] such as newscast transcripts and prisoner-of-war interrogations, to produce a full-fledged report on a particular aspect of an enemy nation's economy, military organization, internal unrest, new technology, and the like.[62] These were published as completed or as needed.

The messages that were kept by Rigby, Shute, and later another lawyer, William R. Perdue, were edited by them and by the two other lawyers under each to produce the daily intercept reports.[63] These were the MAGIC Summary (later called the MAGIC Diplomatic Summary), edited by Rigby; the Military and Diplomatic Supplement to the MAGIC Summary (which began in the summer of 1943 and which from 1 July 1944 was called the MAGIC European Summary), edited by Shute; and the Far Eastern Summary (which began 10 February 1944 and dealt with Japanese military matters), edited by Perdue.[64] For these, the teams studied the day's file of translated decodes from the Signal Security Agency. They passed some to the specialists for information or comment.[65] They scanned other material for background. The editors working on the Far Eastern Summary, for example, looked at the 300 daily Japanese Navy intercepts; at cables from the field; outgoing cables on MAGIC intelligence; diplomatic and commercial intercepts; other MAGIC intelligence such as Japanese order of battle; the codebreaking agency's Weekly Report, giving the status of cryptanalysis of Japanese army systems; at the Boston Series of plaintext German diplomatic messages obtained by the Office of Strategic Services; and at noncodebreaking intelligence material, such as field command intelligence reports, photo interpre-

[59] SRH-185, 003, 013; Shute.

[60] Bell, 52.

[61] SRH-141, 220–23.

[62] SRH-116, 015–19; Bell, 55, 58.

[63] SRH-146, 1.

[64] SRH-062, 087–88; SRH-035, 18; Bell, 50, 52, 56; SRH-062, 049; Perdue, who was in charge from 1942 of what was then called the Japanese Army Supplement; Rigby. Though SRH-062, 038, says that F. A. O'Connell Jr. prepared the Far Eastern Summary, this may refer to a later time.

[65] Fontana; SRH-146, 5.

tations, and Allied operational cables.[66]

From the translated decodes the editors selected the most important messages. That was their basic job. Most of their effort went into deciding what to leave out.[67] In addition, they spoke with the cryptanalysts and translators, often over the secure telephone, to rectify errors or fill in lacunae caused by garbles in the ciphertext and to clarify ambiguities in the Japanese and in the English. Even after this, many messages were, by themselves, scarcely intelligible.[68] To squeeze the maximum out of them, the editors called for files on items mentioned in the intercepts and sought elucidation.[69] They asked research personnel at the Military Intelligence Service to provide information about lesser known people, places, or things, about obscure events or allusions to forthcoming operations. They visited the who's who section, in which each mention of a name or operation was card-indexed, and determined who the person or what the event was.[70] This would be inserted as a footnote to the intercept. The editors conferred with one another on the meaning of a word. Some important intercepts were run verbatim in the summaries, but most were summarized. Rarely, if ever, were comments made beyond identifications; the editors did not evaluate the significance of the information.[71] The editors assembled the several summaries or texts and the identifications into a single writing, which went to Rigby and then to McCormack for final editing. McCormack was an extremely meticulous editor, who often spent a great deal of time making sure that the reports would be quickly and fully intelligible to their readers.[72]

The manuscript then went for typing to one of the two chief secretaries, both male legal stenographers who had worked at Cravath. One of them, at least, Vincent Fontana, followed the Cravath form book in typing the summaries: indent text 15 spaces from the left, indent 25 spaces for a paragraph start, indent 20 fewer for a quotation set apart in the text.[73] Reports averaged ten to twelve pages. The typescripts then were reproduced in bluish-purple ink by the Ditto process in about ten copies—later apparently more. Each was bound in a spiral

[66] SRH-062, 043-44. Anthony Cave Brown, *The Last Hero: Wild Bill Donovan* (New York: Times Books, 1982), 278–286, for Boston Series.

[67] SRH-116, 12.

[68] SRH-062, 044.

[69] Fontana.

[70] SRH-146, 5; Ervin; Shute; Nelson; Fontana.

[71] The summaries themselves.

[72] Fontana; Ervin.

[73] Fontana.

binder, using a device that Special Branch had for this.[74] Then each was put in a locked leather briefcase,[75] called a pouch, and distributed by officer couriers in regularly scheduled staff cars. The diplomatic summary went out early in the afternoon; the military ones were reproduced and bound in the evening and distributed the next morning.[76]

The Army delivered its copies of the MAGIC Summary to the secretary of war (Stimson), the chief of staff (General George C. Marshall), the assistant chief of staff for war plans, the assistant chief of staff for intelligence, the Navy for further distribution, the secretary of state (Cordell Hull) and the assistant secretary of state who followed signal intelligence (probably Adolf A. Berle).[77] The Navy passed the 10 copies it got to the White House, the secretary of the navy (Frank Knox), the commander-in-chief of the U. S. Fleet and chief of naval operations (Admiral Ernest J. King), the chief of war plans, the director of naval intelligence, the vice chief of naval operations, the director of naval communications, the radio intelligence center, the domestic intelligence section and a file.[78] The Far Eastern Summary also went to the president. It was at first distributed in 23 copies; by the end of the war the number of copies rose to 75.[79]

V

Delivery to the president was made by the naval aide. This had been the practice since before Pearl Harbor. For a few months in 1941, Roosevelt's physician, Admiral Ross T. McIntire, acted as naval aide and so brought the intercepts to the chief executive.[80] Once, in the summer, when McIntire was not available, a naval intelligence specialist, who did not want to entrust the pouch to the president's private secretary and who felt that the intercept he had could not wait, brought a message to the president. He was Lieutenant Commander Alwin D. Kramer, a Japanese language expert who headed the translation section of OP-20-G and who, because of his expertise on Japan and MAGIC, carried the intercepts to the high-level naval recipients so that he could answer their questions on MAGIC.[81]

[74] *Ibid.*

[75] Mott; SRH-123, 3.

[76] SRH-132, 3.

[77] SRH-035, 18.

[78] SRH-145, 049.

[79] SRH-062, 049.

[80] *PHA*, 9:3985.

[81] Kahn, 11.

The intercept perhaps concerned Japanese-American negotiations the next day.[82]

When, a little later, Rear Admiral John Beardall became naval aide, Kramer—a methodical and dedicated man whom some called "The Shadow"[83]—took the intercepts to him for delivery to the president.[84] Kramer explained to Beardall points in the summaries about which the president was asking. Beardall, whose office, like Kramer's, was in the Navy Building, would ask Kramer late in the afternoon whether any MAGIC would be coming over for the president. Kramer would say, "There is no MAGIC ready for the president now" or "There might be something later" or "Nothing until tomorrow morning."[85] When there was something, Beardall would receive the intercepts in his office from Kramer and would carry them to the White House, a few blocks away, arriving around 5:30 or 6:00.[86]

About the first week in December 1941, a junior naval officer was assigned to bring MAGIC to the president when Beardall was off duty or at dinner or otherwise unavailable. This was a communications intelligence specialist, Lieutenant Lester R. Schulz.[87] He was assisted, beginning 6 December, by an Ensign Carson.[88] This was the system in effect at the time of Pearl Harbor.

Beardall had a key to the MAGIC pouch and sometimes opened it for the president, sometimes not. Normally, the president returned the sheaf of papers to Beardall when he had finished with it, though on 2 or 3 December 1941 he said he wanted to keep a copy of a Tokyo-Berlin intercept warning that "War may suddenly break out between the Anglo-Saxon nations and Japan ... quicker than anyone dreams." Kramer prepared the copy and gave it to Beardall; he never got it back.[89] Though Beardall never discussed the intercepts with the president, he sometimes called attention to what he considered were important ones. On 4 or 5 December, for example, he said to Roosevelt of an intercept revealing that the Japanese were burning their codes: "Mr. President, this is a very significant dispatch." Roosevelt read it carefully, and then said to Beardall, "When do you think it will happen?" Beardall, thinking that Roosevelt meant

[82] *Ibid.*, 30.

[83] Mott.

[84] *PHA*, 11:5475.

[85] *Ibid.*, 5269, 5277.

[86] *Ibid.*, 11:5277, 5278.

[87] *Ibid.*, 10:4660, 11:5278. He was technically on temporary duty at the White House as communications assistant to the naval aide.

[88] *Ibid.*, 10:4660, 11:5280.

[89] *Ibid.*, 9:4072.

the start of war, said, "Most any time."[90]

The most dramatic case of Roosevelt's receiving intercepts took place on the evening of Saturday, 6 December 1941. Schulz had been ordered by Beardall to stand by and bring some MAGIC expected that evening to the president. It was the first time that MAGIC was delivered after the close of the ordinary working day.[91] Schulz was waiting in Beardall's small office in the corner of the White House's basement mail room when Kramer arrived about 9:15. He was bringing the first thirteen parts of a fourteen-part PURPLE message; the Japanese had deliberately delayed transmitting the fourteenth part, which would break off negotiations with the United States. Schulz obtained permission to bring the MAGIC to the president and an usher accompanied him to the oval study on the second floor and announced him.

Roosevelt, who had excused himself from a dinner party, was seated at his desk. Only his confidante, Harry Hopkins, was with him in the quiet lamplit room. Schulz unlocked the briefcase in which the MAGIC was distributed with the key Beardall had given him, removed the sheaf of intercepts, and handed it to the president. Roosevelt read the thirteen parts in about ten minutes, while Hopkins paced slowly up and down. Then Hopkins read them. The thirteenth part rejected the final American offer and, when Hopkins passed the papers back to the president, Roosevelt said, in effect, "This means war." Hopkins agreed, and for about five minutes they discussed the situation, the deployment of Japanese forces, the movement towards Indochina, and similar matters. The president mentioned a message for peace he had sent the emperor of Japan. Hopkins replied that it was too bad that the United States could not strike the first blow and prevent any kind of surprise in the inevitable war.

"No," the president said in effect, "we can't do that. We are a democracy and a peaceful people." He raised his voice. "But we have a good record." He tried unsuccessfully to get the chief of naval operations on the telephone, deciding against having him paged at the National Theater for fear of causing undue alarm.

The president then returned the papers to Schulz, and, about half an hour after he had entered the study, Schulz left. He returned the papers to Kramer, who was waiting on a bench in the mail room, and soon thereafter went home.[92]

The next morning, Beardall brought the president the fourteenth part, which had been intercepted and solved during the night. It expressed the Japanese government's regret that "it is impossible to reach an agreement through further

[90] *Ibid.*, 11:5284.

[91] *Ibid.*, 11:5278.

[92] *Ibid.*, 10:4659-665.

negotiations." It was about 10 a. m., and the president was in his bedroom. He said "Good morning" to Beardall, read the contents of the pouch, and then commented, in substance, "It looks like the Japanese are going to break off negotiations." There was no further discussion; Beardall took the pouch back to the Navy Department, and was as surprised as everybody else when news came of the attack on Pearl Harbor.[93]

Beardall was replaced as naval aide in January 1942 by Captain John McCrea. He saw or telephoned the president almost every day. To him went the MAGIC intercepts, brought over by an army courier. He also received copies of important naval dispatches that high naval officers thought should be shown to the president. Much of this arrived at his office in the Navy Building. Twice a day, McCrea picked out the naval dispatches he thought would be of interest to the president, added to these all the MAGIC intercepts, stuck them in his pocket or his briefcase, and headed for the White House, where he was expected.

When McCrea arrived in the morning, Roosevelt would usually be either in bed, in which case McCrea would hand him the papers, or in the bathroom, shaving. If the latter, the naval officer would go in, close the toilet cover, sit down on it, and in that inglorious setting would read the leader of the most powerful nation in the world some of the most secret documents of the greatest war in history. In the afternoon, the president, if he were not in the Map Room, a chart-lined office in the basement of the White House, would be seated in a neighboring room in a chair like a dentist's, having his sinuses packed by McIntire or his polio-deformed legs rubbed down by a masseur. Here again, McCrea would read the documents aloud to him.[94]

When the president was at his home in Hyde Park or at his Maryland mountain retreat, Shangri-La, McCrea would tell him about the documents by telephone.[95]

In 1942, the MAGIC documents were reaching McCrea's office in the Navy Department about 2:30 p. m.[96] OP-20-G had, since before Pearl Harbor, flagged the important items by sliding paper clips on them.[97] McCrea commented that although the selection was in general "accurate as to the possibility of their interest to the president," he often spotted other items that only he knew would

[93] *Ibid.*, 5283.

[94] McCrea. As an interesting confirmation, Franklin D. Roosevelt Library, Map Room File, Box 195, Folder Chart Room Log and Standing Orders, p. 67, notes that McCrea saw the president in the doctor's office on 14 and 16 May 1942.

[95] McCrea.

[96] SRH-200, 052.

[97] Kahn, 30.

also interest Roosevelt. After showing the documents to the president, he returned them to the courier for destruction by OP-20-G. In the summer of 1942, when Admiral William D. Leahy became Roosevelt's chief of staff, MAGIC documents may have been addressed to him,[98] though he himself rarely took them in to Roosevelt.[99] That function was still mainly exercised by McCrea, his 1943 successor Rear Admiral Wilson Brown, Brown's assistant, Commander John A. Tyree, and a young naval officer working in the Map Room under Leahy, Lieutenant Commander William C. Mott. All briefed the president orally in the morning and in the afternoon, often when he was getting his sinuses packed and was, Mott said, "a captive audience." They reported the intercept intelligence with the other information that they thought he would want to know and that they had boiled down for his consumption. If Roosevelt wanted to read some papers, they would give them to him; the intercept material was on separate sheets. Mott often underlined on these pages what he thought the president ought to know. The president struck the aides as extremely well informed.[100]

During Roosevelt's trip to the Casablanca conference with British Prime Minister Winston Churchill in January 1943, he was sent messages from Washington that were based on MAGIC Diplomatic Summaries. They were "from Colonel Boone"—almost certainly Marine Corps Colonel Ronald Aubry Boone, an Annapolis graduate who had been in naval intelligence in Washington since February 1939 and had spent eight years in Peking and Shanghai on language, intelligence, diplomatic, and public-relations missions.[101] One such "Colonel Boone" message, of 17 January 1943, identical with MAGIC Diplomatic Summary 296 of the day before, states that "Yamaguchi in Berlin reports to Tokyo that Nazis are upset about American military supplies reaching Vladivostok in Russian bottoms X. They reach Moscow quickly by rail X Germans say until this supply route is cut off, any advance in Russia quite impossible X ..."[102] On 27 January, "Colonel

[98] Adams, 237. Smedberg said that "probably Leahy brought it [MAGIC] around to him [Roosevelt]." Tending to support this is that Leahy knew of MAGIC since, in the late 1930s, he was chief of naval operations, when a cryptanalyst said of his attitudes to signal intelligence that "We couldn't ask for better. He is thoroughly sold" (SRH-355, 259). On the other hand, Elsey doubted that any MAGIC or ULTRA documents went through Leahy, and Mott said that Leahy did not have access to the intercepts he was bringing to Roosevelt.

[99] Mott; inferred from Elsey.

[100] Mott; Tyree.

[101] Service record, Serial No. 093, National Personnel Records Center, St. Louis, Missouri. Clarke said that the colonel delivering to the White House was not in the Army but was a Marine. This would keep delivery in the Navy Department and eliminates a Colonel Boone who was in Army intelligence at the time—Colonel Joseph W. Boone, O15341 (U. S. War Department, Adjutant General's Office, *Army Directory*, April 20, 1943 [Washington], 174). The only Marine Corps Colonel Boone at the time was Ronald Aubry Boone. Department of the Navy, Marine Corps, Headquarters, History and Museums Branch, Reference Section.

[102] Franklin D. Roosevelt Library, Map Room File, Box 15, Folder Casablanca Conference Trip File, Item QMU 171938Z (UTAH 53), and SRS 839, "Magic" Summary No. 490, 19 July 1943.

Boone has indicated for last three days that a Jap movement in strength toward southern Solomons is shaping up X ..."[103] There are about half a dozen such messages to Roosevelt during the conference.

Churchill sometimes sent Roosevelt ULTRA—which the prime minister referred to by his own codename for it, BONIFACE. While they were at the Quebec conference in September, 1944, for example, Churchill passed some to him, which he later said "appeared to me to be of profound significance." More often Churchill gave him documents based on intercepts or referred to information based on them. In a note to the president of 13 August 1943, while both were at the president's home in Hyde Park, New York, Churchill provided an analysis of the situation in Yugoslavia of which he said, "Much of it is taken from BONIFACE sources." In October, he wrote about movements in Italy that "We know from BONIFACE that the enemy is withdrawing to the north.[104]

Likewise, during Roosevelt's visit to Canada for the QUADRANT conference of July and August 1943 and his subsequent fishing trip, he was sent two Colonel Boone messages: One was on 21 August: "Colonel Boone indications that Jap barge traffic from Shortland to Kolombangara and Vella area (in the Solomon Islands) may be proceeding along south Choiseul coast to vicinity Sami Head (7 22 S 156 08 E) then across the channel to destination."[105] The other, four days later, dealt with a report of the Japanese ambassador in Hanoi.[106]

In November of 1943, a Marshall aide and Churchill sent Roosevelt intercepts that they wanted him to see. But he was away at the conferences at Cairo and Teheran when they arrived. He first saw them when he returned on 16 December 1943. After transferring from the battleship *Iowa* to the presidential yacht *Potomac* for the trip up to Washington, he was handed a brown manila envelope in a briefcase that a Map Room watch officer, naval reserve Lieutenant (j. g.) R. H. Myers, had brought with him on the yacht. The intercepts dealt with German forces and forts that would be defending against the Allied invasion

---

[103] *Ibid.*, Item WHITE 21. Cochran, "The Influence of 'Magic' Intelligence," 348, errs in stating that item WHITE 8 in the above folder is identical to MAGIC Diplomatic Summary of 7/29/43; they are different.

[104] Winston S. Churchill and Franklin D. Roosevelt, *The Complete Correspondence*, ed. Warren F. Kimball (Princeton: Princeton University Press, 1984), 3:321 for "appeared to me," 2:389 for "Much of it," 2:502-503 for "We know." Other references to intercepts at 2:4, 389, 502-503, 666-667, 672-673, 677, 3:229, 256, 321, 341, 388. Kimball says that all the intelligence files have not been declassified and suggests that intercepts sent by Churchill to Roosevelt but not yet published might be found therein. This is reinforced by Martin GIlbert's footnote in Randolph Churchill, then Martin Gilbert, *Winston Churchill* (Boston: Houghton Mifflin, 1966-1988), 6 (1983): *Finest Hour, 1939-1941*, at 1260, based on a personal communication, that Churchill sent Roosevelt intercepts via "C," the chief of the Secret Intelligent Service, Sir Stewart Menzies, minuting on them "Make sure the President knows this" or "Make sure the President sees this."

[105] Franklin D. Roosevelt Library, Map Room File, Box 16, Folder Canadian Fishing Trip and Quadrant Conference, WHITE 112.

[106] *Ibid.*, WHITE 128.

of Europe. The Japanese ambassador, General Hiroshi Oshime, and his aides had inspected the Atlantic Wall and reported to Tokyo in great detail on the German measures: "...All the German fortifications on the French coast are very close to the shore and it is quite clear that the Germans plan to smash any enemy attempt to land as close to the water edge as possible.... The coastal defence divisions are distributed as follows: Netherlands Defence Army (covering the Netherlands down to the mouth of the Rhine), 4 divisions; Fifteenth Army (covering the area extending from the mouth of the Rhine to west of Le Havre), 9 divisions; Seventh Army (extending thence to the southern bank of the Loire), 8 divisions...." These messages, all probably decrypted from PURPLE radiograms intercepted at Asmara, Ethiopia, where SSA monitored Berlin-Tokyo traffic in the dun-colored block building of a former Italian radio station, have been called the greatest military intelligence coup of all time—and it certainly gave an insight into enemy plans unprecedented in history in both detail and scope. In July of 1944, Churchill asked that Roosevelt be shown another message. But, Clarke noted in a memo, that one had already been reported in Special Branch's Far Eastern Summary of 20 July 1944 and had been delivered to the White House that day. It was a Japanese message saying that the enemy must not be given any pretext to use chemical warfare.[107]

VI

Roosevelt's confidence in the MAGIC decrypts had been settled well before Pearl Harbor. When he was given the solution of the Japanese ambassador's report of a private audience with himself, he appeared much impressed and commented, as he read the translation, "Yes, I said that" and "That is correct; he said that."[108]

But did ULTRA and MAGIC, which had to compete with so many other messages and problems, capture his interest? The evidence is contradictory.[109]

[107]SRH-111; Carl Boyd, "Significance of MAGIC and the Japanese Ambassador to Berlin: (V) News of Hitler's Defense Preparations for Allied Invasion of Western Europe," *Intelligence and National Security*, 4 (July 1989), 461–481 at 472–474; Hinsley, 3:2:32–33, 771–775.

[108]SRH-305, 24.

[109]Gerhard L. Weinberg, *Reflections on Running a War: Hitler, Churchill, Stalin, Roosevelt, Tojo*, Transcription of the 6th annual Phi Alpha Theta distinguished lecture on history at the State University of Albany on March 20, 1986 (n. p., n. d.), 11, is no doubt correct in observing that Roosevelt got through an enormous amount of paperwork each day and that he "kept track of the 'ultra' materials." Likewise, Cochran, "The Influence of 'Magic' Intelligence," 346, concludes that Roosevelt had "more than a passing interest in intelligence." On the other hand, Adams says, 237, that while Leahy continued to bring ULTRA to the president as it seemed pertinent, "Roosevelt was not much interested in details." None of these comments answers the significant question of whether he much used the intercepts in running the war. On January 1944, Roosevelt, in a message to Churchill, cited an intercept as a possible answer to a puzzling international incident (Kimball, ed., 2:672-673, 677). This might indicate that he was reading the intercepts voluminously and carefully. On

In February 1944, Marshall wrote the president, "I have learned that you seldom see the Army summaries of 'Magic' material." Marshall's remedies—improving the presentation, separating the wheat from the chaff and correlating the items with other information, binding the summaries in a Black Book for convenience of reading—suggest that the president was not particularly interested in the information.[110]

On the other hand, McCormack said of his MAGIC summaries that the president and his advisers "read them avidly."[111] He commented that "the fight for personnel was made easier by high praise which the [Special] Branch received from time to time by General Marshall and other high officials, and occasionally from the President."[112] General Corderman said that although none of the Signal Service Agency's many commendations were ever known to have come from the president, he had the impression that Roosevelt was a great user of the intercepts and a great supporter of those who produced them.[113] Frank Rowlett, the cryptanalyst whose solution of the RED machine had started it all, said that the usefulness of the intercepts to the top leadership was made clear to the cryptanalysts in the most valuable way possible: the cryptanalysts were asked, "Do you have anything more on this? Give it priority!" That request for more of their results was, said Rowlett, "better than any pat on the back."[114]

No doubt the president was glad for the information he got—hence the appreciation sensed by the codebreakers and the evaluators. Nevertheless, the impression remains that the intercepts had only a secondary interest for Roosevelt. The reason may be that—unlike the military and naval men—he could not much use the information they gave him.

During hostilities, intercepts on diplomatic affairs—either of the Axis or of neutral nations—dealt at best with nonbelligerents. But geography and the war situation influenced them far more than Allied pressure. Roosevelt understood this, and so perhaps interested himself but little in the intercepts dealing with these matters. For instance, the Allied invasion of North Africa had given rise to fears that Spain might admit German troops who could close the Strait of Gibraltar and trap Allied forces in the Mediterranean.[115] McCormack later

---

the other hand, his staff might have provided this information to him–an impression reinforced by Kimball's putting a bracketed "WDL?" at the end of the message to suggest that Leahy might have drafted it for him.

[110] National Archives, Record Group 218, Leahy Papers, Folder 126, p. 189.

[111] SRH-185, 013.

[112] *Ibid.*, 004.

[113] Corderman.

[114] Rowlett.

[115] Dallek, 362, 370.

claimed that Special Branch had provided the best intelligence about Spain's postinvasion intentions.[116] But however valuable the information Roosevelt may have gleaned from MAGIC, he seems not to have acted upon it: he sent no instructions to his ambassador in Madrid and did not mention guarding against the German threat in any of his postinvasion messages to Churchill.[117] In August of 1942, however, a MAGIC solution seemed of unusual importance. Marshall pointed out to Roosevelt intercepts showing that Japan had decided not to attack the Soviet Union and urged that Roosevelt tell Soviet leader Joseph Stalin about this. Roosevelt did—without, however, specifying the source of his information, which he affirmed was "definitely authentic."[118]

The MAGIC and ULTRA dealing with military matters was, of course, of extraordinary value to the generals and admirals, and Roosevelt urged it upon them. "Be sure Ernie King sees that!" he would sometimes tell McCrea when shown an intercept. But neither McCrea nor Mott nor Tyree saw cases in which an intercept impelled the president to take a particular action.[119] In part this was because he delegated the detailed running of the war to his military chiefs and seldom interfered.

In this Roosevelt contrasted sharply with Churchill. Soon after messages were solved at the codebreaking center at Bletchley Park, they were sent, together with their interpretations, to MI6 headquarters at Broadway Buildings in London, where a selection was made, put in a special buff-colored box, and taken by messenger to 10 Downing Street, five minutes away. The prime minister, who carried the key to the box on his key ring, sometimes read up to 200 ULTRA intercepts a day, discussed them at his daily meetings with his chiefs of staff, drew the attention of his field commanders to individual items, was told of such technical successes as the solution of a U-boat key, and in general utilized the intercepts in directing the war. But Churchill was much closer to the fighting and bombing than Roosevelt and, as minister of defense, much more intimately involved in the details. The intercepts were therefore of far greater value to him. Not to know what he saw would distort the picture of how he worked during

---

[116] SRH-035, 18–19; SRH-116, 014.

[117] Nothing in Kimball, ed., 2, checking especially 20, 41–42, 75, 76; *FRUS, 1942*, 2:303–315; *FRUS, 1943*, 2:595–97; Willard L. Beaulac (an American diplomat in Spain at the time), *Franco: Silent Ally in World War II* (Carbondale: Southern Illinois University Press, 1986); James W. Cortada, *Two Nations Over Time: Spain and the United States, 1776–1977* (Westport, Conn.: Greenwood Press, 1978); or in Carlton J. H. Hayes (the ambassador), *Wartime Mission to Spain* (New York: Macmillan, 1945).

[118] Franklin D. Roosevelt Library, Map Room Files, Box 167, Naval Aide's File: A/16-Japan and Japanese Islands, 1942–1945, Memo for the President from Marshall, 5 Aug 1942; *FRUS, 1942*, 3:616. I am obliged to Dr. Carl Boyd for this information.

[119] McCrea; Mott; Tyree.

World War II.[120]

Can the same be said of Roosevelt? Not with our present knowledge. Perhaps he used MAGIC and ULTRA extensively in his running of the war, and perhaps he did not. All known sources do not provide enough facts to permit a conclusion one way or the other. He and his advisors cannot be asked, and the chances seem slight that papers yielding direct evidence on the matter will come to light. Possibly comparisons of MAGIC and ULTRA with actions taken in the war will develop a pattern convincing enough to let historians say that the one affected the other. But here, too, the uncertainty even about what intercepts Roosevelt saw and the tenuousness of such influence even with solid data reduce the likelihood of success. Of the two efforts made so far, one fails to convince—even its proposer offers it diffidently—and the other, more plausible, awaits a documentary basis.[121] The almost complete lack of traces of Roosevelt's use of MAGIC and ULTRA despite the great volume of material may indeed suggest that Roosevelt did not much use them. But this is at best a negative argument, and in no way conclusive. In the end, we historians can only say that we do not now know how Roosevelt used MAGIC and ULTRA. And probably we never will.

NOTES

A name alone refers to an interview. Numbers for SRH items beginning with 0 or 00 when under 10 or 100 are the page numbers stamped on the document by archivists; they differ from the document's original typewritten page numbers.

[120] Gilbert, 6:611, 613, and the numerous references under "Enigma decrypts" in 7 (*Road to Victory*); Martin Gilbert, memorandum, 17 March 1987; Hinsley, 2:4 and entries under "Churchill" in the index.

[121] The effort that fails to convince is that of Cochran. He says, at 346, that the decision to get Italy out of the war "could well have been based upon Magic information" about "major strategic differences between the Axis powers, desperate economic and social conditions in Italy, and growing political instability among the fascist elite," citing 17 reports from April to June 1943. But not only did many nonintelligence factors contribute to this decision, not only had the idea been considered even before the North African invasion, but, as Cochran admits, no evidence links the intelligence to the decision, and without evidence his hypothesis remains unproved. The more plausible effort is that of Rohwer. He argues that Churchill knew via ULTRA of Hitler's 21 June 1941 order to U-boats that they not attack large warships whose nationality they did not know, that Churchill passed this information to Roosevelt, and that it encouraged the president to support Britain in the Atlantic more energetically than if he feared German retaliation. Rohwer says that U. S. Navy situation maps show the U-boats dispositions and that this is "clear evidence of the re-transmission of 'Ultra' intelligence from London to Washington." But National Archives, Record Group 457, SRMN-033, Part I, "Cominch File of Messages on U-Boat, Estimates and Situation Reports October 1941-September 1942," which gives individual submarine positions and estimated positions, does not reach back to June, the Navy's Operational Archives does not have any earlier files, and none of the messages in Kimball, ed., 1:210-281 (21 June to 7 December 1941) deal with any such instructions to U-boats. Thus, plausible as Rohwer's suggestion is, it must await further documentation before winning at least my acceptance.

Note

Though this article concludes that cryptologic information was little used by Roosevelt, such material was of enormous value in military operations, including at the highest level. Special Branch prepared the material for use by these high officials. But because this article focuses on Roosevelt, it excludes many details of the workings of the Special Branch. Almost every sentence dealing with that operation could—and some day should—be expanded into an article. The sources offer clues to this.

My claim of exhaustive research requires that I list the sources that I checked for information about Roosevelt and MAGIC and ULTRA. They are:

- Franklin D. Roosevelt Library, Hyde Park, New York. The only records of MAGIC for the president are those for some of his trips away from Washington. These are cited below. The Map Room files have no logs of ULTRA or MAGIC documents or the summaries based on them that went to the president, nor any mention of the couriers who brought him ULTRA or MAGIC documents, nor any record of what the naval aides (as Captain McCrea) brought him in its noting of their visits to him.

- William Emerson, at the time of my inquiry director of the Roosevelt Library. He knew of no unpublished documentary material concerning Roosevelt's receipt of or use of ULTRA or MAGIC.

- Robert Dallek, *Franklin D. Roosevelt and American Foreign Policy, 1932–1945* (New York: Oxford University Press, 1979). The work does not mention ULTRA, and Dallek, who went through all or nearly all the papers dealing with Roosevelt and foreign policy, said on the telephone 14 April 1986 he does not remember seeing anything about it in the documents. He believes that such material would have jumped out at him and that he would have used it in his book as a good story.

- Robert Myers of the White House Map Room staff said that both he and George Elsey of that staff picked up intercepts from the commander in chief, U. S. Fleet (Admiral Ernest King), and Admiral John Tyree, assistant to the naval aide, said that Myers picked up intercepts from the Navy. On the other hand, Elsey said that, although he dealt with messages of Winston Churchill and Joseph Stalin to Roosevelt, he knew nothing about MAGIC or ULTRA documents or summaries going to the president. I have not been able to clear up the contradiction. Asked for any intercepts sent by the Navy to the White House during World War II, the Navy Security Group Command headquarters replied in a letter of 18 April 1988 (Ser GH/4057) that it had only duplicates of Army-originated diplomatic intercepts.

- All of the interviewees were asked about how the Special Branch summaries went to Roosevelt and what use he made of that information; they knew little about the distribution and none knew anything about the president's use of the material.

- Archive of Contemporary History, University of Wyoming. I was told 16 April 1986 by Professor Emmet Chisum that he has never seen anything about distribution of MAGIC to Roosevelt in the papers of Laurance Safford, Dundas Tucker, and C. C. Hiles, all naval officers who worked in or knew about communications intelligence.

- Dundas P. Tucker, "Rhapsody in Purple: A New History of Pearl Harbor," ed. Greg Mellen, *Cryptologia*, 6 (July 1982), 193–228, (October 1982), 346–367, has nothing on MAGIC or ULTRA to Roosevelt.
- Library of Congress, Manuscript Division, Papers of William D. Leahy, Diary. A skim showed nothing dealing with ULTRA or MAGIC. Moreover, these papers became available to the public before ULTRA material was declassified.
- National Archives, Record Group 218, Joint Chiefs of Staff, Leahy Papers. None of the following folders have anything about MAGIC or ULTRA to Roosevelt:
  - Folder 54, Value of OSS
  - Folder 82, Miscellaneous Messages 1945–46
  - Folder 83, Dispatches 1944 Miscellaneous
  - Folder 84, Miscellaneous 1944
  - Folder 89, Conference Teheran 1943
  - Folder 100, Prime-President 1944
  - Folder 101, Prime-President 1943
  - Folder 102, Hurley—"Special"—and Harriman
  - Folder 104, High-Level (not Prime), 1945
  - Folder 105, High-Level Message 1943 1944 not Prime
  - Folder 125, Memoranda to and from President 1945
  - Folder 127, Memos to and from President July 1, 1943 to Nov. 1943
  - Folder 128, Correspondence signed by the President
  - Folder 138, Messages between President and Admiral Leahy
- U. S. Naval Institute, Annapolis, Maryland, "Reminiscences of Vice Admiral William R. Smedberg III, U. S. Navy, Retired," an oral history, July 1979, has nothing on Roosevelt on pages 281–313, which deal with Smedberg's service from September 1944 to March 1946 as combat intelligence officer on the staff of the commander in chief, U. S. Fleet.

I have not credited or used the statements in William Stevenson's *A Man Called Intrepid: The Secret War* (New York: Harcourt Brace Jovanovich, 1976) ascribed to Sir William Stephenson, head of British Security Coordination during World War II, that he brought ULTRA information to Roosevelt. These statements are unsupported by other evidence and my confidence in the book's unsubstantiated statements has been undermined by its basic and numerous errors.

SOURCES

Published Documents

United States. Congress. Joint Committee on the Investigation of the Pearl Harbor Attack. *Pearl Harbor Attack.* Hearings. 79th Congress: 1st and 2nd Sessions. Washington: Government Printing Office, 1946. 39 vols. Cited as *PHA* followed by volume and page number.

———. Department of State. *Foreign Relations of the United States: Diplomatic Papers*. Washington: Government Printing Office. Cited as *FRUS* with year followed by volume and page.

1937, 1: *General*; 2: *British Commonwealth, Europe, Near East, Africa*; 3 and 4: *The Far East* (all 1954).

Japan, 1931–1941, 2 (1943).

1942, 2: *Europe* (1961); 3: *Europe* (1961).

1943, 2: *Europe* (1964).

Unpublished Documents

National Archives. Record Group 457. Records of the National Security Agency.

SRH-061. "Allocation of Special Security Offices to Special Branch, Military Intelligence Service, War Department, 1943–45." 33 pp.

SRH-111. "MAGIC Reports for the Attention of the President 1943–1944." 32 pp.

SRH-116. "Origin, Functions and Problems of the Special Branch, M. I. S." [A memorandum by McCormack to Clarke, 15 April 1943.] 55 pp.

SRH-118. "Incidental Exhibits re Pearl Harbor Investigation" [Collection of Japanese Diplomatic Messages 12 July 1938–21 January 1942. Department of the Army Intelligence Files.] 481 pp.

SRH-141. "Papers from the Personal Files of Alfred McCormack." Parts 1 and 2. 570 pp.

SRH-145. "Collection of Memoranda on Operations of S. I. S. [Signal Intelligence Service] Intercept Activities and Dissemination 1942–45." 293 pp.

SRH-153. "M. I. S., War Department, Liaison Activities in the U. K. 1943– 1945." 20 pp.

SRH-185. "War Experience of Alfred McCormack." 83 pp.

SRH-270. "Army-Navy-F. B. I. Comint [Communications Intelligence] Agreements of 1942." 5 pp.

SRDJ-113,785–114,000. Japanese Diplomatic Translations. Station "C" Intercept Message File. [10 September 1940–7 April 1941.]

SRS. Diplomatic "MAGIC" Summaries 1943. September, October, December.

———. Record Group 218. Joint Chiefs of Staff. Leahy Papers. Folder 126, Memos to the President from General Marshall.

Franklin D. Roosevelt Library, Hyde Park, New York. Map Room File.

Box 15. Presidential Trips 27 September 1942–28 April 1943. Folders Casa- blanca Conference Trip, Presidential Trips Hyde Park 6 February–5 April 1943.

Box 16. Presidential Trips 14 May–7 November 1943. Folder Canadian Fishing Trip and Quadrant Conference 31 July–29 August 1943.

Box 192. Map Room Logs—Incoming Dispatches. 1941–July 1943. Folder Log of Incoming Dispatches April May 1942. Folder Log of Incoming Dispatches December 1941 January February March 1942. Folder Log Sheets September 1942.

Box 195. Map Room Logs: 1942–1945. Folder White House Chart Room—Special Log President-Churchill Dispatches. Folder Chart Room Log and Standing Orders [a log of calls and visitors]. Folder 1-A [log of Map Room].

National Security Agency.

Translated solutions of intercepted Japanese diplomatic messages dealing with the accession of Italy to the Anti-Comintern Pact.

Interviews

In Person

Carter W. Clarke. Head of Special Branch. 6 December 1963.
Thomas Ervin. Liaison officer in Special Branch. 11 March 1986.
Vincent Fontana. Typist and production manager in Special Branch. 12 March 1986.
C. Roger Nelson. Assistant editor of MAGIC Diplomatic Summary. 12 December 1985.
Benjamin R. Shute. Editor of German military summary. 8 March 1986.

By Telephone

Cletus Beard. Head of Asmara intercept post in 1950s. 20 April 1986.
Carter W. Clarke. 10 January and 19 October 1986.
Arthur Compton. Liaison officer in Special Branch. 27 October 1986.
W. Preston Corderman. Head of Signal Security Agency. 20 April 1986.
George Elsey. Deputy to Leahy in Map Room. 12 February 1986.
Forrest McCluney. Head of Special Branch couriers. 9 March 1986.
John McCrea. Naval aide to Roosevelt. 4 and 12 March 1986.
William Mott. Assistant to Leahy. 13 January and 23 March 1987.
Robert Myers. Assistant in Map Room. 26 April 1987.
William R. Perdue. Editor of Far Eastern Summary. 16 February 1987.
William McKinley Rigdon. Assistant to Leahy. 9 March 1986.
Frank Rowlett. Head of cryptanalytical teams that solved RED and PURPLE. 22 April 1986.
W. R. Smedberg. Head of naval combat intelligence. 12 May 1980.
John A. Tyree. Assistant to naval aide. 19 May 1987.

Letters

Carter W. Clarke. December 1985.
W. Preston Corderman. Commander of Signal Service Agency, 1943–1945. 5 December 1983.
John J. McCloy. Assistant secretary of war. 18 July 1984.
Henry W. Rigby. Editor of MAGIC Summary. 10 February 1987.
Lester R. Schultz. Delivered MAGIC to Roosevelt. 3 and 24 January 1986.
J. C. Sharp. Deputy chief, information policy, National Security Agency. 4 November 1986 (Serial J9420A).
Telford Taylor. Head of Special Branch unit at Bletchley Park. 6 January 1985.

Secondary Sources

(Includes only items cited more than once in the notes or of basic importance)

Adams, Henry H. *Witness to Power: The Life of Fleet Admiral William D. Leahy.* Annapolis, Maryland: Naval Institute Press, 1985.

Bell, Ernest L. *An Initial View of ULTRA as an American Weapon*. Keene, New Hampshire: TSU Press, 1977. (A reprint of SRH-005, "Use of [CX/MSS ULTRA] by the United States War Department, 1943–1945," 82 pp., and of SRH-006, "Synthesis of Experiences in the Use of ULTRA Intelligence by the U. S. Army Field Commands in the European Theatre of Operations," 29 pp.)

Cochran, Alexander S., Jr. "The Influence of 'Magic' Intelligence on Allied Strategy in the Mediterranean." In Craig L. Symonds, ed., *New Aspects of Naval History: Selected Papers Presented at the Fourth Naval History Symposium, United States Naval Academy, 25–26 October 1979* (Annapolis, Maryland: Naval Institute Press, 1981), at 340–348.

———. *The MAGIC Diplomatic Summaries: A Chronological Finding Aid*. New York: Garland Publishing, 1982.

Hannah, Theodore M. "Frank B. Rowlett—A Personal Profile." *Cryptologic Spectrum* (National Security Agency), 11 (April 1981), 4–22.

Hinsley, F. H., et al., *British Intelligence in the Second World War* Cambridge and New York: Cambridge University Press, 1979–1988.

Howe, George F. *American Signal Intelligence in Northwest Africa and Western Europe*. [Unclassified]. United States Cryptologic History, Series IV: World War II, Vol. 1. [Fort Meade, Maryland]: National Security Agency/Central Security Service, 1980.

Jones, F. C. *Japan's New Order in Asia: Its Rise and Fall, 1937–1945*. London: Oxford University Press, 1954.

Kahn, David. *The Codebreakers: The Story of Secret Writing*. New York: Macmillan, 1967.

Kesaris, Paul, ed. *The MAGIC Documents: Summaries and Transcripts of the Top-Secret Diplomatic Communications of Japan, 1938–1945*. Index compiled by David Wallace. Frederick, Maryland: University Publications of America, 1982.

Lewis, Graydon A., ed. *A History of Communications Intelligence in the United States with Emphasis on the United States Navy*. Denver: Naval Cryptologic Veterans Association, 1982.

Parrish, Thomas. *The Ultra Americans: The U. S. Role in Breaking the Nazi Codes* New York: Stein and Day, 1986.

Rohwer, Jürgen. "Ultra, the United States and the Battle of the Atlantic 1941." Paper delivered at the 2nd U. S. Army War College International Conference on Intelligence and Military Operations, Carlisle Barracks, Pennsylvania, 12–14 May 1987.

National Archives. Record Group 457. Records of the National Security Agency.

SRH-001. "Historical Background of the Signal Security Agency." Vol. 3. "The Peace." Preface signed by Historian, Army Security Agency. [Assistant Chief of Staff, G-2, 12 April 1946.] 416 pp. Reprinted as *The Origin and Development of the Army Security Agency 1917–1947* (Laguna Hills, Calif.: Aegean Park Press, 1978).

SRH-029. "A Brief History of the Signal Intelligence Service." By William F. Friedman, June 19, 1942. 18 pp. Reprinted in Kesaris, ed., 17–31.

SRH-035. "History of the Special Branch, MIS [Military Intelligence Service], War Department, 1943–1944." 63 pp.

SRH-041. "M. I. S. Contributions to the War Effort." December 1945. 22 pp.

SRH-061. "Allocation of Special Security Officers to Special Branch, Military Intelligence Service, War Department, 1943–45." 33 pp.

SRH-062. "History of Military Intelligence Service. Reports Unit." September 1945. 116 pp.

SRH-117. "History of Special Branch, M. I. S." June 1944–September 1945. Unsigned. 11 pp. Reprinted in Kesaris, ed., 55–67.

SRH-132. "History of the Special Distribution Branch." 7 pp.

SRH-146. "Handling of Ultra Within the Military Intelligence Service (M. I. S.) 1941–1945." [A memorandum by Lieutenant Colonel Thomas E. Ervin, 29 January 1946.] 7 pp.

SRH-149. "A Brief History of Communications Intelligence in the United States." By Laurance F. Safford, Captain, U. S. N., Ret. 22 pp. Reprinted in Kesaris, ed., 1–16.

SRH-152. "Historical Review of OP-20-G." 17 February 1944. [Unsigned]. 13 pp.

SRH-154. "Signal Intelligence Disclosures in the Pearl Harbor Investigations." 47 pp.

SRH-159. "Preliminary Historical Report on the Solution of the 'B' Machine." By William F. Friedman. 14 October 1940. 10 pp.

SRH-305. "The Undeclared War: 'History of R. I.' [radio intelligence]." By L. F. Safford. 15 November 1943. 29 pp.

SRH-349. "The Achievements of the Signal Security Agency in World War II." February 1946. 115 pp.

SRH-355. "Naval Security Group History to World War II." Prepared and compiled by Captain J. S. Holtwick, Jr., U. S. N., Ret. June 1971. Part I, 464 pp.

THE PAPAL CIPHER SECTION IN THE EARLY NINETEENTH CENTURY

David Alvarez

ADDRESS: Department of Government, Saint Mary's College, Moraga CA 94575 USA.

ABSTRACT: In the early 19th century, papal cryptography was moribund. The pope's Secretariat maintained a small cipher section, but this unit was sadly neglected. It shunned cryptanalysis, and limited itself to handling the few secret communications that passed between the Vatican and its diplomatic representative abroad. Papal ciphers were simple, and provided only modest security.

KEYWORDS: Papacy, cryptography, ciphers.

Recent papal cryptography has received little attention from historians. Specialists have illuminated the early history of the subject, namely during the Renaissance, when the papacy was a major force in international politics. But they have not been inclined to pursue the subject into the modern period. As a result, little is known about the cipher practices of the Vatican in the 19th and 20th centuries. This note makes a preliminary foray into a relatively unexplored area.

At the beginning of the 19th century, papal cryptography was the responsibility of the Secretariat of Ciphers, a distinct department in the papal administration. Its staff consisted of three "cifristi" (cipher officers) under the direction of the Cipher Secretary.[1]. The department worked exclusively for the papal Secretariat of State which, under the direction of the Cardinal Secretary of State, supervised the external and internal affairs of the Papal States, the wide band of territory across central Italy under the temporal rule of the popes. In 1814, a reorganization abolished the Secretariat of Ciphers and the office of Cipher Secretary, and moved the cifristi into the Secretariat of State, where they became a separate class of officials. The reorganization also established the new post of "Substitute Secretary of State and Cipher Secretary". As deputy to the Cardinal Secretary of State, the Substitute assumed direction of the day-to-day-affairs of the department, including (in a purely titular capacity) responsibility for codes and ciphers.

[1]The post of Cipher Secretary was established in 1555. For a survey of early papal cryptography, see Aloys Meister, *Die Geheimschrift im Dienste der Päpstlichen Kurie* (Paderborn, 1906)

In this period papal cryptography was closely linked with a single family, the Evangelisti. Giuseppe Evangelisti entered papal service in 1793 as a "scrittore" (copyist) in the Secretariat of State. His father, Francesco, had served the church as a cipher officer in the pontificate of Pope Pius VI (1775-1799). In 1801, Giuseppe followed the paternal example by joining the cifristi. He accompanied Cardinal Ercole Consalvi to the Congress of Vienna in 1814-1815. At his death in 1847, he had been the Secretariat's senior cipher officer for more than four decades. Filippo Evangelisti, Giuseppe's brother, maintained the family tradition by joining the cipher office in 1806. A graduate of the elite Collegio Romano, where he won academic prizes, Filippo served in the cipher section until 1840.[2].

Two other individuals worked in papal cryptography during this period. Vincenzo Scagliosi, of whom little is known, served in the old Secretariat of Ciphers from 1804 until 1814 when that department was incorporated into the Secretariat of State. His successor, Giuseppe Golt, remained in the cipher section until his retirement in 1847.

The cipher section was not an important unit in the Secretariat of State. Papal cryptography had declined since those years in the late Renaissance when the Argenti family, the illustrious predecessors of the Evangelisti, had made the papacy a leader in the field. In the first half of the 19th century, the cryptographic requirements of the Vatican were few, largely because of the modest position of the papacy in international affairs. In 1830 the papacy accredited diplomatic missions to only 11 states. By mid-century the number had increased to 13. These missions generated few secret communications. The record of the nunciature (embassy) in Madrid, one of the more important posts, is revealing. In 1828 the Secretariat of State sent only 12 enciphered dispatches to Madrid, while the nunciature sent no enciphered messages at all to Rome. In 1848 the Secretariat forwarded only 3 enciphered dispatches to Madrid, which responded with 7 of its own. Months might pass without secret correspondence between the Vatican and its nuncio (ambassador) in the Spain.[3]

The volume of enciphered correspondence that passed between the Vatican and its representatives abroad had diminished to such an extent that there was insufficient work for the cifristi.[4] The senior officer, Giuseppe Evangelisti, was

---

[2]Students of the modern papacy owe an enormous debt to Lajos Pasztor, whose researches have illuminated the history of the Secretariat of State in the 19th century. Unless otherwise indicated, the material for this article has been drawn from his magisterial *La Segreteria di Stato e il suo archivio, 1814-1833* , 2 volumes (Stuttgart, 1984)

[3]Giuseppe de Marchi, *Le Nunziature apostoliche dal 1800 al 1956* (Rome, 1957), *passim* ; Vicente Carcel Orti, *Correspondencia Diplomatica del Nuncio Tiberi, 1827-1834* (Pamplona, 1976), xvii and *passim* ; idem. "Los despachos de la Nunciatura de Madrid, 1847-1857," *Archivum Historiae Pontificiae* , 13 (1975), 311-400.

[4]Lajos Pasztor, "La Segreteria di Stato di Gregorio XVI, 1833-1846," *Archivum Historiae Pontificiae* , 15 (1977), 322, 331.

N.° 7.

Udienza di S.M. la Regina

Eminenza Reverendissima

accl N°. 71896
6

11

= 1.° =

Fin dal giorno 2. di questo mese, previa la domanda da me fattane verbalmente, e l'avviso ricevuto per mezzo del Sig.r Ministro di Stato, io ebbi l'onore di essere ammesso all'udienza di S.M. la Regina Isabella, cui presentai la lettera particolare del Santo Padre. Sua Maestà fu con me molto obbligante, e rispettosa, avendo voluto baciarmi la mano nel mio ingresso, e regresso dalla sua stanza, ed avendomi fatto sedere alla sua destra. Il colloquio però fu assai breve, ed alla presenza del così detto Introduttore degli Ambasciatori, che si trattenne sempre sulla porta. Per questa circostanza, ed anche perchè non mi parve prudente, nè d'altronde sarei stato in grado di entrare in alcuna particolarità nella stessa prima visita, io ebbi a limitarmi a poche espressioni miste di gravità, e di ecclesiastica dolcezza per assicurare la Sovrana dei benevoli sentimenti del Santo Padre a di Lei riguardo, del vivo desiderio ond'era animato di provvedere colla maggiore sollecitudine, e stabilità possibile ai riordinamento delle cose religiose nella Spagna, e della ferma di Lui fiducia, di essere all'uopo coadjuvato, e corrisposto dalla

con mezze parole), che mi

parte di Sua Maestà; e del suo Reale Governo. La Regina (4) 1/4 2/8 2/7 2/6 1/6 4/7 4/7 1/6 2/9 1/2 3/5 2/8

avrebbero indicato in essolei un tal quale interno turbamento

2/5 1/6 > 1/4 1/9 1/6 2/6 2/3 1/2 3/9 3/5 1/6 1/3 1/3 1/6 3/5 2/8 2/3 2/7 1/5 2/3 1/4 1/2 3/7 2/8 2/3 2/7 1/6 3/6 3/6 2/8 2/5 1/6 2/3

3/8 2/7 3/7 1/2 2/5 3/4 3/8 1/2 2/5 1/6 2/3 2/7 3/7 1/6 3/5 2/7 2/8 3/7 3/8 3/5 1/3 1/2 2/6 1/6 2/7 3/7 2/8 > se non

prevenuto essersi usata,

fossi stato 2/9 3/5 1/6 3/9 1/6 2/7 3/8 3/7 2/8 1/6 3/6 3/6 1/6 3/5 3/9 2/3 3/8 3/6 1/2 3/7 1/2 > mi rispose, che si rallegrava molto della mia venuta, che vi era Religione in Ispagna, e che tutto si regolerebbe.

Sua Ecz.a Rev.ma

Il Sig.r Card. Gizzi Seg.rio di Stato

Roma

Poscia

The nuncio in Madrid reports an audience with the Queen of Spain, 15 June 1847. Deciphered in the papal Secretariat of State.

usually able to handle all the enciphering and deciphering and still have time for other tasks around the Secretariat. After 1815, Filippo Evangelisti worked less and less with ciphers as he assumed duties in the registry of the Secretariat. It is unlikely that Giuseppe Golt was concerned with secret communications at all. Although assigned to the cipher section, Golt performed archival work and served as secretary to Monsignor Carlo Mauri, who held the post of Substitute for 16 years. After Mauri's death in 1830, Golt assumed responsibility for various legal matters of concern to the Secretariat, although he remained nominally attached to the cipher section.

At any given time, the Vatican had in service a variety of codes and ciphers. The nuncio in The Hague, Monsignor Francesco Capaccini, used several systems.[5] In the simplest procedure, Capaccini would compose a dispatch in which the paragraphs or sections were numbered consecutively. In the even-numbered sections there was no attempt at subterfuge; the sentences meant what they said. In the odd-numbered sections, however, the true meaning of a sentence was the opposite of that written. The statement "our position is strong" was meant to be understood as "our position is weak". The message was then sent in clear. Capaccini also used a jargon code in which an apparently open dispatch contained codewords. In this code, for example, Vatican = MUSEUM, Austria = MR. CERINI, Spain = MR. CARLI, Prince of Orange = MR. FREY, and ambassador = BOOKKEEPER. A third system consisted of two-, three-, and four-digit number groups. Words, listed in alphabetical order, were replaced by codegroups listed in numerical order and beginning with 0 or 1, while individual letters and syllables, listed separately, but again alphabetically, were replaced by groups beginning with 2. Each word could be represented by any of 10 sequential numbers, and proper nouns had, in addition, a corresponding codeword. Austria for example, could be represented by any number in the sequence 090 ... 099 or by the codeword MOROCCO. The Prince of Orange was 1340 ... 1349 or ERNESTO.

Capaccini was an enthusiast who delighted in creating new methods for secret communication. More typical, perhaps, of papal cryptography was a modified two-part nomenclator used by the nuncio in Vienna in 1814. Letters of the alphabet and syllables were represented by two-digit number groups from 00 = F to 99 = to. A list of four-digit groups, each beginning with 7, served for common nouns (7282 = emperor), proper names (7690 = Napoleon Bonaparte), and parts of speech (7070 = this). The number 5 was a null, but there were no homophones. In the 1840's the Madrid nunciature had in service three monoalphabetic ciphers.

[5]Segreteria di Stato, Spogli: F. Capaccini, "Chiavi di cifre imaginate da Mr. Capaccini," Archivio Segreto Vaticano (ASV), Vatican City.

In the first (introduced by the indicator 4), the letters of the alphabet were replaced by ordered pairs of numbers ($a = 1/2$, $b = 1/3$, $c = 1/4\ldots$). In the second system (indicator 6), the letters were replaced by two-digit numerical groups in ascending order ($\underline{a} = 11$, $\underline{b} = 12$, $\underline{c} = 13$, $\underline{d} = 14$, ldots. In the third (indicator 1), the letters were replaced by either letters or numbers in a system in which the letters in the first half of the alphabet were replaced by those in the last half in reverse order ($\underline{a}$ = Z, $\underline{b}$ = Y, $\underline{c}$ = X ...) and the remaining letters by numbers in declining order ($\underline{n} = 12$, $\underline{o} = 11$, $\underline{p} = 10$, $\underline{q} = 9 \ldots$)/. This latter system occasionally substituted symbols for certain proper nouns ($\overset{\infty}{\equiv}$ = France, *C3* = Pope). In all three systems numbers were spelled out, and punctuation was provided by symbols (comma = >, period = $\sqrt{\ }$). None of the ciphers used homophones or nulls.[6]

These simple ciphers show how sadly neglected papal cryptography was.[7] This neglect manifested a pervasive inertia and lassitude that made the papal administration a byword for inefficiency in the early 19th century. In the Secretariat of State the infrequent use of ciphers gave little incentive to improve established practices. Senior officials had no interest in cryptography. One cardinal secretary of state, Giuseppe Albani, considered a cipher section "totally useless" since diplomatic affairs could be conducted more easily through foreign ambassadors in Rome than through the nuncios in distant capitals.[8] Even Francesco Capaccini, an amateur cryptographer and the only official with an appreciation for the subject, considered the cifristi superfluous. Upon his promotion to the dual post of Substitute and Cipher Secretary Capaccini promptly suggested the abolition of the cipher section and the transfer of its few tasks to an ordinary clerk. But the reform was not instituted at the time.

The cipher section was moribund and provided little control or direction. There was no serious attempt at cryptanalysis.[9] Hardly more than clerks, the cifristi performed the mechanical tasks of encipherment and decipherment, but apparently gave little systematic thought to the problems of communications

[6]The Vienna and Madrid systems have been reconstructed by a comparison of the enciphered and deciphered versions of dispatches. For Vienna, see Segretaria di Stato, 1814, rubrica 242, busta 385, fascicolo 4, ASV. For Madrid see Segreteria di Stato (Esteri), 1847, rubrica 11, fogli 43-46, ASV.

[7]The ciphers used by the Madrid nuncio in 1847 were inferior to the cipher prepared for a predecessor by Giovanni Battista Argenti in 1588. Compare Meister, 405-407.

[8]Between 1815 and 1850 several states (Russia, Prussia, Saxony, Hannover, Wurtemburg) maintained diplomatic missions in Rome without accepting a nuncio in their own capitals.

[9]When papal officers intercepted two enciphered letters addressed to Napoleon's entourage on Elba, the Secretariat of State sent them to Cardinal Consalvi in Vienna with the suggestion that he or his cipher clerk, Giuseppe Evangelisti, try to solve them. Consalvi promptly turned the unbroken ciphers over to the Austrians. Ilario Rinieri, ***Corrispondenza inedita dei cardinali Consalvi e Pacca nel tempo del Congresso di Vienna, 1814-1815*** (Turin, 1903), 232.

security. In one dangerous practice, for example, nuncios were allowed to invent their own codes and ciphers. While at The Hague, Monsignor Capaccini used at least 10 different systems in three years. Often a nuncio would use more than one system in a single dispatch; part of the text would be written in a numerical cipher, part in a letter cipher, and part in a jargon code. At times, officials in the Secretariat would despair at their inability to make any sense out of an incoming dispatch. Apparently, security did not compensate for the confusion. In 1801, the vatican's special envoy in Paris had no confidence in his ciphers. At the Congress of Vienna, Cardinal Consalvi correctly surmised that his cipher had been compromised by the Austrians. Fourteen years later, Capaccini complained that the Vatican's codes were open to everyone.[10]

Political decline, administrative enervation, and personal obtuseness all may have contributed to the decline of papal cryptography in the early 19th century. Whatever the explanation, the papacy had clearly slipped to a minor place in the world of secret communications.

BIOGRAPHICAL SKETCH

David Alvarez, PhD, is a professor of government at Saint Mary's College of California. His current research focuses on the role of the Vatican as an intelligence center during the world wars.

[10] Boulay de la Meurthe (ed.), *Documents sur la Négociation du Concordat*, 1 (paris, 1895), xxiii; Alessandro Roveri, *La Missione Consalvi e il Congresso di Vienna*, 3 (Rome, 1970), 545, 567-68; J. P. de Valk (ed.), *Lettres de Francesco Capaccini* (Brussels, 1983), 256.

CRYPTANALYSIS FOR PEACETIME: CODEBREAKING AND THE BIRTH AND STRUCTURE OF THE UNITED NATIONS

Stephen Schlesinger

ADDRESS: 500 West 111th St. #4A, New York NY 10025 USA.

ABSTRACT: This paper examines how the United States used the information it obtained from the confidential diplomatic cable traffic of allies that it intercepted in the six months before and the two months during the conference in San Francisco drafting the charter of the United Nations in 1945 to influence the outcome of that meeting.

KEYWORDS: United Nations, Argentina, U.S.S.R., Poland, Ultra, Magic, peacetime, cryptanalysis, diplomatic, veto, colonies, trusteeship, UN veto, UN Security Council, San Francisco Conference.

[Editor's Note: This is one of the most important articles in the historiography of cryptology. Many are the stories about how codebreaking played a major role in wars. Only two cases are known in which cryptanalysis has significantly affected world events in peacetime. The first was the American solution under Herbert O. Yardley of Japanese diplomatic code messages which enabled the United States, at the Washington naval arms limitation conference of 1921-22, to push Japan into accepting fewer warships than it wanted. This article reports the second.
David Kahn]

The interception of foreign diplomatic traffic by the United States played a major role in enabling America to fashion the United Nations into the organization it wished at the San Francisco conference in 1945. The United States, the primary strategist behind the creation of the U.N., had a war-created cryptanalytic program that included the interception and solution of the embassy cables not only of its enemies, but also of its allies and of neutrals. As World War II wound down, America employed it to uncover the interests of the San Francisco participants in order to mold the organization's charter to its liking.

Secret U.S. files recently released under the Freedom of Information Act reveal how Pentagon operatives eavesdropped on friendly nations in the weeks leading

Total pages--20 TOP SECRET
(Incl. TAB A)

No. 1070 - 28 February 1945

WAR DEPARTMENT
Office of A. C. of S., G-2

By Auth. A. C. of S., G-2
Date 28 February 1945
Initials C. W. C.
GSC

"MAGIC"–DIPLOMATIC SUMMARY

NOTE: No one, without express permission from the proper authorities, may disseminate the information reported in this Summary or communicate it to any other person.

Those authorized to disseminate such information must employ only the most secure means, must take every precaution to avoid compromising the source, and must limit dissemination to the minimum number of secure and responsible persons who need the information in order to discharge their duties.

No action is to be taken on information herein reported, regardless of temporary advantage, if such action might have the effect of revealing the existence of the source to the enemy.

The enemy knows that we attempt to exploit these sources. He does not know, and must not be permitted to learn, either the degree of our success or the particular sources with which we have been successful.

A. MILITARY

1. Japanese reliance on surprise in Indo-China: (169077 168975 169052 DS 15, 26 Feb 45 WCM)

One of the matters in dispute between the Japanese military and civil authorities, in connection with the projected assumption of direct control over Indo-China, has been the policy to be followed toward the native kingdoms (DS 21, 26 Feb 45). A 24 February report from the Embassy at Saigon now clarifies the nature of the argument:

TOP SECRET
ULTRA

up to the San Francisco meeting to find out how they were preparing for it and during the two months of the conference to find out how they were reacting to it. These documents suggest that, in producing a United Nations that the United States envisaged, it was indulging not only in altruism but also in national self-interest. Such revelations indicate that, in retrospect, some revisions in conventional historical judgments on the origins of the United Nations may now be in order.

The 635 pages of diplomatic messages came from the Army's Signal Security Agency, which broke codes and solved intercepts, and its Special Branch, which evaluated, edited, and distributed them as "Magic" Diplomatic Summaries. These summaries reported information obtained from intercepted and solved foreign diplomatic cryptograms. Reproduced in purple ink by the Ditto process and issued daily, they averaged 15 pages each. All were divided into three categories – military, political, and economic – and, where needed, into two others: psychological and subversive and miscellaneous. Sometimes annexes amplified some items or gave the full text of important documents; occasionally maps were included. Within each category, headings described the subject of the intercept or intercepts. The scope was extremely great, ranging from German military plans as reported by the Japanese ambassador to Afghan attitudes to Japanese intelligence activities. The documents do not include any British or Soviet messages.[1]

The "Magic Summaries" reveal that:

- Washington knew in advance the negotiating positions of almost all the 50 or so countries that assembled in San Francisco;

[1]The National Archives and Records Administration, Record Group 457 (Records of the National Security Agency / Central Security Service), " 'Magic' Diplomatic Summaries," 1942-1945. The NSA declassified portions of the summaries for me 28 October 1993 on the basis of my request of 13 August 1993 under the Freedom of Information Act. All are from 1945. The intercept numbers on the declassified portions were found in the full summaries by David Kahn, and sources in each summary are cited below by date (excluding"1945") and page. For the organization and operation of Special Branch, which produced and distributed the summaries, see David Kahn, "Roosevelt, Magic, and Ultra," *Cryptologia*, 16 (October 1992), 289-319.

(Opposite) First page of a "Magic" Diplomatic Summary, showing security restrictions. A.C. of S., G-2 is the army assistant chief of staff for intelligence. C.W.C. are the initials of Carter W. Clarke, General Staff Corps, head of the Military Intelligence Divisions' Special Branch, which produced the summaries. The six-digit numbers in the margin are the numbers of intercepts; the DS 15 may be the number of a special study and the WCM are the initials of the drafter of item number 1. Photo - National Archives

- On key issues – whom to admit to the U.N., decolonization, the Security Council veto, the role of smaller countries, even Soviet views – the U.S. had crucial intelligence beforehand;
- Most nations, including the U.S., sought to push their own interests over those of the world community;
- The U.S. apparently used its surveillance reports to set the agenda of the U.N., to control the debate, to pressure nations to agree to its positions, and to write the U.N. Charter mostly according to its own blueprint.

Whether the U.S. was morally right to make use of Ultra - as the solutions were covernamed - is not an easy question to answer; America found itself in desperate times in the 1940s. Undoubtedly every country that had the capacity to intercept cable traffic was ready to take advantage of this capacity to ferret out as much as possible about the strategies of other governments. And surely the creation of as important a body as the U.N., given past historic failures, also merited special attention. Still, Ultra was not the proper way to treat allies and was of dubious legality.[2]

But President Franklin D. Roosevelt knew he was going to have a difficult time pushing a complicated U.N. structure, even if designed in large part to forestall another League of Nations disaster, through the U.S. Senate. Isolationist sentiment was still strong in the land. To make the United Nations palatable to Congress, Roosevelt, beginning at a conference of the leading Allied nations he convened on 21 August 1944, at Dumbarton Oaks, a study center in Washington, campaigned for a Security Council tightly controlled by the major powers.[3] The United States, Great Britain, the Soviet Union, and China agreed at Dumbarton Oaks to the draft outlines of the United Nations, which provided for a General Assembly with rather modest authority, a Secretariat subject to major-power control, a Military Staff Committee composed of the Big Five (the four nations that met at Dumbarton plus France, which was also invited to be the fifth permanent Council member), and, of course, the all-powerful Security Council, an 11-member body of which the Big Five were to be permanent members with veto powers.

The adoption of the latter veto provision which at first as left in limbo at Dumbarton Oaks reflected F.D.R.'s belief that the Security Council would actually run the United Nations and that, since these five nations were the only

[2]Under the Federal Communications Act of 1934, "no person receiving... any interstate or foreign communication by wire or radio shall divulge or publish the existence, contents, substance ... thereof ... to any person other than the addressee..." (18 United States Code 605).

[3]See Robert Hilderbrand, *Dumbarton Oaks: The Origins of the United Nations and the Search for Postwar Security* (Chapel Hill: University of North Carolina Press, 1990).

ones that possessed the forces to police the world, this prerogative was required. Extending the veto to all nations (as had been done for the Executive Council of the League of Nations) would invite gridlock and inaction. Four months later, at Yalta, Churchill and Stalin, at F.D.R.'s insistence, completed voting procedures reflecting this veto system for the United Nations.

With these building blocks in place, F.D.R. believed in the spring of 1945 that his special conception of the United Nations would have the best chance of acceptance by the other nations of the world. Not only was it in America's national interest, but it basically met the needs of all the earth's capitals. Still, he anticipated that smaller countries would fear the expansive authority of the organizing nations and would probably dispute the draft U.N. Charter. He knew he would probably have to twist arms to get his way.

The intercepted diplomatic notes nonetheless show that in the months leading up to the San Francisco conference, which began 25 April 1945, as well as during the two-month meeting itself, the U.S. used information from its codebreaking to help get its way on the U.N. issues about which it vitally cared.

I

One of the most vexing questions facing the conference from the onset was whether Argentina should be admitted. Soviet Premier Joseph Stalin opposed the admission of Argentina on the grounds that it was a crypto-Nazi country that had helped Hitler during the war. The "United Nations" was, in Stalin's reasoning, originally a wartime alliance of nations opposed to the Axis, not yet a global security organization, and thus like-minded countries should be its members. Most Latin American nations, however, for reasons of hemispheric solidarity, threatened to boycott the U.N. if Argentina's application were turned down. Although the U.S. had objected to Argentina's neutrality during nearly all the war, Washington knew that it had to resolve the question of Argentina's admission or face the possibility that either Moscow or Latin America would refuse to join the U.N., crippling the fledgling organization from the outset.

The general outlines of the Argentine dispute were known to the U.S. through the diplomatic contacts of Roosevelt's coordinator of inter-American affairs, Nelson Rockefeller. But Ultra told the United States precisely what *tactics* the Latin nations were beginning to pursue on Argentina's behalf, and what actions Moscow was, in turn, taking to prevent its admission.

Hemispheric countries, according to Ultra, already were aware that Roosevelt was sidestepping an urgent Argentine proposal in late 1944 to convene a meeting of the Pan American Union on postwar hemisphere problems. Argentina, which had severed its relations with Germany and Japan in January, was going to present its arguments for a continued neutrality at this meeting. Instead, Washington, to isolate Argentina within the Americas to force it to abandon its position, suggested that the Latin republics that had collaborated with Washington in the war effort meet in a conference outside the Pan American Union in Mexico and there consider the Argentine request.[4] The United States may have supposed that, once it induced Buenos Aires to abandon its neutrality and declare war on Germany, it could recast that nation into a more acceptable applicant for U.N. membership.

The Ultra intercepts enabled the United States to track the reaction to F.D.R.'s proposal around the hemisphere. For example, it read an Ecuadoran message of 30 December 1944 to its chargé in Washington, Duran Ballen, instructing him to tell representatives of Chile and Colombia that Ecuador would back Buenos Aires' request. In his answer, Ballen stated that since 16 Latin countries had already accepted invitations to the conference, he advised deemphasizing strong support for Argentina at the Pan American Governing Board. "I therefore suggest that I limit myself to stating Ecuador's point of view without seeking to contradict the plans prepared by nearly all the other countries," he wrote in a message that U.S. officials undoubtedly read with satisfaction. Ballen then did what he proposed doing.[5]

U.S. codebreakers also solved a dispatch of 6 January 1945 from the Argentine chargé in Washington to his government. This reported that the Chilean ambassador was assuring him that the conference in Mexico - now scheduled for February - "should not be considered as an unfriendly act toward Argentina but rather as a means for reaching a solution." The Chilean soothed the Argentinian. Perhaps the meeting would "convert" itself into the Pan American Union conference on postwar problems that Argentina wanted. In a message to its Washington embassy, Chile's foreign ministry said that it had turned down a U.S. request that it demand that the Pan American Governing Board rebuff Buenos Aires' entreaty. And a day earlier, the Chilean ambassador in Brazil informed Santiago that Brazil's acting foreign minister, while not approving of the Mexican parley, nevertheless planned to attend it. In mid-February, shortly before the assemblage in Mexico City, another flurry of messages scanned by Ultra reported on new developments in Latin attitudes on the Argentina problem. Paraguay now proposed to Colombia, Ecuador, and Venezuela that pressure

[4] 6 January, 7-9. See also "The Inter-American Conference on Problems of War and Peace. Held at Mexico City, February 21 - March 8, 1945 (The Chapultec Conference)," in United States, Department of State, *Foreign Relations of the United States: Diplomatic Papers, 1945*, 9: *The American States* (Washington: Government Printing Office, 1969), 1-153; also 223-230.

[5] Ibid.

be put on Argentina directly to join the Allied war effort. Colombia, though, objected to that course as likely to be "badly received" and offered a gentler plan.

All these tidbits may have helped Washington intensify pressure on the other Latin countries, both within and without the Mexico meeting, to compel Argentina to reassess its ideological course.[6] For example, the Chilean envoy in Washington telegraphed Santiago in mid-January about a luncheon with Nelson Rockefeller at the Cuban embassy. He and seven other ambassadors agreed there that Buenos Aires had to be told to "adopt concrete and effective measures to eradicate the conviction of all the American peoples that she has been antagonistic toward the democracies of the continent..." The relentless U.S. blitz against Buenos Aires was gradually generating a hemispheric condemnation that was starting to be felt by Argentine officials - at least those in Washington. Around the time of the Rockefeller lunch, the Argentine chargé in Washington informed his government that, while the Mexico conference would "cover a discussion of the diplomatic misunderstanding" (copying the same word as one of his Latin colleagues), his country might have to think about a compromise.[7] By 21 March 1945, the Argentine military attaché in the U.S. cabled President Farrell of Argentina that it was "decidedly advisable" for Buenos Aires to back an idea recently floated by its foreign minister: the country should hold a plebiscite for or against a declaration of war against Germany and Japan. Ultra enhanced Washington's ability to anticipate objections from, evaluate shifts in the mood of, and calibrate the growing demands upon Argentina by, its neighbors. On 27 March 1945, Argentina entered the war against the Axis powers.[8]

Ecuador now tried to convince the other recalcitrant party to this drama, the U.S.S.R. , to reconsider its opposition to Argentina's admission to the U.N. Ecuador enlisted Peru, Mexico, Cuba, Panama, Venezuela, and Brazil in this campaign. On 7 April, Colombia's foreign minister instructed his envoy in Moscow to tell the Soviet regime that if it were now to oppose Buenos Aires, "we fear it would be a bad beginning for her relations with the American countries, which, for the most part, have established relations with Russia without much conviction and against considerable internal opposition."[9] With this information, the U.S. was in a position in its negotiations with Moscow at San Francisco to persuade Stalin's envoys to admit Argentina, thereby placating the Latins.[10]

[6]8 January, 4-6; 11 January, 9.

[7]19 January, 9-12; 22 January, 13.

[8]26 March, 5-6.

[9]16 April, 10-11.

[10]George McJimsey says in *Harry Hopkins: Ally of the Poor and Defender of Democracy* (Cambridge, Mass.:

II

A second great dispute arose over the U.S. desire to establish a Trusteeship Council at the U.N. Both the British and French feared that such a body might force them to give up their colonies after the war. The U.S. became keenly aware of French apprehensions as early as mid-January, 1945, through Ultra. France's provisional foreign minister, Georges Bidault, telegraphed his ambassador in Washington about reports that the Americans at a preliminary U.N. conference were pressing for an international U.N. committee for colonies modeled on the controversial Mandates Commission of the League of Nations. A French emissary reassured Bidault on the matter: "The American tendency of hastening the normal evolution of colonies toward autonomy, dominion status, or independence was met by objections – and even the protests – of qualified delegates ... including ours."[11] But the issue was not dead.

After the Yalta meeting in early February (from which the French had been excluded), the French representative in Moscow sought out Soviet Foreign Minister Vyacheslav Molotov to obtain additional assurances that the U.N. would not act against Paris's colonial possessions. Molotov responded that Paris should take the matter up with the White House, which, he said, was now assuming primary responsibility over the colonial matter. In any event, he informed the French that a system of "trusteeship" had been "defined only in principle" at Yalta.[12] Solutions of French intercepts revealed these discussions to the Americans.

The French later shared their concerns with the British. After meeting with British Foreign Secretary Anthony Eden in London, Bidault, according to Ultra, expressed his satisfaction with the results: "Mr. Eden explained to me that the idea [of the trusteeship] was an American one and would permit the United States to lay hands chastely on the Japanese islands in the Pacific. The system is not to be applied to any region in Europe nor to any colonies belonging to the Allied countries. The English are determined that no misunderstanding arise in this regard." In their talks, Eden and Prime Minister Winston Churchill approved a continuation of France's "privileged position" in Syria and Lebanon.[13]

But the French, not convinced of the British assurances, undertook a broader offensive to protect their colonies. For example, they convened talks in Paris between General Charles de Gaulle and the foreign minister of the Netherlands,

Harvard University Press, 1987), at 382, that Stalin gave in on Argentina's seating after "both Hopkins and Stalin were willing to concede issues that lay in the other's area of special interest."

[11] 6 February, 5-7.

[12] 24 February, 8-10.

[13] 6 March, 8-11.

which also had overseas territories. Both men agreed, according to Bidault, on "the impossibility of surrendering to an international authority any of their colonies." By early April, Bidault wired all his envoys that France would flatly reject any plan of international control "over all or part of her colonial empire or of the countries placed under her protection." She would, moreover, uphold trusteeships for former Japanese or Italian territories. Bidault told the Greek ambassador that the U.S. was promoting trusteeships simply because it "wants to exercise influence on other people's colonies for selfish political and economic reasons."[14] The codebreaking helped shape U.S. policy. Just as the San Francisco conference was commencing, the U.S. began to reassess its thinking about France. It wanted France to join the Security Council and drop its championship of smaller nations opposed to the Big Five rule. It began to ease its pressure somewhat on the trusteeship issue. America's conciliatory stance and respect toward France's amour-propre had an immediate calming influence, the Ultra documents show.

Paris began displaying more pragmatism on the colonial issue. A French official in the U.S. with experience in territorial problems advised his home office midway through the conference that France should not "turn down a text in which independence is set as the eventual goal for trusteeship, for we would be approximately the only ones to do so." By the end of the San Francisco conference, Bidault himself was imploring his delegates to get U.N. approval for an impartial commission to investigate France's treaty status in Syria and Lebanon. At the conference's conclusion, France's verdict on trusteeship was now cautiously favorable and no longer harshly antagonistic. Its delegate wrote Paris: "The settlement of the trusteeship problem fulfills in broad outlines the instructions of the French government."[15] American diplomacy, guided in part by intelligence from Ultra, had maneuvered France into a satisfactory stand-down.

III

The third serious matter that roiled U.N. members from the outset was the exclusive veto power over U.N. actions that the Big Five as permanent members of the Security Council would hold. Before the conference, Washington had had hints of the profound misgivings of certain mid-sized countries about this allocation of power. France had provisionally turned down becoming the fifth permanent member of the Security Council (after the U.S., the Soviet Union, China, and the United Kingdom) with the accompanying right to the veto because of its ire

[14] 29 March, 9-10; 12 April, 6-8; 14 April, 8-10.

[15] 5 June, 14-15; 25 June, 7-9; 30 June, 4-6.

TOP SECRET

ULTRA

B. <u>POLITICAL</u>

1. <u>Gromyko-Bidault conversation in San Francisco</u>: French Foreign Minister Bidault has reported from San Francisco that on 30 April he had a talk with Russian Ambassador Gromyko at the latter's request. Bidault's message, which is annexed as TAB A, indicates: (181571 HMM)

<u>a</u>. Bidault agreed to a proposal from Foreign Commissar Molotov that the French and Russian delegations should consult together before the various committee sessions.

<u>b</u>. Bidault told Gromyko that he would unconditionally favor admitting the Ukraine, White Russia and Argentina to the Conference, but that France would "abstain from taking a stand" on the question of admitting the Warsaw Government. On the latter point, he "refused to yield" to Mr. Gromyko's "insistence."

2. <u>French delegation's dissatisfaction with its position at San Francisco</u>: A message of 29 April from Fouques Duparc, Secretary General of the French delegation, to General de Gaulle--only parts of which (181[illegible] HMM)

-7-

TOP SECRET

ULTRA

over its exclusion from Yalta. It criticized the veto and held to a romantic notion of becoming the leader of the U.N.'s smaller nations. Ironically, in the months preceding the San Francisco conclave, F.D.R. also privately clashed with Stalin over the breadth of the veto. Stalin wanted the power to bar even *discussion* of issues in the Security Council, not merely to bar *action* by the Security Council, as F.D.R. wished. Whatever the source, F.D.R.'s effort to get the veto in any form was in much more serious trouble from U.N. members than many historians have previously thought.

First, U.S. intelligence obtained a message in mid-March 1945 from the Chilean foreign minister offering his "personal opinion" to his envoys that "the procedure devised at the Crimea Conference [Yalta] for voting in the Security Council is not in accord with the sovereign equality of all peace-loving states and, in operation, would put the permanent members of the Council above the law which will govern all nations." The minister requested his diplomats abroad to find out how other countries regarded the special status of the great powers at the U.N. He received at least five replies – from Costa Rica, Cuba, Italy, Switzerland and the Vatican – agreeing with Chile's reservations. This group was probably representative of the 50 or so states that were to convene at San Francisco, and their reservations suggested that an alarmingly large opposition to the U.S. position was in the making.[16]

In a dispatch in late March, Turkish officials also expressed doubts about the voting procedures. They told French diplomats in Ankara that the setup "seemed destined to make lawful the projects of the large powers against the small – with the system of voting in the Security Council ensuring them impunity." The Turks warned that "the small states are inevitably going to be reduced to the status of satellites of the great." They also feared that bilateral alliances, as, for example, their 1939 mutual assistance pact with England, could be overridden during a crisis, say, with the U.S.S.R., by the veto. But they also conceded the "futility" of modifying the U.N. Charter. Instead they hoped to increase "the number of non-permanent members on the Security Council from six to nine in order to give the Great Powers a less preponderant majority."[17]

---

[16] 11 April, 12-13.

[17] 13 April, 7-9.

(Opposite) Page 7 of the "Magic" Diplomatic Summary for 2 May 1945, reporting on a French intercept. Photo - National Archives

The eavesdropper and his victim, U. S. Secretary of State Edward Stettinius, left, and French Foreign Minister Georges Bidault smile at one another. Would Bidault have been so pleased if he knew that Stettinius was reading his most secret cables? Photo - National Archives

Likewise, France's initial worries with respect to Security Council procedures grew as the conference neared. In early March, Bidault instructed his ambassador in Moscow to advise the Soviets of his concern that, under the veto arrangement, regional and bilateral treaties could be "subordinated to the previous agreement of the Security Council." This is dangerous, he added, because the "automatic nature of regional pacts is ... the essential element of collective security..." The French enlisted the Belgians and sounded out the British on an amendment to "clarify" the section on regional agreements in the U.N. Charter.[18]

With such intelligence data, the United States was able to develop its arguments on behalf of the veto well in advance and thus disable the opposition. Washington contended that, as a matter of realpolitik, there simply was not going to be a viable U.N. organization unless the four or five most powerful nations received the veto. Without it, none of these countries, the U.S. insisted, would entrust any of its sovereignty to an international group, especially not to a resurrected intrusive and powerless League of Nations-like body. The U.S. Congress, for instance, would not ratify the pact and would probably allow America to return to isolationism. Nor would the Soviets join without the veto. And, as a matter of practicality, U.S, delegates added, the other nations in the United Nations would never have enough weapons or influence to impose U.N. edicts on the great powers except with the concurrence of the Big Five themselves. Finally, even with an organization that might distribute authority in a way that would seem unsatisfactory to lesser powers, the superpowers would still always remain in the dock of world opinion. This would stabilize the peace system better than none at all.[19]

Alerted to France's hesitancy by its reading of the French diplomatic transmissions, Washington decided to focus its campaign for the veto on Paris, which had now made itself the leader of the recalcitrant member-states. As the San Francisco conference commenced, America intensified its overtures to France to reconsider its decision to forgo its role as the fifth permanent member of the Security Council. This approach, coming at a time when France was finding it increasingly difficult to act as the champion of the smaller nations, and flattering France's pretensions to being a great power and salving its hurt over Yalta, reignited the Quai d'Orsay's interest.[20] The French soon decided to accept their earlier assigned spot on the Security Council. With France's decision, the campaign to thwart the veto collapsed.

French diplomats saw fresh virtues in the arrangement, Ultra showed. While ostensibly staying above the battle, they were signaling a change of heart. The French U.N. delegate cabled Paris: "However far apart we [the Four Powers and small and medium powers] still are, a conciliatory solution is not impossible, for

---

[18] 7 March, 6-8; 9 March, 6-7.

[19] Clark Eichelberger, *Organizing for Peace: A Personal History of the United Nations* (New York: Harper & Row, 1977), 165; Evan Luard, *Conflict and Peace in the Modern International System* (New York: Macmillan, 1988), 212-215; Harry Truman, *Memoirs: Years of Decision* (New York: Doubleday, 1955), 284-285; Sumner Welles, *Seven Decisions That Shaped History* (New York: Harper, 1950), 185.

[20] Hilderbrand, 40, 120-121; "France Lining Up With Big Powers" (25 April 1945), "France's Position Put by De Gaulle" (26 April 1945), "France Position Still in Doubt" (28 April 1945), "Hints France Asks Major-Power Role" (3 May 1945), all *The New York Times*.

everyone is beginning to realize that the veto is a necessity and that its limits could not be further defined without risks for which no one wishes seriously to assume responsibility." Later, in summing up his country's achievements at San Francisco, the French delegate reflected: "although it [the veto] may in some cases seem an annoyance – and a very grave annoyance – it may also in others be a means of preventing the Council from meddling unduly in affairs which are our own or which we intend to settle through other channels."[21] Finally, discussions at San Francisco had rendered moot the concern of France and Turkey that bilateral and regional pacts would be undermined by the veto. Several Ultra dispatches showed that France no longer worried about this. The American victory was complete.

IV

The fourth event of importance illuminated by the Ultra files were the chronically touchy U.S.-Soviet relations. Intercepted cables – none of them Soviet – contained conflicting assessments of Soviet intentions. On the one hand, they showed the Soviets holding firm to their agreement with F.D.R. forged at Yalta, which had included a commitment to free elections in Poland. On the other hand, Moscow's surreptitious resistance to democratic government in Poland, and its heavy-handed pressures on such border countries as Iran and Turkey, revealed a Soviet Union that was growing increasingly obstinate about its territorial security. Ironically, even after the intercepts disclosed potentially alarming Russian moves, the United States remained puzzled about Moscow's intentions and unsure of how to act toward Stalin. American reactions swung between confrontation and conciliation.

Washington learned from intercepted messages, for example, that Moscow, at least in private conversations with some European powers, was treating the Yalta agreements as sacrosanct. Just days after Yalta, Molotov told the French ambassador to the Soviet Union, according to that envoy's dispatch to Paris on 19 February, that Stalin had expressed great confidence at the outcome of the meeting with Roosevelt and Churchill. And Molotov, according to the official, said that Stalin guaranteed that France would receive "an equal place" with Great Britain, Russia and the U.S. at the U.N.[22]

But Stalin was not willing to jeopardize his agreements with Washington and London to please Paris over the latter's aggressive insistence that the three powers reword the joint summons to San Francisco – an invitation in which

[21] 30 June, 4-6.

[22] 26 February, 8-10.

France had been asked to join by Moscow – to promise preservation of regional and bilateral pacts. The French ambassador messaged home his guess as to why the Soviets refused to rephrase the invitation: "I have reason to believe that at the present stage one of the prime concerns of Soviet policy is carefully to avoid anything that could weaken the assertion of perfect unity of viewpoint and action proclaimed by the Three Powers meeting at Yalta." Why? Because, he concluded, Moscow was "the principal beneficiary of the accord."[23]

Soviet Foreign Minister Vyacheslav Molotov at the podium in San Francisco. His cables, at least, were not read. Photo - National Archives

On the other hand, America was picking up ominous reports from foreign emissaries in Moscow and elsewhere concerning Poland and other Soviet border nations. The crux of the issue with respect to Poland was Stalin's demand

[23] 16 March, 6-7.

that his puppet regime in Lublin be admitted to the U.N. and the U.S. response that Stalin first democratize Lublin by including members of the prewar Polish government-in-exile in London. At Yalta, Stalin had agreed to a three-way British-American-Soviet commission to try to work out a coalition government in Poland. But Washington soon knew from the Ultra records and from its own ambassador in Moscow, Averell Harriman, that Stalin was reneging on that deal. France's delegate to Stalin's puppet regime in Lublin wired Paris that Lublin's policies were gradually turning it into a "quasi-protectorate" of the Soviet Union. Meantime, in early March, France's Moscow envoy reported that the commission was "running into great difficulties" because, according to information from Harriman, the Soviets didn't like "the choice of Polish leaders to be consulted." By mid-March, the French diplomat had had conversations with Poland's pro-Russian emissary to Moscow and informed Paris that the Soviets would probably impose their own regime on Poland.[24]

The impasse over Poland persisted for weeks, gravely imperiling the United Nations' ratification. Stalin was, according to some observers, on the edge of aborting the whole idea of a global organization – a "Western" idea about which he was never enthusiastic – over the matter. Eventually President Harry Truman, in the midst of the San Francisco conference, sent the late F.D.R.'s closest confidante, the seriously ill Harry Hopkins, who knew Stalin from past encounters, as his emissary to the Soviet leader to work out a settlement on Lublin. Hopkins, though terminally ill, arrived in Moscow in late May and conferred with the Soviet chieftain for 10 days.

Hopkins eventually obtained Stalin's agreement to a high-level meeting in Moscow on the beleaguered nation's future to be supervised by the three-way commission. The French liaison in London to the Polish government-in-exile learned of the Hopkins deal from his contacts and, in a message read in Washington, informed Paris in early June that the meeting would include all Polish leadership inside and outside the country. But he noted that the Polish exiles were disappointed that so few representatives of the democratic parties had been invited. Harriman later told the French in md-June that "it is not impossible that a successful conclusion will be reached" at the upcoming Moscow conclave on Poland. By the end of June, the session did produce a Provisional Government of National Unity that nominally included members of all political parties; as a result of this settlement, by early July, the U.S. and Great Britain formally recognized the regime and thereafter it gained U.N. membership. Ultra had warned Washington how stubborn and unyielding Stalin was likely to be on Poland and thereby had encouraged a U.S. backdown on the issue.

[24] 10 March, 8-9; 22 March, 7-9.

During this time, Washington had taken its own action to ward off what it saw as a potential Soviet intrusion in its own sphere of influence in Latin America. It began to caution Latin states throughout the hemisphere against Communist subversion. Ultra revealed that a Venezuelan diplomat telegraphed his home office on 7 May after a session with Rockefeller, the coordinator of inter-American affairs, that "Rockefeller communicated to us the anxiety of the United States government about the Russian attitude." American officials, the envoy wrote, were "beginning to speak of Communism as they once spoke of Nazism, and are invoking continental solidarity and hemispheric defense against it."

Still, as the Ultra documents illuminate, even as it maneuvered its way between various Soviet moves and countermoves, the United States continued to concentrate on the main target at hand – overcoming the Soviet Union's reservations about the United Nations, especially on such matters as Argentina's admission, the breadth of the veto, and Poland. Thus it showed a distinct unwillingness to allow any of these outside events to derail the conference. While F.D.R. at Yalta, and Truman in his first days in office, bargained hard on most issues with Stalin and Molotov, both always judged that the U.S. national interest in establishing the U.N. made it necessary to accommodate Stalin where otherwise they might have challenged him.

At Yalta, F.D.R. accepted the dictator's desire for three votes at the U.N. (one for the Soviet Union, and one for each of its "territories," the Ukraine and Byelorussia). Truman eventually backed a vague settlement of the Polish situation and gave in on a host of other protocol matters at San Francisco. In turn, though, the U.S. gained Stalin's assent to what is now generally regarded as an essential international body. That that assent was a necessity was well understood even then. As the under secretary of state, Sumner Welles, wrote a few years later, without the U.N. "war between the Soviet Union and Western powers would already have been inevitable, and the fate of our civilization would today be trembling in the balance."[25]

The San Francisco conference ended in late June 1945. Helped by Ultra, Washington by and large achieved what it wanted: a Security Council controlled by the five Allies, a weaker General Assembly, and a malleable Secretariat and military commission (though the latter soon vanished in the Cold War mists.) The Ultra intercepts gave the United States advance warnings about problems with members' admissions to the U.N., with decolonization, with the veto and with U.S.-Soviet relations – all essential to America's various bargaining positions. The Ultra intelligence thus gave Washington an edge in its public and in

[25] Welles, xviii.

its behind-the-scenes efforts at the conference.

In the spring of 1945, the U.S. already had a very big edge – Franklin Roosevelt had formally crafted the idea for the United Nations, he had organized the founding meeting in an American city (incidentally making interception easier for the Americans), nearly every country in the world wanted to join it, and the United States was now the most powerful nation on earth, possessing the richest economy and the strongest military. Nonetheless, Washington had to be absolutely certain of gaining its objectives at San Francisco, or the United Nations it desired might have fallen apart. America consequently used every weapon in its arsenal, including one of its most secret - Ultra. And it achieved what it sought.

Stettenius signs the charter of the United Nations for the United States, as President Harry S Truman watches. Photo - National Archives

BIOGRAPHICAL SKETCH

Stephen Schlesinger is currently a visiting scholar at the Taub Urban Research Center at New York University. He recently ended service as director of international organizations in the New York State government. In the 1970s, he was a staff writer at *Time* magazine and editor of *The New Democrat*. A graduate of Harvard Law School, he is also the author of *The New Reformers* (Houghton Mifflin, 1974) and co-author of *Bitter Fruit* (Doubleday, 1982).

ITALIAN DIPLOMATIC CRYPTANALYSIS IN WORLD WAR I

David Alvarez

ADDRESS: Department of Government, St. Mary's College, Moraga CA 94575 USA.

ABSTRACT: After its entry into the First World War in May 1915, Italy established a cryptanalytic unit to attack the military and diplomatic cryptosystems of other governments. This unit retrospectively solved an Austrian diplomatic system and currently read an American system, but it succeeded mainly against the codes and ciphers of minor powers.

KEYWORDS: Italy, diplomatic cryptanalysis, World War I.

1 INTRODUCTION

Although the combatants in the First World War eventually exploited signals intelligence on a large scale, few entered the conflict able to intercept and read secret diplomatic communications. In 1914 France and Russia supported small establishments of codebreakers in their foreign offices and interior ministries, and the Austro-Hungarian secret service purchased the occasional codebook from a venal embassy clerk, but other governments remained indifferent to the possibility of eavesdropping on friends and foes. Italy, which cast its lot with the Entente by declaring war on Austria-Hungary on 5 May 1915, was one of the powers that belatedly recognized the potential of signals intelligence. Though modern cryptology had practically been invented on their peninsula in the early Renaissance, the Italians had long since lost touch with their distinguished heritage and, like most of the belligerents, they entered the war with no codebreaking capability. In 1915 Italy's Foreign Ministry and Interior Ministry had cipher offices, but these units limited their effort to compiling new cryptosystems and encoding and decoding messages from Italy's diplomatic missions and regional prefectures. The cryptologic service of the Italian army consisted of a single officer whose limitations were revealed before the war when he labored two months to solve a correspondence encoded in two commercial systems without being aware that several years earlier a French expert had developed a mechanical solution to such a problem that would have reduced his labor to hours.

With such resources Italy was woefully unprepared to attack the various diplomatic messages that it had begun to intercept even before it entered the war. An appeal to France for assistance was rebuffed, probably because the French cryptanalysts, who had been regularly cracking Italian diplomatic codes since 1887, were reluctant to share their skills with their new ally.[1] Within a year of joining the conflict (circumstantial evidence suggests a date no earlier than the fall of 1915), Italian military intelligence created a three-man cryptanalytic unit and placed it under Luigi Sacco, a lieutenant of engineers who as head of the army's radio intercept service had begun to work on encrypted Austrian messages in his spare time.[2] This *Reparto crittografico* (Cryptographic Bureau), which eventually grew to several score officers, was responsible for both diplomatic and military cryptanalysis. While historians have recorded some of its military successes, such as the solution of Austria's CW and CARNIA codes, systems of 1000 groups for use within a single army, almost nothing is known of its work against the diplomatic cryptosystems of other powers, mainly because of the absence of official records and documentation.[3] It is possible, however, to catch a glimpse of the bureau's diplomatic work in the personal papers of Sidney Sonnino, Italy's wartime foreign minister. Military intelligence passed to the foreign minister various diplomatic intercepts and these documents allow us to sketch the broad outlines of Italian diplomatic cryptanalysis.[4]

2 ENEMIES

Italy eventually declared war on Germany, Bulgaria, and Turkey, but the Austro-Hungarian Empire remained throughout the war its principal antagonist. The bulk of the Italian army was engaged against the Austrians along the common border, which ran through the rugged peaks of the Dolomites and the Julian Alps, and it was from the Austrians that Italy expected to win territory in the Trentino and along the eastern shore of the Adriatic Sea. Italian intelligence

[1] David Kahn, *The Codebreakers* (New York, 1967), 318; Christopher Andrew, "Déchiffrement et Diplomatie: Le cabinet noir du Quai d'Orsay sous la Troisième République," *Relations internationales*, 5 (1976), 37-64, at 42-43.

[2] The birthdate of the Cryptographic Bureau remains obscure. In his memoirs the wartime director of the intelligence service, Odoardo Marchetti, recalled that for a period following the declaration of war the Italian army did nothing but file the cryptograms it intercepted. Then "for a long time" Lieutenant Sacco worked alone on the intercepts before his results convinced his superiors to establish the Cryptographic Bureau. A date in late 1915 or early 1916 seems reasonable. Odoardo Marchetti, *Il servizio informazione dell' esercito italiano nella grande guerra* (Rome, 1937), 88.

[3] Repeated inquiries at the historical office of the Italian army have failed to determine whether the records of the cryptographic bureau still exist.

[4] The Sonnino Papers are in the family archive in Montespertoli (Tuscany), Italy. A microfilm copy is in the Hoover Library, Stanford University, Palo Alto, California.

began intercepting Austrian diplomatic messages as early as March 1915, during negotiations over Italy's neutrality. A member of the Triple Alliance and, therefore, nominally an ally of Austria and Germany, Italy had remained on the sidelines during the crisis of July-August 1914 when its partners had gambled on war. By the spring of 1915 Rome was bargaining in London and in Vienna to determine which alliance was prepared to pay the most for Italian military support. Like all diplomatic missions in the Eternal City, the two Austrian embassies (one accredited to the Italian government and the other to the papacy) used the telegraphic facilities of the Italian postal service to communicate with their foreign ministry. The Austrians, who were old hands at intercepting secret communications, assumed correctly that the Rome telegraph office passed to Italian intelligence copies of all embassy cables, but they relied on their codes to protect the secrecy of their messages. They were not disappointed. Lacking a cryptanalytic capability (the *Reparto crittografico* was yet to be created), the Italians could do nothing but stare in frustration at the rows of numerical groups on the telegraph forms delivered each day by the central post office.

The inability to penetrate Austrian cryptosystems placed the Italians at a serious disadvantage during the negotiations in the spring of 1915 since cryptanalysts in Vienna were reading the encoded correspondence of the Italian embassy.[5] The inability to monitor the exchanges between Austrian diplomats was especially disappointing to Italian policy-makers since they did not entirely trust their ambassador in Vienna (who was a noted Austrophile) to report accurately the true intentions of the Dual Monarchy.[6] For some time after the creation of the *Reparto crittografico* the Austrian diplomatic codes resisted attack. Luigi Sacco's codebreakers did not register their first success until the late summer of 1916 and though this target remains unidentified it was almost certainly not an Austrian diplomatic system.[7] Indeed, the earliest diplomatic decrypts in the Sonnino Papers date from December 1916, and consist of a handful of Vatican telegrams.

The first evidence of progress on the Austrian codes appears in September 1917, when military intelligence sent Sonnino plaintext versions of the encoded telegrams exchanged between the foreign ministry in Vienna and its embassy at

[5]In 1919 the Italian Foreign Ministry learned that the Austrians had been able to read at least four of its cryptosystems. After interrogating an officer in the Austrian cipher service, Italian intelligence concluded, "The Italian diplomatic code [sic] has for a long time been known to Austria which easily decodes any Italian telegram." The report noted that Austria's penetration of Italy's codes began before the war. "Promemoria," Commando della 1 Armata, 23 March 1919, reel 51, Sonnino Papers.

[6]William Renzi, *In the Shadow of the Sword: Italy's Neutrality and Entrance into the Great War, 1914-1915* (New York, 1987), 71, 176-177.

[7]Cesare Amè, *Guerra Segreta in Italia, 1940-1943* (Rome, 1954), 48. Sacco's unit did not achieve a complete solution of an Austrian military cyptogram until August 1917. Kahn, *The Codebreakers*, 318.

the Vatican during the period March-May 1915. Though substantially complete, these solutions contain unrecovered groups and tentative identifications, which indicate that Sacco's cryptanalysts were still reconstructing the code. In October 1917 Sonnino received solutions of four telegrams exchanged between Vienna and its embassy to the Italian government in April and May 1915; military intelligence assured the foreign minister that additional telegrams were forthcoming. Though these four telegrams were also incomplete, the codebreakers had clearly made a significant (though tardy) entry into the Austrian systems of 1915.[8] Surprisingly, there was little further progress for several months. In April 1918 military intelligence sent the foreign minister "corrected" versions of the four cables. Later that month Sonnino received plaintext versions of all telegrams sent or received by the Austrian embassy in Rome during the period March to May 1915, though these decrypts still contained a few unsolved groups and some tentative identifications.[9]

During the war Italian intelligence organized several clandestine incursions into diplomatic missions in various countries in search of secret documents, but the progress against the Austrian diplomatic codes apparently depended upon pure cryptanalysis, though Sacco's codebreakers were able to compare the intercepted codetexts of certain Austrian cables with the plaintext versions which Vienna (in an uncharacteristic indiscretion) included in the "Red Book" of selected documents that it published in the summer of 1915 to justify its policy toward Italy.[10] Of course the success against the 1915 messages was of limited value in 1917. Sacco later claimed that during the war his unit solved a superenciphered Austrian diplomatic [11] Without further details it is impossible to determine whether he is referring to the 1915 messages or to a second effort against the Austrians. There is no evidence of a second penetration of Austrian diplomatic codes in the papers of Sidney Sonnino.

Though modest, Italy's achievement against Austrian diplomatic codes starkly contrasted with its even worse record against the cryptosystems of the other Central Powers. The Sonnino Papers offer no evidence that Italian cryptanalysts were able to crack the diplomatic codes of Germany, Bulgaria, or Turkey.

---

[8] "Raccolta di telegrammi scambiati tra il Ministro degli Esteri Austro-Ungarico e l'ambasciatore A.U. presso la Sante Sede nel periodo 13 marzo 1915 - 22 maggio 1915," reel 52, Sonnino Papers; Chief of Section R to Sonnino, 22 October 1917, reel 47, Sonnino Papers. The characteristics of the Austrian system(s) remain unknown.

[9] Chief of the Intelligence Service to Sonnino, April 1918, reel 47, Sonnino Papers.

[10] When Italy entered the war Italian intelligence did not possess any foreign codebooks or cipher tables. Marchetti, Il servizio informazione, 87.

[11] Kahn, *The Codebreakers*, 319.

3 ALLIES

One of the ironies of international relations is that a government may expend as much effort to collect intelligence on friendly powers as it does to secure information about hostile ones. Because Italy was constantly concerned that Britain, France and Russia might abandon her (as she had abandoned Germany and Austria-Hungary in 1914) by arranging a separate peace with Austria-Hungary or reneging on secret agreements for the distribution of the territorial spoils of victory, she kept a wary eye on her allies throughout the war. The papers of the foreign minister contain reports from Italian army and naval intelligence on such subjects as Allied politicians, French morale, British secret service operations, and alleged contacts between British, French or Russian diplomats and their Austrian and German counterparts. This discreet intelligence effort included attempts to monitor the secret communications of the Allied governments. The *Reparto crittografico* apparently had no success against the encrypted communications of Britain, France, and Russia, but it was able to crack at least some of the cryptosystems of Greece and the United States. For example, a cable from the Greek ambassador in London to his foreign ministry, dated 12 June 1918 and sent to Sonnino by the *Reparto crittografico* on 5 July ("Deciphering delayed by a change of cipher") revealed that British intelligence was assisting a Bulgarian revolutionary named Bonbolof to return to his country to foment a coup that would remove the king and separate Bulgaria from the Central Powers.[12] In the last days of the war an intercepted telegram from the State Department in Washington to the American embassy in Rome exposed Anglo-American concerns over financial conditions and currency stabilization in Archangel and Siberia, Russian territories occupied by or under the influence of Allied military forces, which had invaded the areas as an anti-Bolshevik measure. When passed to Sonnino, this intercept contained 22 undeciphered groups, suggesting that Italian cryptanalysts were still reconstructing this American code at the end of the war. By the time the peace conference convened at Versailles, however, the codebreakers had sufficient command of American cryptosystems to allow Sonnino to eavesdrop as the State Department and its embassy in Rome discussed the prospects for replacing him as foreign minister with an individual more amendable to American plans for the postwar world.[13]

---

[12] "3 telegrammi decifrati," reel 52, Sonnino Papers. Italian intelligence dated this telegram 12 July 1918, but since it was already decrypted and in Sonnino's hands by 5 July, the later date is clearly a slip of the pen.

[13] "5 telegrammi decifrati," reel 51, Sonnino Papers; Denis Mack Smith, *Italy and its Monarchy* (New Haven, 1989), 234.

4 NEUTRALS

The communications of neutral powers also attracted the attention of Italian intelligence. During the war these governments maintained their diplomatic representation in Berlin, Vienna, Constantinople, and Sofia, and the confidential telegrams of their embassies offered Italy an opportunity to gain some insight into the political, military and economic affairs of the Central Powers. Also, the reports cabled to their respective foreign ministries by neutral embassies in Allied capitals could usefully supplement the information reported by Italian representatives in those cities. Finally, the surveillance of neutral communications might alert Italian authorities to the activities of the army of spies, revolutionaries, secret emissaries, and confidence tricksters who found Berne, Madrid and Stockholm secure locales for their intrigues.

By the end of the war the *Reparto crittografico* had penetrated at least some of the diplomatic cryptosystems of Spain and Switzerland and were reading messages on the Madrid-Rome and Berne-Rome circuits. Spanish codes (which seem to have been read by almost everybody) were broken as early as September 1917. A decrypted cable to Madrid from the Spanish embassy to the Vatican alerted Sonnino that Pope Benedict XV was pressing British Prime Minister David Lloyd George to remain open to negotiations in the face of Berlin's unsatisfactory response to the pope's peace proposal of August. The telegrams of the Spanish and Swiss embassies in Rome were intercepted at the local telegraphic office, but some Spanish radio messages were plucked from the air by Italian listening posts. On 2 July 1918, for example, the Italians intercepted and decrypted a radiogram from the Spanish embassy in Vienna. It was forwarding to Madrid a message from the Spanish embassy in Constantinople concerning Turkey's interest in a separate peace.[14]

The Vatican, the smallest of the states, provided the *Reparto crittografico* with its greatest success and its richest harvest of diplomatic intelligence. Throughout the war Italy considered the Vatican a major intelligence target; indeed, Rome's intelligence effort against the papacy was matched only by its effort against its principal military opponent, Austria-Hungary. The preoccupation with the Vatican reflected, in part, a continuing concern with the "Roman Question." Ever since the occupation of Rome by the forces of the House of Savoy in 1870 and the incorporation of the Papal States into the Kingdom of Italy, the papacy and the Italian government had glared at each other across the symbolic boundary of the Tiber. The popes had never renounced their claims to temporal power and the Italian authorities suspected, with some reason, that they were alert

[14] "3 telegrammi decifrati," reel 52, Sonnino Papers.

for any opportunity to reassert those claims. With the outbreak of war Italy became especially apprehensive that the Vatican would find an opportunity in the political dislocations of the conflict and its aftermath to seek a restoration of at least some of its former territory. These apprehensions were sufficiently great that Rome insisted that the 1915 Treaty of London, which formalized its adhesion to the Allied cause, include a secret clause (Article 15) pledging London, Paris, and St. Petersburg to support Italy in opposing any role for the Vatican in a future peace conference. For the Italian government suspected the political sympathies of the papacy. The Vatican had long considered the Austro-Hungarian Empire its strongest supporter in Europe, while its attitude toward anti-clerical France was exceeded in coolness only by its feelings toward Italy itself. The Italians feared that the Vatican might place its considerable influence at the service of the Central Powers in return for a commitment from Berlin and Vienna to support the restoration of the popes to some form of temporal power at the expense of Italy.[15]

The earliest Vatican decrypts in the Sonnino papers, a few telegrams between the papal Secretariat of State and its nuncio (ambassador) in Vienna, date from the end of 1916, but by 1918 Italian intelligence was routinely intercepting and reading the communications of nuncios in Austria, Belgium, Germany, Holland, Spain, Switzerland, Turkey, and the United States. The *Reparto crittografico* had such a command of papal codes that it was able to read messages which, as a result of mistakes by papal code clerks, were gibberish to their intended recipients. When the Secretariat of State mistakenly sent a telegram to the nunciature (embassy) in Germany in a system reserved for the nunciature in Austria, Sacco's cryptanalysts immediately caught the mistake and were able to read the message while papal officials were still trying to sort things out. On another occasion the codebreakers expressed their frustration with the sloppy cryptographic practices of papal diplomats by complaining at the end of one decrypt: "This telegram contains so many errors, omissions and repetitions that it is in some parts indecipherable."[16] The effort against papal communications faltered only in the last year of the war when the Vatican improved its cryptographic procedures and introduced superencipherment for its more confidential messages.[17]

Through the surveillance of papal communications, Italian intelligence was privy to all aspects of the Vatican's secret diplomacy. The cryptanalysts must

[15]David Alvarez, "A German Agent in the Vatican: The Gerlach Affair," *Intelligence and National Security* (forthcoming).

[16]Pacelli to Gasparri, 11 October 1917, and Ragonesi to Gasparri, 18 August 1917, reel 48, Sonnino Papers.

[17]David Alvarez, "Faded Lustre: Vatican Cryptography, 1815-1920," *Cryptologia* (forthcoming).

have smiled as they decoded the warning, "Destroy this message and maintain absolute secrecy in this matter," in a telegram from the Cardinal Secretary of State to the nuncios in Munich (where the nunciature to Germany was located) and Vienna concerning an opportunity to detach Italy from its Entente partners.[18] Intercepts proved especially useful in the late summer of 1917 as the belligerent and neutral capitals reacted to Pope Benedict XV's seven-point peace proposal of 1 August. Decrypts of traffic on the Munich-Vatican circuit revealed that the Vatican had discussed the terms of the proposal with Berlin before presenting it to the Allies. Decrypts also revealed that in its effort to sell the proposal to the Allies the Vatican exaggerated the extent to which Berlin and Vienna were prepared to make political and territorial concessions.[19] Surveillance of papal communications also alerted the Italian foreign ministry to the Vatican's efforts, despite Article 15 of the Treaty of London, to secure representation at the Versailles peace conference by mobilizing Catholic opinion and lobbying statesmen such as U. S. President Woodrow Wilson and British Prime Minister David Lloyd George. Forewarned, Foreign Minister Sonnino counteracted such papal maneuvers as a plan to stimulate a press campaign against Article 15 and an effort to enlist the sympathies of the British monarch, George V.[20]

Occasionally, the Vatican's diplomatic messages contained items of military interest. This was especially true in the fall of 1917 when Austro-German forces smashed the Italians at Caporetto, advanced across northeastern Italy, and threatened Venice. From Berne, Munich, and Vienna, papal diplomats cabled to the Vatican daily reports on the offensive. Gleaned from official bulletins, contacts in the host government, and conversations with colleagues in the local diplomatic communities, these reports included references to the situation at the front ("Colle di Lana lost, province of Belluno threatened, Italians retreating along a front of 150 kilometers"), Italian losses ("From 24 October to 1 November, 250,000 prisoners and over 2000 pieces of artillery"), the direction of the attack ("On the lower Piave German troops are nearing the Vidor bridge."), and the intentions of Italy's enemies ("Probably the plan is to separate Italy from France in order to aggravate Italy's economic and military situation.").[21] Delivered each day to the Foreign Ministry, these intercepts were especially welcome

[18] Gasparri to Aversa and Valfre, 23 February 1917, reel 48, Sonnino Papers.

[19] Pacelli to Gasparri, 27 September 1917, and Valfre to Gasparri, 25 August 1917, reel 48, Sonnino Papers; Italo Garzia, *La Questione Romana durante la prima guerra mondiale* (Naples, 1981), 161-164.

[20] Ragonesi to Gasparri, 14 January 1918, Gasparri to Ogno, 15 December 1918 and 20 April 1919, reel 48, Sonnino Papers; Dragan Zivojinovic, *The United States and the Vatican Policies, 1914-1918* (Boulder, 1978), 134-138.

[21] Marchetti to Gasparri, 6 and 11 November 1917, and Pacelli to Gasparri, 7 and 12 November 1917, reel 48, Sonnino Papers.

to Sonnino since the Italian High Command deliberately issued misleading or false reports to civilian authorities in order to obscure the scope of the debacle and the incompetence of the general staff.[22]

The papal messages, especially those from Berne, Munich, and Vienna, produced a trove of information which permitted Italian intelligence to monitor not only the activities of the Vatican, but also political developments in Austria and Germany. The telegrams of the Munich nuncio, Monsignor Eugenio Pacelli, allowed the Italians to chart the military's growing influence in German politics and to follow the political maneuvers in Berlin that led to the replacement of Theobald von Bethmann-Hollweg as chancellor by Georg Michaelis in July 1917 and the latter's removal in favor of Georg von Hertling in November of that year.[23] In November 1918 the reports of the Munich and Vienna nunciatures described the deteriorating economic and political conditions in Germany and Austria and provided early indications of the overthrow of the Wittelsbach dynasty in Bavaria and the creation of a Bavarian socialist republic.[24]

5 CONCLUSION

Any appraisal of Italian diplomatic cryptanalysis that depends upon the Sonnino Papers must acknowledge the limitations of that source. The decrypts in the collection probably represent only a portion of the diplomatic messages intercepted and decrypted by the *Reparto crittografico*. Some messages may have been destroyed or returned to the intelligence service, while others may have remained in the Foreign Ministry's files when Sonnino left his post in the summer of 1919. Also, the foreign minister may have received and retained only those messages relevant to his principal diplomatic concerns. A selection process can be inferred from the Vatican decrypts. The majority of these messages originated in the important nunciatures in Munich and Vienna, the heart of enemy territory. There are relatively few telegrams from the nunciatures in Berne, Brussels, Madrid, and Washington, and none at all from papal missions in such capitals as Lima and Buenos Aires. Since it is unlikely that the *Reparto crittografico* would have been able to read the cryptosystems used by the papal Secretariat of State to communicate with its most important missions but remain unable to penetrate the less sophisticated systems in service at minor papal missions, the absence of

[22] Smith, *Italy and its Monarchy*, 220, 227.

[23] For Pacelli's numerous reports on German domestic politics, see reel 48, Sonnino Papers. Monsignor Pacelli capped a distinguished career in the papal diplomatic service by serving as Cardinal Secretary of State for nine years (1930-1939). In 1939 he was elected to the papacy as Pope Pius XII (1939-1958).

[24] Pacelli to Gasparri and Valfre to Gasparri, 8 November 1918, reel 48 Sonnino Papers.

decrypts from the Latin American nunciatures probably reflects the interests and priorities of the foreign minister rather than the limitations of the Cryptographic Bureau.[25]

Nonetheless, the intercepts in the Sonnino Papers remain the only primary evidence of Italian diplomatic cryptanalysis during the First World War. This evidence indicates that Italy was a latecomer to wartime signals intelligence and that its codebreakers never equaled the achievements of their counterparts in Britain and France.[26] The *Reparto crittografico* succeeded, at best, only moderately in reading the diplomatic communications of other governments. The retrospective reading of an early Austrian code and the current reading of an American code were probably the highpoints of a cryptanalytic effort whose success was mainly limited to the cryptosystems of a few minor powers.[27] Of course, much of the cryptanalytic history of the First World War remains unwritten, but the available evidence suggests that when the full record is revealed, Italy will not stand in the first rank.

BIOGRAPHICAL SKETCH

David Alvarez is a professor at Saint Mary's College of California. His current research focuses on SIGINT and the European neutrals in World War II.

[25] The Secretariat of State tried to give its most important missions more secure systems. When, in 1918, it introduced superenciphered codes, the new system was initially reserved for the nunciatures in Munich, Vienna, Berne, and Madrid.

[26] During the war French cryptanalysts solved the diplomatic cryptosystems of Austria-Hungary, Bulgaria, Germany, Greece, Italy, Spain, and Turkey, while British codebreakers read the systems of Austria-Hungary, Denmark, Germany, Greece, Japan, Norway, Spain, Sweden, and the United States. Andrew, "Déchiffrement et Diplomatie," 58-59; A. G. Denniston, "The Government Code and Cypher School Between the Wars," *Intelligence and National Security*, 1 (January, 1986), 53-54.

[27] It is possible that the *Reparto crittografico* cracked British, French or German codes but that no trace of such successes remains in the Sonnino Papers. It is, however, unlikely that Sonnino would save the odd Greek or Swiss intercept but retain not a single intercept from the major belligerents, especially in view of the fact that his papers contain many intelligence reports on Britain, France and Germany drawn from sources other than intercepted communications.

FADED LUSTRE: VATICAN CRYPTOGRAPHY, 1815-1920

David Alvarez

ADDRESS: Department of Government, St. Mary's College, Moraga CA 94575 USA.

ABSTRACT: The Vatican, whose cryptology was once perhaps the finest in the West, lagged behind other states in responding to advances in cryptography in the period 1815-1920. It depended upon nomenclators long after other states had abandoned that system, and it did not begin to encipher its codes until decades after the practice had become common in the foreign ministries of other governments. Only during the First World War did the Vatican seriously attempt to upgrade its diplomatic cryptosystems.

KEYWORDS: Vatican, cryptography, nomenclators, polyalphabetics, enciphered codes

I. ORGANIZATION

Papal cryptography has a long and often distinguished history. The popes were among the first in modern Western civilization to recognize the utility of codes and ciphers. The first modern code, a short list of single letter substitutions for nouns, can be found in the archives of the Vatican. This code dates from the first half of the fourteenth century, a period during which the papal curia (central administration) was already using simple ciphers for some of its correspondence with cardinals and archbishops. In 1379 a papal secretary compiled a set of cryptographic keys that are the oldest extant and that include the first nomenclator, a system that combines a cipher system of substitutes for individual letters with a code list of equivalents for entire words. The post of cipher secretary was created in the curia in 1555, and in the late 1580s the office fell under the control of a remarkable family, the Argentis. As cipher secretaries to six popes, Giovanni Batista Argenti and Matteo Argenti, his nephew and successor, elevated papal cryptography to a dominant position in the late Renaissance. Their accomplishments included the first use of a mnemonic key to mix a cipher alphabet, an early solution of a polyalphabetic cipher, and the preparation of a manual of cryptography that represented the state of the art at the time.[1]

[1] David Kahn, *The Codebreakers* (New York, 1967), 106-107, 113; Aloys Meister, *Die Geheimschrift im Dienste der Päpstlichen Kurie* (Paderborn, 1906).

At the beginning of the nineteenth century, papal cryptography was the responsibility of the Secretariat of Ciphers, a distinct department in the papal administration. Its staff consisted of three *cifristi* (cipher clerks) under the direction of the cipher secretary. The office worked exclusively for the Secretariat of State, the pope's foreign ministry. In 1814, a reorganization abolished the Secretariat of Ciphers and the position of cipher secretary and moved the *cifristi* into the Secretariat of State, where they became a separate class of officials responsible for ciphering and deciphering dispatches, maintaining secure custody of the codes and ciphers, and creating new cryptographic systems. The reorganization also established the new post of substitute secretary of state and cipher secretary. As deputy to the cardinal secretary of state, the substitute was responsible for the day-to-day operation of the department, including the supervision of cryptography. From 1814 to 1840, the total staff of the Secretariat of State, which for most of the period supervised both the external and internal affairs of the Papal States, consisted of the cardinal secretary of state, the substitute, four or five *minutanti* (chief clerks), three *cifristi*, four or five *scrittori* (copyists), and two *archivisti* (archivists).[2]

Throughout this period, papal cryptography was closely linked with another family, the Evangelisti. Giuseppe Evangelisti entered the Secretariat of State in 1793 as a copyist. His father and uncle had served as cipher clerks in the old Secretariat of Ciphers in the pontificates of Clement XIV (1769-1774) and Pius VI (1775-1799). In 1801 Giuseppe continued the family tradition by joining the *cifristi*. He accompanied Cardinal Ercole Consalvi to the Congress of Vienna (1814-1815), where he was responsible for the cardinal's voluminous secret correspondence. At his death in 1847, he had been the Secretariat's senior cipher officer for four decades. Filippo Evangelisti, Giuseppe's younger brother, followed the expected path by joining the *cifristi* in 1806. A graduate of the elite Collegio Romano, where he won several academic prizes, Filippo was included in the small suite of officials who followed Pius VII to Genoa in March 1815 when the pontiff, reacting to Napoleon's escape from Elba and the invasion of the Papal States by King Murat of Naples, sought refuge in northern Italy. Filippo Evangelisti served in the cipher section of the Secretariat until 1840.

Two other individuals worked in papal cryptography during this period. Vincenzo Scagliosi, of whom little is known, served in the old Secretariat of Ciphers from 1804 until 1814, when that department was incorporated into the Secretariat of State and he transferred into the ranks of the *minutanti*. His successor,

[2]For the organizational history of the Secretariat of State in this period, see Lajos Pásztor, *La Segreteria di Stato e il suo Archivio, 1814-1833* (Stuttgart, 1984).

Giuseppe Golt, remained in the cipher section until his retirement in 1847.[3]

II. SYSTEMS, 1815-1870

In the first decades of the nineteenth century, the Vatican had in service a variety of codes and ciphers. Francesco Capaccini, for instance, used several systems while serving as nuncio (papal ambassador) in the Low Countries between 1828 and 1831. In the simplest procedure, Capaccini would compose a dispatch in which the paragraphs were numbered consecutively. In the even-numbered paragraphs there was no attempt at subterfuge; the sentences meant what they said. In the odd- numbered sections, however, the true meaning of a sentence was the opposite of that written. The statement "our position is strong" was meant to be understood as "our position is weak." Such a message was sent in clear. Monsignor Capaccini also used at least two jargon codes which substituted codewords for the names of certain individuals or places in an otherwise open dispatch . In one such code, which had more than 70 equivalents and also included a small cipher, *Rome* = AUNT, *France* = MR. NANTINI, *Spain* = MR. CARLI, *Amsterdam* = SAN ORESTE, and *Prince of Orange* = MR. FREY. A third system consisted of two-, three-, and four-digit numerical groups. Nouns could be represented by any of 10 sequential numbers, and many had, in addition, a corresponding codeword. Austria, for example, could be represented by any group in the sequence 090 ... 099, or by the codeword MOROCCO. The Pope was 1240 ... 1249 or MERCURIO. Capaccini also used a null code in which the first words of certain lines in an apparently innocuous letter would, if separated from the text and combined in the correct order, form a message. The reader could identify the relevant words in the appropriate lines because the last letter of one was connected to the first letter of the next without the pen having been lifted from the page. The correspondents agreed ahead of time on the order in which the words would be read.[4]

Capaccini's cryptographic repertory was probably atypical among papal diplomats, since he was an enthusiast who delighted in creating methods of secret communication. In the first half of the nineteenth century the papacy's basic system for secret communication was the nomenclator, that system, half code and half cipher, which had dominated western cryptography since the fifteenth century. The Argentis had discussed the use of nomenclators in their famous cipher manual, and in an institution as tradition-bound as the Vatican, the Evangelisti were

[3]Ibid., 78, and chapter 5.

[4]Vatican City. Archivio Segreto Vaticano (ASV). Segreteria di Stato, Spogli: F. Capaccini, busta 3, fasc. A.

not inclined to challenge the authority of their illustrious predecessors.

The system prepared in 1856 for Monsignor Alessandro Bianchi, the last nuncio to the Grandduchy of Tuscany before its absorption into the Kingdom of Italy, is representative of the papal nomenclators of the period.[5] It was a modified two-part system with three-digit numerical equivalents (from 242 to 989, skipping any group with a 0, 1, or 3) for the letters of the alphabet and almost 200 common bigrams and trigrams. Four-digit groups (3008 to 3989) served for nouns, proper names, parts of speech, and commonly used prefixes and suffixes. For punctuation there were three-digit groups each beginning and ending in 1 (*comma* = 121, *period* = 161). In the alphabet, each vowel was represented by three homophones, as were the consonants n, r, and t. The other consonants, except for h, l, the combination qu, and the (in Italian) rarely used j, w, x, and y, each had two homophones. In Italian, a preposition will have different endings depending upon the gender and number of its object. In Franchi's nomenclator, the compiler has given each variation its own group; thus, the preposition da (by, from) is represented by the group 857, while its variants *dagli*, *dai*, *dal*, *dalla*, and *dalle* are represented, respectively, by the groups 542, 986, 758, 752, and 928.

A section labeled "Precautions" provided directions on the use of the system.

> The dispatch, or the portion of it written in code, should be entirely in numbers, without paragraph divisions. It should not be divided by commas, periods, and accent marks, and even less by [unenciphered] words, even if they are insignificant. All must be done with numbers, which should be clear, intelligible, and always placed at an equal distance from each other.
>
> The numbers 1 and 0 are nulls, except a 0 which has value as part of a four-digit group that always begins with the number 3, a number that appears nowhere else.
>
> A 1, separated from another 1 by a single number, serves as punctuation, and, by itself, serves to distinguish one word from another to avoid confusion in deciphering. For this reason do not omit to use it. The 0 can also be used for the purpose of dividing words, except in cases where 0 is used at the end of a four-digit group. This will facilitate decoding.

In the first half of the nineteenth century, the Vatican also made occasional use of ciphers. In the 1840s the nunciature in Madrid had in service three monoalphabetic substitutions. The first system substituted ordered pairs of numbers

[5] ASV. Segreteria di Stato, Spogli: A. Franchi, v. 1.

($a = \frac{1}{2}$, $b = \frac{1}{3}$, $c = \frac{1}{4}$, ...) for the letters of the alphabet. In the second system the letters were replaced by two-digit numerical groups in ascending order (a = 11, b = 12, c = 13, ...). In the third system the letters in the first half of the alphabet were replaced by those in the last half in reverse order (a = Z, b = Y, c = X, ...), and the remaining letters by numbers in descending order (n = 12, o = 11, p = 10, ...). The latter system occasionally substituted symbols for certain proper nouns (*France* = F, *Pope* = ∞). The three ciphers used symbols for punctuation (*comma* = $\succ$, *period* = $\sqrt{}$) and spelled-out all numbers. Apparently, there were neither homophones nor nulls. Provincial delegates (governors) who administered various cities and territories within the Papal States, that broad swath of territory curving across central Italy from Rome in the southwest to Ravenna in the northeast, also used simple ciphers to protect their telegraphic communications with Rome. In the spring of 1860, the delegate in the Adriatic port of Ancona had a system that substituted two-digit numerical elements for letters of the alphabet (a = 28, b = 69) and common Italian bigrams and trigrams (*al* = 58, *co* = 49, *pre* = 64). In the same year, the governor of Pesaro depended upon a cipher that merely mixed the alphabet in a system of reciprocal substitution (a = U, u = A, b = H, h = B).[6]

Given the relative simplicity of these ciphers, it is rather surprising to discover that the Vatican seriously experimented with systems of polyalphabetic substitution. Such a system had been conceived in the fifteenth century by the great Renaissance polymath Leon Battista Alberti, but it had never caught on in the government chancelleries and embassies, where the professional cryptographers much preferred the nomenclator. While recognizing that a polyalphabetic cipher avoided that fixed replacement of a letter with the same cipher symbol throughout a message which is the particular weakness of monoalphabetic systems, the professionals felt that the more complex system required too much time and attention. The frequently changing alphabets could too easily confuse a code clerk for whom the traditional nomenclator seemed more straightforward and less difficult to use.[7]

Papal cryptographers were no strangers to polyalphabetic ciphers. Such ciphers were known to and used by the Argentis in the sixteenth century. Matteo Argenti considered them "the noblest and the greatest [ciphers] in the world" and discussed them in his famous manual. While reluctant to adopt polyalphabetics for general service, the Argentis occasionally prepared one for the use of a partic-

[6]The ciphers of the Madrid nuncio and the Ancona and Pesaro delegates have been partially reconstructed by the author from their dispatches. For Madrid, see ASV. Segreteria di Stato, Esteri, 1847, rubrica 11, fogli 43-46. For the delegates, see ASV. Segreteria di Stato, 1860, rubrica 165, fasc. 2-3.

[7]On the development of polyalphabetic substitution, see Kahn, *The Codebreakers*, 125ff.

ular papal official.[8] Subsequently, the papacy, like other European governments, came to depend upon the nomenclator to protect its secret communications, but the memory of polyalphabetics apparently survived.

The private papers of Francesco Capaccini, a career diplomat whose service between 1815 and 1844 included senior posts in the Secretariat of State, as well as ambassadorial assignments in Holland and Portugal, include two polyalphabetic systems.[9] In the first, the letters of the alphabet, in column, are numbered from 1 to 26 and, again, from 27 to 52. A central column lists the alphabet again, this time with each letter associated with a key word (a = PARAVAMI). To encipher the phrase, *Il Re di Marocco* (The King of Morocco), the cipher clerk would write below it a keyword, e.g. PEPELAGRO. Referring to the first column, he would add the numerical value of each plaintext letter to the numerical value of the keyword letter below it. The sum would then be converted, again by reference to the first column, to a letter, creating the ciphertext YQHJPJTSGEHST. The clerk would place at the front of the ciphertext the letter identifying the keyword.

plain	i	l	r	e	d	i	m	a	r	o	c	c	o
+	9	12	18	5	4	9	13	1	18	15	3	3	15
key	P	E	P	E	L	A	G	R	O	P	E	P	E
	16	5	16	5	12	1	7	18	15	16	5	16	5
cipher	25	17	34	10	16	10	20	19	33	31	8	19	20
	y	q	h	j	p	j	t	s	g	e	h	s	t

To decipher a message like QEROMGSB, the process would be reversed. The first letter of the ciphertext would indicate which keyword was in use. This keyword would be written under the ciphertext. The numerical value of each letter in the keyword would be subtracted from that of the cipher letter above it, and the result converted into a plaintext letter to reveal the words *i Masoni* (the Masons).

indicator Q = VENTRES

cipher	e	r	o	m	g	s	b
-	31	18	15	39	33	19	28
key	V	E	N	T	R	E	S
	22	5	14	20	18	5	19
clear	9	13	1	19	15	14	9
	i	m	a	s	o	n	i

[8] Ibid., 150-151.

[9] ASV. Segreteria di Stato, Spogli: F. Capaccini, busta 3, fasc. A.

The second polyalphabetic in the Capaccini papers consists of two tables, one for enciphering and one for deciphering. The former has 26 columns, one for each letter of the alphabet. Within each column, the letters of the alphabet are listed in order with each letter associated with a number from 50 to 78. These numbers are in mixed order, and the numbers 55, 66, and 77 are not represented. Along the bottom of the table are 52 key phrases, in groups of four, each of the first 26 identified by a letter of the alphabet, and each of the remainder by a number between 10 and 38 (skipping 11, 22, and 33). To encipher "abbey" with the keyword CABA, the writer would copy the plaintext under the key, search down the column represented by the first letter of the keyword to find the first letter of the plaintext, and then enter the corresponding number as the first element of the ciphertext. He would proceed in the same way with the second letter and so on to the end of the message.

A	B	C		Y	Z
a+a = 63	b+a = 62	c+a = 74	...	y+a = 51	z+a = 68
a+b = 62	b+b = 65	c+b = 53	...	y+b = 60	z+b = 67
a+c = 74	b+c = 53	c+c = 64	...	y+c = 70	z+c = 62
a+d = 69	b+d = 52	c+d = 73	...	y+d = 71	z+d = 50
a+e = 78	b+e = 56	c+e = 51	...	y+e = 63	z+e = 73
⋮	⋮	⋮	⋮	⋮	
a+y = 51	b+y = 60	c+y = 70	⋮	y+y = 74	z+y = 56
a+z = 68	b+z = 67	c+z = 62	⋮	y+z = 56	z+z = 57

key	C	A	B	A	C
plain	a	b	b	e	y
cipher	74	62	65	78	70

To indicate which keyphrase he had selected from the 52 available at the bottom of the table, the writer would place at the head of the ciphertext the letter or number that identified the keyword. Two-digit groups served to denote punctuation (*period* = 11, *comma* = 22) and to indicate a change in the key (88).

The presence of these ciphers in the papers of an important papal diplomat is intriguing in view of established opinion that in the early nineteenth century governments avoided polyalphabetics and depended almost exclusively upon nomenclators for their diplomatic correspondence This opinion is, at first glance, reinforced by the fact that Capaccini, distrusting the official cryptographic systems of the Secretariat of State, often created codes and ciphers for his private use. Were the papal polyalphabetic ciphers aberrations concocted by Capaccini for his personal amusement or private use? Circumstantial evidence suggests that the polyalphabetics may have seen more general service in the Secretariat of State.

The first of the two polyalphabetics is handwritten on a single sheet of paper, and bears the notation, *Questa chiave Mgr. Leardi* ("Monsignor Leardi's key"). This is a reference to Paolo Leardi who was the papal nuncio in Vienna from 1816 to 1823, and it indicates that this polyalphabetic cipher was in service at the Vienna nunciature during his tenure. Indeed, as late as 1825 the nunciature was still using this cipher to communicate with the Secretariat of State.[10]

The second of the polyalphabetics is even more suggestive. It is printed on long sheets of paper. The keywords along the bottom have been written by hand next to the printed letters and numbers which serve as indicators. The Capaccini papers contain several "blank" tables; that is, tables where the keywords have not been entered at the bottom. This indicates that the tables had been prepared ahead of time for future use, and that someone anticipated using the tables with sufficient frequency to justify printing multiple copies. There are two "complete" tables; that is tables in which the keywords have been filled in by hand. A marginal notation on one table indicates that it was prepared for Antonio Tosti who was the pope's diplomatic representative in Turin from 1822 to 1828. In the second complete table, the keywords differ from the first. What is interesting about this second table is the fact that one cluster of four keywords has the notation *Parigi* (Paris), another cluster of four has *Lisbona* (Lisbon), another has *Vienna*, while two clusters are marked *Comune* (common). It was not unusual for the Secretariat of State to issue two ciphers to a nuncio; one, the *cifra particolare* (private cipher), was for confidential communication with the Vatican, the other, the *cifra comune* (common cipher), was for communication between nuncios. One can conclude that the Secretariat of State was using the polyalphabetic cipher to communicate with papal diplomatic missions in Austria, France, and Portugal, and that the same cipher was in use among the nuncios in those capitals. In 1831 Capaccini used this system to report to the Vatican from his post as nuncio in the Low Countries.

The presence of these polyalphabetic ciphers in the personal papers of Monsignor Capaccini lends additional credence to the argument that such systems were, at least for a time, in general service in the papal Secretariat of State. From 1815 to 1824, Capaccini was the *minutante* (chief clerk) responsible for drafting the Secretariat's correspondence with the nuncios in foreign capitals. Apparently the polyalphabetics were used for the most confidential messages, and only

[10] See the correspondence in ASV. Segreteria di Stato, 1823-1825, rubrica 247.

Capaccini and Luigi Armellini, his successor as *minutante* for diplomatic affairs, were allowed to encipher and decipher dispatches in these systems. Reports in more routine systems, such as nomenclators, were left to the personnel of the cipher section.[11]

It is even likely that Capaccini designed the polyalphabetic ciphers, for in June 1829 the amateur cryptographer, then on a diplomatic mission to Holland, recalled an interesting encounter for Cardinal Albani.

> Since our old ciphers were known to everyone, when I was in the Secretariat of State I composed, with much effort, a new one which is the one I now use. Pope Leo XII laughed at me, saying that no cipher is indecipherable, but when he saw that a professional cryptologist, who was then in Rome, could not break mine and had to give up, he authorized me to use it for correspondence. This cipher, however, is very difficult and requires much time to cipher, and perhaps even more to decipher, as Your Eminence will have already discovered ...[12]

The cryptographic systems of the papacy were clearly of uneven quality. The polyalphabetic ciphers (to the extent that they were used) certainly represented the best effort of the Secretariat of State. At a time when other governments depended upon the familiar and more easily managed nomenclator for their diplomatic communications (even though they considered the polyalphabetics unbreakable), these papal ciphers were probably as good as any system then in use in Europe.[13] As for the Vatican's nomenclators, the one in service at the nunciature in Florence in 1856 (supra) was probably indicative of the best in the repertory of the Secretariat. Although rather short (fewer than 900 plaintext elements in a period when nomenclators of 3,000 elements were not unknown), it was a respectable system. It was as good as some of those used by France in the first half of the century, and in its use of homophones and nulls was superior to a contemporary American system. On the other hand, it was less sophisticated than a Spanish system which, in addition to homophones and nulls, utilized such devices as special groups to indicate a doubled letter or to cancel a preceding group.[14] Unfortunately, the nomenclator of the nuncio in Florence

[11] Pásztor, *La Segreteria di Stato*, 207-208.

[12] J. P. de Valk (ed.), *Lettres de Francesco Capaccini* (Brussels, 1983), 256. Leo XII was pope from September 1823 to February 1829.

[13] A general solution for polyalphabetic ciphers was not available until 1863. Kahn, *The Codebreakers*, 191.

[14] Ralph Weber, *United States Diplomatic Codes and Ciphers, 1775-1938* (Chicago, 1979), 478-489. The French and Spanish systems are among the papers of Britain's Decyphering Branch in the British Library (London), Add. MSS. 32265 and 32301.

was an anachronism on the day it entered service since, in response to the expansion of telegraphic communications, European foreign offices at mid-century were rushing to replace their nomenclators with full-fledged codebooks.[15] The flimsy substitution ciphers used by papal governors and, at least for a time, the nuncio in Madrid offered scant security and would not have withstood even the most cursory professional scrutiny. Of course, in the case of the governors, the danger of interception was practically nil since their telegrams and dispatches moved along internal lines. Their schoolboy ciphers probably sufficed to protect a message from the casual curiosity of a clerk, messenger, or cable operator.

III. DECLINE

Whatever their merits, the cryptographic systems of the Secretariat of State engendered little confidence among papal officials. The Vatican's special envoy to Paris in 1801 did not trust his ciphers. At the Congress of Vienna, the pope's representative, Cardinal Consalvi, correctly surmised that his nomenclator had been compromised by the Austrian intelligence service. Fourteen years later, Francesco Capaccini complained that the Vatican's ciphers (except, of course, those of his own devising) were open to everyone, and Pope Leo XII, who before his election to the Throne of St. Peter had led diplomatic missions to France, Germany, and Switzerland, expressed his conviction that no cryptographic system was secure. As late as 1859, the nuncio in Vienna warned the Secretariat of State that his cipher was insecure.[16]

Several factors, aside from the inherent weaknesses of some of the systems, undermined the integrity of the Vatican's ciphers. Cryptographic materials were a natural target for the foreign agents who occasionally insinuated themselves into the households of nuncios.[17] Official dispatches might easily be intercepted and copied in transit, thereby providing foreign cryptanalysts an opportunity to study and attack papal ciphers.[18] This threat was not diminished by the poor communications practices of papal diplomats. To save time and effort, nuncios

[15] On this process, which was substantially completed by the 1860s, see Kahn *The Codebreakers*, 190.

[16] Boulay de la Meurthe (ed.), *Documents sur la Négociation du Concordat* (Paris, 1895), xxiii; Alessandro Roveri (ed.), *La Missione Consalvi e il Congresso di Vienna* (Rome, 1970), v. 3, 545, 567-68; Carla Meneguzzi Rostagni (ed.), *Il Carteggio Antonelli-DeLuca*, 1859-1861 (Rome, 1983), 61.

[17] During the Congress of Vienna the Austrian secret police obtained documents from an agent inside the papal nunciature. One nuncio in Madrid discovered that his valet was reading his confidential correspondence. A confidant of Francesco Capaccini during the latter's mission to the Low Countries may well have been an agent of the local Neapolitan ambassador.

[18] Lacking a regular courier service of its own, the Secretariat of State often depended upon the diplomatic couriers of Austria, France and Spain. The Austrians were especially skillful at opening dispatches entrusted to their care.

usually enciphered only those portions of a dispatch which they considered the most sensitive. Appraisals of sensitivity varied, naturally, from individual to individual. Often a nuncio would encipher only the names of people and places; sometimes only one or two words in an entire message might be enciphered. Such practices (especially the latter) immeasurably aided the efforts of anyone seeking to break the cipher. With most of the message in clear, the cryptanalyst had a context of subject matter, time, and locale in which to place the unknown text.

Communications security was further compromised by the low priority of cryptography in the Secretariat of State. This situation reflected a decline in the use of codes and ciphers. In the first half of the nineteenth century, the cryptographic requirements of the Vatican were few, largely because of the modest position of the papacy in world affairs. In 1830, for example, the papacy accredited diplomatic missions to only 11 countries. By midcentury the number had increased slightly to 13. These missions generated relatively few secret communications. The record of the nunciature in Madrid, one of the more important posts in the Vatican's diplomatic network, is revealing. In 1828 the Secretariat of State sent only 12 enciphered dispatches to Madrid, while the nunciature sent no enciphered messages at all to Rome. In 1848 the Secretariat forwarded only three enciphered dispatches to Madrid, which responded with seven of its own. Months might pass without secret correspondence between the Vatican and its nuncio in Spain.[19]

The volume of enciphered correspondence that passed between the Vatican and its representatives abroad had diminished to such an extent that there was insufficient work for the *cifristi*.[20] The senior officer, Giuseppe Evangelisti, was usually able to handle most of the enciphering and deciphering and still have time for other tasks around the Secretariat. After 1815, Filippo Evangelisti worked less and less with ciphers as he assumed duties in the registry of the Secretariat. The third member of the section, Giuseppe Golt, was not concerned with secret communications at all. He performed archival work and served as secretary to Monsignor Carlo Mauri, who held the post of substitute for 16 years. Upon Mauri's death in 1830, Golt assumed responsibility for various legal matters of interest to the Secretariat, although he remained nominally attached to the cipher section.

Papal cryptography was rather neglected in the last decades of the Papal States. For some time, perhaps more than a century, the cipher section had

[19] Vicente Carcel Orti (ed.), *Correspondencia Diplomatica del Nuncio Tiberi, 1827-1834* (Pamplona, 1976), xvii and passim; idem, "Los Despachos de la Nunciatura de Madrid, 1847-1857," *Archivum Historiae Pontificiae*, 13 (1975), 311-400. The totals for enciphered messages, low as they are, still exaggerate the use of ciphers by the Madrid nunciature since only portions (at times only a few words) of a dispatch would be enciphered.

[20] Lajos Pásztor, "La Segreteria di Stato di Gregorio XVI, 1833-1846" *Archivum Historiae Pontificiae*, 15 (1977), 322.

been coasting, propelled (ever less forcefully) by the memories of past accomplishments. By the middle of the nineteenth century it had coasted almost to a stop. The Secretariat of State and its cipher section were not immune to that malaise marked by inertia, lassitude, and a "qualitative loss of energy" that pervaded the papal administration in the 18th and 19th centuries, and that made the Roman bureaucracy a byword in Europe for inefficiency and incompetence.[21] Having obtained their posts through family connections, favoritism, or bribery, papal administrators in general had little training or even interest in their duties. They performed their assigned tasks with varying degrees of industry and competence, but they resisted any hint of reform or innovation which threatened traditional routines and prerogatives. Describing the papal bureaucracy, one Austrian representative concluded, "In general, they are individuals with little or no knowledge of the duties entrusted to them," while a second agent bemoaned the "total lack of all system and method."[22] In the Secretariat the *cifristi* performed the mechanical tasks of enciphering and deciphering the few dispatches which required their attention, but gave little systematic thought to communications security. There were certainly signs of laxness in cryptographic matters. In one dangerous practice, nuncios were allowed to invent their own codes and ciphers. Francesco Capaccini, alone, used at least 10 different systems during his three years in The Hague. Often a nuncio used more than one system in a single dispatch; part of the text would be written in a numerical cipher, part in a letter cipher, and part in a jargon code. Officials in the Secretariat would sometimes despair of their ability to make any sense of a particular dispatch.

Inertia led the cipher section to adopt a conservative (not to say antiquarian) approach to cryptography. Furthermore, the relatively infrequent use of ciphers provided little incentive to improve established practices, especially in an institution as tradition-bound as the Roman Curia. The result was a tendency to continue tried- and-true procedures and to eschew innovation. The experiment with polyalphabetic ciphers in the late 1820s is the exception that confirms the rule. The temporary adoption of polyalphabetic systems reflected the personal enthusiasm of the senior *minutante*, Francesco Capaccini, rather than the considered judgment of the cipher section. Polyalphabetics were notoriously laborious and prone to errors in ciphering and deciphering as correspondents tried to keep track of the constant shift in alphabets. There is evidence that Capaccini's systems caused some confusion in the Secretariat of State. The margins of

---

[21] Hanns Gross, *Rome in the Age of the Enlightenment: The Post-Tridentine Syndrome and the Ancien Regime* (Cambridge, 1990), 10-11.

[22] Alan Reinerman, *Austria and the Papacy in the Age of Metternich*, v. 2 (Washington, DC, 1989), 182, 187, 193; *Carte segrete e atti ufficiali della Polizia austriaca in Italia* (no editor), v. 1 (Capolago, 1851), 382.

dispatches bear such notations as "Because of a mistake in enciphering it isn't possible to read the name of the person indicated" or "Armellini gave up trying to decipher this dispatch." Several lines of ciphertext might remain unintelligible, a sure sign of confusion over the proper key.[23] It is easy to imagine officials cursing these new systems as a needless complication and urging a return to the familiar systems which had served well in the past. With the exception of some of Monsignor Capaccini's creations (which apparently did not survive his departure from the Secretariat of State), the cryptographic systems in service at the Vatican in the early nineteenth century differed little from those in service 300 years earlier. The simple substitution cipher in which a nuncio to Spain composed some of his reports in the 1840s ($a = 11$, $b = 12$, $c = 13$, $d = 14 \ldots$) was similar to a cipher used by a nuncio to France in the 1570s ($a = 1$, $b = 3$, $c = 4$, $d = 6 \ldots$), with the latter, because of its broken sequence of numerical equivalents, revealing a marginally greater imagination. Indeed, in an interesting case of historical regression, the substitution ciphers of the Madrid nuncio were greatly inferior to the cryptographic systems prepared for his predecessors by Giovanni Battista Argenti in the late sixteenth century.[24]

The lethargy of the cipher section might have been overcome by strong leadership and direction from the senior officers in the Secretariat of State, but none was forthcoming. These officers had no interest in cryptography. Ercole Consalvi, the great cardinal secretary of state who skillfully navigated the papacy through the shoals of the post-Napoleonic years, was sensitive to the problems of communications security, but thought that trustworthy messengers, cover addresses and clandestine mail routes provided greater security than ciphers. Consalvi retired in 1823, and his successors as cardinal secretary of state exhibited even less interest in secret communications. Cardinal Giulio Della Somaglia was an octogenarian when he assumed direction of the Secretariat and had little energy for even the most routine tasks of his office. Cardinal Tommaso Bernetti was a skillful diplomatist, but he paid scant attention to the administration of his department. One Secretary of State, Cardinal Giuseppe Albani, actually saw no need for ciphers. Discussing the office of cipher secretary, Albani argued,

> Today this title and responsibility of Cipher Secretary is totally useless. At other times there was a separate cipher department, because the Secretariat of State conducted all of its affairs with foreign courts through its nuncios, and it was necessary to make use of the cipher in correspondence, but now that the number of foreign ambassadors

[23] de Valk, *Lettres de Francesco Capaccini*, xxvi, 100, 179.

[24] Pierre Hurtubise, *Correspondance du nonce en France, Antonio Maria Salviati, 1572-1578*, v. 1 (Rome, 1975), viii; Meister, *Die Geheimschrift im Dienste der Päpstlichen Kurie*, 355-357, 405-407.

> to the Holy See has increased and all prefer to conduct affairs here in Rome, the cipher is used little or not at all ...[25]

Unfortunately, indifference at the top was not balanced by interest at the next level of the Secretariat. Ideally, the substitute, who after 1814 also held the title of cipher secretary, should have manifested such interest since he was administratively responsible for all cryptographic matters. In practice, however, this responsibility was purely nominal. With one exception the substitute/cipher secretaries during this period knew nothing of codes and ciphers, and without exception they relegated cryptography to a priority far below more pressing problems of departmental finance, personnel, and general administration.

Monsignor Carlo Mauri, the first official to combine the offices of substitute and cipher secretary, had little understanding of communications security. During the Congress of Vienna, his naive faith in the inviolability of papal correspondence drove Cardinal Consalvi to distraction. Mauri would send letters directly to Consalvi rather than to the pre-arranged cover address in Vienna, believing that the cardinal's nomenclator would protect the confidential messages from prying eyes. The cardinal secretary of state had a higher opinion of Austrian cryptanalysis, but his repeated warnings were of no avail. In a letter to his friend Cardinal Pacca, Consalvi exploded, "I confess that when reading the cipher messages that are sent to me directly through the mail, I would like to slam the head of monsignor Substitute against the wall."[26] Mauri preferred to devote his attention to various legal questions of interest to the Secretariat of State, and it was on such matters that he spent most of his time until he died in office in 1830.

His successor as substitute and cipher secretary, Francesco Capaccini, seemed prepared by appetite and experience to assume the responsibilities of both offices. Unlike most of his colleagues who came to the Secretariat of State with a background in theology or law, Capaccini had studied mathematics and astronomy, and had been teaching the latter science at the University of Naples when he was recruited into the Secretariat as a *minutante* in 1815. He was a man of many gifts, and when he temporarily left the Secretariat in 1824 the Prussian ambassador commented that the department had lost the only official who understood diplomatic affairs. He might have added that it had lost the only official who understood cryptography. Capaccini had a keen interest in codes and ciphers, and he considered the Vatican's cryptographic systems hopelessly outdated or compromised. It was almost certainly because of his influence that the Secretariat of State, at least for a time, adopted polyalphabetic ciphers for

[25] Pásztor, *La Segreteria di Stato*, 208.

[26] Roveri, *La Missione Consalvi*, 568.

its most confidential messages.

Capaccini came to his high post from a stint as nuncio to the Low Countries where he exhibited an obsessive concern with the security of his communications. Surprisingly, this concern abated when he returned to Rome in 1831 to take up his duties as substitute/cipher secretary. He turned his energies to administrative matters, in particular plans to reorganize the Secretariat of State. When he thought of the cipher section, it was only to consider ways to reduce its presence in the department. This rather unexpected attitude in one previously so sensitive to the importance of cryptography was almost certainly a reaction to financial exigencies which were pressing upon the Secretariat. By 1831 the financial affairs of the papacy had reached such a dismal state that the Prussian minister in Rome informed his government that "bankruptcy is at the door."[27] Revolutionary disturbances in the pope's northern provinces had required extraordinary expenditures while impeding the collection of taxes in the region. This condition, combined with "normal" maladministration in the pontifical Treasury, created a situation where there were doubts that the pope could pay the salaries of his officials. In this climate Capaccini, who as substitute was the chief administrative officer of the Secretariat, concluded that the cryptographic needs of the department had declined to such an extent that a cipher section was an extravagance. In fact, there was no longer enough work for even a full-time cipher clerk. As early as 1823, Capaccini, while still a *minutante*, had advised the cardinal secretary of state that

> the two most senior and most faithful copyists would be able to learn in half a day the very simple art of ciphering and deciphering, so that when there is an occasion to write in cipher or to decipher, it could be done as quickly as now, but without the expense of over 100 *scudi* a month for the Cipher Section.[28]

As substitute, Capaccini continued to argue for the abolition of the Cipher Section. Shortly after his appointment to the second-ranking position in the Secretariat, he proposed the elimination of the *cifristi* and the transfer of their duties to one of the officials responsible for the correspondence with the nuncios. Capaccini's proposal was not formally adopted, but its purposes were achieved in time. In 1840 the *cifristi* were reduced to two when Filippo Evangelisti retired. On 1 March 1847 Giuseppe Evangelisti died after forty-six years of service as a cipher officer for the papacy. That same day Giuseppe Golt retired, and with his departure the *cifristi* effectively disappeared as a separate group of officials

[27] Reinerman, *Austria and the Papacy in the Age of Metternich*, 178, 374(n.1).

[28] Pásztor, La Segreteria di Stato, v. 2, 378.

in the Secretariat of State. The post of *prima cifrista*, which had been in the Evangelisti family for over half a century, was abolished, and responsibility for cryptography passed to a junior official in the Secretariat.[29]

Between 1841 and 1844, Francesco Cappacini was absent from Rome on special diplomatic missions to Holland and Portugal. Although he retained his office of substitute and cipher secretary, he clearly could not fulfill its responsibilities from The Hague or Lisbon, so that the Secretariat was again without the services of the one man who had both an appetite and aptitude for cryptography. In 1844 he was made a cardinal and transferred to the largely honorific post of Auditor General of the Apostolic Chamber. He died in 1845. His immediate successors in the Secretariat of State were more or less competent administrators, but they demonstrated little appreciation of the requirements of secret communications. The experience of one substitute, who in the midst of the crisis surrounding the Roman revolution of November 1848 could not determine which cipher was in use at the Brussels nunciature, suggests how far affairs had been allowed to slip.[30]

IV. CONSOLIDATION

On 7 September 1870, after a brief skirmish with papal troops at the northern gates to the city, an Italian army occupied Rome and brought to a close a thousand years of papal temporal power in central Italy. The incorporation of the Eternal City into the Kingdom of Italy concluded a decades-long effort by the Piedmontese House of Savoy to unify the various duchies and principalities of the peninsula under a single crown. The new rulers of Italy had no wish to undercut the pope's position as an independent spiritual leader, and they promised to guarantee his sovereignty over the Vatican complex of palaces and gardens and his right to enter into diplomatic relations with other states. But, inconsolable over the loss of his territories and convinced that a government that granted rights could repeal them, Pope Pius IX refused any accomodation with the new regime. The uncompromising pontiff registered his protest by withdrawing into the vastness of his palace as the self-styled "prisoner of the Vatican," and refusing any contact with the "usurpers."

The loss of the temporal power wrought a sea change in papal diplomacy. For centuries the papacy had been a significant presence in European affairs, but after the occupation of Rome this presence was much reduced. With the

[29] As late as 1848 the roster of the Secretariat of State still included two *cifristi*, but one performed only archival duties, and his title was merely titular. By the 1850s, a certain Monsignor Emidio Gentilucci was responsible for codes and ciphers. Lajos Pásztor, "La Segreteria di Stato di Pio IX durante il triennio 1848-1850," *Annali della Fondazione italiana per la storia amministrativa*, 3 (1966), 308-365.

[30] Michele Fatica, *Le Relazioni diplomatiche fra lo Stato Pontificio e la Francia* (Rome, 1971), 389.

pope a self-proclaimed prisoner in his palace, and his Secretariat of State reduced to hurling a string of protests against the "usurpation" by the House of Savoy, the Vatican quickly slipped to the margins of international affairs. Its diplomatic establishment contracted accordingly. By 1880 the entire staff of the Secretariat of State, from the cardinal secretary to the humblest underarchivist, numbered only 11 individuals supervising a diplomatic network shrunk to eight nunciatures. This establishment remained fairly stable through the pontificate of Pope Leo XIII (1878-1903) despite that pontiff's aggressive efforts to project a more visible and important role for the Holy See in world affairs. In 1904, when Leo's successor, Pius X, had been on the papal throne for a year, the strength of the Secretariat declined to 10 officials, although the number of nunciatures had increased by one to nine.

In 1908 Pius X reorganized his diplomatic machinery by incorporating into the Secretariat of State certain hitherto independent offices whose responsibilities partook of a diplomatic character. While the cardinal secretary of state remained the pope's foreign minister, the reorganization divided his department into three sections. The First Section (Extraordinary Affairs) was responsible for issues arising from the Vatican's relations with foreign governments, especially issues relating to concordats (treaties). This section was supervised of a secretary who was assisted by an undersecretary. The Second Section (Ordinary Affairs) handled the routine administrative tasks of the Secretariat, such as accounts, personnel and communications. The substitute, who retained the nominal title of cipher secretary, now directed only this section. The Third Section (Apostolic Briefs) prepared papal messages of a formal or ceremonial nature. Its director had the title of chancellor. This enlarged Secretariat still numbered scarcely 16 or 17 individuals, and the level of activity remained modest. In 1910 the undersecretary for extraordinary affairs, in a letter to the cardinal secretary of state asking to be relieved of his position, complained that though he diligently appeared in his office each day, "nine times out of ten, there was nothing to do."[31]

The cryptographic requirements of the Secretariat of State changed significantly in the period following the loss of the temporal power. The volume of encoded traffic declined with the number of nunciatures and with the more modest diplomatic agenda of the Holy See, which required fewer secret communications. Even after the turn of the century an entire year might pass without an exchange of coded messages between the Vatican and one of its less important diplomatic outposts.[32] As mentioned, the Secretariat of State, in the 1850s, had

---

[31] Benigni to Merry del Val, 21 June 1910, ASV. Segreteria di Stato, Spogli: U. Benigni, busta 31.

[32] Falconi to Merry del Val, 28 July 1906 and 1 January 1907, ASV. Archivio Nunziatura Stati Uniti (ANSU),

abandoned any effort to maintain a distinct cipher office. After the surrender of Rome, the Secretariat was content to continue the practice of assigning responsibility for cryptographic matters to one of the department's junior officers under the general supervision of the substitute secretary of state, who also held the purely formal title of cipher secretary. Until the end of the century that junior officer was Monsignor Filippo Frassenelli, a *minutante* who was also listed on the department's roster as cipher officer (*incaricato della Cifra*). After Frassenelli's retirement, the Secretariat no longer bothered to distinguish the cipher officer with a separate title or notation on the rolls. Rather than a security measure to hide the identity of this official, the new practice probably reflected the insignificant status of cryptography in the department. Frassenelli's replacement was another *minutante*, the layman Camillo Barluzzi, who served until his retirement in 1903. His immediate successor cannot be identified. After the reorganization of the Secretariat in 1908, cryptography fell within the purview of the Second Section (Ordinary Affairs). A junior officer in this section, Monsignor Domenico Spada, seems to have been assigned the responsibility for codes and ciphers.

The cipher officer encoded and decoded messages, and prepared new ciphers and codes for the nunciatures. The latter was not an onerous task since papal cryptography underwent a certain consolidation after 1870. In the first half of that century, the Secretariat of State had used an array of cryptographic systems including jargon codes, null codes, nomenclators, and monoalphabetic and polyalphabetic ciphers. By the 1870s, however, the foreign ministry of the Vatican had abandoned diversity for standardization. For the next 40 years the standard was that familiar standby, the two-part nomenclator, which the Secretariat issued to each of its diplomatic missions. Though these codes differed from each other in certain details such as the equivalencies of specific codegroups or the size of the plaintext repertory, they all followed the same model. This is well illustrated by the system used by the Vatican's special envoy to Mexico at the turn of the century.[33]

This nomenclator consisted of 729 three-digit groups from 000 to 888 (skipping any trio that included the number 9) and 100 four-digit groups from 9900 to 9999. Except when part of a group introduced by 99, 9 was a null as were the groups 854, 855 and 856. There were groups for letters of the alphabet (*a* = 857, *b* = 310), numbers (*one* = 555, *fifty* = 331), punctuation (*comma* = 9978), common Italian prefixes, suffixes, and syllables (*mente* = 766, *er* = 9914), and complete words (*Paris* = 648, *cardinal* = 9930). The presence of codegroups for such user-specific terms as Yucatan (330) and Vera Cruz (240) confirms that

sezione 1, fasc. 62.

[33] ASV. Segreteria di Stato, Visita Apostolica nel Mexico, 1896-1900, busta 1.

each diplomatic mission had a version of the standard system adapted to its particular need. Scattered throughout were numerical groups with no plaintext equivalents so as to add to the lexicon as required. There were no homophones.

Between 1870 and 1914, similar systems (two-part nomenclators of three- and four- digit numerical groups with a null digit that was doubled to introduce the four-digit groups) were in service at papal diplomatic missions around the world, although toward the end of the period the large folded sheets of paper which were the distinctive format of the nomenclator began to give way to small codebooks. Occasionally, one mission would borrow the code of another. This was especially true for temporary missions or those whose arrangements had to be expedited. When Monsignor Alessandro Franchi, temporarily recalled to Rome from his post as nuncio to Spain, was sent as the pope's special envoy to Constantinople in 1871, he carried a copy of the nomenclator then in service at his Madrid nunciature. When, in 1900, the archbishop of Havana, then engaged in delicate negotiations with the secular authorities over the status of church lands in Cuba, asked the Secretariat of State for a code, he was authorized to use the system of the apostolic delegation in Washington.[34]

The Secretariat also continued its practice (dating back to the 16th century) of issuing two codes to each nunciature: a *cifra particolare* for correspondence with the Vatican, and a *cifra comune* for correspondence with other nuncios. The practice, however, sometimes proved imperfect as the distinction between "private" and "common" codes proved too much for the pope's ambassadors. In 1893 the new nuncio in Madrid, Monsignor Serafino Cretoni, used "Cifra X," the designated *cifra comune*, in a letter to his more experienced colleague in Paris, Monsignor Domenico Ferrata, seeking advice on how to prepare for the visit to Madrid of the Queen of Portugal. Monsignor Ferrata returned Cretoni's letter unread, saying that the Paris nunciature did not have a code corresponding to that used by the Madrid nuncio. Cretoni then sent the apparently useless "Cifra X" to the Secretariat, and asked for a system which would allow him to communicate with Paris. A rather puzzled Secretariat promptly returned the same code to Cretoni, insisting that "Cifra X" was still in service as the *cifra comune*, but leaving unanswered the question why the Paris nunciature seemed never to have heard of such a code.[35]

Although the two-part numerical code dominated papal cryptography until the First World War, the Secretariat of State occasionally used other systems.

[34] For Franchi's code, compare the encoded correspondence from Madrid and Constantinople in ASV. Segreteria di Stato, Spogli: Artibani, busta 1, fasc. 2 and Franchi, busta 1. For the archbishop of Havana, see Sbarretti to Martinelli, 21 March 1900, ASV. ANSU, sezione 1, fasc. 17.

[35] Cretoni to the Secretariat of State, 13 July 1893 and Barluzzi to Cretoni, 29 July 1893, ASV. Archivio Nunziatura Madrid (ANM), scatola 585.

During his tenure as nuncio to Spain (1879-1881), Monsignor Angelo Bianchi used a homophonic substitution in which the letters of the alphabet were replaced by two- digit numerical groups from 00 to 99 in mixed order.[36]

a	b	c	d	e	f	g	h	i	k	l	m
07	14	11	09	02	08	01	70	04	52	54	20
28	24	38	35	19	15	16	82	15		67	41
39	30	65	49	25	55	83	93	23		88	75
48	59	71	63	44	98	91		42		97	99
86	87	89	85	79				60			
95		92		96				76			

n	o	p	q	r	s	t	u	v	x	y	z
21	05	33	45	06	27	17	00	43	03	69	37
61	12	40	74	13	31	26	34	66	51		57
80	22	56		36	84	64	47	77			72
90	46	68		62	94	81	58				
	78										

period(.) 10
nulls 29, 32, 50, 53, 73

Except for the letters K and Y (both rarely used in Italian) each letter had several cipher equivalents, with the more frequently used letters having up to six homophones. The system included five null groups which could be inserted in the middle of an enciphered word as well as between enciphered words. Instructions also directed the cipher clerk to begin and end each line of ciphertext with any single-digit number he pleased. The first and last numbers of each line would then be disregarded during deciphering. Using this system, the message "Arrived Madrid" could be transformed (with nulls) into the following rather unwieldy cryptogram:

-	A	R	R	I	V	E	D	-	M	A	-	D	R	I	D	-
6	07	62	13	23	77	02	09	73	41	28	50	85	36	04	49	9

Arrived Madrid = 6076213237702097341285085360449 9

In its continued reliance on nomenclators and cumbersome monoalphabetic ciphers, the Vatican lagged behind other governments in responding to recent

[36] ASV. Spogli: A Bianchi, busta 2. The author has not been able to discover whether similar ciphers were in service at other nunciatures at this time.

advances in cryptography which made such traditional systems hopelessly obsolete. Unlike the papal Secretariat of State, many foreign ministries, by the 1870s, had abandoned the small, two-part nomenclators that had served for centuries in favor of large, one-part codebooks with lexicons of up to 50,000 entries. With their numerical or alphabetic equivalents for thousands of words and phrases, these codes shortened messages sent by telegraph, which had come to be widely used for diplomatic communications, and so lowered message costs. In 1876 (three years before Monsignor Bianchi carried his unwieldy cipher to Madrid) the U. S. State Department had introduced its Red Code of more than a thousand pages of words and phrases, each of which had a corresponding codeword and five-digit codenumber. With this, a clerk could reduce a 16 word plaintext to as few as five codewords.[37] Many governments also sought further security by superenciphering their coded messages, either by transposing the digits of the codegroups or, more commonly, by adding to each codegroup a keynumber, the so-called "additive." By the end of the century enciphered codes were in service at almost every foreign ministry. The Red Code (1876) and Blue Code (1899) of the State Department, for example, used several different methods of encipherment including transposition, additives and, in the Blue Code, subtractives. Even minor powers adopted the practice. The foreign ministry of the Netherlands communicated with its diplomatic missions by means of a system based on the commercial code developed by F. J. Sittler, a former director of French posts and telegraphs. In Sittler's codebook each word or phrase was represented by a four-digit numerical group in which the first two digits indicated the page and the second two the line on which the word or phrase could be found. The Dutch adapted this system by scrambling the page numeration and reversing the order of the codegroup pairs so that the first two digits now indicated the line and the last two the page. A cipher table was then used to transform each numerical codegroup pair into a letter pair, producing an enciphered code of four-letter groups. [38] Such procedures, which were increasingly widespread in western chancelleries at the turn of the century, clearly provided more security than the old-fashioned nomenclators and ciphers of the Vatican.

V. TOWARDS REFORM

Though slow to abandon traditional methods, the Vatican was not insensitive to the need to improve its cryptographic systems. The Secretariat of State was

[37] Ralph Weber, *Masked Dispatches: Cryptograms and Cryptology in American History, 1775-1900* (Fort George Meade MD, 1993), 203.

[38] Ibid., 213, 229; David Alvarez, "A Dutch Enciphered Code," *Cryptologia*, 19(3) (July 1995), 261-264.

even willing to consider systems invented by private cryptographers. In 1907, for example, it asked one of its nuncios to evaluate an enciphered code offered to the Vatican by one Ricardo Candriani, an Argentine of Italian extraction.[39] Invariably, however, the Secretariat returned to its own resources. In 1902 the substitute instructed Camillo Barluzzi, the official then responsible for the ciphers of the Secretariat, to solicit from the nuncios "opinions and suggestions that would be useful in the event that a new cipher is adopted."[40] The response to this solicitation is not recorded, but in the spring of 1907 the Secretariat of State introduced a new system in which the codegroups combined letters and numbers. The characteristics of this system cannot be determined from the available evidence, but the code quickly proved impractical. Some cable companies would not accept a mixed group such as 4KA14 as a single group for transmission purposes; instead, the companies insisted on treating it as three groups, 4, KA, 14, and charging for three groups rather than one. Shortly after the new system went into service, nuncios began to complain that their cable expenses had increased dramatically. Then in the summer, the Italian parliament enacted new regulations which prohibited the combination of numbers and letters in encoded telegrams. Since the Vatican depended upon Italian cable facilities for the transmission and reception of all its telegrams, this legislation doomed a system whose utility was already in question because of financial considerations. After only a few months, the Secretariat of State withdrew the system and reintroduced the old two-part numerical codes.[41] The fact that the Secretariat had to return the old codes in the same packet that contained the letter announcing the recall of the new system suggests that each nunciature had, at any given time, only one system with which to communicate with the Vatican.

In 1913, the Secretariat of State had the idea of improving its cryptographic systems through superencipherment. In July of that year, the nuncio in Brazil acknowledged receipt of a new code in a "perfectly sealed package which bore no signs of tampering." The package had been entrusted to one Bruno Chaves (a priest returning to Brazil from Rome), and the nuncio informed the Secretariat that the old code (which had been in service since 1902) would be returned by hand of the bishop of Bahia, who was travelling to Rome on church business.[42]

[39] The nuncio judged the system "ingenious" and "most secure," but noted that it was cumbersome to use since encoding required three separate operations. The Secretariat declined Candriani's offer. Locatelli to Merry del Val, 19 June and 21 October 1907, ASV. Archivio Nunziatura Argentina (ANA), scatola 61, fasc. 7.

[40] Barluzzi to Marchetti, 3 May 1902, ASV. ANSU, fasc. 18.

[41] Dolci to Merry del Val, 26 June 1907, ASV. Archivio Nunziatura Peru, busta 55, fasc. 3; Locatelli to Merry del Val, 24 July 1907 and Della Chiesa to Locatelli, 20 August 1907, ASV. ANA, scatola 61, fasc. 7.

[42] Aversa to the Cardinal Secretary of State, 10 July 1913 and 18 August 1913, ASV. Archivio Nunziatura Brasile, n. 136, fasc. 681.

The new code, which the Secretariat considered "much more complete, serviceable, and economical" than the old, incorporated two important innovations.[43] As we have seen in the case of the code of the Vatican's envoy to Mexico, the standard papal code of this period identified a number (e.g. 5) which, when appearing alone, was a null, but which could be doubled to introduce a legitimate four-digit codegroup (e.g., 5512). The new code for the Brazilian nunciature revised this feature by allowing the doubled null to complete as well as introduce a four-digit group. The new procedure doubled the number of four-digit groups by adding, for example, groups from 0055 to 9955 to the usual sequence of groups which ranged from 5500 to 5599. A more significant innovation introduced superencipherment tables consisting of 100 numerical pairs from 00 to 99 with letter bigrams (always beginning with a consonant) corresponding to each numerical pair (*00* = SE, *01* = PO ... *99* = HO). A message would be encoded in the numerical code and then assembled into five-digit groups (e.g., 38306 36005 ...), as required for telegraphic transmission. These groups would then be divided into numerical pairs (38 30 63 60 05 ...). Each pair would then be replaced by its corresponding bigram from the cipher table (*38* = TO, *30* = KU, *63* = DA, *60* = ME, *05* = FO ...) and the letter pairs combined to form "words" of 10 letters (TOKUDAMEFO ...). International telegraphic conventions authorized the transmission of such artificial words, each of which represented two of the original numerical codegroups, as a single group. Although the procedure of superencipherment enhanced security, it appealed to the Secretariat of State primarily because it cut cable costs by half. For some unknown reason, this system proved short-lived for it was not in service when, the next year, an assassination in Sarajevo set off a world war.

VI. WARTIME CRYPTOGRAPHY

The Vatican did not replace its cryptographic systems when World War I began. As had long been the custom in the Secretariat of State, each papal diplomatic mission had its own code. The creation and distribution of unique codebooks for each of almost two dozen missions were tasks clearly beyond the capacity of the small Secretariat where, among the staff of 14 in 1914, a single officer was responsible for cryptography.[44] Consequently, for some time the nuncios had to depend upon their prewar cryptographic systems. These systems were seriously

[43] Merry del Val to Aversa, 4 June 1913, Ibid.

[44] Monsignor Domenico Spada, who joined the Secretariat in 1909, was responsible for codes and ciphers during the war. He was assisted by Count Stanislao Caterini, a lay employee of the department since 1912. Scapinelli to Gasparri, 21 May 1915, ASV. Segreteria di Stato, Guerra 1914-18, rubrica 244, fasc. 4.

deficient for wartime use. The lexicons of 700 to 800 plaintext elements were actually smaller than those of many papal codes of the 19th century, and the codes' utility was further restricted by the omission of many secular words such as *offensive*, *armistice*, *belligerents*, *submarine*, and *blockade* and their code equivalents. The limitations of this papal cryptography are suggested by comparison to a German diplomatic code in use, early in the war, at German missions in Central and South America. This code had 25,000 plaintext elements and used many homophones. The wartime director of the cryptanalysis unit of the Italian Supreme Command believed that a good diplomatic code should have at least 10,000 elements. Even the U. S. State Department (hardly a leader in cryptography) began the war with a codebook of almost 1500 pages, much harder to reconstruct than the paltry 14 pages of the codebook in service at the Lisbon nunciature in August 1914.[45]

Thus the Vatican's cryptosystems at the start of the war provided illusory security against the scrutiny of foreign governments seeking information about papal attitudes and intentions. One government was a particular threat. Suspicious of the Vatican's political sympathies and concerned that the papacy would use the dislocations of the war and its aftermath to reassert its claims to temporal sovereignty in the Eternal City (the so-called "Roman Question"), the Italian government was especially eager for intelligence on the Vatican.

Communications intelligence became an important instrument of that surveillance. Under the leadership of Luigi Sacco, an enthusiastic young lieutenant of engineers, a cryptanalysis unit had been established in the intelligence division of the army high command. In addition to attacking the cryptographic systems of the Austrian army, this unit also monitored the diplomatic traffic passing in and out of Rome. In the course of the war Sacco's unit read, with more or less regularity, the secret diplomatic messages of several countries including Greece, Spain, Switzerland, and the United States. By the end of 1916 (at the latest), Italian intelligence was routinely intercepting and reading the encoded communications between the Secretariat of State and papal diplomatic missions in Austria-Hungary, Belgium, Germany, the Netherlands, Spain, Switzerland, Turkey, and the United States. Plaintext transcriptions of such communications were often delivered to Italy's foreign minister, Sidney Sonnino, within 24 hours of their transmission.[46]

From the available evidence it is difficult to determine whether Italy's success

---

[45] Kahn, *The Codebreakers*, 289; Luigi Sacco, *Manual of Cryptography* (1936, trans. Laguna Hills, CA, 1977), 58; Weber, *United States Diplomatic Codes and Ciphers, 1775-1938*, 246. For the Lisbon codebook, see David Alvarez, "A Papal Diplomatic Code," *Cryptologia*, 16(2) (April, 1992), 174-176.

[46] David Alvarez, "Italian Diplomatic Cryptanalysis in World War I," *Cryptologia*, 20(1) (January 1996), 1-10.

in reading the Vatican's secret communications stemmed from cryptanalysis or from the clandestine acquisition of papal codebooks. Sacco's solution in 1917 of an Austrian diplomatic code suggests why theft may be likely. The texts of the messages, which passed between the Austrian foreign ministry and its two embassies in Rome (one accredited to the Italian government and one to the Vatican), include tentative identifications and undeciphered groups. These imply that the code is being reconstructed.[47] The intercepts of the papal telegrams, on the other hand, are completely deciphered from the start. This suggests possession of a codebook. In fact during the war Italian agents burglarized the diplomatic missions of neutral as well as enemy powers in search of secret documents including codebooks. Given Italy's concern over Vatican policy, the Secretariat of State would have been a prime target for covert operations. At least one papal nunciature was infiltrated: Italian intelligence had an agent inside the nunciature in Madrid, which was a collection center for papal diplomatic mail to and from South America. The nunciature in Berne was under especially close surveillance.[48]

While Italy was the principal transgressor, other governments successfully attacked the Vatican's codes during the war. In 1917 the Swiss were reading at least one of the systems in use at the Munich nunciature. Although direct evidence is lacking, it is highly likely that Britain's skilled cryptanalysts in Room 40 (who successfully attacked cryptographic systems far more difficult than those of the papacy) were able to read encoded papal messages. They perhaps paid special attention to those that passed through the transatlantic cables, which the British closely monitored. Germany may have achieved the most dramatic success. Through Matthias Erzberger, the head of the Catholic Center Party and a familiar figure in Vatican circles, the German Foreign Ministry apparently sought, in the spring of 1915, to recruit the services of a junior officer in the Munich nunciature. The Germans hoped to convince this monsignor to provide, among other items, the codebook of the nunciature, ostensibly to allow Berlin to communicate directly with the Vatican in the event of war with Italy. A successful approach would have paid unexpected dividends, for the papal diplomat was soon promoted to a senior position in another European nunciature. While it is impossible from the available evidence to determine conclusively that the papal diplomat was a German informant, it seems that he received covert payments from Franz von Stockhammern, the director of German espionage in Switzerland, and that these payments continued at least until the summer of 1916. Whatever

[47] Ibid.

[48] Marchetti to Sonnino, 10 January 1918, reel 52, Sidney Sonnino Papers (microfilm), Hoover Library, Stanford University, Palo Alto CA.

their success with this monsignor, German intelligence closely monitored the communications of the nunciature in Munich throughout the war and read some of the encoded traffic on the Munich-Berne-Vatican circuit.[49]

Various factors undermined the integrity of papal codes. In the early years of the war, each nunciature had a single, unique code. All encoded communications beween the nunciature in Madrid and the Secretariat of State, for example, were in one system. This practice would ease the work of foreign cryptanalysts seeking to break into the Madrid-Vatican circuit. Lacking cable or wireless facilities of its own, the Vatican was compelled to use the facilities of the Italian telegraph service. Italian military intelligence was thus able to obtain copies of every telegram sent or received by the Secretariat of State, thereby enhancing its effort against papal codes. In foreign capitals the nuncios were also dependent on local facilities. Messages to and from distant posts often passed through several sets of hands. Cables for the papal representative in Warsaw, Monsignor Achille Ratti (the future Pope Pius XI), for example, moved through Italian, Swiss, German, and (after the armistice) Polish telegraph offices. In 1920, when he was on a special, fact-finding mission in Upper Silesia (then a disputed area under a temporary Inter-Allied Plebiscite Commission), Monsignor Ratti actually sent and received encoded messages through the facilities of the Polish consulate in that territory.[50] Since Poland claimed parts of the disputed territory, the consular officials would have been most interested in the nuncio's confidential reports, and they perhaps passed copies of such reports to the fledgling Polish cryptanalytic service, the *Biuro Szyfrów* (Cipher Bureau), which had already achieved notable success against the cryptographic systems of the new Bolshevik regime in Russia.

Papal representatives were sometimes lax in their communications practices. The Secretariat of State specifically warned the nuncios against repeating in their reports the plaintext of encoded cables, but there were lapses. Once the Vatican's representative in Washington acknowledged, in an open dispatch, the receipt of a coded message from the Secretariat, the contents of which he then summarized. This same officer often repeated verbatim the plaintext of coded messages from the Vatican in his open letters to Cardinal James Gibbons of Baltimore. Monsignor Bonaventura Cerretti, a senior officer in the Secretariat on a special mission to Paris, entrusted to the British diplomatic pouch an open letter to the cardinal secretary of state which included the dates, reference numbers and summaries of three coded messages he had recently cabled to the Vatican.

---

[49] Hort to Pacelli, 10 May 1922, ASV. Archivio Nunziatura Baviera (ANB), busta 307, pos. 1; "Fidelis" to the Undersecretary of State of His Holiness (undated, but probably the summer of 1917), ASV.'Segreteria di Stato, Guerra 1914-18, rubrica 244, fasc. 6. Wilhelm Patin, ***Beiträge zur Geschichte der deutsch-vatikanischen Beziehungen in den letzten Jahrzehnten*** (Berlin, 1942), 51.

[50] Farolfi to Ratti, 14 November 1920, ASV. Archivio Nunziatura Varsavia (ANVa), scatola 190, fasc. 1.

Trusting in the inviolability of the diplomatic pouch, papal diplomats usually did not encode dispatches entrusted to couriers. Since the security of the pouch was, in fact, problematic, this was a potentially dangerous practice. Nuncios often prepared dispatches to follow and elaborate upon developments which they had briefly reported by encoded telegram. Any power which surreptitiously gained access to the Vatican bag could, after comparing dates, use the nuncios' open reports as cribs to attack the intercepted cables. Some nuncios would encode only the more sensitive passages of a dispatch despite an explicit warning from the Secretariat of State that such practice greatly facilitated the efforts of foreign cryptanalysts. In one extreme case, the nuncio in Berne sent a long message in which only two words were in code.[51]

The Secretariat of State, itself, was not beyond reproach. In May 1915 it prepared a new code for the nuncio in Vienna, and naively gave it to a homeward bound Austrian diplomat for delivery. While there is no evidence that this messenger betrayed his trust, it would have been an exceptional diplomat, indeed, who would have, in wartime, resisted the temptation to sneak a look at the cryptographic system of an important neutral. The following month the Secretariat gave one of its codes to the Prussian ambassador to the Holy See in order to facilitate communication after the ambassador was compelled by Italian belligerency to move his mission from Rome to Switzerland. Since all papal codes in this period were variants of the same system (two-part, numerical codes with three- and four-digit groups and a null number), this courtesy helped not only the German diplomat but the German cryptanalysts as well.[52]

The experience of the nuncio to Bolivia suggests that, as late as the final year of the war, a degree of amateurism still tainted the Secretariat's attitude toward cryptography. In the spring of 1918, Monsignor Rudolfo Caroli, concluding that the codebook of the La Paz nunciature was hopelessly inadequate, decided to pass the time at his distant and quiet post by creating a new system. The result was a four volume code of more than 30,000 words and phrases with provision for a daily superencipherment key. Caroli proudly offered this work to the Secretariat, graciously noting that he would understand if the department wished to retain its current codes, even though none were as clever and efficient as his "marvel." The Secretariat of State was underwhelmed. Cardinal Gasparri eventually thanked Monsignor Caroli for his "colossal" efforts, but hastened to remind the nuncio

---

[51] Bonzano to Gasparri, 9 June 1915, ASV. ANSU, sezione 1, fasc. 62; Cerretti to Gasparri, 27 November 1918, Vatican City. Archivio della Congregazione per gli Affari Ecclesiastici Straordinari (AAES), Stati Ecclesiastici, 214, Questione Romana, III, 1917-20; Marchetti to Gasparri, 20 June 1917, AAES, Guerra Europa, 1914-18, Iniziative Pontificie per la Pace.

[52] Scapinelli to Gasparri, 21 May 1915, ASV. Segreteria di Stato, Guerra 1914-18, rubrica 244, fasc. 4; Tedeschini to Muhlberg, 6 June 1915, ibid., fasc. 2.

that the Vatican's codes were the product of long experience and careful consideration. Left unspoken was the implicit suggestion that the good monsignor (who, before his appointment to La Paz in 1913, had presumably sharpened his cryptographic skills as the bishop of a small Italian town and, before that, as an expert on religious ceremonial) might seek other diversions and leave cryptography to the Secretariat of State. The cardinal secretary of state, however, was reluctant to disappoint the nuncio after so much labor, so he authorized him to use his code to communicate with the Vatican. Allowing diplomats to invent their own codes was a dangerous practice, since only experienced cryptologists could judge the security of a new system. Fortunately, Caroli's "marvel" proved too inconvenient for the clerks in the Secretariat, and after a few months the nuncio was instructed to return to his standard system.[53]

The Vatican was not completely oblivious to the threat to its confidential communications. In October 1915, the Secretariat of State warned its diplomatic missions that it was aware of "grave abuses" concerning the custody of confidential papers, "some of which, though most sensitive and important, had come into the possession of foreigners." Diplomatic personnel were informed that anyone, of whatever rank, who removed documents from the archive of a nunciature or delegation without authorization would be immediately suspended not only from his post but also from the priesthood. To lend substance to this warning, the first secretary in the nunciature to Belgium was sacked for leaking to an Italian journalist information concerning the Vatican's interest in a separate peace between Belgium and Germany.[54]

The Secretariat of State was aware that its encoded cables were intercepted and studied by foreign governments, and at least some of the nuncios worried about the integrity of their cryptographic systems. As late as the spring of 1916, the nuncio in Germany hesitated to commit his most secret messages to the telegraph office because he doubted the security of his code.[55] To protect its cryptographic systems, the Secretariat of State tightened its procedures for handling encoded messages. Previously, it had been common for the code clerk to write the plaintext of an incoming dispatch or telegram above the codetext or on an attached sheet. Both the codetext and the plaintext would then go into the department's files where they could be examined (and compared) by anyone who gained access to the files. By the middle of the war, the Secretariat had abolished this practice in favor of a method by which the clerk decoded a cable onto a separate piece of paper. Only this plaintext would enter the files, while the

[53] See the Caroli-Gasparri correspondence in ASV. Archivio Nunziatura Bolivia, fasc. 108.

[54] See the circular in ASV. Archivio Nunziatura Vienna (ANV), scatola 190, fasc. 148.

[55] Frühwirth to Gasparri, 14 May 1916, ASV. Segreteria di Stato, Guerra 1914-18, rubrica 244, fasc. 29.

codetext would be destroyed. In dispatches where only a portion of the message was in code, the plaintext might still be written above the codetext, but now the latter was obliterated. The goal was to make it impossible for unauthorized eyes to compare the codetext of a message with the plaintext. Unfortunately, the nunciatures did not always follow the new procedures. The papal mission to Poland, for example, continued to file the plaintext with the codetext of both outgoing and incoming messages well into 1920.[56]

The Secretariat of State also responded to the threat against its communications security by changing its codes more frequently. Whereas before 1914 a nunciature might have used the same codebook for more than a decade, during the war (especially after Italian belligerency) the more important posts received new codes as often as every six months. For delivery of new codes, the Secretariat decided that it could not trust the ordinary mail or the diplomatic couriers of friendly or neutral states. Special couriers were now employed for this sensitive task– usually papal diplomats traveling to their posts, as when the nuncio in Madrid received a new codebook from a monsignor returning to the nunciature in Argentina, or when the secretary to the nuncio in Munich returned from a leave in Rome with a new codebook for the nunciature. Nuncios were now required to confirm that the sealed packets containing cryptographic materials arrived "absolutely intact and without any sign of tampering." The more important posts were given additional codes to allow them to vary the cryptography of their messages. When Monsignor Eugenio Pacelli (the future Pope Pius XII) was appointed nuncio to Germany in the spring of 1917, he departed for his post with not one, but two fresh codebooks in his baggage.[57]

Like other states during this period, the Vatican concluded (somewhat belatedly) that its cryptographic systems were inadequate for the challenges of the war. In the spring of 1916, the Secretariat of State distributed new codebooks that were "larger and more complete" than those previously in service. These were two-part codes of four-digit numerical groups, a single null number and an expanded plaintext dictionary.[58] In January 1918 the Secretariat took a more significant step by introducing a superencipherment for use by its missions in Berne, Madrid, Munich, and Vienna. In a letter to the nuncio in Munich, Car-

[56] ASV. ANVa, scatola 190, fasc. 1 and 5.

[57] Cardinal Gasparri was annoyed by Monsignor Pacelli's practice of signing some of his cables "Pacelli" and others "Pacelli, Nunzio Apostolico." The nuncio explained that when he had taken custody of the two new codebooks he had been instructed by the Secretariat's cipher officer to indicate the use of the first by signing the message with his surname, and to indicate the use of the second by adding "Nunzio Apostolico" to his name.

[58] For an example from the Vienna nunciature, see Valfre to Gasparri, 20 September 1917, AAES, Guerra Europa, 1914-18, fasc. 12.

dinal Gasparri explained the Secretariat's action:

> The need to guarantee the secrecy of the code has led the Holy See to introduce certain modifications in its normal use.
>
> It is reasonable to assume that copies of encoded telegrams are retained by government agencies, and that these copies are carefully studied by experts in cryptology who, by means of careful comparison of texts and attention to information from newspapers and other sources, succeed, more or less quickly, in exposing the codes of the Holy See.
>
> To deal with this serious problem, it is necessary to confuse these cryptologists by undermining the basis of their studies. This can be achieved by the use of "transformation tables" that do not replace the system currently in use, but complement it by transforming and distorting coded telegrams before they are consigned to cable offices for transmission.[59]

Each mission received a new code along with so-called "transformation tables" consisting of two columns, each containing 100 pairs of digits from 00 to 99. In column A these groups were arranged sequentially (00, 01, 02 ...), while in column B they were arranged randomly (29, 43, 09 ...). To prepare a secret communication, the nuncio would first use his codebook to encode his message in the usual way. Next he would divide the codetext into groups of two digits. He would then find each of these bigrams in column A of his cipher (transformation) table, substitute the corresponding bigrams from column B, and arrange the latter into the standard five- digit group for transmission.[60] The codetext 16042 97831, for example, would be divided into 16 04 29 78 31. The cipher table would indicate that

16	corresponds to	08
04	"	31
29	"	88
78	"	54
31	"	17

The table enciphers the original codetext 16042 97831 into 08318 85417.

By early spring Monsignor Pacelli in Germany was using three different cipher tables with his codebooks. Each table carried identifying indicators: the first, the prime numbers 1, 5, and 7; the second, the number 2 and its multiples 4 and

[59] Gasparri to Pacelli, 26 January 1918, ASV. ANB, busta 307, pos.1.

[60] Ibid.

8; and the third the number 3 and its multiples 6 and 9. Whoever enciphered a message would indicate which table was in use by making the first group in the enciphered text consist of four rather than five digits. He would then expand this into a five digit group by prefixing one of the three table indicators. An enciphered message from Munich beginning with 2, 4, or 8, therefore, would mean that Pacelli had used his second table.

Only the most secret communications were to be sent in this *cifra transformata*, while less sensitive material would be sent in straight code without encipherment. The Secretariat of State was confident that this new system would guarantee "full and absolute" secrecy.[61] The available evidence suggests that this confidence was not entirely misplaced. The Secretariat's enciphered codes immediately began to cause problems for Italian cryptanalysts. While continuing to read many coded telegrams on the important Munich-Vatican circuit, they were unable, well into the summer of 1918, to break the new system in which the more sensitive messages had been enciphered. This lapse, which was in marked contrast to earlier success in reading practically all of Pacelli's communications, was probably due to frequent changes of code and the selective use of the *cifra transformata*. The nuncio in Germany used the new system to discuss with his superiors at the Vatican Austria's interest in a negotiated peace, discussions which apparently escaped the scrutiny of Italian cryptanalysts. Similarly, reports from the papal representative in Switzerland concerning Austrian overtures in September 1918 for a separate peace remained hidden from Italian eyes. By the last month of the war, Italian intelligence was reading some of the cables from Berne, but one of the nuncio's codes remained unbroken, as did one of the codes of the nuncio in Vienna.[62]

Despite the apparent success of the *cifra transformata*, the Secretariat of State did not put the system into general service. The unenciphered, two-part numerical code remained the workhorse of the papal diplomatic corps right through the end of the war. When Monsignor Ratti departed in the summer of 1918 to become the Vatican's new representative in Poland, he carried a codebook that exemplified the "larger and more complete" systems first introduced by the Secretariat of State in the spring of 1916. This was a two-part, four-digit code with groups for nouns (*Russia* = 6281, *information* = 0036), the alphabet (*A* = 5004, *Z* = 6210), common Italian syllables, prefixes, and suffixes (*es* = 3641, *st* = 3487, *mente* = 0679), parts of speech (*without* = 2816), numbers (*3* =

[61] Ibid.

[62] See the notations on the intercepts from the Munich-Vatican circuit for 31 March, 31 May, and 27 June 1918; from the Berne-Vatican circuit for 24 and 25 September 1918; and from the Vienna-Vatican circuit for 7 and 28 October 1918, reel 48, Sonnino Papers.

6523, *100* = 5425), and punctuation (*period[.]* = 7074). Similar codes in service at other papal diplomatic missions included some sort of null feature, but Monsignor Ratti used no such feature when encoding his reports. As late as the summer of 1919, this code was not enciphered. Still, the plaintext repertory of several thousand elements was a marked improvement over the small codes of several hundred elements with which papal missions began the war. In 1920 the Secretariat of State put into service a new code on the Vatican-Warsaw circuit. This was again a two-part code, but of five- rather than four-digit groups. It was notably more efficient than its predecessor. The latter, for example, required three groups for the word *corriere* (mail/courier): *corr* = 4608, *i* = 5310, *ere* = 3608, while the new code used a single group: *corriere* = 08902. It too was routinely used without superencipherment. Sometime in early 1920, however, the Warsaw nunciature began to superencipher its telegrams, but from the available evidence it is impossible to determine if this represented the introduction of the *cifra transformata* or a system based on additives.[63]

The Vatican occasionally employed jargon codes in which an apparently open message contained code words or phrases. As early as January 1915, the nuncio in Vienna was using a short code in which *Italy* = MONSIGNOR G, *Austria* = MONSIGNOR T, *Germany* = ABBOT S, *Pope* = MISTER PRIMO, and *Emperor Franz Joseph* = MISTER T. During secret exchanges with Austria concerning a separate peace in the fall of 1918, Monsignor Pacelli sought additional security by employing a jargon in which the Austrian emperor was the ABBOT PRIMATE, Vienna was the CENTRAL ABBEY, and the world war was the WELL-KNOWN CONTROVERSY, while the entire communication was composed in such a manner as to suggest that its subject was ecclesiastical business. Such a message would then be encoded by the usual two-part code. It is unlikely that such a transparent cover would have fooled experienced cryptanalysts who almost certainly would have guessed what "well-known controversy" was so secret as to require encoding. But it may have deterred them by eliminating from the plaintext words they would look for, such as Charles, the new emperor of Austria.[64]

Monsignor Bonaventura Cerretti carried a similar jargon code on his special mission to the Versailles Peace Conference which opened in January 1919. Cerretti was ostensibly at the conference to ask the victorious Allies to refrain from

[63] Monsignor Ratti's codes have been partially reconstructed by the author from the telegrams in ASV. AnVa, scatola 190, fasc. 1 and 5.

[64] Scapinelli to Gasparri, 19 January 1915, ASV. Segreteria di Stato, Guerra 1914-18, rubrica 244, fasc. 28; Pacelli to Gasparri, 12 September 1918, AAES. Guerra Europa, 1914-18, fasc. 18-22. Occasionally the Vatican used codes to protect the secrecy of internal church administration, but usually codes were used only for diplomatic affairs.

expelling German missionaries from former German colonies in Africa, China, and the Pacific islands. In fact, he was the Vatican's unofficial representative to the peace conference, where he met with various world leaders and opened secret negotiations with Italian Prime Minister Vittorio Orlando on a reconciliation between the Vatican and the Kingdom of Italy. To communicate with the Secretariat of State, Cerretti had a jargon code of two sections.[65] In the first, various individuals or subjects were identified by ecclesiastical terms. For example,

Woodrow Wilson = ABBOT GENERAL
Colonel House = BENEDICTINE ABBOT
Lloyd George = VICAR GENERAL
Vittorio Orlando = MASTER GENERAL
Sidney Sonnino = ASSISTANT GENERAL
Georges Clemenceau = LAY BROTHER
League of Nations = GENERAL CHAPTER
Peace Conference = PROVINCIAL CHAPTER, etc.

In the second, a series of code phrases allowed Cerretti to report secretly the progess of his confidential discussions on the Roman Question with Orlando. The code phrases included such statements as,

ENRICO HAS BEEN ADMITTED TO THE EXAMS.
Orlando has recognized the necessity of a territorial solution.

ENRICO PREFERS THE MILITIA.
Orlando agrees to negotiations between Italy and the Vatican.

ENRICO HAS MADE HIS FIRST COMMUNION.
Orlando will not accept a [Vatican] outlet to the sea.

Whatever the benefits of jargon codes or the *cifra transformata*, there can be little doubt that throughout the war the Holy See was plagued by poor communications security. The problem complicated the administration of a global church and seriously undermined papal diplomacy. Italy, the state for which the relevant cryptanalytical evidence is most readily available, provides a case in point. A power with which the Vatican's relations were cool and distant when not actually hostile, Italy was privy to most of the confidential communications of the Holy

[65] "Missione di Mons. Cerretti a Parigi," AAES. Asia (1919), fasc. 86.

See. In February 1917 it listened as Cardinal Gasparri cabled the nuncios in Vienna and Munich about an opportunity to detach Italy from its Entente partners. Rome's cryptanalysts must have smiled at the cardinal's warning to the nuncios to "destroy this message and maintain absolute secrecy in this matter."[66] When intercepts in the summer of 1917 revealed the papal Secretariat of State discussing with Berlin (but not London, Paris, Rome or Washington) the terms of Pope Benedict's peace proposal before its presentation to the Allies on 9 August, Italian suspicions that the papacy favored the Central Powers could only have been fueled. Reluctant to pursue any settlement that did not address its territorial ambitions, the Italian government was fortified in its opposition to the pope's peace proposal by cryptanalytical evidence that the Secretariat of State was exaggerating the extent to which Berlin and Vienna were prepared to make concessions. In late September 1917, Cardinal Gasparri assured an emissary of the Italian government that Germany was prepared to reestablish Belgium and pay an indemnity to Brussels and that Austria was open to some of Italy's territorial aspirations. The Italian foreign minister, however, knew from the intercepts that the nuncio in Germany had informed the Vatican that Berlin's position on Belgium remained vague and equivocal, while the nuncio in Austria had reported that Vienna had excluded any territorial concessions.[67]

VII. CONCLUSION

Between 1815 and 1920 the Vatican lagged behind other states in responding to developments in cryptographic practice. Throughout most of the period it entrusted its confidential communications to outdated cryptosystems of uncertain security. Particular factors may account for the Vatican's failure to keep pace with advances in the field of diplomatic codes and ciphers. For reasons of economy the Secretariat of State lacked professional cryptologists and delegated cryptography to junior officers who picked up what knowledge they could on the job, often while performing other duties within the department. An exaggerated respect for tradition and a preference for proven and long-established procedures (characteristic elements of the administrative culture of the Vatican) discouraged innovation in all areas, including cryptography. Finally, the Vatican's modest profile in international diplomacy reduced the demand for secure communications and provided little incentive to change established practices. Without more research into the cryptographic practices of minor powers in the

[66] Gasparri to Aversa and Valfre, 23 February 1917, reel 48, Sonnino Papers.

[67] Pacelli to Gasparri, 27 September 1917 and Valfre to Gasparri, 25 August 1917, reel 48, Sonnino Papers; Italo Garzia, *La Questione Romana durante la prima guerra mondiale* (Naples, 1981), 161-164.

same period it is impossible to determine whether the experience of the Vatican was common to small states. It is possible, however, to conclude that the reputation of papal cryptography, once so lustrous, had sadly faded.

BIOGRAPHICAL SKETCH

David Alvarez is a professor at Saint Mary's College of California. He is completing a study of German espionage at the Vatican during the Second World War.

BLETCHLEY PARK 1941-1945

William F. Clarke

[William F. Clarke, barrister, received a commission as an Assistant Paymaster RNVR at the beginning of 1915. He was the son of Oscar Wilde's counsel and had been educated at Uppingham and Magdalen College, Oxford, and was fluent in German. In 1916, the Director of Intelligence, William R. " Blinker" Hall, recruited him for Room 40. Clarke was only an average cryptanalyst, but his legal training made him a good Intelligence Officer. After WW I he and Frank Birch, another cryptanalysts, were assigned to compile a secret record of the activities of the German fleet. Clarke was one of the few people who remained with Room 40's successor, the Government Code and Cypher School (GC&CS) from its inception through the end of WW II. He wrote extensively about the history of the organizations in which he served, his experiences, views and co-workers.

This is the third of four sections from Clarke's papers and it was written nine years after he ended a 29 year cryptologic career which began in 1916. Clarke's papers are held by the Churchill Archives Centre, Churchill College Cambridge, with reference identification CLKE 3. The copyright to these papers is held by Clarke's son, Judge Edward Clarke QC, and all excerpts are printed with his permission.]

Forsan et haec olim memisse iuvabit.

[Perhaps someday it will be pleasant to remember this.]

The receipt yesterday of a letter from my oldest and most valued co-worker asking for some information for a very old colleague of mine from whom circumstances have estranged me about a detail relating to our common work in the 1914 - 18 war has led my thoughts back to these years which were certainly the most unhappy of a very happy life so I will try to write my account of them as fairly as I can. I recognise [*sic*] that the events which made them so unhappy for me may have been partly my own fault and partly the result of a sequence of events over which none of us had complete control but compelled as I was to be in the role of an onlooker without any power or responsibility, I may be able to judge fairly.

At the beginning of the year all seemed going well; the Italian section which was under my control was going well despite difficulties; it is true that thanks largely to the stupidity of the detached section in Egypt, which had been established in spite of my protests while I was away for an operation, thanks to an intrigue by Titterton, more care had been taken by the Italians with regard to their codes which were now much more difficult but valuable work by two junior members of my staff Grey and Clarke had shown that a most important part of their communications were ciphered on a Swedish machine, obtainable in the market whose possible use by them had been betrayed by a plain language message. I had handed over to the Head of the German section Bodsworth, my most capable cryptographer and the others were fully employed on other Italian codes. I urged that work on the machine material should be undertaken at once by our general cryptographic section under Tiltman but it was some time before I could get this done.

In the meantime important developments had occurred with relation to the German codes; their machine had come into our hands and great progress was being made. Also the entry of Italy into the war had made the task of Naval section one especially as combined action between the two allies was developing. So in March 1941 I suggested the two sections should be amalgamated under Birch, he was in favour [*sic*] and I was quite willing to work under him as I had in 1919 but at first the proposal did not seem to find favour [*sic*] among the high-ups. At a meeting in April I managed to get my way re the machine and the cryptographic section took on the task which was beyond the powers of my heavily reduced staff whose work was fatally hampered by the retention in Egypt of an Italian captured code book which was in current use by Italian submarines but which in their inexperience they had put aside as useless; it finally reached us after peace with Italy. Early in June the machine secret was solved and a large part of the Italian traffic became legible. This was decisive and on June 15 the merger of the German and Italian sections took place; Birch very properly was to be head and Bodsworth his second, I was to be on the staff in a rather undefined capacity. Bodsworth's position was, I thought and still think was a mistake; he was very bad at managing staff; worse still he took over the management of the Italian section with a Professor Vincent as No 2. Vincent was Professor of Italian at Cambridge, a fine scholar but like many of his class good at his own classics but very poor at the language used by naval officers; when he was working under me my first task on arrival at the office when he had been on night duty was to see how many mistakes in translation he had made and send in the necessary corrections.

Bodsworth was too busy on his own special work to devote much time to the Italian part and was, I think suffering from an inferiority complex as regards Vincent whom as his elder at one time, I think, tutor or at any rate superior at Cambridge, he regarded with some awe. As a result Vincent ran one section with a protegee of his, Crichton, who was useless and idle.

However all went well for a time; the drastic change produced by the solution of the important German and Italian codes necessitated much increase of staff and reorganisation [*sic*]; in this I seemed to be useful. Kinsley, extremely capable had my old 1917 - 18 job of intelligence officer in chief which he kept till the end of the war. A few young R.N.V.R. special branch officers were brought in as I.O.s to maintain constant watch on current traffic; the head of these was Gorley Putt who was excellent in every way. In their early days I used to visit them early in the morning, before Birch arrived and try to train them in their job which after a little experience they carried out most successfully. At first the decode sections worked separately but on August 21st the German and Italian decode sections were merged and all their decodes were passed to Admiralty under one head, Walter Ettinghausen, a very able Oxonian who now I believe is a prominent member of the Israeli government under another name; he was excellent and one pleasant memory of unhappy years is the continued good terms which continued until I left, between him and myself.

The next memorable event which took place was a visit from the Prime Minister on September 6th. This had for me some humourous [*sic*] sides. I had known Winston Churchill for many

The Victorian mansion, Bletchley Park, which was the headquarters of the G.C.&C.S. in WW II.

years, my father was a great admirer and defender of his and I had helped him in the writing of his well known books. The day before his visit I was told by our chief that it was proposed to introduce to him some of the senior staff, including myself and Jimmie Turner. He duly arrived and was shown a sort of composite section, of all kinds, including Vincent who under the guidance of one Wall, a very competent cryptographer of my staff in the pre– amalgamation days, had managed to solve a submarine Italian code which should have been solved a long time before if Vincent had done his work properly. Turner and I were not presented. Winston made us a very nice speech of thanks and went off to lunch at Blenheim. He had arrived in the usual cortege of cars with flags flying and must have been spotted by the local inhabitants; for lunch the B.P. staff crowded out into the town and no doubt talked about the visit to all and sundry; late that afternoon an order came round to say that his visit was to be kept secret, another instance, if more were needed of our wonderful security which had allowed motors bearing the name of a well known tabulating firm to ply to and from our premises.

Late in October our I.O.s were put into watches.

Then came the first big row; it concerned one of my most able helps for many years; she was a Miss White who sent originally to be my secretary, I had found to have the makings of a quite passable cryptographer. She had done remarkably good work helping me; when De Grey came to bring up to date his knowledge in that line, I handed him over to her, so I did Lord Lytton who at one time was anxious to join us but whom I persuaded that he would be better employed elsewhere. On the reorganisation [*sic*] she had been left to work under Vincent and Crichton both of whom were her enemies and who were jealous of her, realising [*sic*] she was better at the job than they; in fact she often had had to teach them their job and point out their mistakes in the past so having attained power they were bent on revenge. In this they were helped by her temperament which was not easy. In spite of my efforts to prevent what I regarded as a calamity, they persuaded Birch to ask for her transfer; Bodsworth collaborated in this in spite of the fact that he had previously told me he could not spare her when I asked she might be lent to me for what I thought was a very important job. Another section was delighted to have her but her work was lost to the section. The full correspondence on this matter will be found in my files labeled The White Conspiracy.

This most unfortunate episode was my first break with Birch who had preferred others advice to mine; I think it shook his previous good opinion of me, certainly his attitude to me in after years showed a great change. All this happened in November and very soon afterwards I was taken ill with a complaint that was to lead to a severe operation a few years later.

Things however apparently settled down and our work went on but during my rather prolonged absence many changes took place. It should be stated here that, acting in accordance with the lessons of the previous war, it had always been my practice to circulate appreciations of messages decoded rather than the text; our experience in the early days of that war had shown that we were not merely a decoding but an intelligence section and that the higher ups made too many mistakes by their incorrect deductions. This was more easy in the days of the old type of codes when some words had to be guessed but when machine codes came in all the words were normally legible, only the language difficulty remaining.

While the two sections were separate I kept to the old method but as it involved a little more work, it was unpopular with my staff and they had while I was away adopted the easier course and the actual text was sent to the O.I.C. whose functions are described in "The War at Sea" Vol. I, p18. I personally think this was a mistake; during the prewar years I had always urged that we should be in close proximity to Operations Division as had been the case at the end of the previous war; the only result had been as I have stated in previous chapters that my German section had been removed to the Admiralty where it was completely useless and I had to build up a new one as the head of which I was lucky to get Birch. This development had taken from me all control over the Italian traffic circulated which I had exercised before my unfortunate illness.

However I found plenty to do, in spite of the fact that relations between me and many of the section were for ever strained. The mass of deciphered intercepts necessitated a largely increased staff and the establishment of a system of indexing and arranging the intercepts to ensure rapid ascertainment of previous information on matters dealt with in a current message. This needed a great deal of supervision as the staff entrusted with these tasks was of necessity inexperienced and many mistakes were made.

Another amusing task I had was to compose differences between Dilly Knox and Birch; Dilly was a genius in cryptography as well as in other matters but he was very temperamental and had to be handled tactfully or else he refused to play; I happened to be one of the few people who understood how to manage him and when these clashes occurred I was called into restore peace and ward off the resignation which he threatened. He died soon after and his death is one of the many for which the high ups at B.P. were responsible, another was Leslie Lambert alias A.J. Alan. Both were sick men and no consideration was shown them. This was characteristic; in the years before the war I had a valuable colleague, Rees, who had had his lungs injured by shrapnel in the First World War; in consequence when there was fog he was grievously distressed and not fit to work but when at these times I sent him away I was reprimanded by Denniston. The same thing happened when I varied the hours of one of my staff who was only temporary and was preparing for an examination; in spite of my protests she was dismissed only to be employed in an important job by the S.I.S.

In the meantime in March 1942 great changes had taken place at B.P. The diplomatic section under Denniston had been moved back to London; if any part were to be moved to this danger area it should have been Naval section or some part of it so as to be nearer Operations Division. This meant that most of my old and experienced colleagues went with it and so I lost most of my friends. Travis was now in supreme command under Quex's successor, a very unworthy heir, named Menzies who disliked me as he was connected by marriage with the Wilson family who had been the subject of bitter comment by my father in the famous Baccarat case. Of my real friends only Strachey was left and he was a sick man. Quex's death early in the war was a real calamity, he was a very brilliant chief who trusted his subordinates and to whom one could go knowing that if your application was wise, and he always investigated fully, you would get what you wanted, be it staff or improvements in organisation [*sic*].

B.P. now consisted of the three service sections and the usual competition started as to which was to be the largest, one had seen this in the former war and its evil results; staff was added to do useless jobs and thanks to the poor material enrolled the work was badly done. Advertisements even appeared in the Times asking for recruits. Naval section was no less to blame in this respect than the other sections.

Quite early on tabulating machines had been brought into use by us and while they were of great use in certain circumstances, they did, in my opinion, a good deal of harm as many idle people thought they relieved them of personal indexing; this, I think, led to the failure to solve many of the hand codes, as contrasted with the machine types.

The increase of staff led to the building of more commodious quarters and a large new block was built for naval section to which we duly removed.

During 1942 the war at sea was chiefly concerned with the submarine menace, apart from the escape of Scharnhorst and Gneisenau from Brest and very important the interception of Italian transports carrying men and supplies to Africa. Thanks to the solving of the Swedish machine this last was very successful. The work of the section had become largely a matter of routine, the Swedish messages were dealt with most successfully by a small section and the German naval traffic outside, the only really active work being in the hands of the circulating party under Ettinghausen and intelligence appreciations by Hinsley and his aides. The larger part of the staff was employed on record keeping. This required constant overseeing and a large part of my time was devoted to this. I was officially shown on so called Naval Section state as part of the Staff but Staff papers were carefully kept from me and most of my activities were sub rosa; however I found most of the subsections glad of my suggestions. I had learnt that my formal submissions were either ignored or turned down so except in more important matters I did not indulge in them. The section was now approaching the 1000 mark as compared with the 24 who had gone down in August 1939.

The Italian section had made no progress with the main codes; the famous submarine code displayed to the P.M. had been changed within a week of his visit, a strange coincidence which might have seemed curious but it had only been solved after being compromised by long use. Work on the German naval traffic had been halted for a while by their employing additional wheels but this had been soon overcome.

So matters went on in the early part of 1943 but the course of events was changing in our favour [*sic*]; things were going badly for our enemies all along the various fronts and our political leaders had started to talk about "Unconditional Surrender". Early in the year a copy of an Italian code "Sigma" came into our possession and helped our anti Italian operations. But Italy was already at its last gasp, the Axis forces had been driven out of Africa by early May and in July Italian territory was invaded. Negotiation for her surrender started and in August the Italian section personnel were dispersed to other duties; on September 9th came the armistice and charge of its remnants was given to me; these consisted of 4 or 5 and their job was to keep a watch on such Italian communications as were permitted and the more interesting task of investigating old traffic which had never been deciphered with the aid of a number of old code books which had come into our hands; these were very incomplete as no condition had been imposed as to their surrender; this was because it was an armistice and not a complete surrender so we had the same situation as we had had in 1918. It was however a quite amusing job and gave me work to do which was really worth doing, a welcome change from the preceding period. The Spanish and French sections were also put under me so again I had plenty to do and as the staff transferred was mostly old staff of mine less the senior members who had got themselves more showy posts, I was happier than I had been for some time. It was nice to be doing cryptographic work again even if it was of very minor importance.

The Spanish section had carried on bravely with its rather monotonous task as the traffic was of no great import but it had been recognised [*sic*] that there was always a possibility of Franco taking sides against us and in this work continuity was of the utmost importance. Once the codes of a nation are put on one side as of no moment it is very difficult to pick up the threads. This had been shown in the case of the French; in 1937 we were reading their traffic with comparative ease but they then had recourse to a system which, though we found the system, was too hard to decipher on the very small amount of traffic intercepted. They found it too complicated and reverted to an easier one but no effort was made to deal with it except as a very part time job of one or two of the diplomatic staff. My staff which the Admiralty kept on reuseing the moment world affairs looked calmer, had their hands too full especially as in 1938, just before the Munich affair I had managed, thanks to Quex's intervention, I having bypassed Denniston, to start a German section. This sounds hardly credible but it is a fact. This section had a narrow escape from extinction, as Quex, impatient of the little information we had been able to get at that time sent a minute to Denniston, asking if he thought our work was going to be of any use in view of the increasing difficulty of foreign codes. Denniston always a pessimist said " No" but I wrote a very strong minute, taking the opposite view; in it I said that machines were in use and this type of ciphering would spread and that in war there was a good chance of their capture. This turned out to be an accurate prediction. When war broke out this work stopped but was resumed shortly after the collapse of France; I was not told of this though I had studied the subject for many years. The staff employed had practically no experience of French methods and made little progress and the section did not last long. Now in 1944 we started again when we had the advantage of dealing with material that was that of a fleet that was allied with us; the traffic was scanty and progress was slow though now the Americans were collaborating but we had seldom found them of great assistance; the staffs they employed were very much larger than ours but our smaller numbers generally produced better results.

But in the meantime we knew of the proposed landing in France and attention was naturally focused on this; it took place in June and success after success followed with only a very occasional setback. The end seemed very near. My mothers death in October upset me and within a fortnight my old trouble returned; it is curious to note that these attacks of mine seemed to have some strange connection with my mental state; an immediate operation was averted; I wanted badly to be in at the death but I had to be away for over two months during which the last desperate German offensive after some initial success was repulsed and Allied progress resumed. My prolonged absence had of course put me out of touch with things and it took me a long time to pick up the threads and Birch's Staff had never liked me and resented my occasional interferences, they treated me with scanty courtesy, and took every opportunity to subject me to minor inconveniences.

V.E. day soon came. We had a thanksgiving service conducted by the local parson; I was deputed to make the arrangements as I had become a sort of office church warden when I arranged every Sunday for the conduct of the very nice little services which the padre, a charming person called Milne who could not do too much for us, always willing to arrange extra services at hours to suite our convenience. Some amusement was caused by Nigel de Grey reading a message to us from Travis whom he described as " Our beloved chief" a description I think Travis himself would have been the first to disclaim; able he was especially in working his way up to get the top job, pushing out Denniston than whom he was a much more efficient organiser [*sic*]; he was always very nice to me.

After this the work of Naval section really came to an end, after almost exactly 21 years of existence during 16 of which I had been its head and its founder; it had started with two or three members, never exceeded 30 before the war and finished up over 1000 strong; I sent a message of my thanks to Hope (N.S.L.), Rees and Miss White who had been with me in the early days, all three were now in London with the diplomatic show. Now two jobs remained, the first the future of the whole organisation and the second the preparation of a history.

The first was considered by a small committee on which there was a lamentable absence of those who had had experience of peace conditions which are so different from those of war; I gave evidence before it but my views carried little weight, I fear, though I never learnt what recommendations were made though I did last long enough to see some of the first steps in reorganisation with which I profoundly disagreed.

The second, as to which I can only speak of the Naval Section side, was complicated by the masses of captured documents which had fallen into hands. These were in the first place entrusted to a section which was quite unsuited for the task and my efforts to have the material divided into historical and cryptographical sections was only considered by of the junior Staff with no real knowledge either of history or cryptography and so remained without success. The task of writing the history of Naval section during the war was entrusted to the various existing subsections who apparently on the assumption that all had started from scratch in 1939 or even later; thus the identification of call signs, a very important job was stated to have begun with a Miss Linahan who had not had anything to do with it until very late in the war it had actually been very important from the very start of Naval Section in 1924. The Roman historian Livy says somewhere *Semper enim quod postremum adjectum est, id rem totam videtur traxisse*, a translation of which sage remark is "For the last stroke to a business seems to get the credit for the whole performance". Of course the question of a history on the lines of that in which I had assisted Birch in 1919 was under discussion and now (October 1954) nine years afterwards seems from the query to which I have referred in the opening lines of this chapter of events to be still in course of preparation was another matter for consideration and Birch at any rate hinted that I might be of use to him in this; he was given the job but I never heard any more about it; as it has turned out this is fortunate for me at least as I have enjoyed nearly nine years of peace in retirement. Travis informed me of the fact that I was to be retired on September 26th 1945, I received the formal notice from the F.O. on October 8th, a couple of days after I had made the acquaintance of Mr. Ernest Bevin, then Foreign Minister, at my London club of which somewhat surprisingly he had recently become a very popular member.

My days in G.C. & C.S., our official description, were numbered and on October 17th my old complaint flared up again and this time an operation was necessary; I left B.P. in a hurry that day and never went back; some of my papers were sent to me as they happened to be locked up in a box of which I had the key but most of them were taken by Birch and I shall never see them again; I expect most of them have either been destroyed or placed with the records of Room 40 O.B. which Birch and I among others so carefully put in order in 1919. Anyway they have gone and these reminiscences have had to be compiled without them which must be my excuse for such misstatements as have been made. I have tried to be fair even to my most bitter enemies to all of whom I must be grateful for nine years of peace.

Commenced October 24th 1954.
Finished October 28th 1954.

WHY WAS SAFFORD PESSIMISTIC ABOUT BREAKING THE GERMAN ENIGMA CIPHER MACHINE IN 1942?

Louis Kruh

ADDRESS: 17 Alfred Road W., Merrick NY 11566 USA.

ABSTRACT: In 1942, Captain Lawrence F. Safford doubted that the German Enigma would be broken. This implied that he was unaware that the British had already broken the Enigma Cipher. Various authorities and colleagues of Safford speculate on his assertion.

KEYWORDS: Safford, Enigma, Weeks, Currier, Rosen, Sinkov, Rowlett, Kahn.

This article is reprinted, with permission, from the Winter 1988 (Volume 9, Number 2, pp. 3, 11), issue of *Cryptolog*. *Cryptolog* is a publication of the Naval Cryptologic Veterans Association, 3421 Stack St., Eugene OR 97404 USA.

In a memorandum dated March 18, 1942, Captain Lawrence F. Safford wrote, "Our prospects of ever breaking the German 'Enigma' cipher machine are rather poor." (See section 2 in the complete memo which follows.)

Safford's assertion implies that he was not aware that the British had solved German Enigma ciphers. But, he wrote this memo a year after Navy Lieutenant Robert H. Weeks and Ensign Prescott Currier along with Captain Abraham Sinkov and Lieutenant Leo Rosen of the Signal Intelligence Service had visited Britain's Government Code and Cipher School at Bletchley Park.

Some authors have speculated that the British may not have informed the American mission of the extent of their success with the Enigma.[1, p.312] Also, the April 11, 1941 report prepared by Sinkov and Rosen on their trip did not mention the Enigma. (Information is unavailable on any Weeks-Currier report.) The Signal Security Agency official history, however, points out that for security reasons the omission of Enigma was deliberate. The British had obtained a special pledge of secrecy from the American officers who agreed to reveal what they learned to just three persons: the Assistant Chief of Staff, G-2; the Chief, SIS; and Mr. William F. Friedman. Only then did the British disclose "the existence of their 'E' operations (a successful attack upon traffic enciphered by

Op-20-Q MEMORANDUM FOR OP-20. March 18, 1942

Subject: Security of information obtained from enemy communications.

1. During World War I (1914-18), the British Admiralty was very successful in reading German naval messages and tracking German submarines by D/F. This was a well-guarded secret during the war. After the war, the British let down the bars of secrecy and told the world what they had done and how good they were. Apparently, it never occurred to the British that the Germans would profit by these revelations.

2. In World War II, the German ciphers and communication procedure are such that the information obtainable by radio is substantially zero. With the exception of "weather codes" and certain other "minor systems" the only German naval messages which have been read have been the result of captures. Our prospects of ever breaking the German "Enigma" cipher machine are rather poor.

3. We are much more fortunately situated with regard to the Japanese Navy. Our solution of Japanese naval systems began back in 1925. We are able to read most of the messages in the most important Japanese naval ciphering system and have completely solved one of the secondary systems. This is the result of sixteen years of unremitting effort along these lines. We are also getting considerable information by direction finding. We are faced with constant danger that news of our success will leak out, whereupon the Japanese will adopt steps which will nullify our efforts.

4. In 1938 and 1939, the Japanese Navy used a system which we have never succeeded in solving. They abandoned this system in 1940 and adopted a system which we can solve. The reason for abandoning the secure system is not known, but we surmise it was due to its slowness and complexity. There is nothing that will stop their return to it if they find out that we are reading their messages.

5. What is even worse, they may adopt the German Navy cipher machine if news of our success leaks out. If this happens, our source of information will be reduced to direction finding and traffic analysis.

6. This memorandum is submitted for the purpose of impressing upon all concerned that we must not compromise our one sure source of enemy information by a too wide dissemination of information.

Respectfully,
L. F. Safford,
Captain, U. S. Navy, Op-20-Q.

the German high- security cryptographic machine known as the Enigma ...)."[2, pp.12-13]

Presumably, Weeks and Currier made similar pledges – the Signal Security Agency history does not discuss Navy details – but surely Safford, head of the Navy's cryptographic unit, OP-20-G, would be among the Navy people permitted to receive the information. However, if Safford knew about British success with Enigma ciphers, what is the explanation for his seemingly uninformed comment.

Frank B. Rowlett, a key member of SIS who knew Safford, was asked for his opinion on the apparent discrepancy. He wrote,

> I do not consider it possible that Safford was not aware of the British success with the Enigma at the time he wrote the memorandom in 1942. As you know, an American team of cryptologists visited England in 1941 and were told of the UK work on the Enigma intercepts. I do not see how the information on the German Enigma brought back by Weeks and Currier could have been kept from Safford, considering the position he held in the Navy cryptologic organization.

One possible reason for Safford implying that the Enigma had not been broken is perhaps best described by the word "disinformation." The subject of Safford's memo was "Secrecy of information obtained from enemy communications" and what better way to protect Enigma derived intelligence then to highlight dismal prospects of ever breaking Enigma ciphers. Or as Winston Churchill said in 1943, "In war-time, truth is so precious that she should always be attended by a bodyguard of lies."[3, p.338]

Another view comes from David Kahn, author of *The Codebreakers* and the leading historian on cryptologic matters. He says that Safford was referring only to the Naval Enigma, which had been decrypted with the assistance of clues gathered from captured U-boats and other German ships. Kahn also suggests that Safford may have been alluding specifically to the then current inability to crack the new 4-rotor Enigma Shark cipher which was introduced 1 February 1942. (It was not solved until early December 1942 with the help of a codebook taken from the captured U-559.)[4, pp.749-750]

The opinions of naval officers who had worked with Safford were also solicited. None of the replies had a definitive answer but two of them had some pertinent information.

A retired Lieutenant Commander wrote:

> Prior to the Pearl Harbor incident, Captain Safford, though chiefly concerned with the Japanese problem, did devote considerable time

> with the embryonic German Naval section, discussing at length with Mrs. Driscoll matters associated with the German Naval problem.
>
> Around July or August 1940, the section was comprised of four persons. Since Mrs. Driscoll was severely handicapped in walking because of an earlier auto accident, it was my job to relay the various requests and questions directly to Captain Safford - this occurred frequently and I spent much time in his office.
>
> Such was the climate of our operations until either late 1940 or early 1941, when Commander Wenger became our boss and we no longer had contact with Safford. I must stress that I am not certain of these dates - it could have been later, but I do clearly recall that this shift did take place.
>
> By March 1942 (the date of the memorandom in question) the German Naval section had expanded considerably.
>
> It may have been around June 1941 when the first of the English visited us - and by the fall of 1941, plans had already started for the development of the U.S. Navy version of "Bombe". Again, I'm pretty vague as to exact months.
>
> I have no knowledge as to why Captain Safford made the statement you quoted from the memorandom - but one may draw some conclusions:
>
> 1. Since he had been out of touch with the German section since either late 1940 or early 1941, he may not have been up to date in the progression of things since the development of Bombe plans in mid 1941, though he was aware of the information that Weeks and Currier brought back from England.
>
> 2. The memorandom was written in March 1942, 3 months after Pearl Harbor. Things were in a great state of flux, and the matter of the Winds message must have placed a great stress and inner turmoil on his over-worried mind, so that he may have overlooked or forgotten the actual facts.

Another retired officer explained that "My personal knowledge of the naval Enigma was in February and March 1941. At the time the British had several copies of the machine and sets of rotors and umkehrwalz, some of which, at least, had been taken in the Lofoten raid. They had only the key/starting setting material for a few days, and their big problem was trying to solve the coding of each message's starting settings. As you are aware from the extensive literature on the Enigma, they apparently did solve this, but I do not know when."

In regard to the Currier-Weeks trip to Bletchley Park, he said they returned with,

> ...several days' settings and a "paper" Enigma, with all the rotor and umkehrwalz laid out. With this material we were able to decipher messages taken from our intercept logs of the days covered. Captain Safford's "except by capture" seems to bear this out. What the exact situation was by March of 1942, I do not know. I feel sure he would have known of any recent successes in the matter.

One final conjecture is that Safford's pessimism was based on the apparently remote possibility of a purely cryptanalytical solution of the Naval Enigma cipher in contrast to one assisted by capture of rotors, key lists, etc., which had been the situation up to that time.

Supporting or contrary views are welcome.

REFERENCES

1. Bamford, J. 1982. *The Puzzle Palace*. Boston: Houghton Mifflin Co.
2. U.S. Army Security Agency, History of the Signal Security Agency, Vol. Two. 1947. *The General Cryptanalytic Problems*. Washington, D.C.: Army Security Agency.
3. Churchill, W. 1949. *The Second World War*, Vol. V. Boston: Houghton Mifflin Co.
4. Hinsley, F. H., et al. 1981. *British Intelligence in the Second World War*, Vol. II. London: HMSO.

BIOGRAPHICAL SKETCH

Louis Kruh is an indefatigable researcher and collector of cryptologic memorabilia, particularly cipher devices. His interest in cryptology spans more than 40 years and he responds knowledgeably and promptly to all queries regarding literature and equipment. He is co-author with Dr. C. A. Deavours, of *Machine Cryptography and Modern Cryptanalysis*, (Artech House, Inc. 1985). He retired as Director of Advertising for New York Telephone in June 1990 and he is setting about practicing the law he studied.

S. I. S./CB

Henry D Ephron

ADDRESS: 800 Woodworth Ave., Missoula MT 59801 USA.

ABSTRACT: From Fort Monmouth to Vint Hill Farms to SWPA: S.I.S./CB, Brisbane (Col. Abraham Sinkov, CO).

An introduction to the Japanese method of superenciphermant, by means of additives, of their encoded messages and a detailed description of some of our (successful) efforts at solving this superencipherment – by means of step-by-step solution of an imaginary worksheet. Later the Japanese improved the security of their communications, superenciphering by passing the digits through a numerical square (a 10×10 table) instead of by mere addition. Encipherment is demonstrated in detail by means of a step- by-step solution of an English encoded tect superenciphmered by this method.

KEYWORDS: Japanese, codegroups, superencipherment, additives, encicode groups, cryptanalysis, *dendai*, redundancies, shift to squares (10×10 tables), solving the square.

PART I

It was the fateful day of May 7, 1942 that a dozen American SIS men arrived in Melbourne, Australia, and joined a handful of Aussies working in a beautiful mansion in a fashionable suburb under the command of Captain A. W. Sanford, A.I.F. (that is, Central Bureau, just activated, April 15, 1942). Two of the four American officers stayed with the outfit: (then) Captain Harry L. Clark and Lt. Charles E. Girhard. The enlisted men were Cpl. Joseph Fitzpatrick and seven Pfcs: Lester Truex, Tris Johnson, Leo Tepperwein, DeWitt Barnett, Chet Young, William Estep, and John Dalas. Who would have guessed then that from this bit of seed would grow and flourish an outfit of well over two thousand officers and men (including WACS), generously sprinkled with awards, citations, and promotions, and including even the Legion of Merit?

Major Abraham Sinkov (later Colonel) arrived to take charge of the entire outfit, with a few others, on July 1, and a few more personnel reached the southern port on July 22. It would be presumptuous on my part to attempt to point out the genius of Col. Sinkov. Whether as a cryptanalyst or as a Commanding Officer, this scholar and mathematician seemed able to keep a finger on every facet of the work that went on in the outfit from the interception of the radio communiques, through the collateral channels and collateral information and the grinding by the IBM Business Machines which were the predecessors of the computer, to the individual efforts and results of the personnel engaged in various cryptanalytic activities. I will merely say that I considered myself and my colleagues lucky to have him at the head of the outfit. To attempt to comment on and name the personnel as the outfit grew would be an impossible task, of course. Perhaps, however, I should mention T/4 Joseph E. Richard whose genius made the first successful cryptanalytic break in the continual, agonizing efforts of the group and finally started the outfit off to happier results. For which Joe, as a deserved reward, became Chief Warrant Officer Richard. At the other extreme I cannot refrain from mentioning, without naming him, one permanent army man who, when it became possible for men over a certain age to ask for a discharge, immediately did, saying to me, "I didn't join the army to fight in any wars."

According to an article in the *Sigaffe Clamor* (October 14, 1944) the "Melbourne idyll" ended with the move to Brisbane on September 25, 1942, a move "greeted by no less a personage than 'Tokyo Rose' from Radio Tokyo. 'So you have moved to Brisbane,' she said. 'We'll be over soon to visit you with bombers.' We are still waiting, lady."[1]

It was not until sometime in 1943 that the small select group which included myself joined the outfit, but because it happened before the great increase in size began, we too were always numbered among the *anasazi*, "the ancient ones," and were included in the picture taken at the (original Brisbane) Henry Street Headquarters. I have used the word "select" advisedly, but since we are dealing with the army way of doing things, the word is not necessarily used here in an ameliorative sense. To appreciate the irony of our becoming members of this excellent outfit for the duration it is necessary to point out that we were indeed carefully selected, but not, I repeat "not," for this purpose.

I had enlisted in 1942 at the urging of the Office of the Chief Signal Officer, War Department, Washington ("The Chief Signal Officer directs me to advise that * * * *"), with an assignment at his request from The Adjutant General to the Cryptographic Division of the Signal Corps School at Fort Monmouth, New

---

[1]For two years the SIS organization had been shunted from pillar to post before it was allowed to join the army as a separate organization with its own quarters (July 25, 1944): the Signal Intelligence Service, United States Armed Forces in the Far East (USAFFE). Permission was then granted to publish the *Sigaffe Clamor* twice monthly: "by and for the personnel of Signal Intelligence Service (USAFFE), Col. A. Sinkov, commanding." Of particular interest was this additional sentence: "The *Clamor* may be mailed home." It was about this time that the USAFFE censorship regulations changed our official address from "Somewhere in Australia" to "Brisbane, Queensland." The information I have given here can be found in greater detail in various issues of the *Clamor* and/or the *S.I.S. Record*, which was published in 1946 with the added longer title, *Signal Intelligence Service in the Far East 1942-1946*, "An Historical and Pictorial Record." I have checked details with these publications as far as possible.

Jersey (actually, in plain language, the school for Cryptanalysis). It was my understanding that the purpose was for further, efficient training in the art, but, of course, I had not yet been initiated into army ways and "Catch 22." When the school moved on October 5, 1942 to its new quarters in the new army camp at Vint Hill Farms in Virginia, we found that the attitude of those in charge was that the school was a leftover of the old days and no longer of any value. For the most part grammar school dropouts were put in charge, apparently with instructions to put those "Ph.D.s" in their place. It was an extremely unfair situation to put this practically-illiterate cadre (they couldn't even count correctly), up against some of the army's most brilliant and educated personnel. It took the school about a month to get rid of all these men and replace them with its own staff. However, I myself heard the C.O. of the whole camp, speaking to another officer, say that since cryptanalysis was passe and the school really served no purpose, we were to serve as KPs and do guard duty for the much larger radio school. This practice was carried to the point where on one occasion of which I happened to be aware, someone came into a German class and took the instructor away to put him on KP! Fortunately for us the army, playing its usual game of musical chairs with its officers, soon replaced its unfortunate choice of a C.O.

Although I was on the staff technically as a "teacher," I also ranked as a student–no matter what your duties or rank you continued your studies during whatever hours were available to you. Along with other "students" I was given a refresher course in German. There were classes in various European languages as well as a class in Japanese. In early 1943 a request came to the school from Brigadier General Akin, Chief Signal Officer, SWPA, to "select six of your best men." We were to join five enlisted men from Arlington Hall and be attached as a group of eleven men (and four officers) to the CSO himself. The experience became another great lesson in the army way of doing things: "Hurry up and wait!" It was to be done "pronto," for we would in fact be flown across the Pacific, almost unheard of then for enlisted men in view of the shortage of space even on one of the cargo planes (C-47), such as we were crammed into along with a number of officers. But before that we were relieved of all duties and spent more than two weeks doing such things as running the obstacle course, hiking, policing the area, and other busywork.

I noted with puzzlement that all six of the men chosen were from those who had had refresher or cram courses in European languages. I did not know that we were slated for the Philippine underground, and Military Intelligence on one of its periodic visits looking for personnel had promised me that they would not insist if I preferred to remain in the work I was doing–although they had the clout to take me if they wanted to. I therefore had no objection to being one of those "selected," but out of curiosity, I naively asked why "they" were sending men from the European language classes to SWPA instead of taking them from the Japanese language class. The answer was that they had a good Japanese class going and didn't want to do anything to disturb it. It made sense of a sort–but I learned afterward that the men who had studied Japanese in that class all ended up in the European theater.

There were in the school at this time three men who were considered nuisances, silly, undependable, "gold-brick," or "Good Soldier Schweik" types. All three were selected for inclusion in our group of six, and thus in one fell stroke the school got rid of the three men who were considered the greatest detriment to the school. Poor fellows! In the field they got short shrift. Anybody for the front lines?

Of course, the destiny of our own little group lay forward of the front lines, for once inside the Philippines we would presumably not have been heard from again unless and until our forces successfully retrieved the Islands at some future date. And even then, of course, it would have been "iffy." However, it was fated that I should remain in cryptanalysis, even if it was just by luck. I've always insisted that regardless of what you have enlisted for and regardless of what the army has promised, you must depend on luck to get into and to remain in the field you wanted. Had a call come in for a large number of permanent KPs, I have no doubt that my "spec" number would have been changed to that of a permanent KP. You do not tell the general that you are unable to fill his order.

When I arrived at the School at Fort Monmouth with my assignment from The Adjutant General I found that insofar as the School was concerned, it was a worthless piece of paper and I was immediately shipped off to basic training without any promise of being accepted. The School made its own decisions as to whom to accept. One of my friends, an excellent professional photographer enlisted in the Signal Corps for photography. Once in the Corps he too was sent to basic training and from there was picked for cryptanalysis school just as I was–and a number of others–willy-nilly.

What happened was that after hurrying up and waiting, and then hurrying up again by flying across the Pacific Ocean, we found it necessary to wait again. The Chief Signal Officer, SWPA, Brigadier General S. B. Akin, was not there to welcome us. He had returned to Washington in what proved a successful effort to get himself another star. Meanwhile the Base Commander, who had been desperately trying to get a few additional men for Major Sinkov's hard-pressed little unit on Henry Street, found himself with an unexpected gift, a group of eleven (trained) men. He accepted five of the group for About-to-be-Major General Akin's special project and sent the other six to Major Sinkov without ever mentioning what he had done to the CSO. Four of us fitted happily into the SIS outfit, and thus, purely by luck, I ended up doing the very work I had enlisted for. Meanwhile the other five, having hurried up, began their next wait.

We found a small but efficient, well-integrated group of enlisted men and officers working side by side, and enthusiasm making up for experience, it was not long before we were helping them quite effectively in what they were doing. They were working on one of the Japanese systems mentioned by Kahn in his chapter on "The Scrutable Orientals."[2] The messages, which, like all radio-sent material, were available to anyone who wished to intercept them, were encoded with four-figure code groups which were then superenciphered by means of four-figure additive groups.[3]

Although four-figure groups allowed for as many as ten thousand code groups, the Japanese knew as well as we did that the frequent repetition of many of the groups used, especially those necessary under war conditions, would allow our cryptanalysts easy entry into their secret code. Kahn, on page 402, explains the use of such additives and demonstrates how effectively they conceal the repetitions of the plaintext codegroups. As in the case of the codegroups the four-figure numerical additives allowed for a series of 10,000 different groups without a single repeat.[4] All the cryptographer had to do was to write the additives in order over the codegroups of his message and add each pair of groups without writing or carrying the tens digits (e.g., 5986 + 7259 = 2135): a simple task and very effective.

Before our decoders could even begin to try to solve the meanings of the codegroups we had to solve these keys, i.e., the additives, and subtract them from the four-figure encicode groups of the message. And before we could even begin to think of trying to do that we had to find a number of instances of consecutive text enciphered by the same string of additives. Such instances of text were typed on extra-wide sheets of paper with the coinciding groups underneath each other so that all groups superenciphered by the same additive were in the same column. The rest of the texts of the messages were also written in before and/or after these groups so arranged. Messages incorrectly placed wasted time but did no further harm–they merely yielded no results.

---

[2]David Kahn, *The Codebreakers* (The Macmillan Company, 1969), p. 587 (Kahn).

[3]I shall go into greater detail in this article, but I touched upon the subject in "An American Cryptanalyst in Australia," *The Australian Rationalist Quarterly*, No. 2, Apr-May-Jun 1980, pp. 11 f., and No. 9, Jan-Feb-Mar 1982, p. 7; also *Cryptologia*, Vol. IX, October 1985, pp. 337 ff. (Crypt-Australia).

[4]Joseph E. Richard, mentioned earlier, reminded me in a personal letter dated 1 February 1988, after reading my Australian article in *Cryptologia* (see note 3), that additive books were composed of several books each of 10,000 such groups: "A few had only 3 books, i.e., 30,000, but most had 5 books, i.e., 50,000."

On my first morning with the outfit, my personal secret came out: I could type! Apparently in the army, if on a typing test you did forty words a minute, you were rated a typist, but at the crypt school at Fort Monmouth where you were tested not on words but on code groups, a mere twenty such (errorless) groups of five characters per minute gave you the rating of a typist. I was immediately placed in front of a typewriter with the longest roller I had ever seen and asked to prepare the superwide worksheets of anywhere from five to ten messages per sheet.

Officially I did that for several days, but actually I spent most of my time learning how this superencipherment worked and how others went about seeking to extract the hidden additives and the meaning of the codegroups. Since none of our little group had had any training at all in Japanese, the decoding aspect of the work was not for us. Instead we concentrated on the cryptanalysis of the superencipherment, for which a knowledge of the Japanese language was not a requisite, although I personally believe that the ability to think like a Nipponese would have been of considerable help in the cryptanalysis. Only rarely did we use terms in their language, but rather a rough translation of the meaning in English. For example, although we did say and write such expressions as *den dai*, roughly "Message No.," we did not bother putting the number into Japanese, saying and writing instead, "one, two"–rather than *ichi, ni*.

We used every military term we could think of as probable words in our decipherment attempts, as well as collateral information concerning events which might affect the content of individual messages, such as, for example, the fact that Okinawa had just been bombed about the time we intercepted a flurry of messages radioed from there. One might then look for words like, "planes" or "pilots," and even such face-saving expressions as "damage slight," or "damage none," which were almost mandatory. But identification was absolutely mandatory for the benefit of the recipient, and as I got into the work I found that of particular interest and help.[5] Whatever Col. Sinkov's reason for choosing me for the task shortly after I arrived, it was he that saw the possibilities in having a complete record of all *dendai*-information as it became available from the intercepted messages. Keeping that record became my highest priority, and I was permitted for a time to work swing shift all by myself in order to be able to bring the *dendai*-record book up to date during hours when no one was using

[5]See Kahn, p. 753, on the sliding ratio between the element of time and need for security, and especially, on p. 587, on how disastrous a special attempt at extra secrecy by the Japanese actually turned out for them. Even with the relatively simple methods used in the cryptosystem we were working on, the intended recipients of the messages (cryptographers) must often have found their decipherment difficult or even impossible, in whole or in part, because of errors in encoding, in enciphering, in transmission, in reception–not to mention mere typographical errors.

the worksheets.

As I have written elsewhere,[6] in those pre-computer days I became a sort of living computer with the information stored in my memory bank, and some of my co-workers, instead of looking up the information in the record book would call out to me, e.g., "Dendai, I've got Blank Shipping on ______," and ask for the *dendais* on that date. If the information was available, I would give it to them; for example, 125 to 127. Putting the correct number into the text (by digits) gave us two or three more additives to add to the string already recorded.[7] With his Australian fondness for Gilbert and Sullivan, an Aussie cobber from CB (Central Bureau), punning on Deadeye Dick, dubbed me Dendai Dick.

To illustrate in a bit of detail how we worked with such a worksheet I have created a portion of an imaginary sheet, using words, phrases, and kana strings typical of those we encountered in our work, such as *bu ra tsu ku ui do* (= "black widow"), "[*bi ta mi n B*]" (= "vitamin B"), (Blank) "Shipping Company *den dai*" (+ numerals), "to be cont.," "the end," and so forth. (See Figure 1.) Assume that we are at an early stage of our work and that the majority of the plaincode groups cannot be translated even after the additive has been removed, since we did work under this handicap during the period I am describing. Even they could help, however, if they were groups that had been met several times before, for obtaining recognizable groups could show that we were on the right track. I will refer to them as "friendly" groups. Note that the messages in lines 6 and 8 begin in the middle of this portion of the worksheet and those in lines 3 and 7 end in the middle. This phenomenon was a constant on a full sized worksheet, and because of it only a limited number of columns were likely to have any real depth to work with.

It must be understood that I have deliberately brought together on one portion of a worksheet too many redundancies, and so forth. In actual practice one would have had to work on several worksheets to find so many obviously helpful codegroups. However, the time was to come when we were able to follow the travels of a frustrated Japanese general for six months from island to island. He had been ordered to report to Tokyo, but because of the sad shape of the Imperial Navy he was finding it virtually impossible to get there. After he had been marooned on an island for a time, a submarine might arrive and he would get aboard and travel to another island, but at the end of the six months of vicissitudes of which I was personally aware from the deciphered messages he

[6] Crypt-Australia, 1980, p. 12; 1985, p. 340.

[7] This was important because, if you obtained an additive which was one of a known string, you could try out the remainder of the string in the adjacent columns in the hope that the additive was from the same book of ten thousand. Thus a good deal more of the messages on the worksheet could be (at least) deciphered.

had not yet succeeded in reaching Tokyo.

To help the reader follow the demonstration of the cryptanalysis, I give a small list of (assumed) plaincode groups whose meanings have already been decoded. By assumption no plaincode group used in these messages, which is not in the list, has been decoded. For realism a few undecoded groups of more frequent occurrence ("friendly" groups) have also been assumed.

LIST OF PLAINCODE GROUPS

								friendly groups
0068	in	2865	(Eng.) B	5565	colon (:)	7959	two	
0140	three			5752	box	7984	began	1234
0229	daybreak	3443	the end	5878	nine			1597
0631	KU	3661	four	5939	DO	8166	boat	2148
0978	RA	3871	none	5948	BI	8216	(kana) N	3701
		3974	DAI			8403	zero	3851
1357	part			6170	metal	8437	]	6489
1648	damage	4003	general	6295	six	8570	attacked	6942
1713	five	4012	Kyoto	6530	planes	8651	TSU	7340
1726	and	4241	DEN	6711	TA	8686	HQ	8250
1733	bombed	4412	MI					8639
		4504	seven	7315	[	9101	company	8756
2036	eight	4927	shipping	7324	injured	9122	officer	9165
2332	to be cont.	4942	water	7409	at	9213	codebook	
2457	slight			7658	men	9327	one	
2486	locked	5393	pilot	7802	UI	9705	sunk	
2550	BU	5439	killed					

The worksheet has now been prepared, and I can have considerable confidence in the line-up of the messages considering the number of repeated groups it has yielded. Presumably, as a relative beginner, I work at first with various redundant and routine expressions which have yielded entry in the past, but somehow, for a long time nothing works. It often happens that the expert scans the fresh worksheet, casually tries several "hunches," and almost immediately succeeds in correctly placing that all-important first plaincode group, and the time would come when that could happen to me too. But, for whatever reason, that doesn't usually work for the beginner or the less gifted. Of the old-timer it may sometimes be said that he uses the sense of smell, or ESP, or just a common old-fashioned hunch, but never of the tyro. Somehow hunches do work, but even then it should be noted that they never work if you deliberately try to use them. A thought occurs to you unbidden, you try something and it works. You had a hunch! But try deliberately to use them, and nothing works. And at first nothing worked for me. No luck! The gifted expert, of course, creates his own luck.

SAMPLE WORKSHEET

	1	2	3	4	5	6	7	8	9	10	11	12	13	14
1.	4694	5717	8023	5237	5664	6779	7017	2510	2981	7503	6633	5752	2851	6395
2.	4694	2118	3874	1378	8724	0550	6844	7149	4232	5933	3699	0430	6023	9087
3.	2305	7512	5792	0476	5552									
4.	9184	7197	5792	3762	0475	4863	7153	1138	9870	0784	8920	6677	0248	3164
5.	7754	1149	9176	0509	7604	5891	3457	3197	1017	5694	6470	3853	3501	9617
6.										6308	6470	4277	4650	8066
7.	4543	8471	5597	7722	8194	5228	5727	7746	2408	7904	5463			
8.				7860	7938	5421	4589	3623	0870	0793	7778	4849	5742	1276
9.	8482	2704	3197	4619	5664	8130	1937	1208	8600	4005	9439	8568	9172	8905

Figure 1. Text superenciphered by means of additives.

To some extent it perhaps only seemed to take so very long. The tyro is easily discouraged. Sooner or later one of the words that should work, does work. In this case let us say that I finally try out the plaincode groups 3443, "the end," and 2332, "to be continued." I run "to be continued" through the first message, and lo and behold at the very end of line 1, I strike paydirt, and I learn how easily the feeling of depression can be dispelled. In line 1, column 14, 6395 - 2332 = 4063, which is then tried as additive for the whole of column 14. Line 2, 9087 - 4063 = 5024 (?). But line 4, 3164 - 4063 = 9101, plaincode for "company." Lines 5 and 8 yield 5654 and 7213, both unknown and therefore meaningless, but line 6, 8066 - 4063 = 4003, "general," and line 9, 8905 - 4063 = 4942, "water." Keep in mind (as I have indicated above) that we are not "carrying" in our addition and subtraction (for example, 6 + 4 = 0).

There was always a good deal of comment, both humorous and serious, between co-workers in the hutment, although each of us worked alone on a worksheet in his own table-space. This rather uninhibited conversation stimulated thought and prevented dullness from setting in, and along with an occasional friendly look over a shoulder at the results being obtained on a worksheet, often led to quite helpful suggestions. I myself, let us assume, looking at the three isolated words, company, general, and water, am wondering, "What next?" A co-worker, however, looking over my shoulder at this point suggests that I ought to try "shipping" in line 4, column 13, before the word "company." 0248 - 4927 gives me the possible additive 6321 and that in turn subtracted from 2851 in line 1, yields 6530, "planes," a good possibility. However, in the rest of the column the plaincode results yield no recognizable words. And now I am unable to proceed further with these words. I must begin the process over again of trying to break into the worksheet from scratch.

I begin trying DEN in line two (without results) but notice the shortness of line 3 (only five groups), and a streak of laziness overcomes me. I shift from line 2 to line 3, and my "silly" hunch pays off. Encicode 2305 minus plaincode 4241 yields the possible additive 8164. None of the rest of the column below yields anything except a "friendly" group in line 7: 4543 - 8164 = 6489, but the repeated group in lines 1 and 2 (4694 - 8164 = 6530) yields the word "planes," a corroborating coincidence for column 13, where line 1 also was deciphered (at least tentatively) to 6530, "planes." Suddenly the world looks brighter. I now have something to work with in adjacent columns. And we may now, with some confidence, inscribe the additive 8164 in column 1 along with the additives 6321 and 4063 in columns 13 and 14.

I enter DAI, 3974, after DEN, in line 3, column 2: 7512 - 3974 = 4648, the additive for column 2. In line 2, 2118 - 4648 = 8570, "attacked." In line 9, 2704 - 4648 = 8166, "boat" (note "water" in column 14, obtained earlier).

Presumably a number follows, but since I do not know the sender, my specialized work on "dendais" does not help me here. For the repeated group in column 3, 5792, we may begin by testing 9327, the numeral "one": 5792 - 9327 = 6475, which subtracted from 5597 in line 7, gives 9122, "officer," as well as 7409, "at," in line 2. As I have indicated, the army method of sending those trained in Japanese to the European Theatre and those in the European languages to the Southwest Pacific, forced us to work with English translations or rough approximations of the various groups' meanings. So "at" was good enough here, as was "Message No." for "dendai." At any rate, we can now enter the additive 6475 in column 3.

Assuming that the message number is not simply "one," I now try for another digit in column 4, beginning again with "one": 0476 - 9327 = 1159. In line 2, 1378 - 1159 yields 0229, "daybreak" (note, "planes attacked at daybreak"), and in line 8, 7860 - 1159 = 6711, kana "TA." With the aid of IBM machines we had been able to list all words previously encountered spelled out in kana–in (English) alphabetical order by first syllable, as well as by second syllable, third, and so forth. We therefore now have the possibility of trying out all such listed words with TA in line 8, as well as the number we have been working on.

Having entered the additive 1159 in column 4, let us assume a message number above one hundred for line 3. Trying the digits in order in column 5, when we reach "eight," we get 5552 - 2036 = 3526, which as the additive gives us the friendly group 2148 in lines 1 and 9, 0475 - 3526 = 7959, "two," in line 4, and more importantly as it turned out, 7938 - 3526 = 4412, kana "MI," in line 8. We may enter 3526 with full confidence as the additive in column 5.

With TA and MI in columns 4 and 5 we find only one listed word to test in the IBM alphabetical run-off: [BI-TA-MI-N B], "vitamin B," giving us, if correct, kana "N" (8216) in column 6, the English letter "B" (2865) in 7, and the closing bracket, "]" (8437) in 8. In column 6, "N" gives us "men" in line 4 after "two," and "HQ" in line 5. We may therefore inscribe the additive 7215 in column 6.

"B" (2865) in column 7 yields 5393, "pilot," in line 1, 5439, "killed" (after "men"), in line 4, 1733, "bombed" (after "HQ"), in line 5, and 9213, "codebook," in line 9, and we may enter the additive 2724 for column 7.

"]" (8437) in line 8 yields 7324, "injured" (after "pilot") in line 1, the friendly group 6942 in line 4, 2550, kana "BU" in line 7, and we are warranted in entering the additive 5296 in column 8. For kana "BU" the IBM alphabetical list gives as one possibility "BU-RA-TSU-KU UI-DO," "Black Widow," the name of the plane, which seems like the best choice.

Kana "RA" in column 9 yields the friendly group 7340 in line 4, and in line 9, 6170, which has been assumed to be the plaincode for "metal," perhaps insufficient reason for confidence. We try TSU in column 10 and again get a friendly group, 8250, in line 1, and 5752 in line 9, a possible group for "box." More careful consideration would add to our confidence, but for the moment we try KU in column 11, the last group in that message, hoping for stronger corroboration. (As for its being the last group in the message, please remember that these messages have been cut in two at some arbitrarily-chosen intermediate point, and that then the end has been placed before the beginning.)

Placement of KU gives us 1648, "damage," for the repeated group, 6470, in lines 5 and 6, an encouraging result since the word is frequently encountered in these messages because of the importance of face-saving in the Nipponese culture. I first try 3871, "none," in column 12, line 5, without satisfactory results, but when I try it in line 6, I get 2457, "slight," in line 5: "Damage none" in one message and "Damage slight" in the other.

Returning to columns 9 and 10, I find that my results for the two columns when considered together were quite corroborative of my assumptions, for we have a metal box in line 9 associated with the plaincode group 9213, "codebook" in column 7, altogether an interesting and suggestive combination. We may enter the additives 2530, 9353, 5832, and 1406 in columns 9-12 and thus complete this mini-illustration of our work in the field with the additives. But I must repeat my cautionary reminder that in actual practice, in the early days of the work, we met so much success only working with a number of worksheets–not on a single one, and certainly not on only part of one.

Once in the field when this context of a codebook locked in a metal box appeared in a message, it turned out that the boat (see line 9, column 2) was on fire and doomed and that the message was intended to inform the authorities that the codebook had been placed in a metal box which was both locked and carefully bound before being thrown overboard. For like ourselves the Nipponese knew that the best way to solve a code was to get hold of the codebook one way or another! By placing it in a strong metal box, binding the box and locking it, and then throwing it overboard in very deep water, they were able to assure themselves that we could not possibly get our hands on that copy of the codebook.

Conversely, since we knew as well as they did that the most successful method of solving a code was to get hold of the codebook, we actually had in the hutment in which I worked at Ascot Park, a large map of the area stuck with colored pins, each denoting a spot where we knew that a codebook had been sunk or where a ship had gone down which might have contained a codebook. Unfortunately the ocean in the area was much too deep as a rule for any of our "frogs" to be able to go down and bring up one of these codebooks. As a rule! But I do remember a conference with the Chief Signal Officer, SWPA, in which we studied all these pins and their locations and found one where the water was only a little too deep for a diver. The CSO pointed to it and said, a bit dramatically to be sure, "We'll get that one." And a diver did go down in a successful bid for that codebook, which he brought up locked inside the metal box as described!

It turned out that even though the "fireproof" box had kept the code and additive books from burning up in the fire that sank the ship, the intense heat of the flames had charred the pages into solid black sheets on which not a thing was visible. The technology, of course, had long been available for bringing the print on such pages back automatically, and one of our technicians demonstrated how with a camera (infra red, I presume) he could photograph an entire page and bring the contents out quite clearly. However, we did it painstakingly by rubbing the "now-you-see-it-now-you-don't" print with a piece of cotton soaked in an alcoholic liquid. I remember doing my share of this work to some of the additive pages, one group at a time. As long as the spot was wet the numerals remained quite clear, although they became invisible again quickly enough–but not before I had called them off to a T-5, who recorded them in the additive book we were re-creating.

	1	2	3	4	5	6	7	8	9	0	= plaincode
1 /	4	0	3	5	9	8	1	7	6	2	
2 /	1	3	6	8	0	9	7	5	2	4	
3 /	5	7	9	2	4	1	8	3	6	0	
4 /	0	2	5	6	4	3	7	8	1	9	
5 /	9	1	4	0	6	8	5	2	7	3	= cipher
6 /	3	6	1	4	9	0	2	4	8	7	(encicode)
7 /	1	7	9	5	2	4	3	8	0	6	
8 /	3	8	4	6	7	9	5	0	2	1	
9 /	4	1	2	3	8	0	7	6	9	5	
0 /	2	4	0	9	3	7	6	1	5	8	
Key											

Figure 2. Square used in encipherment.

PART II

As I have pointed out, the Nipponese knew that we would be making every effort to read their (radio) mail and did what they thought necessary to keep us from being able to do so. At some point they had begun to realize that we were being successful in our efforts, that the superencipherment by means of additives was insufficient to render us impotent. The "Yanks" had turned out to be a harder nut to crack than they had expected and, it appeared, smarter too! They would have to work out a more difficult and safer system–and they did. We became aware of this only when the additive system suddenly no longer worked for us.

The additive books were still being used for the superencipherment, it turned out, but the digits of the numerical groups were now being passed through a square, through a sort of numerical Vigenere. (See Figure 2.) Each row had all ten digits arranged in random order (if we may take the liberty of calling zero a digit). The rows and the columns were each numbered in order from zero to nine (or one to zero). The column numbers were the plaincode numerals, the row numbers those of the keys (the former additives), and the digits inside the square (at row-and-column intersection) the encicode numerals.

It was a simple enough matter for the Japanese cryptographer to take a plaincode group, such as 8352, and encipher it with a keygroup, such as 9716, thus: row 9, column 8 = 6; row 7, column 3 = 9; row 1, column 5 = 9; row 6, column 2 = 6: encicode 6996. It was equally simple for him to decipher encicode 6996 with the key 9716, thus: 6 in row 9 = 8, 9 in row 7 = 3, 9 in row 1 = 5, 6 in row 6 = 2: plaincode 8352. However, it was much more of a nuisance for us working without the key to attempt to decipher the encicode, for as you can see by looking down the columns in Figure 2 almost every column has repeated digits, with the result that even assuming the correct plaincode group will not necessarily give you the key, but may rather give you several possible keys. If you look for the cipher digits 6, 9, 9, 6 in columns 8, 3, 5, 2 respectively, you will find two 9's in both columns 3 and 5, and have to work with the possible keygroups, 9316, 9366, 9766, and 9716. Often there were many more choices than that.

The technique involved in the cryptanalysis remained essentially the same as I described it with the additives, except for this new method of finding and using the key. The use of the square did not stop us although it was a nuisance which did slow us down considerably. In addition to that, the enemy had foisted upon us the additional task of solving the square itself before we could work with it. And showing that they knew well enough what we were up to, they began to introduce new squares at a fast pace.

I did not myself ever work on the solving of the squares themselves but did spend a great deal of time with them deciphering messages and extracting keys. There is nothing additional I can offer of actual material from the war period, and it would take a whole cryptanalytic outfit to demonstrate methods of solution working with several books of 10,000 keys and a 10,000 group codebook. I do, however, want to demonstrate, insofar as I can, the decipherment and cryptanalysis involved in working with squares of this sort. I think I can illustrate that with a simple problem based on one which someone else prepared for me, one simple enough so that I was able to solve it.

In 1946, with my World War II military service and cryptanalytic activity well behind me, I met someone in Chicago whom I scarcely knew, but who seemed to know me well enough, and who had been engaged in similar work. He sent me some humorous (English) encicode problems he had constructed using numerical code groups superenciphered in the manner I have described. I left Chicago shortly afterward and never heard from him again. Included were some explanatory materials and a quasi-code book of perhaps a thousand numerical groups with meanings he had assigned them.

Insofar as these problems in English used additives, they were quite simple to solve. Of two problems, however, using squares for the superencipherment, I solved only one, which, I must admit, was made easier for me because I knew what sort of thing to expect from my earlier solutions of the simpler problems. This problem is the only one I have been able to locate–nor have I found the quasi-code book.

Almost a decade ago when I decided to discuss this work for the benefit of the American Cryptogram Association I studied this worksheet in an attempt

9259 7156 5138 4084 9139 6232 8240 1574 6500 0650 3652 0363 1503
3988 2037 5707 8023 3246 8540 8526 7818 8957 7667 6221 1329 4441
4247 1123 0531 7171 0383 3270 3729 2992 2614 5243 2855 3224 9453
6656 7112 7677 8209 9279 1974 0048 3159

3310 8309 1057 0323 5616 2235 4220 0179 9031 2304 5061 8229 2557
3706 0628 6285 6672 2616 6787 2097 1228 3009 2940 9485 9318 5951
8208 3892 8035 4106 5721 9616 5729 5820 0202 4108 9673 9293 1029
5719 2841 6927 9074 9856 0192 2162 8602

9262 4257 6507 1560 8987 0160 4720 7914 9658 0003 8587 8049 7751
2097 3394 2764 2427 7097 1897 4267 0840 3374 5510 9517 5404 1246
6266 8737 0167 6636 8358 9861 7672 5286 6856 4556 3801 4187 2083
1760 2137 9487 7096 8529 7483 6556 9646 0323

3440 3688 4288 7303 3413 8162 6059 3136 0767 3033 1984 9631 0014
5243 1250 7516 9361 6510 0168 6434 7251 7226 8479 1040 0686 4939
2411 8844 7724 0581 9377 6523 4663 3651 1283 2050 5479 0776 9119
4533 7026 9582 7768 8900 0650 9172 5925 5025

1003 3869 5222 4034 6748 1176 0014 5739 0179 7215 8114 7597 1729
0698 7448 1010 4208 8587 0077 0734 2814 3659 0392 0164 0844 4731
0742 6780 0690 8465 5968 8820 1203 8179 8610 0723 7712 5266 0919
5980 3009 0766 3919 2492 3785 0212 7121

4557 1950 3598 6764 1139 7190 1903 1980 0521 6735 2344 4881 9689
5488 0735 9459 6768 7822 2110 0603 1762 2082 5612 0414 7816 6795
6812 0882 3770 2551 2515 2357 8384 1787 6046 9253 6139 9742 6125
1401 6179 0473 4012 6688 0137 6550 6656 0217

0179 6011 3344 9124 8357 5383 0187 6688 4644 4555 5761 3491 5996
0478 0615 4126 1987 6120 1326 5814 8386 4798 2110 0767 8472 5921
6737 2331 7751 1801 8191 4942 4430 9147 7929 5721 8194 8023 4230
5986 4620 7537 4026 2523 9919 3357 0122 2207

5779 7684 6941 5929 0089 6536 4051 6392 7099 5358 2428 1564 9104
8353 2743 4981 4350 7984 2351 2552 6035 9153 0661 0159 8286 9784
0343 7620 9750 7614 2001 3330 6829 2781 2511 0567 3252 1462 1374
0743 2084 6615 4586 5688 5939 1784 7312 0092

2229 2821 6656 4395 5286 3847 6927 7608 3219 1187 6528 2930 1421
4469 4897 3310 2720 6146 7324 4720 7256 6012 4438 9874 0989 4410
7243 3068 5579 3979 0615 6290 6013 0343 7121 8524 6737 2122 7983
8776 6768 6917 0690 4648 9982 9547 0846 5305

4620 9758 8286 7637 0138 2931 9076 4557 1950 4412 1511 5000 7471
6285 5586 5968 3090 8285 0877 1287 9985 5836 2127 0040 5249 3642
6167 9568 1290 3061 9935 4642 8224 0261 9654 2971 3655 0332 2350
0323 3023 5658 7878 8403 3276 8004 9233 7848

Figure 3. The mixed bag. Ten parts of a message in encicode.

	1	2	3	4	5	6	7	8	9	10	
1)	3440	3688	4288	7303	3413	8162	6059	3136	0767	3033	(1
2)	9856	0192	2162	8602				3310	8309	1057	(2
3)	8194	8023	4230	5986	4620	7537	4026	2523	9919	3357	(3
4)	3061	9935	4642	8224	0261	9654	2971	3655	0332	2350	(4
5)	5612	0414	7816	6795	6812	0882	3770	2551	2515	2357	(5
6)	5707	8023	3246	8540	8526	7818	8957	7667	6221	1329	(6
7)	3219	1187	6528	2930	1421	4469	4897	3310	2720	6146	(7
8)	7684	6941	5929	0089	6536	4051	6392	7099	5358	2428	(8
9)	3785	0212	7121				1003	3869	5222	4034	(9
10	4187	2083	1760	2137	9487	7096	8529	7483	6556	9646	(10

	11	12	13	14	15	16	17	18	19	20	
1)	1984	9631	0014	5243	1250	7516	9361	6510	0168	6434	(1
2)	0323	5616	2235	4220	0179	9031	2304	5061	8229	2557	(2
3)	0122	2207			0179	6011	3344	9124	8357	5383	(3
4)	0323	3023	5658	7878	8403	3276	8004	9233	7848		(4
5)	8384	1787	6046	9253	6139	9742	6125	1401	6179	0473	(5
6)	4441	4247	1123	0531	7171	0383	3270	3729	2992	2614	(6
7)	7324	4720	7256	6012	4438	9874	0989	4410	7243	3068	(7
8)	1564	9104	8353	2743	4981	4350	7984	2351	2552	6035	(8
9)	6748	1176	0014	5739	0179	7215	8114	7597	1729	0698	(9
10)	0323			9262	4257	6507	1560	8987	0160	4720	(10

	21	22	23	24	25	26	27	28	29	30	
1)	7251	7226	8479	1040	0686	4939	2411	8844	7724	0581	(1
2)	3706	0628	6285	6672	2616	6787	2097	1228	3009	2940	(2
3)	0187	6688	4644	4555	5761	3491	5996	0478	0615	4126	(3
4)		4620	9758	8286	7637	0138	2931	9076	4557	1950	(4
5)	4012	6688	0137	6550	6656	0217			4557	1950	(5
6)	5243	2855	3224	9453	6656	7112	7677	8209	9279	1974	(6
7)	5579	3979	0615	6290	6013	0343	7121	8524	6737	2122	(7
8)	9153	0661	0159	8286	9784	0343	7620	9750	7614	2001	(8
9)	7448	1010	4208	8587	0077	0734	2814	3659	0392	0164	(9
10)	7914	9658	0003	8587	8049	7751	2097	3394	2764	2427	(10

	31	32	33	34	35	36	37	38	39	40	
1)	9377	6523	4663	3651	1283	2050	5479	0776	9119	4533	(1
2)	9485	9318	5951	8208	3892	8035	4106	5721	9616	5729	(2
3)	1987	6120	1326	5814	8386	4798	2110	0767	8472	5921	(3
4)	4412	1511	5000	7471	6285	5586	5968	3090	8285	0877	(4
5)	3598	6764	1139	7190	1903	1980	0521	6735	2344	4881	(5
6)	0048	3159				9259	7156	5138	4084	9139	(6
7)	7983	8776	6768	6917	0690	4648	9982	9547	0846	5305	(7
8)	3330	6829	2781	2511	0567	3252	1462	1374	0743	2084	(8
9)	0844	4731	0742	6780	0690	8465	5968	8820	1203	8179	(9
10)	7097	1897	4267	0840	3374	5510	9517	5404	1246	6266	(10

	41	42	43	44	45	46	47	48	49	50	
1)	7026	9582	7768	8900	0650	9172	5925	5025			(1
2)	5820	0202	4108	9673	9293	1029	5719	2481	6927	9074	(2
3)	6737	2331	7751	1801	8191	4942	4430	9147	7929	5721	(3
4)	1287	9985	5836	2127	0040	5249	3642	6167	9568	1290	(4
5)	9689	5488	0735	9459	6768	7822	2110	0603	1762	2082	(5
6)	6232	8240	1574	6500	0650	3652	0363	1503	3988	2037	(6
7)			2229	2821	6656	4395	5286	3847	6927	7608	(7
8)	6615	4586	5688	5939	1784	7312	0092			5779	(8
9)	8610	0723	7712	5266	0919	5980	3009	0766	3919	2492	(9
10)	8737	0167	6636	8358	9861	7672	5286	6856	4556	3801	(10

Figure 4. The worksheet. The parts aligned under their keys.

to recall how I went about solving the problem. In the process I was able to reconstruct a part of the codebook, but after that I was too familiar with the worksheet to be able to demonstrate how one might go about solving the problem from scratch, since I would have been remembering the decipherment of many of the numerical groups. To illustrate a step at a time what might happen in attempting a solution I needed material I would not recognize. To obtain such material, the steps of whose solution might integrate with those of my successful solution in 1946, I re-enciphered the plaintext with modifications in the encoding as well as in the text itself, and, of course, with a totally new set of keys. And in order to keep the numerical groups meaningless, I enciphered only one of the digits at a time, as, for example, all the 9's, wherever I found them. In other words, I did not work at any time with complete groups, either key or code.

After the enciphering was complete, I copied off the encicode text of the messages on separate sheets, mixed them up, and put them away for several months. (See Figure 3.) Thus, before I could begin to re-solve the messages, I was faced with the task of lining them up correctly, as I had been in 1946, with all groups of each column under the same (though unknown) key. (See Figure 4, and note the repeated groups within the columns.) I then proceeded to re-solve the encicode messages without resorting to any of their special context–which I might have remembered. Instead of assuming the sort of text I had assumed for the Japanese example, of course, I assumed English terms, but only of the sort I had assumed in 1946, when I had begun with no idea of the special content. Undoubtedly it was an easier task this second time around, and perhaps that was fortunate for my special purpose, for which success was mandatory. It is interesting to note, however, that any memory I might have retained of the fifty keys I used would have proved absolutely worthless since insofar as I know there is no way to reproduce the original order of the rows of the square, and so any keys obtained by cryptanalysis would bear no resemblance whatever to the original keys used in the enciphering. (The same phenomenon, of course, held true for additives too as long as every bit of our knowledge of the system had been obtained by cryptanalysis and there had been no opportunity of correcting our original assumptions via an authentic [original] code or additive book.)

I began then, as I had in 1946, by looking for some of the inevitable redundancies, for such common words as "the, and, of, to," and such grammatical endings as "-ing, -ly, -s, -d," etc. And although the bisection technique helped hide the redundancy of the beginning of each message, the weakness inherent in the use of groups for identification purposes proved as helpful to me here as they did for us in the Southwest Pacific during World War II.[8]

[8]See Crypt-Australia, 1980, p. 12, or 1985, p. 340. See also Kahn, s.v. "Redundancy."

It may be thought that working with a small codebook (about 1000 groups instead of 10,000) would make the problem simpler, and to some extent that is undoubtedly true. But I rendered the question moot in 1946 by not even working with the thousand. Instead I went through the list picking out the commoner "words" and prepared a list of only a few hundred. It was with this and even smaller lists that I worked, rather than with the thousand, looking for words most likely to appear in any text–especially during the earlier, more difficult stages, of the solution. And I lined up these smaller lists in numerical order by first, second, third, and final digits–for obvious reasons. For example, when I noted that in column 19, rows 1 and 10, I had 0168 and 0160, I wasted little time finding in my shorter list the excellent fit, plaincode 2882, "-ing" and 2888, "-ment." In column (C) 10, rows (R) 3, 4, and 5, I had 3357, 2350, 2357, and here too my list suggested 2882 and 2888 for the latter two, although it showed more than one possibility for the first (but did include the correct plaincode group). I merely made a note of this at the time, but I mention it because it turned out that "-ing" and "-ment" were correct both times.[9]

I will not waste time on the tentative assumptions I made which proved futile after I had lined up the ten messages as shown in Figure 4. What I sought to begin with were two (or more) assumptions which seemed to corroborate each other, and which would enable me to place two or more digits in the same row (K) and/or the same column (PC) of the (so far empty) square. I paid especial attention to columns containing repeated groups and this finally paid off in column (C) 45. Note 0650 in rows (R) 1 and 6 of C45, 6656 in R7, and 0040 in R4. Compare especially 6656 and 0040. I assumed that the key number in positions 1, 2, and 4 was identical and that one of these groups was plaincode (PC) 9999, "the." Experimenting with the three groups, I found that if 6656 were PC9999 and 0040 were PC2222, ":" (colon), 0650 would automatically become plaincode PC2992, "-ly." Curiously 6656 reappeared *as a repeat* in C25, R5 and 6. I assumed that here too we had PC9999, "the," and that since presumably no two key groups were identical, at least one key number must be different of the key groups of C25 and 45. The odds were, of course, 3 to 1 on the 6, and so I assumed a 6 in two rows (K) of the square in column (PC) 9 (see Figure 5A).

Note carefully which are the key groups, which the encicode groups, and which the plaincode groups (column 45):

---

[9] **In order to avoid confusion, since I will be referring constantly to the rows and columns of both the messages and the square, I will refer to the rows of the *square* as "keys" (symbol K) and the columns as plaincode (symbol PC). Rows (symbol R) and columns (symbol C) will refer to the aligned messages on the worksheet (Figure 4). The solved groups will also be referred to as keys (K) and plaincode (PC). EC will designate encicode (numeral) groups.**

Key (K):	2232	2232	2232
Encicode (EC):	6656	0040	0650
Plaincode (PC):	9999	2222	2992

Note again that the numbering of my rows is arbitrary. And since even if our assumptions are correct we do not know which 6 is in row (K) 1, we may write the key for C25 as 1/2 1/2 3 1/2, where the solidus or slash mark may be read as "or."

	1	2	3	4	5	6	7	8	9	0	= plaincode = (PC)	1	2	3	4	5	6	7	8	9	0	
1 /									6						2					6	0	/ 1
2 /		0							6		encicode		0			9				6		/ 2
3 /		4							5		(EC)		4			3			1	5	8	/ 3
Key																						/ K
					A											B						

Figure 5. Genesis of the square; first three rows.

In C45, R9, we have 0919 = 2 x y x. With one digit known there were not many groups to fit the pattern in my high frequency list–actually seven all told: "the, colon, E, G, -ing, -(e)D, -ment," but with all plaincode 2's and 9's already used in C45 I could eliminate these four: "the, colon, -ing (2929), E (2525)." Since the two best possibilities (PC2585 = "(e)D" and PC2888 = "-ment") have a common 8 in the third position, I added 1 in K3 under PC8. Assuming PC2585 = "(e)D," I pencilled 9 lightly in K2 under PC5. I was, of course, at all times ready to erase an assumption as necessary and used my eraser freely.

In C25, R1 and 2 we have the same pattern: EC0686 and 2616, but this time we know (by assumption, at least) that the repeated 6's are PC9's: x9x9. I found six possibilities: 0909 = "begin," 4989 = "P," 9919 = "red," 2929, 1999, and 9999. The last three have PC2 or 9 in the third position (K3), and so may be eliminated. 1 in K3 is PC8 and so PC4989 is "P" if our previous assumptions are not at fault. That leaves the likely PC0909 = "begin," for EC0686 (or 9919 = "red"). In either case the first key number cannot be 2, for 0 in K2 = PC2. Therefore change the key for C25 to 1 1/2 3 1/2. So for PC4989, "P," I placed 2 in K1 under 4, and assuming 0909, "begin," for EC0686, I placed a lightly penciled 0 in K1 under 0, and 8 in K3 under 0. For the inclusion of these assumptions in the square see Figure 5B.

Since in 1946 I had found the search for "the" fruitful I had tried PC9999 in other places before meeting with the success already described. C11 and 15 contained the only double repetitions, EC0323 and 0179 respectively, and I had assumed 9999 for 0323 without success. I had, however, experimented with a (blank) square and filled in a number of cells based on the assumptions that 0323 (C11) = PC6505, "to," and 0179 (C15) = PC9999, "the." Returning to this square now, where EC0323 = PC6505, "to," had led to EC7324 = PC7502, "seven," in R7, corroborating my placement of the seventh part of the message, I found that I could rearrange the K-rows of this square to mesh with the other square without any conflicts.

A

PC	1	2	3	4	5	6	7	8	9	0
1 /		4			3				5	
2 /		7							6	
3 /		2							1	
4 /		0							6	
5 /									0	
6 /									7	
7 /									9	
8 /										
9 /										
0 /										
Key										

B

PC	1	2	3	4	5	6	7	8	9	0
1 /		7		2					6	0
2 /		0			9				6	
3 /		4			3			1	5	8
4 /									0	
5 /									9	
6 /		2							1	
7 /										
8 /										
9 /										
0 /									7	
Key										

C

PC	1	2	3	4	5	6	7	8	9	0
1 /		7		2					6	0
2 /		0			9			7	6	2
3 /	2	4			3			1	5	8
4 /		7							0	6
5 /		1				0	7		9	
6 /	0	2						8	1	
7 /										
8 /										
9 /										
0 /									7	3
Key										

Figure 6. The square: first assumptions.

I assumed, therefore, R2, C11 and 12, EC0323 5616, to be PC6505 9999, "to the," and R7, EC7324 4720, to be PC7502 2222, "seven, colon" (a standard redundancy), along with the assumption that the triple repeat in C15, EC0179, was PC9999, "the." Omitting a few of my riskier assumptions, and assigning the K-rows arbitrarily, this gave me the results shown in Figure 6A. Comparison with Figure 5B shows that this row K1 is our previous row K3, K4 is K2, and K2 is K1. Combining the two squares we obtain the square shown in Figure 6B.

The new keys then are C11, K[5]323, C12, K3162, and C15, K4605. Comparison of EC1564 (C11, R8) = PCx992 and EC7171 (C15, R6) = PCx99x, shows that the assumption that in K5, EC1 = PC2, will give us PC2992 = "-ly" for both. We may add EC7 under 2 in K4 and EC1 under 2 in K5. I also assumed for EC6139 (C15, R5) = PCx9x9, PC0909, "begin." That added EC6 under 0 in K4, and EC3 under 0 in K0. We may also add the 0 under 6 in K5 and 2 under 0 in K2, both from our PC6505, "to," in C11, as well as 7 under 7 in K5 from our PC7502 = "seven" in R7.

In C11-12, R3, EC0122 2207 = PC680x x4xx, the assumption of EC2 = PC1 in K3 gives us PC6801 14xx, "when," followed by the possible excellent context PC1418, "ever," which I assumed, placing 0 under 1 in K6 and 7 under 8 in K2. For EC1787 (C12, R5) = 82x8, I assumed 8288, "possible," and placed 8 in K6 under 8. (See Figure 6C.)

For R4, C12, EC3023 = PC502x I confidently assumed "an" after "to." Usually three sure digits out of four provided an easy choice, but obviously not always, for "an" turned out to be wrong. I mention this failed assumption because of the confidence with which I assumed it. Actually I frequently had to try out and drop assumptions for which at the time I could not find sufficient corroboration, including some which later turned out to be correct. Note R7, C10, where I assumed EC6146 = PC4609, "part," preceding "seven," but dropped it when I found no way to place the assumed values in their K-rows even though EC2720 in C9 worked well for PC0909, "begin." "Begin part 7" was, of course, a type of redundancy which we found very helpful in the field during the war.

We have only one place where two consecutive groups are repeated, C29 and 30, R4 and 5, EC4557 1950. Experimenting with these two groups I found that if I assumed 1950 = 9999, "the," and the key for C30, K6534, EC2940 (C30, R2) became PC2929, "-ing." EC1974 = PC99xx (R6) compared with EC0164 = 12xx (R9) showed that if EC4 = PC6 in K4, the first could be PC9966, "without," and the second PC1276, "period," with the placement of 7 under 6 and 6 under 7 in K3, without creating a conflict in the now well-filled row K3. (Obviously the odds were that wrong assumptions would produce one or more conflicts in a well-filled K-row.)

In R10, for EC2427 = 2x12, I assumed PC2112, "his," and placed 4 in K5 under 1. At this point I had a bit of serendipity. Debating whether to assume that EC4126 = PCx210 (R3) = 5210, "situation," I tried the group preceding it in C29, EC0615, as 9999, "the," for "the situation," and found that this assumption gave me K4163 for C29, which in turn gave me PC2222, "colon," in R1 for EC7724, and PC2992, "-ly," in R8 for EC7614. In R2, preceding "-ing," I needed a verb for EC3009 = x01x and assumed PC7014, "love," which gave me 3 under 7 in K4, 9 under 4 in K3, without a conflict, and, of course, 0 under 3, completing the row K3.

I refrained from making a final decision on C29, R4, EC4557 = 6xx6 = PC6136?, "of," preceding "the", and on C30, R1, EC0581 = 1x0x = PC1001?, "according," R3, "situation," R7, EC2122 = 221x = PC2215?, "de," and R8, EC2001 = 263x = PC2631?, "pro-." Instead I went back to C9 and 10, R7, to the possible assumption for EC2720 6146 of PC0909 4609, "begin part," preceding "seven." As indicated earlier it was in C10, R3, 4, and 5 (EC3357, 2350, and 2357) that my list suggested PC2882 and 2888, "-ing" and "-ment." K2 in the fourth position (EC6 = PC9) gives EC0 = PC2 and EC7 = PC8, and so we can assume for R3, 4, and 5, PCx888, 2882, "-ing," and 2888, "-ment." EC2 = PC2 gives us K6 in position 1, and so we can place 6 under 4 in K6. With "-ment" eliminated, I assumed R3, EC3357 = PC6888, "haste," and placed 3 under 6 in

K6.

I now had EC4257 = PC62x7 (C15, R10), for which I assumed PC6277, "on," and I placed 5 under 7 in K0. This gave me EC1250 = x276 in R1, and I assumed PC1276, "period," and placed 1 under 1 in K4. C10, R2, EC1057 = 90x8 became PC9088, "example," since EC5 = PC8, although not yet assigned to K7. I did now assign the third position to K7, placing 5 under 8 and from EC6146 = PC4609, "part," 4 under 0 in K7. For R10, EC9646 = x909, I assumed PC0909, "begin" (before "to"), and placed 9 under 0 in K6.

After C9, R7, "begin" (EC2720 = PC0909), I assumed K2024 for C9. Then R6, EC6221 = 9x01 = PC9801, "(e)D," yielded 2 under 8 in K0; R10, EC6556 = 97x0 = PC9740, "-ence," 5 under 4 in K2. R9, EC5222 = 480x = PC4805, "be," placed 2 under 5 in K4; R8, EC5358 = 404x = PC4048, "mistake," 8 under 8 in K4. For R5, EC2515 = PC07x 3/4 (i.e., 3 or 4) I assumed PC0774, "arrange," before "-ment" in C10. I placed 1 under 7 in K2, 5 under 4 in K4, and added 9 under 3 to finish K4. For R4, EC0332 = 20x5 I assumed PC2035, "refuse," before "-ing" and placed 3 under 3 in K2.

In C10, R1 (K6172), I now had EC3033 = 60x3 and assumed PC6023, "one," giving me 3 under 2 in K7. But in C9, R2, I had EC8309 = x023, and so I recognized "one" again: 6023. I placed 8 under 6 in K2 and added 4 under 1 to finish K-row 2. (See Figure 7A for the current state of the square.)

I now returned to the colon in message 1 (C29). If my placement was correct, C27 and 28 should be (EC2411 8844), "part one": PC4609 6023. Positions 2 and 3 of the key for C28 seemed to be K3, but positions 1 and 4 gave me trouble. I finally put 4 under 3 in K8 (position 4). EC8 = PC6 in K2 gave me a conflict, but compared with 8844 and using Kx338, EC8524 in R7 gave me PC6913, "this." EC8209 in R6 = 613x gave me a satisfactory PC6136, "of," and I placed 9 under 6 in K8. Also EC2931 in R4, C27, compared with EC2411, gave me 43x9, which I assumed to be PC4319, "were." So I accepted "part one" as correct and placed 3 under 1 in K8.

Then, assuming that the repeated group in C27, EC2097 = 49xx (with the assumption EC7 = PC7 in K6) is PC4967, "and," I labelled the two keys (C27 and 28) K1486 x338. PC4609, "part," gave me 1 under 0 in K8. C27, R3, EC5996 = x364 I assumed as PC1364, "they," and placed 5 under 1 in K1. In R7, EC7121 = 21x9 gave me PC2199, "how," and I placed 2 under 9 in K8. R9, EC2814 = 480x I assumed as PC4805, "be," and placed 4 under 5 in K6, and then added 5 under 3 to finish K6.

EC3394 in C28, R10, = x543, I assumed as PC0543, "so," and so EC8 = PC6 should probably have been placed in K0 for position 1 of this column (C28) rather than in K2. Checking all the values down this column with K0 in the first position, I found no conflicts and so assumed 8 under 6 and 9 under 1 in K0. K0338 for C28 allowed me to assume EC9076 in R4 = PC1364, "they," already shown to be the subject in R3, C27, and placed 6 under 4 in K8 as well as corroborating the 9 under 1 in K0.

If we go back to C25, K1 1/2 3 1/2, we can now correct the key to K1131. In R4, EC7637 = PC2952, "other." In R3 we may assume EC5761 = PC1276, "period," and place 1 under 6 in K1. The EC7 = PC2 in the second and fourth positions corrected the 1/2 in the key to 1 in each case. So it turned out that while K2232 gave us EC6656 = PC9999, "the," in C45, all three 6's in C25 derived their 9's from K1, and the complete key for C25 turned out to be K1131. In R7 EC6013 = 908x became PC9088, "example," and I placed 3 under 8 in K1.

In C25-29, R1, we now have PC0909 xxxx 4609 6023 2222, "begin xxxx part one:". For EC4939 in C26 I decided to try "message" (PC1600), giving me K5/2 806 for the key to C26. This yielded EC0343 = PC6136, "of," in R7 and 8, correcting the key to K5806, and placing 4 under 3 in K0. EC0734 in R9 gave me PC6505, "to," placing 7 under 5 in K8. EC3491 in R3, becoming PC4319, "were," placed 3 under 4 in K5. "Were they" in R3, C26-27, helped corroborate and make firm the "were they" in R4, C27-28. I was, of course, usually aware of such corroboration before I added a new value to the square, since I was frequently working on more than one column at a time–as well as trying out and discarding alternate values here and there.

Returning to C15, R5, PC0909, "begin," I assumed that it was probably followed in C16-18 by PC4609 5174 2222, "part five:". The first corroboration I found was in C18, where EC1401 = 2222, ":," gave me K5325, which in turn gave me EC9124 = PC9801, "(e)D," in R3, and EC3729 = PC4609, "part," in R6. I could assume then in R6, C17-20, EC3270 3729 2992 2614 = PC0909 4609 0499 2222, "begin part six:". (Lest this be considered as something which deliberately made this problem easier, let me point out that in the field during the war the phenomenon of redundancies multiplying redundancies proved a constant help, because of the necessity of the sender, as I have shown in the earlier part of this article, to identify the system, the sender, and the message and part numbers, and to use such groups as "begin, message, the end, to be cont.," necessary punctuation, and so forth. Note that this problem has the redundancies of English grammatical verbiage.)

EC3270 = PC0909 gave K0794 (C17), for 7 under 0 can by this time be placed only in K9. EC9361 (R1) gave PC 1221, "very," placing 6 under 2 in K9. EC6125 in R5, already assumed as 5174, "five," placed 6 under 5 in K0, 1 under 1 in K7, and 2 under 7 in K9. (See Figure 7B for the current state of the square.)

= PC = (columns); = EC (rows)

Key	1	2	3	4	5	6	7	8	9	0
1 /		7		2					6	0
2 /	4	0	3	5	9	8	1	7	6	2
3 /	2	4	0	9	3	7	6	1	5	8
4 /	1	7	9	5	2	4	3	8	0	6
5 /	4	1				0	7		9	
6 /	0	2		6		3		8	1	9
7 /		3						5		4
8 /										
9 /										
0 /							5	2	7	3

A

Key	1	2	3	4	5	6	7	8	9	0
1 /	5	7		2		1		3	6	0
2 /	4	0	3	5	9	8	1	7	6	2
3 /	2	4	0	9	3	7	6	1	5	8
4 /	1	7	9	5	2	4	3	8	0	6
5 /	4	1		3		0	7		9	
6 /	0	2	5	6	4	3	7	8	1	9
7 /	1	3						5		4
8 /	3		4		7	9			2	1
9 /		6					2			7
0 /	9		4		6	8	5	2	7	3

B

Figure 7. Square recovery.

EC8114 in R9 (C17) = PC6136, "of," placed 1 under 3 in K9. In C16, R5, EC9742 = PC4609, "part," produced K3378?, which was corrected to 3377 by R2, EC9031 = PC4321, "well," and R6, EC0383 = PC3542, "to be cont." I placed 8 under 4 and 2 under 9 in K7. Note that "to be cont." immediately precedes "begin part 6." Obviously what we have here is the beginning of the message immediately following its last group. This phenomenon was a constant in our work during the war because of the bisection technique used to hide the redundant groups by cutting the message at some arbitrary point and placing the latter part before the beginning.

C16, R8, EC4350 = PC2585, "(e)D," placed 0 under 5 in K7. R10, EC6507 = PC7957, "a," placed 7 under 7 in K7. EC2992 = PC0499, "six," in C19, R6, gave me K235 7/8 for C19. This was corrected to K2358 by R1, EC0168 = PC2882, "-ing," and R10, EC0160 = PC2888, "-ment." This enabled me to place 6 under 8 in K5, 8 under 2 and 0 under 8 in K8. R2, EC8229 = PC6136, "of," placed 2 under 3 in K5, and R8, EC2552 = PC0909, "begin," placed 5 under 0 in K5 and enabled me to add 8 under 5 to finish the K-row. We can also assume here in R8, EC6035 9153 0661 = PC 4609 3285 2222, "part eight colon."

C20, R6, EC2614 = PC2222, ":," and R8, EC6035 = PC4609, "part," supplied K6903 for C20 and placed 1 under 2 in K0 and 0 under 6 in K9. I added 0 under 4 in K0 to finish that K-row. Continuing with C20, EC5383 in R3 = PC3165, "place," and EC0473 in R5 = PC1495, "would," placed 3 under 1 and 4 under 4 in K9 without creating a conflict. Moving on to C22, R8, EC0661 = PC2222, ":," gave K2995, with which the repeated EC6688 became PC9295, "upon," and placed 8 under 9 in K9. R6, EC2855 = PC0980, "design," yielded 5 under 8 in K9, and I added 9 under 5 to finish K9.

In C30 (K6534), R1, I had EC0581 = PC1001, "according." Moving to C31, EC9377 = PC6505, "to," was an obvious assumption and could now give the manageable K 7/8 398. Several groups in the column corrected the key to K7398, such as R5, EC3598 = PC2952, "other," and I placed 9 under 6 in K7 (and 6 under 3 to finish K7). In C10 (K6172), R8, EC2428 = PC2596, "N," added to the preceding "mistake," gave me "mistaken," and I placed 4 under 5 in K1. Then moving back to C25, K1131, R8, EC9784 = PC3205, "habit," gave me 9 under 3 in K1, making it possible to finish that K-row with 8 under 7. Both of these values were then corroborated by R10, EC8049 = PC7023, "with," placing 9 under 3 and 8 under 7 in K1. C28 (K0338), R4, EC9076 = PC1364, "they," tentatively assumed earlier, placed 6 under 4 in K8. EC5 under PC7 then finished the row K8 as well as the square. See Figure 8 for the completed square, keeping in mind that this square, which resulted from the decipherment, is quite different from the square which was used in the encipherment, for which see above, Figure 2. Note, however, that it is only the numbering of the rows which is different, not the numbering of the columns.

	1	2	3	4	5	6	7	8	9	0
1 /	5	7	9	2	4	1	8	3	6	0
2 /	4	0	3	5	9	8	1	7	6	2
3 /	2	4	0	9	3	7	6	1	5	8
4 /	1	7	9	5	2	4	3	8	0	6
5 /	4	1	2	3	8	0	7	6	9	5
6 /	0	2	5	6	4	3	7	8	1	9
7 /	1	3	6	8	0	9	7	5	2	4
8 /	3	8	4	6	7	9	5	0	2	1
9 /	3	6	1	4	9	0	2	5	8	7
0 /	9	1	4	0	6	8	5	2	7	3

Figure 8. Square as deciphered.

ADDENDA

Without any intention of guessing what sort of experimentation one or another reader of this article may wish to try, if any, I have decided to include here a moderate amount of the materials which might be necessary for such experimentation. First the fifty keys obtained by decipherment, arranged consecutively and divided and numbered in groups of five.

```
9627 2656 4435 1591 9371  #5  1609 8531 8729 2024 6172 #10  5323
3162 9158 1447 4605 #15  3377 0794 5325 2358 6903 #20  4073 2995
7401 9792 1131 #25 5806 1486 0338 4163 6534 #30  7398 0722 6245
3719 5067 #35  8838 3529 1040 7497 2811 #40  9474 6533 4201 7659
2232 #45  1946 4723 3889 0585 1467 #50
```

Following are the first three messages as deciphered and still bisected, with the individual groups shown separated by a dash (–). C8 (etc.) indicates the column in which the enciphered message began.

```
C1: which--go--to--make--up--a--marriage--name--ly--one--man--and
--one--woman--period--this--very--surprise--ing--state--ment--
concerning--marriage--to be cont.--begin--message--part--one--
colon--according--to--recent--important--find--ing--s--by--men--
of--science--there--are--two--absolute--ly--necessary--part--s--
C8:  period--one--example--to--the--point--is--the--well--known--
case--of--j--a--c--k--s--p--rat--and--his--love--ing--wife--
period--to becont.--begin--part--two--colon--may--not--at--once--
seem--self--evident--but--let--us--examine--it--care--ful--ly--in
--detail--  C15: the--cook--dare--d--to--place--lean--upon--her--
plate--period--were--they--single--the--situation--would--have--
been--grave--for--to becont.--period--begin--part--three--colon--
now--j--him--self--could--not--bear--the--very--thought--of--fat
--while--mrs--s--retire--d--in-- haste--when--ever--
```

For anyone who might wish to try out the deciphering technique with or without consulting the list of keys given above, I add, in numerical order, a list of every plaincode group which appears in messages four through seven. Seven includes the end of the J.S. portion of the whole message.

```
0145, or.   0239, same.   0266, social.   0404, refer.  0415,
certain. 0471, general.   0499, six.   0608, own.   0774,
arrange.   0867, fashion.  0909, begin.   0925, men.   0980,
design.   0990, there.   1045, continue.   1191, serve.   1218,
up.   1276, period.   1348, finally.   1364, they.   1495, would.
1760, no.   2014, y.   2035, refuse.   2057, al.  2112, his.
2199, how.   2215, de.   2222, colon.   2458, now.   2525, e.
2585, (e)d.  2743, com.   2824, long.   2829, new.   2877, se.
2882, ing.  2888, ment.   2952, other.   2992, ly.   3003,
actual.   3100, match.   3296, x.   3366, strong.   3422,
practical.   3474, meat.   3481, such.   3509, take.
3542, to be cont.   3631, being.   3769, success.   4070, as.
4101, application.   4319, were.   4598, ies.   4609, part.
4693, marriage.   4774, us.   4805, be.   4910, mr.   4967, and.
 4989, p.   5021, an.  5023, work.   5043, but.   5083, for.
5102, have.   5133, individual.   5173, in.   5174, five.   5230,
```

human. 5342, waste. 5390, develop. 5555, is. 5562, could.
5791, regard. 5843, ful. 5997, it. 6008, hand. 6023,
one. 6136, of. 6242, fare. 6277, on. 6411, insist.
6505, to. 6566, that. 6650, let. 6789, might. 6913, this.
6927, share. 6996, war. 7022, dispose. 7040, will. 7311,
cook. 7375, variety. 7491, vice. 7502, seven. 7528, good.
7660, put. 7704, imagine. 7957, a. 8012, hold. 8111,
paragraph. 8242, urge. 8288, possible. 8323, i. 8434,
difficult. 8479, L. 8544, er. 8727, women. 8781, all.
8888, can. 8902, any. 9088, example. 9134, house. 9295,
upon. 9740, ence. 9751, ar. 9779, less. 9801, (e)d.
9833, both. 9843, better. 9876, each. 9895, set. 9919,
red. 9930, four. 9966, without. 9977, prefer. 9999, the.

BIOGRAPHICAL SKETCH

Henry D Ephron is co-editor with his wife Marguerite of Xenocrypts (foreign-language ciphers) for the American Cryptogram Association. He was introduced to cryptanalysis in the 1920's by a pulp magazine's "Yardleygrams," and one of Yardley's captains in WW I, Latin Professor Charles H. Beeson of the University of Chicago. Educated there in Classical Greek and Latin, Professor Ephron then spent the war years in cryptanalysis and later did research in Mycenaean (pre-alphabetic) writing, a task which shared the difficulties and the fun of cryptanalysis. His cryptanalytic mentors in the Army Signal Corps were Lambros Callimahos, William Bundy, and, of course, his CO in the field, Abraham Sinkov.

AGNES MEYER DRISCOLL

Lt. Susan M. Lujan, USNR

[This paper was submitted as an entry in the Rear Admiral Don H. McDowell Essay Contest, established in 1986 by the Commander, Naval Security Group Command. The Contest seeks to stimulate critical thinking, to improve writing skills, and to foster an appreciation for the historical development of the cryptologic profession among junior cryptologic officers. Papers are written on technical, operations, or historical topics. Submission of a paper is a graduation requirement for the Cryptologic Division Officer Course (CDOC), a formal nineteen week school taught in Washington DC.

Lt. Susan M. Lujan, USNR, submitted the paper to the CDOC on 30 October 1987.]

INTRODUCTION

As history is rewritten with the declassification of previously classified information, people who were once *behind the scenes* are often revealed to be heroes who provided information and services to military and political leaders which allowed those leaders to make decisions based on insightful judgments. Mrs. Agnes May (Meyer) Driscoll spent forty-one years of devoted government service from 1918 to 1959 in the field of cryptology. She was often the first in a group of skilled analysts to break new systems and codes considered milestones in the history of the field, including solutions to the Hebern cipher machine, the Japanese Orange Grand Naval Maneuvers cipher, the Japanese *Red Book* and *Blue Book* superencipherments, and more. Due to the sensitive nature of her work, Mrs. Driscoll led a private lifestyle characterized by anonymity. Thus, time and lack of documentation have obscured some aspects of her life, making her as mysterious in some ways as the cryptanalysis she performed. This paper is a tribute to this remarkable historical character.

I. EARLY LIFE

Agnes May Meyer was born on 24 July 1889 in Geneseo, Illinois and was raised there by her parents, Dr. Gustav Frederick Meyer (PhD) and Lucy Andrews Shaw Meyer. She attended Osterbein University from 1907 to 1909 and later received an A.B. degree from Ohio State University in 1911; her main undergraduate subjects were mathematics, physics, music, and languages. Interestingly, in the context of the Victorian era, Miss Meyer was an atypical woman since she pursued technical subjects. "As recently as a hundred years ago, it was assumed that because men are larger and heavier their brains must be larger and heavier too."[1]

After college, Miss Meyer moved to Amarillo, Texas, where she worked at Lowry Phillips Military Academy as Director of Music from 1911 to 1914. She was later the Head of Mathematics Department at Amarillo High School from 1914 to 1918.

II. WORLD WAR I

Woodrow Wilson won his second term on the slogan, "He kept us out of war." Shortly after his inauguration in 1917, however, the United States entered World War I.

Rapid wartime expansion rocked the country which had already been experiencing a growing labor shortage. "While the general citizenry was not aware in 1916 of how great was the probability of the United States being drawn into the war in Europe, the military in this country was already involved in preparations that had to be made if we were not to be caught short in an emergency. This was particularly true of the leadership of the Navy."[2]

The shortage of personnel to do work that was immediately necessary was a problem that the Secretary of the Navy, Josephus Daniels solved with a far-reaching decision:

> 'Is there any law that says a yeoman must be a man'? I asked my legal advisers. The answer was that there was not but that only men had heretofore been enlisted. The law did not say 'male.' 'Then enroll women in the Naval Reserve as yeoman,' I said.[3]

---

[1] Tobias, Sheila. *Overcoming Math Anxiety* (1978), p. 80.

[2] Hancock, Joy (Captain, U.S. Navy, Retired). *Lady in the Navy, A Personal Reminiscence* (1972), p. 22.

[3] Daniels, Josephus. *Our Navy At War* (1922), pp. 328-329.

Although the enrollment of women in the Navy was shocking, women from all points of the compass answered the *Call to Colors*. Agnes May Meyer, a first generation American, was one of those answering the call, and she was recruited in the highest rating obtainable for a woman in World War I, Chief Yeoman (F). Recruiting in such a rating was only done in a few instances. "Other qualifications being equal, in the selection of chief yeoman, preference was given to those who had a working knowledge of Stenography."[4] Stenography was but one of Miss Meyer's talents. In addition, she had a background in statistics, math, physics, engineering, clerical skills, teaching, and was proficient in German, French, Latin, and Japanese. She enlisted on 22 June 1918 and began active duty on 23 July 1918. She worked in communications, where she began her career as a cryptographer and cryptanalyst.

With demobilization, the government gave the yeoman (F) an option: "All Yeoman (F) on duty in the Navy department who do not desire to accept civil service appointments will be released by July 15th or as soon thereafter as your office can handle the releases.... Those who accept civil service appointments will be released from active duty as reservists on July 31st, and their civil service appointments will date from the day following."[5] Miss Meyer's records show that her active duty ended on 31 July 1919 and that she was honorably discharged on 5 February 1920.

III. EARLY WORK IN CRYPTOLOGY

The Office of the Director of Naval Communications (DNC) hired Agnes Meyer as a stenographer, and in 1920 she was given the title *clerk*. "It was during this time that a turning point came. Miss Meyer was invited in 1920 to join an elite group of people studying at the Department of Ciphers at Riverbank Laboratories in Chicago, Illinois, owned and overseen by millionaire George Fabyan. Fabyan paid her expense and her salary and the Navy gave its blessing to the venture."[6] In a letter of 1 March 1920 to Commander Milo F. Draemel, Officer in Charge, Navy Code and Signal Section, Colonel Fabyan wrote about Miss Meyer, "We were very favorably impressed with the young lady and should the condition arise that it is deemed the part of wisdom to dispense with her services at Washington, I want to say, for your personal information, that you are authorized to send her to Riverbank at my expense and I will give her employment

[4] Dessey, Eunice C. *The First Enlisted Women 1917-1918* (1955), p. 30.

[5] *Ibid.*, p. 66.

[6] *Pioneers In U.S. Cryptology, Part II. The NSA Family 35 Year Celebration.* History and Publications Division, National Security Agency (July 1987), p. 13.

in the Cipher Department, at salary which is equal to that which she is now receiving and, furthermore, should occasion arise that you wanted her returned to the Department, I will then release her for that purpose and, in the meantime, she has been kept in touch with that sort of work and would be available."[7]

IV. CIPHER MACHINES

Early in 1921, the inventor of the first cipher machine to employ the rotor principle, Edward Hugh Hebern, "advertised an 'unbreakable cipher' in a marine magazine, but Miss Agnes Meyer, a cryptanalyst in the Navy's Code and Signal Section, solved the sample message."[8] Whether Miss Meyer had actually returned to Washington or remained at Riverbank at that time is unknown.

Between 1920 and 1922 the sequence of events and activities in Miss Meyer's life were not documented. Evidence does show, however, that she was "attached to the (Yardley's) New York Office for five months in 1920."[9] Under the *Education* heading on applications for government service submitted after 1920, she did list a course in cryptanalysis at New York Laboratories, a place which became known as the *Black Chamber*.

On 15 January 1923, Miss Meyer resigned, and in February 1923 Edward Hebern hired her as Technical Advisor and Liaison with the Navy at The Hebern Electric Code Company which he had incorporated in 1921 with encouragement from the Navy. Although the Navy was very interested in 1923 in purchasing the Hebern cipher machine once it was perfected, the Navy had not provided Hebern with any financial assistance. He was unable to bear the burden of the initial investment (including building a factory, machine dies, molds, patterns, etc.), and "in the spring of 1924, the firm defaulted on the interest on its $100,000 mortgage."[10]

In August 1924, Mrs. Driscoll returned to the Navy department, having married a Washington lawyer, Michael Bernard Driscoll, in the interim. When or where they met cannot be determined, but they may have known each other for some time since Mr. Driscoll was from Cairo, Illinois.

During the mid-1930's, Mrs. Driscoll testified before the Senate Naval Affairs Committee that in 1922 she had helped invent a device called the *Communication Machine* or *Cipher Machine*. She worked with Lieutenant Commander William F. Gresham on the development of this device. "Later after GRESHAM's death

[7]Colonel George Fabyan's letter of March 1, 1920 to Commander Milo F. Draemel.

[8]Kahn, David. *The Codebreakers, The Story of Secret Writing* (1967), p. 415.

[9]Holtwick, J. S. (U.S. Navy, Retired) *Naval Security Group History To World War II*, Part I, p. 35.

[10]Kahn, *Codebreakers*, p. 418.

in August 1935, a claim was made by his widow for recompense .... In the course of hearings on this bill, Mrs. Agnes DRISCOLL, the Navy's principal civilian cryptanalyst for many years, claimed a share in the invention. A clipping from the Manila Daily Bulletin of Tuesday, June 8, 1937, datelined Washington (by airmail), quotes Senator David L. WALSH, Chairman of Senate Naval Affairs Committee, that 'The Navy now possesses a device which prevents decoding of its secret radio communications....' "[11] Mrs. Driscoll was awarded $15,000 for her part in inventing the device.

V. THE RED BOOK

"In the spring of 1920 the secret ONI [Office of Naval Intelligence] fund had financed a major undercover operation that involved not just one but a whole series of break-ins at the Japanese Consulate in New York City."[12] The Japanese Fleet code book was photographed page by page and then translated over the course of four years by Dr. Emerson J. Haworth and his wife. This translation was called the *Red Book* because it was bound into two volumes with red bindings.

Lieutenant Lawrence F. Safford, who was destined to become the *father* of naval cryptology, was appointed head of the newly established Cryptographic Research Desk when he arrived at the Naval Code and Signal Section in January 1924. After she returned to DNC, Mrs. Driscoll began to teach Lieutenant Safford the science of cryptology. He reportedly had a natural talent; however, if he is known as the *father* of naval cryptology, then *Miss Aggie* (as she was known) would be the *grandmother* of cryptology.

The Research Desk assumed responsibility for deciphering and analyzing messages sent by Japanese naval radio stations. In 1925, Lieutenant (j.g.) Joseph J. Rochefort joined the group and also came under the tutelage of Miss Aggie. After Safford had rotated to sea duty in February, Rochefort took charge of the research desk and "actual cryptanalysis of the 'Red Book' messages was first undertaken. Mrs. Driscoll was responsible for the initial solution and for the solution of most of the new ciphers and 'transposition forms' used with this code."[13] "This was no small achievement because the two-volume Red Book contained some 100,000 code groups, with as many as 3 groups assigned to each

[11] *Ibid.*, p. 18.

[12] Layton, Edwin T. (Rear Admiral, U.S. Navy, Retired), and Pineau, Roger (Captain, U.S. Naval Reserve, Retired). *"And I Was There"* (1985). p. 31.

[13] Safford, L. F. (Captain, U.S. Navy). *The Undeclared War* ("History of R. I.") (15 November 1943), p. 05.

Japanese word or expression."[14] Mrs. Driscoll received a promotion and a raise in 1925.

Miss Aggie's diligence paid off in other ways. According to Safford, information gained from this deciphered code included:

> "(1) Form, phraseology, and subject matter of secret Japanese naval messages.
>
> (2) Various accidents and casualties of Japanese men-of-war.
>
> (3) General knowledge that Japanese naval maneuvers where much more realist that ours, particularly in night torpedo attacks.
>
> (4) Exact knowledge of Japanese fuel supplies (oil, coal, and gasoline).
>
> (5) Early knowledge of Japanese advances in naval aviation."[15]

VI. GRAND MANEUVERS

In the spring of 1930, the Japanese Imperial Navy conducted exercises known as the Grand Fleet Maneuvers. "The Japanese introduced a new cipher system for these maneuvers plus a daily change of key, but used the good old 'Red Book.' All hands turned to these messages, Mrs. Driscoll got the first break as usual, and the various daily keys were solved without too much effort."[16] Her work resulted in an intelligence windfall which revealed how the Japanese Fleet would conduct combat operations against the United States. Even more significantly, further analysis of the data indicated that the Japanese had successfully exploited a weakness in U.S. operations. "Their [Japanese] naval intelligence had apparently succeeded in obtaining a very accurate picture of the latest modifications in our war plan Orange."[17] The use of cryptology to discover what the enemy knew about our own forces probably gave senior officials a new appreciation for the importance of cryptology.

The Research Desk (Op-20-G) received machines during this time "for relieving the cryptanalytic section of a good deal of their purely mechanical labor. Apparently Mrs. DRISCOLL, the principal cryptanalyst, was not too enthusiastic about this approach, having for more than 10 years operated with pencil

[14] Layton, *"And I Was There"*, p. 34.

[15] Safford, *The Undeclared War*, p. 05.

[16] *Ibid.*, p. 08.

[17] Layton, *"And I Was There"*, p. 35.

and cross-section paper."[18]

VII. THE BLUE BOOK

In the fall of 1931, Mrs. Driscoll was the first to notice that the Japanese had instituted a new code. This new code, known as the *Blue Book*, superseded the Red Book. Faced with solving code and cipher simultaneously, the OP-20-G team worked to recover the code by cryptographic analysis and reconstruction. Using an IBM tabulating machine facilitated the task of keeping track of the nearly 85,000 code groups and changing cipher keys. No other cryptologists had ever broken a complex code and cipher system without assistance from information obtained by compromise of either the code or system. Devoting all her time to pure cryptanalysis, Mrs. Driscoll was the first to break the code. Thanks to her efforts, in 1936 important information was learned:

> The most important and certainly the most dramatic incident connected with the 'Blue Book' was the message reporting [the battleship] NAGATO's post-modernization trials.... The NAGATO's new speed was 26+ knots.... This created consternation in the higher echelons of the Navy Department because...our new battleships (then only in the blueprint stage) were going to have a speed of only 24 knots."[19]

As a result of this information, our battleships were redesigned to have the edge over their Japanese counterparts with a speed which ultimately was to reach 28 knots. Safford was quoted as saying "that this coup 'paid for our peacetime RI [Radio Intelligence] organization a thousand times over.' "[20] The Blue Book was not superseded by another code until 1938.

VIII. TO CATCH A SPY

In 1936, while working on intercepted Japanese foreign office traffic, Mrs. Driscoll consulted the Japanese linguist, Lieutenant Layton, and identified a word *T-O-M-I-M-U-R-A*, which Layton said could be rendered as Tomison or Thompson. Mrs. Driscoll responded, "Lieutenant Layton, I want to thank you. You helped me to solve part of a cipher that has been bothering me for some time. The

[18]Holtwick, *Naval Security Group History To World War II*, Part I, p. 79.

[19]*Ibid.*, p. 90.

[20]Layton, *"And I Was There"*, p. 58.

name is Thompson."[21] After investigation, Thompson was revealed to be a Pacific Fleet radioman.

> The American who was his accomplice and referred to as Agent K in the same message turned out to be John S. Farnsworth, a cashiered navy officer. It was he who masterminded an espionage ring that turned over engineering, and gunnery secrets.... Farnsworth and Thompson were arrested in 1936. Although the evidence that unmasked Harry H. Thompson and led to the pair's apprehension could never be presented in court, they were convicted the following year of conspiring to violate federal espionage laws and sentenced to four to twelve years in prison.[22]

IX. AGGIE, WE HARDLY KNEW YE

The brief character sketch of Agnes May (Meyer) Driscoll which follows is an excerpt from the late Rear Admiral Edwin T. Layton's book, *And I Was There*:

> I had been warned not to patronize Madam 'X,' as her colleagues sometimes referred to her, because she was sensitive to her role as a woman in a man's world. Because of this she kept to herself as much as possible and none of us was invited to socialize with her and her lawyer husband. While she could be warm and friendly, she usually affected an air of intense detachment, which was heightened by her tailored clothes and shunning of makeup. It was surprising to hear Miss Aggie curse, which she frequently did – as fluently as any sailor whom I have ever heard.
>
> In the navy she was without peer as a cryptanalyst. Some of her pupils... were more able mathematicians but she had taught cryptanalysis to all of them, and none ever questioned her superb talent and determination in breaking codes and ciphers. She understood machines and how to apply them.... But her principal talent was her ability to get to the root of a problem, sort out its essential components, and find a way to solve it. Among her uniformed colleagues, she was held in the highest esteem throughout her long career....[23]

In October 1937, Mrs. Driscoll was in an automobile accident in which two of the other passengers were killed. "She suffered two broken jaws and a bro-

[21] *Ibid.*, p. 59.

[22] *Ibid.*, p. 59.

[23] *Ibid.*, p. 58.

ken leg...not returning fully to duty until September 1938."[24] When she did return with the use of a cane, she was reportedly "physically changed, but her personality and attitude had apparently been affected as well."[25]

Agnes Meyer Driscoll (r) at the retirment of her assistant, Mrs. Talley (c) with General Samford (l) in 1956.

Photo Department of Defense.

X. DESIGNING THE PERFECT MOUSETRAP

By 1 June 1939 Japan was using an operations code cipher (JN-25) which was radically different from any other Japanese navy cipher.

[24]Holtwick, *Naval Security Group History To World War II*, Part I, p. 160.

[25]*Pioneers In U.S. Cryptology, Part II. The NSA Family 35 Year Celebration.*

> It was a five-numeral, two-part code, consisting of 33,333 groups in the 'dictionary book,' to each of which were added by false arithmetic (no borrowing or carrying) a five-digit number selected in sequence according to the start key from a second book of random additives. The resulting encipherment consisted of a series of five-digit numbers divisible by three to provide a convenient check against garbles.[26]

Mrs. Driscoll recognized the traffic as machine generated, and she developed a manual decryption for this Japanese machine ("M1"). Working from known characteristics of the cipher, Lieutenant Jack S. Holtwick built an electric coding machine which decrypted the Japanese Navy's machine cipher. "This analog worked well enough... until it [the cipher] was abandoned in 1938. But the real value of the 'M1' machine was that it gave Friedman's team a headstart in constructing the device that would imitate the 'M3' machine which was then being used by the Japanese foreign ministry in 1935 to encode its most confidential communications."[27]

XI. CONCLUSION

From this point until her retirement at age 70 in 1959, there is no documentation of Agnes Driscoll's specific achievements. What we do know of her as a cryptologist is that her most significant contributions were to the development of Naval cryptology and to the Navy's victory in World War II. She modestly described her job in cryptology as *scientific duties*; she probably approached her job with quiet loyalty and dedication (forty years of government service) grounded in a solid understanding of the importance of cryptology. In addition to this unpretentious dignity, Mrs. Driscoll had a certain drive to accomplish wholeheartedly any task at hand, as evidenced by her proficiency and some devoted achievements in her life. Her unique character and the important role she played as a pioneer in the burgeoning field of cryptology in the early part of this century immediately appealed to me as a topic for the Rear Admiral McDowell Essay.

---

[26]Layton, *"And I Was There"*, p. 77.

[27]*Ibid.*, p. 79.

LETTERS TO THE EDITOR

The following two letters relate to the article, "Why Was Safford Pessimistic About Breaking the German Enigma Cipher Machine in 1942," by Louis Kruh which appeared in *Cryptologia*, (1990) 14(3): 253-257.

From Colin Burke, 6319 Red Haven Rd, Columbia MD 21045 USA.

October 1990

Dear Editors

I enclose some general comments re the Safford memo, p. 254, and my conclusions.

a) The political setting: When Safford wrote this memo he was well on the way to being permanently shoved out of any significant role in OP20 but was not yet put totally aside. And, he may have been continuing his attempt to return to a role in "G". The Redman group was on its way in and was beginning to forge its new policies.

   I think Safford was giving policy advice to the new group. The memo has to be placed in a political as well as a technical context to appreciate it:

   1) he is advising the new group to stand against political pressures and to keep Magic to as few people as possible—that might have been one of his responsibilities as 'Q'
   2) he is warning the new group, which included Wenger, who had always been an advocate for the use high speed equipment, not to count on a quick way into Enigma, a system he had been tackling since early 1941
   3) he is perhaps warning the new men not to give Britain too much about or from Magic

b) Conclusions: I agree with your final conjecture, p. 257. Placed in the appropriate contexts, Safford's assertions that only "captures" had allowed the previous entries into Naval Enigma and that prospects for re-entry were rather poor, seem reasonable. Given what England had probably told us about methods by early 1941, his history was well grounded. The importance of the later capture to the re-entry into submarine Enigma at the end of 1942 validated his prediction. If we add the helpful but not critical assumption that OP20G had not been informed of the "bombe" and its allied methods by early 1942, then Safford's memo is explained.[1] And, even if the assumption about the bombe is dropped, Safford's statements were those of a well informed man. Note that the British bombes were in operation long before the mid-1941 break into the 3 wheel Naval Enigma which suggests that some captures might have been needed to initiate bombe usage, even for the simplest 3 wheel systems.

Finally, the materials in this article reinforce my beliefs about how much and how soon Britain shared the Ultra secrets with us. Again, why were there no references to the bombes, even guarded ones, by Safford or the two Navy men you quote?

Comments on Particular Points

I) p. 253: "Our prospects of ever breaking the German "Enigma" cipher machine are rather poor." There are several ways to interpret this. Let me cite some that are not emphasized in your article.

   1) "our" could well refer to OP20G, not to both Britain and America—thus he could have known of the earlier 3 wheel break by Britain and yet made his assertion. The following support this.

      a) Safford did not trust England nor did he approve of sharing American secrets with it

      b) Safford thought ("Rhapsody in Purple") that Britain had not kept its promises concerning the February-March 1941 exchange

      c) he wanted an independent American capability for all systems for all nations

   2) "ever breaking"

      a) the Weeks-Currier trip came before England had cracked even the three wheel Naval Enigma so that it could not have told the Americans how to enter it via any general method (also see your article, p. 257)

[1] Joe Eachus' surprise when he discovered the use of the diagonal board in the Turing bombe when he went to England in late summer, 1942, is only one among many small indications that Britain had not told OP20G all about its Enigma capability.

b) I have seen no hard evidence that Britain told OP20G the necessary details to allow it to enter the 3 wheel system by itself after England's entry in mid-1941 when more 'pure' techniques were available (also see you article, p. 257)

c) I have not seen any evidence that Britain gave OP20G knowledge of more general techniques (pure cryptanalytic) for Naval Enigma systems before 1942

d) Safford's reference seems to be to "naval" Enigma — he could have known of the more pure or general British approaches to air force or other Enigma but was correct that Naval systems awaited captures

e) certainly, in March 1942, the prospects of either England or America cracking the 4 wheel system were low <u>and</u>, I believe, Safford wanted an independent American capability

  i) I have some reason to believe that when the Americans started their bombe project (Feb-March) 1942, the design was very different from Turing's and the logical approach to a solution was also very different

  ii) the U. S. surely knew that Britain had been locked out of the 4 wheel system and that it could not promise a speedy re-entry

II) p. 255: "A Retired Lieutenant Commander wrote...." This man seems to have his dates wrong, he is consistently one year too early

1) the official history or OP20G states that no Enigma work was begun until early 1941 (NOT 1940)

2) Wenger did not return from "sea duty" until late 1941 and probably did not begin any OP20G operational work until early 1942

3) Safford did not leave (get kicked out) as head of G until 1942

4) although the Americans decided to build some sort of bombe on their own in early 1942 it was not until the fall of <u>1942</u> that anything like a firm commitment to a design was made

5) because of the above we should conclude that Safford was well aware of what was going on in March, 1942

III) p. 256: "Another retired officer...." This man also seems to be somewhat confused about dates, but is perhaps a few months rather than a year off the mark. Despite that possible 'slip' he says much that seems acceptable and much that is very important to the questions of how much and when Britain shared it methods, and the Safford memo, as well.

1) Didn't the Lofoten raid come after the conclusion of the Weeks-Currier visit (and if this isn't one of those men someone else who I should know of was at GC&CS in 1941)? [Editor's note: The Lofoten raid occurred on March 4, 1941, and it is known that the two Army members of the U. S. intelligence mission to England remained there through the end of March; presumably so did Weeks and Currier.]
2) if this man was either Weeks or Currier and if England sent over the techniques which allowed them into the three wheel Enigma in mid-1941, he should have known more about all this

IV) Your comment, p. 257: "One final conjecture...." Much of my wandering depends upon a meaning for the elusive concept, "a purely cryptanalytic solution" and this section is directed more to the general question of how much Britain told us than to the Safford memo in your article.

1) In conjunction with the "several days' settings" quote Safford's "captures" assertion strongly suggests that England did not tell us about their "bombe" and its use of 'cribs' in 1941
    a) why wasn't the bombe mentioned along with the mention, p. 257, of the paper Enigma given the Americans?
    b) why didn't Safford include something about the Turing bombe approach in the memo, at least to directly discount the possibilities of such an approach?

From Ralph Erskine who can be reached c/o *Cryptologia*, Rose-Hulman Institute of Technology, Terre Haute IN 47803 USA.

10 October 1990 - Belfast NORTHERN IRELAND

Sir

In a recent article, Louis Kruh asked for views on Captain Lawrence Safford's knowledge of the British breaking of Enigma.

Safford's memorandum was expressly intended to emphasise to others the paramount need to ensure that the US Navy breaking of, in particular, the JN 25 series of codes was not revealed to the Japanese. He was anxious that this might occur due to the resulting intelligence being disseminated too widely. His passing reference to Enigma, occupying only about one line, is clearly not an exhaustive statement of the Allied capability to break any form of Enigma.

Safford was right to be concerned. His memorandum is dated 18 March 1942. On 30 August 1942, 3 Skl II Staffel in the Seekriegsleitung (the German Navy's war staff) is apparently recorded as stating –

> When the Japanese were presented with some captured documents from the prize ship "Nankin", the [Japanese] Navy suspected that its signal methods were no longer unknown to the enemy and it requested German assistance in revising its coding systems.[2]

The Australian ss *Nankin*, bound from Freemantle, had been captured by the German raider *Thor* on 5 May. Documents seized from the *Nankin* were transferred to a German supply ship, *Regenberg*, and reached the German naval attache in Japan, Rear-Admiral Paul Wenneker, only on 18 July. It would seem that the papers seized included intelligence summaries which could only have been derived from the Allied breaking of Japanese naval codes. On 29 August, after being authorised by Berlin to do so, Wenneker showed some of the reports to the Japanese — hence the remarks by 3 Skl quoted above.[3]

The further event that Safford feared — the supply of Enigma by the Germans to the Japanese also took place, although perhaps not until around 1943. Fortunately, it would appear that the Germans did not hand over the plugboard version of Enigma. If they had done so and Enigma's rotors had been rewired (as elementary security would have required), the US Navy would have had the utmost difficulty in penetrating "Japanese" naval Enigma — if the Japanese could have produced it quickly and distributed it in sufficient numbers.

Safford was writing purely in a naval context and not referring to Luftwaffe or Army ciphers, machine or otherwise. Moreover, since naval Enigma ("Enigma M") presented special problems to the cryptanalysts, it was almost irrelevant whether Luftwaffe or Army Enigma ciphers were being broken by the British or not. In the last sentence of paragraph 2 and the first of paragraph 3, Safford is clearly referring to US Navy cryptanalytical work. It is therefore reasonable to read the penultimate sentence of paragraph 2 (stating, in effect, that Enigma M — "German naval messages" which were not "weather codes [or] other minor systems" — had been read only as a result of captures), as also referring to American naval work on Enigma, and therefore only to Enigma M.

The first effective breaking of naval Enigma ("Enigma M") by Bletchley Park ("BP") started on 12 March, after the capture of Heimisch (the main Kriegsmarine Enigma cipher at that time) keylists for at least part of February on 4

[2]Report on documents from 3 Skl II Staffel (NID 24/T50/45): PRO ADM 223/255.

[3]See James Rusbridger, "The Sinking of the "Automedon", the Capture of the "Nankin," *Encounter* 64(5) (May 1985), 8.

March 1941 during a commando raid in the Lofotens.[4] However, the British had not then recovered the bigram tables used to encode the message key on Enigma M, which held BP back badly, despite the fact that the keylists gave the main settings — another indication just how difficult it was to penetrate Enigma M.[5] BP read or broke 26 Heimisch daily keys for dates between 1 February and 9 May, which led to the decipherment of about 340 signals — probably less than 10 per cent of the naval traffic intercepted during that period. The British captured further important Enigma M material in late April and mid-May, including keylists and a weather short signal book, the Wetterkurzschlüssel, which was immediately to prove a crucially important source of cribs.

According to Safford, and the second retired US Navy officer quoted in the article, the US Navy also read some Enigma M traffic, probably around May 1941 (after Lieutenant Weeks and Ensign Currier returned to the States).[6] The British must, at the minimum, have supplied the US Navy with —

> a "paper" copy of Enigma M (the wiring of its 8 rotors and its reflector);
> some keylists (probably those for part of February 1941).[7]

BP's possession of this material was one of the United Kingdom's most closely kept secrets. Only a select few on each side of the Atlantic knew about the captures, but Safford was clearly one. And since the US Navy in Washington would have realised, from at least September 1941 on, that the British were breaking naval Enigma, it would be most surprising if Safford, in view of his position in Op 20 G, was not aware of that fact.[8]

Because the US Navy was primarily concerned with escorting the Atlantic convoys, it was really only interested in the radio traffic to and from the Atlantic U-boats, and not in the signals to surface vessels. On 1 February 1942 the Atlantic U-boats had begun to use four-rotor Enigma (M4) on Triton (known to the Allies as "Shark") — a cipher introduced specially for the Atlantic U-boats on 5 October 1941. Worse still, a second edition of the Wetterkurzschlüssel had been introduced on 20 January 1942. Lacking that edition, BP was therefore without its only source of Triton cribs for the bombes (fast key-finding aids) and, for the first time, blinded by the Triton traffic. In the result, BP broke Triton on only three days between 1 February and 11 December 1942.

[4] Ralph Erskine, "Breaking Heimisch and Triton," *Intelligence and National Security*, (1988) 3: 162.

[5] *Cf.* the retired officer quoted in Kruh, 256.

[6] Kruh, 254 (paragraph 3).

[7] Kruh, 257.

[8] See e.g. Ultra signal 1334A/8 September 1941 to CNO Washington, giving U-boat positions: PRO ADM 223/102.

Since there was no prospect of BP reconstructing the new Wetterkurzschlüssel cryptanalytically in 1942, the Allies were dependent on capturing it, probably from a U-boat, without the Germans realising that the code had been compromised. The codebook was printed using water soluble ink, so that recovering it intact from a sunk U-boat would have been virtually impossible.

The capture of the Wetterkurzschlüssel (and another short signal book, the Kurzsignalheft) required something approaching a minor miracle, and the supreme sacrifice by Lieutenant Tony Fasson and AB Colin Grazier (both from *HMS Petard*) on 30 October 1942. They had boarded the sinking U 559 in the Eastern Mediterranean, but were unable to escape from the submarine when it suddenly went down. For their bravery, which undoubtedly helped significantly to shorten the war, so saving many thousands of lives, they were awarded the George Cross (the second highest British medal). Tommy Brown, a 16 year old canteen assistant who helped them, received the George Medal for his part in the capture. But it was not until 2 December that BP became convinced that it could break into Triton again.[9] And that happened because, by great good fortune, the U-boats sent weather short signals on Triton using M4 in three-rotor mode, so that the absence of four-rotor bombes was not too much of a handicap; the three-rotor bombes sufficed — just about. Given the run of luck required by the Allies before M4 could be broken, it is little wonder that Safford was pessimistic in early 1942.

A further factor in Safford's thinking may have been that the knowledge then available to the Allies possibly suggested that four-rotor bombes would also be required to break Triton on M4. And he would have probably have known that there was little prospect of those bombes entering service before July 1943. In March 1942 who, then, would have rated the chances of breaking Triton as higher than, in Safford's words "rather poor"?

So far as is known, the US Navy did not break Enigma M cryptanalytically in 1941 — or indeed in 1942. All of Safford's statements are entirely consistent with his referring in March 1942 only to Enigma M4. Read in that light, they are extremely accurate, showing Safford to have been very well-informed indeed, since breaking M4 was truly a problem of horrifying proportions.

Safford refers to the breaking of the German naval meteorological cipher by the US Navy. This once again highlights the lack of liaison between the US Navy and the US Army on cryptanalytical matters. The German navy met cipher was in fact a code which was super-enciphered using trigram tables. It was broadcast by Norddeich (call sign DAN) on 125 and 150 kHz, and included reports from numerous met stations in occupied Europe, Portugal and other

[9] See Erskine "Breaking Heimisch and Triton", 170.

countries. Apparently, although the US Navy had broken the met cipher before March 1942, the US Army meteorological cryptanalytical section (under Captain Edmund J. Wrigley) was unaware of that fact in mid-1942, notwithstanding that the Army was interested in breaking it. It would be interesting to know whether the US Navy was still breaking the met cipher at that time.

Perhaps the most significant part of the memorandum is Safford's statement that "German [non-minor — that is Enigma M] naval messages ... have been read". As we have seen, it means that the British must have given the US Navy Enigma M's wiring, and some captured keylists. It is confirmed by the remarks to the same effect (by the second retired US Navy officer) quoted in the article.[10] This directly controverts Safford's post-war claim that the British reneged on a bargain to supply Enigma material in exchange for the Purple machine handed over to BP by the Sinkov/Weeks/Currier mission in January 1941.[11] Since the memorandum was written less than a year after the events and is supported by other evidence, it is to be preferred to Safford's post-war account. Further evidence on the issue would be welcome, in view of the seriousness of Safford's charges of bad faith.

---

[10] Kruh, 257.

[11] Dundas P. Tucker "Rhapsody in Purple: A New History of Pearl Harbor — Part I," *Cryptologia* (1982) 6: 214-7.

MY "PURPLE" TRIP TO ENGLAND IN 1941

Prescott Currier

ADDRESS: Prescott Currier died in 1995.

ABSTRACT: The United States brings the Purple machine to Britain and gets information in return.

KEYWORDS: "Purple," Bletchley Park, Agnes Driscoll, Leo Rosen, Robert Weeks, Abraham Sinkov, Alastair Denniston, John Tiltman, Edward Travis, Humphrey Sandwith, Clive Loehnis, Frank Birch, Don Seiler, Enigma.

[Introduction to Currier's talk at a National Security Agency historical meeting: Captain Prescott Courier was a cryptanalyst in the U. S. Navy's Op20G from 1936 to 1941. He was a member of the negotiating team sent to GC&CS at Bletchley Park [in January 1941], our first mission of cooperation with the British. In February, 1943, he moved to Nebraska Avenue [headquarters of naval communications intelligence] and was assigned to Op 20GC as a senior watch officer, where he stayed until 1945. From there he went to the Russian Language School at Boulder; 1946 to '48, he was the naval representative to Britain's Government Communications Headquarters; 1949 director of research at the Naval Security Group; in 1951 office of Naval Intelligence; 1952 to '54 officer in charge of the Naval Security Group stationed at Kamiseya, Japan; 1955 to '59, senior U. S. liaison officer, London; 1959, director of operations Naval Security Group and then back at NSA. This is one of the real giants in the business. Captain Currier.]

I'm no giant. It took me two breaths to get up those stairs. I'm reminded of something – before I start – some years ago I served on a petit jury along with a friend of mine who was a retired professor from the University of Maine. I complained to him after we got about halfway through that I was never selected for the final 12. I never sat on a jury and tried a case. And he said, you know why? He said it's that bow tie of yours. He said the lawyers don't trust you. They're afraid you know too much.

I thought that some of you might be interested in knowing what life was really like between 1936 and 1940, what we did and how many of us there were. Op 20G in those days was in the back end of the first floor of the sixth wing of the Navy Department. There were eight rooms, total. In the back room were three Japanese linguists, all civilians, Phil Cate, Fred Woodrough, and Dr. Hoffman, who had been a missionary. In the middle room there was a great hodgepodge of everything under the sun including files, papers and that sort of thing. There were probably on the order of about eight or a dozen people in this room at various times, on and off, crypto clerks, cipher people, floor-sweepers and an occasional cryptanalyst – me. In the next room, there were two of us – Mrs. [Agnes] Driscoll and myself. We were the only acknowledged cryptanalysts there at the time. There had been others. There were other naval officers who had been taught, some by Mrs. Driscoll, some who had learned it on their own, but at that particular time – early on, '36, say - there were only two of us. There were two and sometimes three language officers in the front room. On the righthand side of the corridor was our machine section, which was composed of two card punchers, one collator, one sorter and one printer. We rented them from IBM for about, gee, I don't know, $50 a month, something – anyway it was not very much. And that was it.

By December, 1940, our personnel had tripled and we had moved out of the back of the sixth wing, but we were essentially the same kind of an organization. I had gone on leave home. I'd only been married a few months. The day after Christmas I got a note from Joe Wenger who said, come back immediately. So I got back to Washington as quickly as I could in the wintertime and when I arrived I was told that I was going to be making a trip to England. They didn't know exactly when or under what circumstances. I was told almost nothing else. It later turned out in about a week that I was to be one of two Navy representatives [the other was Robert Weeks] who, along with Abe Sinkov and Leo Rosen from the Army, were going to England to bring with us two copies of the Purple machine, what we had done on the JN25 codes, a copy of the typewriter that the U. S. Navy intercept operators used and an enormous amount of additional material on DF and various other items that the Navy considered to be important in the coding business on this side of the Atlantic. I can't remember precisely what was crated because I never saw it before I left. It turned out – I didn't know this, by the way, before I left – but it turned out that there had been something called the Tizard commission [a British unit], which had been debating a proposition that would have allowed the U. S. and the U. K. to exchange information on intelligence, along with other technical things. Originally this arrangement did not include comint but it was eventually decided, not long before I was asked to return, that we would in fact include an exchange of comint material. Bill Friedman was supposed to have been the man who led the group, but about a week before – which occasioned my being called off leave – had gone to the hospital with a breakdown and he was replaced by Abe Sinkov.

Now, strangely enough, I was given no instructions. I wasn't told where I was going. I knew roughly what we had with us. I was not told what to ask for. I wasn't instructed how to behave or what to offer or what to say or even who I was to see, which certainly is rather strange now, but it didn't [seem so] at the time. I didn't really give it a second thought. In any event, the four of us one cold day in January – about the 15th, I guess – eventually were told we'd be leaving. We went down to Annapolis, waited on the dock down there for several hours. It turned out that we were going to England on the *HMS King George V*, one of the two new battleships that the British had just built. The *Prince of Wales* was her sister ship. The *Prince of Wales*, by the way, was sunk by the Japanese a little bit later. We waited there for a longish length of time and eventually us and all our gear were loaded into a motorwheeler and we went out into the road where the *King George* was anchored. When we got there we couldn't go aboard because she also had aboard Lord Halifax, who was to be the ambassador to the U. S. He had not seen fit to debark so we couldn't go aboard until he left. So we sat out there in a bumpy sea in an open boat full of a lot of gear that we didn't want to get wet waiting until the future ambassador to the United States came ashore. He eventually did, of course. We immediately went aboard. All the gear was taken care of beautifully. Arrangements obviously had been made and they not only expected us but they knew precisely who we were, what we had and how much room it would take to store it. The trip was relatively uneventful. We went out east, joined a beef convoy coming up from Argentina, and it took something like 12 or 13 days, we arrived at Scapa Flow on the 7th of February, I believe, with no trouble on the way, not a shot fired in anger.

At Scapa Flow we were told we would be flown from there south in a Short flying boat, of which there were two there waiting for us. It turned out that the people who ordered the flying boat hadn't measured the hatch through which all our gear was supposed to go and it was only about half wide enough. A cruiser, the *Newcastle*, I believe, was there and her captain let the captain of the *King George V* know that he was going south and he'd be very pleased to take us along. So, we got alongside and we loaded all our gear, transferred it to the *Newcastle* – sorry *Neptune*, wasn't the *Newcastle*, the *Neptune* – and we left the following morning alone. We were going down the swept channel on the east coast of England where all of the convoys traveled. The following noon we approached a convoy – we were going a lot faster than they – and so our captain obviously elevated to get by and get south as soon as he could. In the course of doing this, a reconnaissance bomber came overhead and saw the convoy and saw us and we all knew what this meant – that in about 20 minutes it would begin to be noisy. True enough, it did, and we all went below decks and sat there. I was trying to eat some soup but my mouth was so dry I couldn't swallow it. I've never been so scared in my life. We heard a bomb landing on one side and one on the other side and the ship would bounce out of the water and back down again. Then I heard something that sounded like someone dragging chains along the deck. The German were making strafing runs and eventually when it was all over, we went topside. All our gear, by the way, which was in large wooden crates, had been stowed topside. We went to look at them and fortunately all of the ammunition was antipersonnel which exploded on contact and when it hit the heavy wood crating, it exploded and left a little hollow but it never penetrated so that none of the gear that we brought with us was damaged in the least. We put into Sheerness that night and we were met by two staff cars and a truck. We were introduced to the – I've forgotten who it was, Sandwith, I think, Navy captain, plus the drivers. We were driven through the night up to Bletchley, which was about 120 miles to the north. We arrived there well after dark – eight or nine o' clock probably – and were driven through the gates to a large, forbidding-looking brick building with all the blackout curtains drawn. [We were taken] into a room and there standing behind the desk were four men. One of them, [was] Alastair Dennison [the director of GC&CS], one of the former members of Room 40 in the Admiralty. Jumbo Travis was also there. He later became director. John Tiltman, whom many of you know, was there in his full regimentals with his hands behind his back. We were introduced and we made a few inconsequential remarks and we were driven off to our billet. Now our billet was the country home of the chairman of the board of the, at the time, Anglo-Persian Oil Company, and an extremely wealthy man. He had turned over his entire house to these four traveling Americans, fully staffed, the larder full, with a butler and three upstairs maids and a cook to take care of us for the time that we were there.

The British split us up. The Navy group and the Army group never traveled together. We each had a car and our own driver, so [the Navy] never went to the same place as they [Army] did – at least [not] at the same time. We weren't intentionally kept apart. We saw each other almost every night, but we often did our own thing and saw the things that we, as service representatives, were interested in. I was struck by the real friendly feeling of complete cooperation which belies some things that have been said about this little trip of ours by others who've written about it since then. We went everywhere, including Hut 6. We watched the operation of the bombe.. We were told in great detail the solution of the Enigma, who worked on it, how the bombe worked, what they had to do in order to get a readable text. Bob Weeks and I were not Enigma mathematicians or cryptanalysts but Abe and Leo Rosen were and it was left to them to do whatever would be done in making the necessary notes and records that we would take back with us.

We spent a good deal of time at Bletchley, Bob Weeks and I, principally in the Naval Section but around in various others also. We spent many days traveling around to the north of England and to the south visiting Royal Navy intercept stations, up at Scarborough, for instance, where we were introduced to – we had tea in the Wrens barracks – and we were introduced to the immobile Wrens and the mobile Wrens. The mobile Wrens were the ones who could be sent to Singapore. We also spent time at Flowerdown, at Winchester, at the intercept station there and several days traveling around the south of England, one of them which I think is quite memorable, really, we went down to Dover Castle and we went up on the cliffs where the British had installed their radar, which all of you have seen in pictures, I'm sure. The radar was manned by [the women's auxiliary of] the Royal Air Force – alleged to be the prettiest women in the armed services, and they'd never let anyone forget it. There were, as a matter of fact. We went in and watched the screen of CRTs, watched the screens, and then of course while we were there, a German bomber group took off just over the Channel and one of the girls came up to me and tapped me on the shoulder and said, in about three minutes, go outside and look up. So I went outside – all of us did – looked up and here came the German squadron of bombers, contrails very obvious. And over this side, the Spitfires taking off. And we just watched them as they came together. And then they got rather mixed up for a bit and – you could follow this whole thing from the ground – a whole squadron of bombers suddenly turned around and back they went and the Spitfires came back and landed. That was during the time when daylight bombing was still being undertaken by both the British and the Germans, but it was a very expensive way of doing business and in that particular case the Germans decided they'd try it again another day.

The extent of what we actually learned and what we were told was as much as we wanted to know and we were told a good many things that we didn't know we wanted to know until after we were told that the information existed. Now of the material that we brought the Purple machines of course, were the most important. That was the sort of thing that the British wanted the most. One of the things that we wanted from them the most was all the information that they had on the Enigma, the structure and manufacturing of the bombe, which was an Enigma analog actually. They did not have a spare Enigma machine to give us. But they did allow us to spend as much time as we wanted to in Hut 6 watch the operation of the bombe, have someone there to explain precisely what was going on, how the menu was fed in, what happened when the red lights went on and when they discovered that they had a match. All extremely interesting and very valuable. Even though we did not in fact bring back an Enigma, we brought back all of the information that we really wanted and there was never any question that anyone was holding anything from us.

We spent a lot of time visiting around. For instance, I spent a lot of time over in the Naval Section which was run, by the way, by a man named Frank Birch, who was a movie director. His wife was an actress. He ended up, by the way, by writing the [GC&CS] history, some eight or nine volumes, an enormous work. I got to know him fairly well. In fact I stayed in his house a couple of nights. They had, by the way, a lot of material that I thought might be interesting to the Navy, all German material, such as coastal charts and recognition codes for the E-boats, mostly captured. There were no copying machines in those days. Cameras, yes, but not Xerox. So I spent something on the order of two weeks tracing maps of the Channel and copying recognition codes. It would have taken possibly 20 minutes total if we had had a copier at the time.

The Admiralty, in the form of NSD9, which was the equivalent of Op20G, [was] commanded by Captain Humphrey Sandwith at the time. His No. 2 was Clive Loehnis, who later became the director of GC&HQ and whom we met then and whom we've kept in touch with all these years. But I was advised never to ride with him. He drove what was called a flying bedpan and he offered me a lift down to Winchester to see the intercept station and I was strongly advised by several other people in NSD9, don't take him up on it. Make your own way. Which we did. However, he later, by the way, when he was director of GCHQ, drove his Jag up to Cheltenham and ran off the road and injured himself to the point where they weren't quite sure he was alive. But that's the only bad accident he ever had.

The feeling that I got when I was there was only one of extreme friendship and willingness to produce and provide to us anything we saw that we wanted. And they did. We could go absolutely anywhere and they were very good at answering questions and providing the documents that they had.

There were various things, such as learning how to play rounders. Barbara Kenworthy taught me how to play rounders. She married Joe Eachus by the way. He went over there later. But Barbara was my introduction to rounders and she taught me how to play and eventually it turned out that I could hit more home runs than anybody else and I wore myself out running around the bases because I couldn't stop. See, once you'd hit a ball, you had to run until you got back to the home plate.

There were several other very pleasurable instants. I must mention that there were two girls who said they had a car but no petrol and if we could get some petrol, we could go for a ride up in the Cotswolds. So, I went to my Scottish driver and said, Do you have a hose. Can you get me a gallon of petrol out of the car? He said, Sure. So I got a gallon and we gave it to the gals and took a lovely ride up into the Cotswolds.

On another visit, our driver was a Scottie with a very strong Highland accent and none of the English could understand it. We were driving up north and I don't remember where we were going at the time, but we were stopped at a road block by local police. This was done all over England. And here they had a [military] vehicle with two men in civilian clothes in the back seat, obviously not British, a Scottish driver that they couldn't understand, and he got out there and stamped his feet – I cannot imitate him. "See, ya conno do this, they're on a secret mission." And the sergeant nodded his head and off we went to the police station. But it did get cleared up. We did get out and we did come back.

The only other trip that we took where the Navy gathered some material which the Army either didn't know about or didn't want was a new DF [direction finder], some new DF equipment being built by Marconi out in Norfolk. We went out to get it and we were given a complete rundown on its good points and how it was to be set up, all the instructions and that sort of thing, and it was all crated and ready to go and we arranged to have it sent. About two weeks later we finished what we were supposed to do – more of the same that I've been telling you – and we headed north to Helensbrough, up in the Clyde, where we were to board the HMS *Revenge* for a trip home. We got there and we went aboard and we checked all of the gear that had been put aboard and the only thing that wasn't there was all of this DF equipment. So Bob Weeks and I started at one end of Clydeside and walked through all the buildings and warehouses. It probably was only about a half a mile, but it seemed like three miles. And we found it under an old blanket in the corner of one of the warehouses, alongside some I-beams that nobody knew what to do with. How it got there, I don't know, but we found it. We told them about it and we got someone to come and get it and they put it aboard. We were told that we were in a hurry, that we had to get it aboard that night because they wanted everything squared away. They were leaving early the next morning with a convoy.

Well, early the next morning dawned and we weren't going anywhere. The convoy wasn't ready so we were told that we could take a couple of days off. So we walked up to Loch Lomond, had a couple of good lunches, walked around, came back and the following morning we left. It was the biggest and fastest convoy that had left the Clyde since the beginning of the war. There were 20 liners, all over 20,000 tons, all capable of at least 20 knots, loaded with troops for North Africa. We accompanied them out to the middle Atlantic, where we left. Now the *Revenge* was a battleship built in 1911, I believe, for duty in the North Sea, very low freeboard, and it was bumpy during that trip coming back and more often than not there was green water over the bow, so we didn't spend a lot of time out walking around.

But we were very fortunate in having aboard the next high commissioner to Canada, [Malcolm] MacDonald [M.P.], the premier in exile of Poland with his staff, and a couple of other people who were never identified to me and I don't know who they were, but they were important because they all had their own staffs, too. The prime minister of Poland on his staff had a young musician who was a private major in the British Army. Whenever the weather was fair enough, he would move the little piano that was in the wardroom out onto the deck and have a concert for the entire ship's company. Then in the evenings MacDonald – provided the weather was good – would have a discussion group on international politics including the prime minister of Poland and his group, the high commissioner to Canada's group and this other character whom I never identified but was important. And we spent several hours, many evenings, being lectured to by these giants. So coming back was more fun than going over.

Going over, by the way, there's just one thing I might mention, I forgot. In the course of our trip across, eastward, I became friendly with the wireless officer who was a young man about my age and so he took me up into their communications center and took me through the area and into the radar room. That night at dinner the young RNVR [Royal Navy Volunteer Reserve] officer, wavy Navy [sleeve stripes] type, got up. He had had more than his share of drinks before dinner and he called on the captain of the ship. He said, You've allowed these American spies to see British radar. I want you to do something about it. Well, they took him below and stuck him in his bunk. But that shows you how at that time the feeling – at least in the British Navy – was that their radar was the only instrument that was keeping them from being beaten during the war and they damn well weren't going to release the details of it to anyone, including the Americans. I didn't get that feeling once I was in England, but this young man obviously felt that way and he called upon his captain to do something about us, not allow us to see any more British radar and how it worked.

The outcome of all of this was when we got back to Halifax – we hadn't been there more than an hour – when a U. S. destroyer pulled up alongside, a four-stacker, the USS *Overton*, Bob Weeks and I were on deck and a man got off and walked across. It was a man named Stefanic, a lieutenant commander whom Bob Weeks and I knew, and I thought, Well, isn't this lucky, here we now have a ride back to Washington. I didn't realize at the time that it had all been planned and that he was up there to meet us and take us back to Washington. We got all our gear loaded into the destroyer, and we lashed it down on the deck, lashed it into the bunks as best we could, and we left that night after dark, running without lights. As we got out to the entrance of Halifax harbor, two or three trawlers were coming into the harbor. The captain didn't see them until the last minute because they were also running without lights and we did a quick 180 and everything that we had so carefully laid out and lashed down in the captain's cabin came crashing down onto the floor, everything, all over the place. It was a mess. The deck was covered. Fortunately, there was nothing breakable so we got it all back. It was a bad trip going back. It was bumpy all the way. A young seaman came up to me just before we got back to Washington and wanted to know whether there was anything I could do to get him off the *Overton*. He didn't like it. But we pulled into Washington Navy Yard in the middle of the day. We were tired. We probably looked a bit bedraggled because we'd been up a lot of the night. We did have a lot of gear that we had to get ashore and as we came ashore, Don Seiler, who had a machine shop down in the Navy Yard – he was the one, by the way, who made the first analog of the RED machine – he just happened to be there. He saw us coming off and he made a remark to one of his friends, Here they come, they look like whipped dogs and they don't have anything. This got round and you can imagine what some writers made of this remark of his that later got out, got in print and the first thing you know that became gospel: We went over there, we gave them everything we had, we came back emptyhanded with nothing. It took a long time before that view of our trip and the results of our negotiation ever got put to rest, but eventually it was. I trust that this [talk] has let some of you know precisely what went on in those days, how we did it. We were not giants. The problems were simple.

Introduction to the Technology section of the third collection of *CRYPTOLOGIA* articles

Cryptologic Technology
by Cipher Deavours

It seems only yesterday that Artech published the first collection of *CRYPTOLOGIA* articles. But, it was 10 years ago. In that time, much information has come to light concerning the history of cryptologic technology.

The reader will find this section of *Selections from CRYPTOLOGIA: History, People, and Technology* comprised of 19 articles of which 14 are concerned with the all important World War II era, and of these, six articles deal with various aspects of the Enigma cipher machine, including the English translation of Gilbert Block's meticulously researched Enigma chronology, *Enigma Avant Altra*. As more details are revealed about cyptologic development between 1920 to 1945, it becomes truly difficult to believe that so much could have happened in such a relatively short span of years. Machine cryptography was born during this time and matured rapidly during the catastrophe known as World War II. What might be called the world's first digital computers were developed during this war to cope with intricacies necessary to unravel the increasingly complex ciphers generated by the new cipher machines. Many of those early cryptanalytic devices performed computational feats with speeds that leave modern microprocessors in the dust, as has been demonstrated repeatedly.

Interest in the German Enigma cipher machine remains as high as it was 10 years ago. At least one site on the World Wide Web is devoted exclusively to “Enigmatology.” Historians and cryptologists keep finding new and significant information about the Enigma story. The “autoscritcher,” an interesting and powerful derivative of the British bombes, is described in one article of this series. Even more recently, details concerning the Enigma variant used by the Abwehr have been disclosed. The Enigma chronology is important because we now know so much about the politics, technology, and use of that machine that we can see more clearly, than in most other instances, just how these factors work together and contribute to the security or insecurity of a cryptographic system.

Gordon Welchman, the author of the *Hut 6 Story* and inventor of the bombe’s diagonal plugboard, often remarked that he believed that hard lessons concerning communications security learned with the Enigma were being lost. There are many people in the cryptologic field whose lack of knowledge about the history of their subject causes them to repeat the same errors made decades ago—only this time with a computer. College and university courses in cryptology abound these days, and, in these courses, “classical” cryptology is often dismissed with a few waves of the hand as interesting but largely irrelevant to the computer era. The fault in this case is with the instructors who sometimes do not have enough background in this subject to see connections between the past and present. Old techniques and methods are rediscovered and thought to be new. History is important. Without the anchor of historical knowledge, technological advancement is not assured and one is adrift in a lonely sea of disconnected results that cannot be synthesized into useful cryptologic knowledge. *CRYPTOLOGIA* and Artech House are making available an archive not likely to be matched anywhere else. It is a treasure to read, appreciate, and use.

Although the importance of the World War II era to cryptologic technology can hardly be overestimated, there exists much material from earlier days which has gone largely unnoticed. The Victorian age was an especially remarkable one. Two articles in this section present new material on the ingenuity and efforts of Victorian cryptographers to construct secure cipher systems. Charles Dodgson (a.k.a. Lewis Carroll), the author of *Alice in Wonderland,* had an intense interest in cryptography and some of his work with ciphers is presented in this volume.

We have only one article related to the cryptology of the Vietnam era. The cryptologic side of that war is an interesting and varied story that remains to be told. All types of cryptographic systems were in use during the Vietnamese war, ranging from simple cardboard devices, to commercial cipher machines, to high tech electronic systems. This fact may come as a shock to those who believe that the electronic age has relegated hand and mechanical systems to the history books. As the years pass, we hope to know more about the cryptologic history of that conflict.

So, we recommend that readers consider these samples from the history of secret writing and draw whatever lessons seem relevant to their current pursuit. Lambros Callimahos, who was, in many respects, William Friedman's successor, was once asked if he thought that something had been lost when cryptology became electronic. The coldness of computer algorithms and the mysteries of the circuit board had replaced the older mechanical and hand systems where one could achieve a good grasp of how things worked. Callimahos immediately made a remark which, I for one, did not understand until years later. He said, "Well, somebody has to run the machines, right?" Indeed, he was right.

VIET CONG SIGINT AND U.S. ARMY COMSEC IN VIETNAM

Lt. Gen. Charles R. Myer

[Very little information has come to public notice about Viet Cong successes in reading the messages of the United States Army and the Army of the Republic of Vietnam during the Vietnam War, or about U.S. Army attempts to secure its messages. The following extract may be regarded as a first attempt to at least partially fill that gap. It tells, for example, about how Viet Cong interception enabled those forces to avoid the devastation of B-52 bombings, and, on the other hand, about American difficulties in getting soldiers to secure communications and in getting equipment to work. The extract consists of chapters 7 and 8 of an official history in the series Vietnam Studies, *Division-Level Communications 1962-1973*, by Lieutenant General Charles R. Myer, a career signals officer who served twice in Vietnam. The work was published by the Department of the Army in 1982. – David Kahn, Ed.]

Chapter VII COMMUNICATIONS SECURITY THREAT

Few communications subjects can arouse more controversy among professional soldiers than that of security. The argument centers on security versus responsiveness. The points under debate usually involve the techniques used to try to gain some degree of communications security, such as changing call signs and frequencies and using codes for the coordinates of locations and other critical data. To many people, the ideal solution to all but the frequency problem would probably be a small, lightweight, durable, unclassified black box which would automatically scramble and unscramble transmissions so that only those people with a similar black box could understand the transmission. The black box was not available when U.S. troops were committed in Vietnam.

The security measure of changing call signs was very controversial on two points. One was the confusion created by changes. Colonel Sid Berry, the commander of the 1st Brigade, 1st Infantry Division, from June 1966 to February 1967, spoke for many commanders when he said, "It simplifies communications for units and individuals to keep the same frequencies and particularly call signs. Frequent changing of call signs confuses friends more effectively than enemies." The other point was that new call signs frequently cast a poor image or were too long. Although there were many complaints about the denigrating image reflected by call signs, one of the most vivid was provided by Lieutenant Colonel Norman E. Archibald when he recalled his 1970 experiences as division signal officer of the 1st Cavalry Division: "The signal battalion still had a difficult time selling call signs like 'Supreme Capon' (... the commander said, 'What? You want everyone to call my troops castrated chickens?') ..." Brigadier General William S. Coleman, in 1967-68 assistant commander of the 1st Infantry Division, expressed both objections: "Another change that I hated to see was the fancy call signs that the division was ordered to adopt. Imagine being on the radio and saying, 'Sailing Gerta Delta One One Juliet, this Mister is Taboo Four Four.' Why, by the time you get the call signs out, you're forgotten what you wanted to say."

Although these and other complaints were understandable, a communications security problem did exist in Vietnam and had to be dealt with. Early in the war, many people apparently had assumed that the enemy was unsophisticated and that communications security did not warrant much concern. That may have been true when American troops first arrived, but the enemy quickly adjusted.

It soon became evident that he was using poor communications security to good advantage: U.S. troops found their own radio equipment when they swept enemy positions; the enemy might disappear from a location just before a planned U.S. attack; B-52 bomber strikes did not produce expected results because the enemy apparently anticipated them. Some plans probably leaked to enemy agents operating within the Vietnamese political structure at the province and district levels, where Americans had to get political clearance for operational areas to ensure that there was no conflict with local defense forces. But even these suspected leaks could not explain the problems in established free fire zones and already authorized areas of temporary operations. There had to be another source from which the enemy was obtaining critical information.

There was. On 20 December 1969 elements of the 2nd Battalion, 2nd Infantry, and of the 2nd Battalion, 28th Infantry, both from the 1st Brigade, 1st Infantry Division, overran an enemy installation about four kilometers north of Ben Suc in the II Corps zone and captured twelve enemy soldiers. They also seized an assortment of documents and communications equipment, including three U.S. Army FM radios of the AN/PRC-25/77 variety; one Chinese Communist AM receiver, compatible with the U.S. AM radios of the AN/GRC series; seven Sony transistor radios; one Panasonic receiver; one homemade receiver; and one homemade transmitter. All the equipment was in excellent operating condition, and the homemade receiver and transmitter reflected a very high quality of workmanship. With this equipment, this Viet Cong and North Vietnamese Army radio intercept unit could monitor and exploit virtually all nonsecure voice and manual Morse code communications among U.S. and allied tactical units within receiving range. Interrogated prisoners indicated that, along with building up equipment, the enemy had instituted a large number of English linguists. These linguists became an integral part of many Viet Cong and North Vietnamese units such as the intercept unit.

Among the captured documents were several booklets containing extensive instructions on proper intercept techniques and detailed analyses of the communications procedures and exploitable weaknesses of U.S. and allied units. Specific comments were included on the communications procedures of the 1st Infantry Division, 1st Cavalry Division, 4th Infantry Division, 25th Infantry Division, 11th Armored Cavalry Regiment, 5th Vietnamese Division, and the various Military Assistance Command advisory teams. These booklets were current, lengthy, and very detailed. On point of origin codes the instruction was, "... they usually use landmarks or a PO (point of origin) from which they use LEFT, RIGHT, UP and DOWN to designate a position." In one journal entry a 1st Infantry Division unit message had been copied: "Presently my one six element is at CPT Coutine (actually Checkpoint Canteen) R.6U2.2." The actual coordinates were written above the entry. That decoding points of origin and shackle codes appeared frequently throughout the journal indicated that deciphering such codes apparently posed no major problem.

The unit also noted the importance of obtaining information from warning nets. By monitoring these nets the enemy could get data on artillery and air attacks and transport of wounded. He could extract all intelligence concerning the units engaged, fire bases, landing zones, and air reconnaissance. He could also get information that would help analyze the traffic intercepted from U.S. infantry communications. The journal identified U.S. advisory and South Vietnamese unit nets as very productive for intelligence data. It further noted that the South Vietnamese made monitoring easy because they never changed call signs or encrypted messages. The journal gave examples of B-52 warning messages. It instructed the monitors to be sure to get the coordinates because they represented the box, or target area, which had to be reported for the security of their own troops.

The captured radio intercept unit was one small team targeted on several American and Republic of Vietnamese units. A logical assumption is that by 1969, when this unit was captured, the enemy could field a number of such teams. This attribute of the enemy in Vietnam must be accorded any future enemy; that is, his ability to adjust to American operational techniques and to improve his capacities. The enemy in 1969 was not what he had been. As the situation changed, he changed.

These examples are but a few of how the enemy used U.S. communications means and procedures to gather intelligence and to assist his operations. There are numerous instances on record of the enemy jamming radio frequencies and sending false messages. These bogus transmissions used imitation to try to turn fires or forces to an area chosen by the enemy. In one case the enemy tapped the internal telephone lines of a defensive base and diverted reserve forces from the area where he attacked.

General Abrams, then head of the Military Assistance Command, after examining the captured documents and being briefed on the radio intercept team incident, summed up his reaction: "This work is really rather startling; the attention to detail, complete accuracy, and thorough professionalism is amazing. These guys are reading our mail, and everyone will be informed that they are."

Chapter VIII
SECURITY RESPONSE

In the history of strategic military operations, it has long been public knowledge that Allied operations in the Pacific during World War II were indebted to the breaking of the Japanese operational communications codes. More recently it has been revealed that the British were able to break the strategic communications code by which Hitler received information and issued directives to his senior commanders. High command levels should have and did get top priority in American attempts to safeguard communications against similar breaches. With the men at the tactical level not really wanting signal security gear, there was a relatively modest attempt in the U.S. Army to develop new equipment for that level until the need became more evident.

Signal security equals a combination of people, equipment, and systems. People participating in tactical operations have a natural dislike for signal security measures; they prefer to conduct business in the simplest, most direct manner possible. Early in Vietnam, *possible* in most cases meant whatever superiors would permit in the tactical situation involved. The people part of the equation did not improve until the chain of command became convinced of the need for strict communications discipline. The conviction grew in proportion to information about the threat, and interest and participation grew as equipment and systems improved. The signal security problem was never fully resolved during the Vietnam era despite many improvements and much hard work. It was, and still is, one of the most difficult problems facing tactical communicators.

Along with the need to communicate as rapidly and directly as possible during combat, another factor worked against effective communications security: a soldier in the thick of battle feels the need for friendly assurance. This assurance can come from talking—to anyone else who has a radio. When this happens, chances are that little thought is given to what information is actually being transmitted.

Before secure voice equipment was in general use, oral communications requiring any secrecy had to be painstakingly coded and decoded manually. All too often coding was circumvented in favor of simplicity, speed, and assurance. The tactical communicator needed some type of voice encryption device.

In 1965, when he was the U.S. Army, Vietnam, signal officer, Colonel Kenneth Ring recalled that there were some eight hundred KY-8 security devices in a stateside depot. These devices had no mounting brackets or connecting cables because their application had not been settled. Colonel Ring requested, on behalf of U.S. Army, Vietnam, that a new equipment training team be sent to Vietnam to demonstrate the devices and discuss their tactical uses; the team arrived in August 1965. From this start, the security device applications and improvements progressed through some tortuous evolutions.

Voice security devices were issued to the field units in Vietnam beginning in 1965 with the KY-8 for stationary or vehicular use; the KY-8 was fully distributed by the third quarter of fiscal year 1968. The KY-28 was issued for use in aircraft beginning in 1967 and ending a year later. The KY-38, for manpack or mobile use, also was issued initially in 1967 and was fully distributed in 1968.

The voice security gear, like most newly developed equipment, had its problems. The main cause of failure was heat. The KY-8 had to be kept in well ventilated surroundings and away from direct sunlight. During a 1969 presidential visit, an overheated KY-8 temporarily stopped secure voice communications between II Field Force headquarters and Tan Son Nhut.

The 199th Light Infantry Brigade station at Xuan Loc, in the II Field Force commanding general's secure FM voice net, continually broke down. Many pieces of equipment were generating heat in a poorly ventilated bunker and causing the KY-8 to fail. When it was moved to a cooler location, the KY-8 operated normally. Commanders and communicators became aware that, for reliable communications, the security equipment had to operate in the coolest possible environment.

In at least one case the ingenuity of a well intentioned communicator backfired. First, he placed a filled Lyster (water) bag over the KY-8 to cool it, but the device still overheated. Next, he removed the cover of the KY-8 to increase ventilation. That improved the operation of the KY-8 but violated security by exposing the equipment to view and giving the enemy an opportunity to intercept intelligible signals.

In a successful innovation, the 125th Signal Battalion engineered and installed a secure land line to expedite spot reports between the 372nd Radio Relay Company and the 25th Infantry Division G-2. The battalion, operating two KY-8's in essentially a back-to-back test mode, replaced the short back-to-back cable with spiral-four cable of the required length. This system worked with almost no trouble and with very high quality.

During September 1968, the PRC-77 and KY-38 combination was issued down to the infantry battalion level, providing for the first time secure voice communications to infantry companies and in the battalion command nets. At the company level plans and operations could be discussed rapidly, safely, and explicitly without fear of enemy detection. Field units used two methods to offset the weight disadvantage (over 50 pounds) of the PRC-77/KY-38 package. One was to use the equipment fully by having one man carry the PRC-77 radio and another the KY-38. The second was to move the field unit as normal, then have supporting aircraft fly in the equipment for use in static or night defensive positions.

In the fall of 1969, reports on NESTOR (code name given narrow-band secure voice equipment) utilization were revealing. Many units under the operational control of the II Field Force were not using their KY-38s as much as desired. The PRC-77/KY-38 when carried by one man, as intended, was just too heavy. Other units reported that they needed VRC-12/KY-38 interconnecting cables. They wanted to use the smaller KY-38 in lieu of the KY-8. The KY-8 was prone to overheat and ran on generated power. Power was limited at fire support bases, and the field troops thought it unwise to run the noisy generators at night. The KY-38, on the other hand, was battery powered. Acting on this feedback, Electronics Command Laboratories went into an emergency production of interconnecting cables for the VRC-12 and the KY-38 and by late 1969 had produced three hundred. A substitution of the KY-38 for the KY-8 in vehicles was also agreed upon.

One of the most serious equipment problems in the field was the lack of kits and special cables for installing NESTOR gear in aircraft and vehicles. The installation kits and most of the cables was a supply responsibility of the Electronics Command. There was a variety of kits which adapted NESTOR equipment to the tactical series radios used in vehicles and aircraft. The initial supply of kits, once they had been developed, was adequate early in fiscal year 1969. As the use and uses of the equipment increased, however, kits and cable components became increasingly hard to find. The so-called X-mode cable for the KY-8 was also in short supply. Another aspect of the kit shortage came to light during the U.S. withdrawal from Vietnam. Although retrieving the NESTOR equipment from vehicles was no problem, redistributing it was often haphazard. A similar situation existed with KY-28's in aircraft, principally helicopters. Losses and evacuations of aircraft which had been fitted for the equipment made the shortage worse, since most of the replacement aircraft did not come with the kits installed.

The number of KY-28 keying devices for the NESTOR family of equipment was adequate when units were operating in one locale; when divisional units were dispersed (the normal practice), more keying devices were needed. The 101st Airborne Division recommended that keying devices for the KY-28 and KY-38 be supplied as required. In July 1969, II Field Force established a common NESTOR key list for all units operating in the III Corps zone. To maintain compatibility, key changes had to occur simultaneously in all units. The time chosen for this change was midnight, tactically the worst possible time because the greatest number of enemy contacts occurred from 2200 to 0200. Moreover, where several units shared the same keying device, having to move at night to change key settings was inconvenient and dangerous and added to the reasons for not using the equipment. Later the time of the daily NESTOR key change was moved to 0600.

As the use of communications security equipment increased in Vietnam, so did the need for logistical support. A reorganization was approved by the commanding general of U.S. Army, Vietnam, in mid-1967. The revised structure consisted of the Communications Security Logistics Support Center, Vietnam, which functioned as a general support facility serving U.S. Army, Vietnam, with six subordinate units strategically located in the four corps areas. The center was also assigned the mission of organizing, training, and deploying seven contact detachments to augment direct support to the combat divisions. Support began to improve.

Security codes were another problem. The early units had not been very conscious of communications security, and the unauthorized practice of using homemade communications security codes and shackles eventually became widespread. This dangerous practice, which persisted throughout the conflict and which gave users a false sense of security, was probably more dangerous than communicating in the clear. As the number of units burgeoned and the need for compartmenting code systems grew, U.S. Army, Vietnam, requested and received from the National Security Agency a whole series of preprinted codes to cover the expansion. These codes, while not perfect, were a great improvement in both the employment and distribution of such systems. Getting the infantryman under fire, or the helicopter pilot supporting him, to employ the codes was another matter.

The continuing need for easier authentication and more reliable means for passing limited classified traffic led to a new development in the later stages of the conflict: the KAL-55B authentication wheel, or "whiz wheel," a circular authentication table in a plastic covered disc. This innovation simplified communications security measures to a level acceptable to most users.

Signal security, particularly in voice radio transmissions, was a major problem area throughout the period of combat operations in Vietnam. Army Regulation 380-40, *Department of the Army Policy for Safeguarding Communications Security Information*, assigns responsibility for this type of security to both the commander and the individual. All users of communications facilities were more or less aware of their vulnerability to enemy intercept, analysis, and decoding, and of the need for authentication and encoding. The gap between this knowledge and actual practice was immense, and in Vietnam it seemed at times an insurmountable problem.

SOME VICTORIAN PERIODIC POLYALPHABETIC CIPHERS

Francine Abeles and Stanley H. Lipson

ADDRESS: Department of Mathematics, Kean College of New Jersey, Union NJ 07083 USA.

ABSTRACT: Charles L. Dodgson, better known as Lewis Carroll, invented several polyalphabetic ciphers by 1868. The two that he published are put into historical perspective in this paper. His Alphabet Cipher produces a Vigenère enciphered text, while his Telegraph Cipher is equivalent to a Beaufort cipher. In constructing the Telegraph Cipher, Dodgson used two sliding alphabets, similar to the St. Cyr slide invented by Auguste Kerckhoff fifteen years later. A program is available to explore these ciphers and another, the Variant Beaufort.

KEYWORDS: Dodgson, Carroll, cryptography, cipher, computer, Beaufort, Vigenère.

BACKGROUND

In 1843 the first electric telegraph was opened from Paddington to Slough on the Great Western Railway. The telegraph increased the volume of communications, particularly signal communication – command and reconnaissance messages. The telegraph allowed commanding officers, for the first time, the opportunity to control large armies in the field. But as military communications became more effective, the possibility of message interception increased. Tactical messages required protection – telegraph wires could be tapped. The older methods of secret code reliant on code books became outmoded because they were too easy to capture and too hard to reissue rapidly to the many and widely dispersed telegraph stations. The solution lay in a forgotten branch of cryptography – the cipher: simple, easily acquired, easily read, and secure. Ciphers could be printed at low cost on one sheet of paper and readily distributed. By using different keywords secrecy was guaranteed and even if one of the keys was learned, the entire system would not be compromised. So the field cipher was born.

The first system to be used by the military already existed. It was the Vigenère cipher, a form of polyalphabetic substitution invented in the 16th century. It fell out of favor because it was comparatively slow, difficult to use, and prone to error compared with nomenclators, the prevailing system, consisting of lists of names, words, syllables and numeric replacements. This cipher was considered *unbreakable* (until 1917) when in reality it had been broken in 1863 by the Prussian Major Friedrich Kasiski. His 95 page book, *Die Geheimschriften und die Dechiffrir-Kunst* went unnoticed, however, so the world did not know that all periodic polyalphabetic ciphers (those with repeating keywords) were not secure!

THE VIGENÈRE CIPHER

The cipher known by this name in the 19th century is actually not the one invented by Blaise de Vigenère (1585) but a simpler and less secure form of it. Nevertheless, even in its degraded form, it is the archetype of the most famous cipher system in history.

	A	B	C	D	E	F	G	H	I	J	K	L	M	N	O	P	Q	R	S	T	U	V	W	X	Y	Z
A	A	B	C	D	E	F	G	H	I	J	K	L	M	N	O	P	Q	R	S	T	U	V	W	X	Y	Z
B	B	C	D	E	F	G	H	I	J	K	L	M	N	O	P	Q	R	S	T	U	V	W	X	Y	Z	A
C	C	D	E	F	G	H	I	J	K	L	M	N	O	P	Q	R	S	T	U	V	W	X	Y	Z	A	B
D	D	E	F	G	H	I	J	K	L	M	N	O	P	Q	R	S	T	U	V	W	X	Y	Z	A	B	C
E	E	F	G	H	I	J	K	L	M	N	O	P	Q	R	S	T	U	V	W	X	Y	Z	A	B	C	D
F	F	G	H	I	J	K	L	M	N	O	P	Q	R	S	T	U	V	W	X	Y	Z	A	B	C	D	E
G	G	H	I	J	K	L	M	N	O	P	Q	R	S	T	U	V	W	X	Y	Z	A	B	C	D	E	F
H	H	I	J	K	L	M	N	O	P	Q	R	S	T	U	V	W	X	Y	Z	A	B	C	D	E	F	G
I	I	J	K	L	M	N	O	P	Q	R	S	T	U	V	W	X	Y	Z	A	B	C	D	E	F	G	H
J	J	K	L	M	N	O	P	Q	R	S	T	U	V	W	X	Y	Z	A	B	C	D	E	F	G	H	I
K	K	L	M	N	O	P	Q	R	S	T	U	V	W	X	Y	Z	A	B	C	D	E	F	G	H	I	J
L	L	M	N	O	P	Q	R	S	T	U	V	W	X	Y	Z	A	B	C	D	E	F	G	H	I	J	K
M	M	N	O	P	Q	R	S	T	U	V	W	X	Y	Z	A	B	C	D	E	F	G	H	I	J	K	L
N	N	O	P	Q	R	S	T	U	V	W	X	Y	Z	A	B	C	D	E	F	G	H	I	J	K	L	M
O	O	P	Q	R	S	T	U	V	W	X	Y	Z	A	B	C	D	E	F	G	H	I	J	K	L	M	N
P	P	Q	R	S	T	U	V	W	X	Y	Z	A	B	C	D	E	F	G	H	I	J	K	L	M	N	O
Q	Q	R	S	T	U	V	W	X	Y	Z	A	B	C	D	E	F	G	H	I	J	K	L	M	N	O	P
R	R	S	T	U	V	W	X	Y	Z	A	B	C	D	E	F	G	H	I	J	K	L	M	N	O	P	Q
W	S	T	U	V	W	X	Y	Z	A	B	C	D	E	F	G	H	I	J	K	L	M	N	O	P	Q	R
T	T	U	V	W	X	Y	Z	A	B	C	D	E	F	G	H	I	J	K	L	M	N	O	P	Q	R	S
U	U	V	W	X	Y	Z	A	B	C	D	E	F	G	H	I	J	K	L	M	N	O	P	Q	R	S	T
V	V	W	X	Y	Z	A	B	C	D	E	F	G	H	I	J	K	L	M	N	O	P	Q	R	S	T	U
W	W	X	Y	Z	A	B	C	D	E	F	G	H	I	J	K	L	M	N	O	P	Q	R	S	T	U	V
X	X	Y	Z	A	B	C	D	E	F	G	H	I	J	K	L	M	N	O	P	Q	R	S	T	U	V	W
Y	Y	Z	A	B	C	D	E	F	G	H	I	J	K	L	M	N	O	P	Q	R	S	T	U	V	W	X
Z	Z	A	B	C	D	E	F	G	H	I	J	K	L	M	N	O	P	Q	R	S	T	U	V	W	X	Y

The example that follows is a version of this polyalphabetic substitution system called The Alphabet Cipher, authored by C. L. Dodgson.[1, p. 1156-57] It was *reinvented* by Dodgson probably in 1868. The keyword, `VIGILANCE` is repeated over the message to be enciphered, called the plaintext. The intersection of the column headed by `V` and the row headed by `M` gives the cipher-letter `H`, etc. This encipherment can be represented as

Key + Plain = Cipher.

Keyword	VIGILANCEVIGILANCE
Plaintext	MEETMEONTUESDAYEVE
Ciphertext	HMKBXEBPXPMYLLYRXI

Note that different cipher letters could result from the same letter of plaintext, e.g. `M` is encoded by `V` and by `L` to give `H` and `X`. Geometrically, the cipher alphabet is translated behind (to the left of) the plain alphabet a number of places determined by the key letter. This can be seen by using the same letter, say `D`, encoding the entire alphabet, and observing the three place shift.

Key	DDDDD
	01234
Plain	ABCDE
Cipher	DEFGH
	34567

So Cipher = Plain + 3, e.g. `E` = `B` + 3.

THE BEAUFORT CIPHER

In 1857, another polyalphabetic substitution cipher was *reinvented* by retired Admiral Sir Francis Beaufort of the British Royal Navy. Its origin goes back to the early 18th century and is ascribed to Giovanni Sestri (1710). Like the Vigenère, it uses an alphabetic square which differs only in the repetition of the alphabet on all four sides, placing an "A" in all four corners.

WARWARWARWA	Keyword
MEETMEATSIX	Plaintext
KWNDONWHZOD	Ciphertext

To encipher the first letter of this example, the table is entered at row `M`, the plaintext letter, and followed to the tabular entry of the keyletter `W`; then the column heading above that entry gives the cipher letter `K`.

Again, a different cipher letter may result from the same letter of plaintext, e.g. `E` is encoded as both `W` and `N`. Geometrically, the cipher alphabet is a reversed alphabet translated forward (to the right) of the plain alphabet. Using the same keyletter, `D` and encoding the plaintext we obtain

Key	DDDDD
Plain	ABCDE
	01234
Cipher	DCBAZ
	321025

So Cipher = Plain - 3; e.g. `A` = `D - 3`. This encipherment can be represented as

$$\text{Key - Plain = Cipher},$$

which in decipherment becomes

$$\text{Key - Cipher = Plain}.$$

Another cipher, derived from the Beaufort cipher, and called Variant Beaufort can be given by its encipherment scheme,

$$\text{Plaintext - Key = Cipher}.$$

If we compare this with deciphering using the Vigenère cipher:

$$\text{Cipher - Key = Plain},$$

we see that each of these two processes is the inverse of the other.

THE TELEGRAPH CIPHER

In his diary entry of April 22, 1868 Dodgson wrote, "sitting up at night I invented a new cipher, which I think of calling The Telegraph Cipher."[3, vol 2 p.268-9] Both the Alphabet Cipher and the Telegraph Cipher were published anonymously and undated. There is no mention of the Alphabet Cipher in his diary. Dodgson's biographer and relative, Stuart Collingwood, and his bibliographer, Falconer Madan, ascribed 1868 to both ciphers. But we know that Dodgson had produced two ciphers ten years earlier, in February, 1858 – both appear in his unpublished diaries. Roger L. Green had this to say about their omission:

> The cipher is too long and complicated to reproduce here. Moreover, Dodgson printed an improved and much shorter method, The Alphabet Cipher, in 1868...and another variant, The Telegraph Cipher, in the same year.[3, vol 1, p, 140]

Apparently Green was unaware that there were actually two different ciphers in the entries of February 23 and 26 and although these are also periodic polyalphabetic ciphers, they are unrelated to the Alphabet and Telegraph ciphers. Because of their novel construction, each will be the subject of a separate paper.

From Dodgson's statement, it would seem that the Telegraph Cipher followed the Alphabet Cipher chronologically.[2] Dodgson communicated his invention to George Ward Hunt, First Lord of the Admiralty. No response is known. Dodgson's enciphering scheme is interesting. Unlike the Vigenère and Beaufort tables, it uses two sliding alphabets: a key letter alphabet A-Z and a plaintext alphabet A-Z,A.

KEY-ALPHABET

```
a b c d e f g h i j k l m n o p q r s t u v w x y z

a b c d e f g h i j k l m n o p q r s t u v w x y z a
```

MESSAGE-ALPHABET

Dodgson's instructions for using the cipher are to find the keyletter on the key alphabet, slide the plaintext alphabet under it so that the plaintext letter appears under the Keyletter. Then using the appropriate "A' on the plaintext alphabet, the cipher lettter appears on the key alphabet above that "A". This scheme produces a Beaufort enciphered text. It is entirely probable that Dodgson was aware of the 1857 cipher. What is curious, however, is that the identification of the polyalphabetic tableau and this cryptographic slide (as we will call the sliding alphabet) was not made until 1883 by Auguste Kerckhoffs, a Dutchman, in a book entitled *La Cryptographie Militaire*, considered one of the best books on cryptology. He named his cryptographic slide, the St. Cyr system, after the French military academy where it was taught. Kerckhoff's instructions showed the equivalence of the cryptographic slide and the Vigenère tableau. It was not until 1910 that Felix Delastelle, by shifting the position of the key, plaintext, ciphertext and index letters in the St. Cyr slide showed that other codes like the Beaufort cipher would result. He published his work in a book entitled, *Traite Elementaire de Cryptographie.*

CONCLUSION

Like Beaufort and Charles Babbage, Dodgson was a dilettante in cryptography. Although he did not systematically study ciphers, his inventions show that his knowledge of them was *state of the art* for his time. Codes produced by periodic polyalphabetic systems were considered best from the mid-nineteenth century until 1939. Indeed the German Enigma machine (WW II) was just an automatic (electric) encipherment of periodic polyalphabetic ciphers yielding incredible complexity! The deciphering techniques used to break it were critical to the birth of the computer. Mathematically, cipher systems enjoy group properties – indeed they are early applications of abstract groups; but more than this, they have philosophical importance. Darwin had shown earlier in the century that the determinism of natural selection depends on the disorder, the random mutation of individual genes. Cipher systems demonstrate the opposite – that disorder in the form of ciphertext can arise from a determinate system, plaintext. By the end of the century it was becoming clear that science recognized many faces of determinism. The march toward the determinism of the computing machine had begun.

NOTE ON SOFTWARE

The authors have written a program, "CIPHER_PLAY," in Pascal for compilation with Borland's Turbo Pascal 4.0. Information about the program is available from Dr. Lipson.

REFERENCES

1. Dodgson, C. L. (Lewis Carroll). 1982. *The Penguin Complete Lewis Carroll.* Harmondsworth, England: Penguin Books.

2. Dodgson, C. L. "The Telegraph Cipher." Unpublished Manuscript. Berol Collection, New York University.

3. Green, R. L., ed. 1954. *The Diaries of Lewis Carroll*, v.2. New York: Oxford University Press. 268-9.

4. Kahn, D. 1967. *The Codebreakers.* New York: Macmillan.

BIOGRAPHICAL SKETCH

Both Francine Abeles and Stanley Lipson hold doctorate degrees in mathematics from Columbia University and have recently acquired masters degrees in computer science from Stevens Institute in New Jersey. Both teach mathematics and computer science at Kean College of New Jersey. Dr. Abeles is the editor of *The Mathematical Pamphlets of Charles L. Dodgson* (in progress.)

APPENDIX

Readers will probably have little difficulty in finding Dodgson's errors and in recovering both the plaintext and the key for the following[1]:

```
"Jgmu qjl vgrv x ugemdt pupdeto?" wxxl x ugmh vj f jji,

"Ge'n ijsk tukcbb qfds fb qrug eq xud eyk cxdmfit ddjdef:

Fbu cgkskg mglb gf mstutt, had vbj, okc Ljudz jgmu nt jxxmh,

Pa st'ok krv eykgb vj dogoq grukc vqb zmfvtn as tuemak exjmh

Wrs-brs, brs-brs, wrs-brs, brs-brs; "Vrct fdjsi!" fgu Ugmh.

Njj-sjj, sjj-sjj, njj-sjj, sjj-sjj; Jji wxxl "Urbe ek fxdmh."

"Ljs'v efdn ma lb as zapsi hlgvv," wxxl Srr, "zru qrsmxbr yadsl,

Lh jtbv xok urbt jgey zxmkkm imlk, eyoe nvpdef mhtt vj mht zxjlbu,

Xv pn cz rsk xbcbbexak -- na Urdmh btblk'm wvrdu,

nht dok'm gkkkkm o ifvtsv ruut mhxm'w brxey genstxiqm gkzamc."

Njj-sjj, sjj-sjj, njj-sjj, sjj-sjj; "G urbe dogb jjo Ljudz!"

Eab-jab, jab-jab, eab-jab, jab-jab; Ugmh wxxl "Eyoe'n jjudz."
```

[1]From an unpublished manuscript by C. L. Dodgson in the *Houghton Collection* of the Pierpont Morgan Library: Houghton Alice dep. 529. Cited with the permission of the Estate of Charles L. Dodgson.

CHAOCIPHER ENTERS THE COMPUTER AGE WHEN ITS METHOD IS DISCLOSED TO CRYPTOLOGIA EDITORS

John Byrne[1], Cipher A. Deavours[2] and Louis Kruh[3]

ADDRESS: (1) c\o *Cryptologia*, Rose-Hulman Institute of Technology, Terre Haute IN 47803 USA, (2) Department of Mathematics and Computer Science, Kean College of New Jersey, Union NJ 07083 USA, and (3) 17 Alfred Road West, Merrick NY 11566 USA.

ABSTRACT: John F. Byrne invented Chaocipher in 1918 and tried unsuccessfully for almost 40 years to interest the U.S. government in his cipher system. He offered a reward to anyone who could break his cipher but the reward was never claimed. In 1989, John Byrne, son of John F. Byrne, demonstrated Chaocipher to two *Cryptologia* editors to determine if it had any commercial value. After making some improvements and providing additional information they jointly issue a new challenge to would-be solvers.

KEYWORDS: John F. Byrne, Chaocipher, Silent Years, perfect accuracy, computer algorithm, John Byrne.

BACKGROUND

In his autobiography, *Silent Years*[1], John F. Byrne, a lifelong friend of James Joyce, devoted the last chapter to Chaocipher which he had invented in 1918. Byrne described his attempts starting in 1920 to interest the State, War, and Navy Departments in his indecipherable cipher and his frustration with the disinterest shown by William F. Friedman and other cryptanalytic experts after he had demonstrated his machine.

In a letter to Friedman, February 17, 1957, who had written earlier about "why you put in that book the completely extraneous matter of ciphers when you had so much of interest to tell of Joyce," Byrne admitted that "my main reason for writing the book was to be on record in the matter of my 'Chaocipher'." Byrne was still so convinced of its indecipherability that he included an enciphered message in his book and offered a $5,000 reward to anyone who solved it within three months after the book was published. He specifically challenged Bell Laboratories, where he had also demonstrated his machine, all "cryptanalytic and cryptographic" experts, and the American Cryptogram Association and its local affiliate, the New York Cipher Society, to solve the cipher. The reward was never claimed.

Since then, however, many attempts have been made to solve the Chaocipher. (For an excellent analysis by G. E. Mellen see [4].)

Over the years many hypotheses about the cipher have been suggested and most of them were demolished in Mellen's article, e.g., it is a crude rotor system, contains an element of transposition, autokey cipher, Vernam tape system, and a form of polyalphabetic cipher with a random nonrepeating key.

An overlooked, possibly valuable clue is in *The Editor's Notebook of the American Cryptogram Association 1952-1956*. This is a detailed chronological diary kept by Henry E. Langen who was editor of the ACA's magazine, *The Cryptogram*, during that period. The diary contains a letter from Byrne, December 10, 1953, in answer to Langen's request for a demonstration of Chaocipher. Byrne said he would be delighted to demonstrate his apparatus and after many postponements, Byrne, accompanied by his wife, visited Langen in New Jersey on May 25, 1954. For two hours he discussed his book and Chaocipher but he did not bring the cipher machine "explaining that it was too heavy and cumbersome." He did bring blueprints which Langen did not understand. According to Langen, "He did explain that the machine is made up somewhat like a typewriter with two revolving disks with the alphabets arranged along the periphery in a complete disorder." Langen commented parenthetically that "With only two disks used; I am a bit confused as to how this can result in such utter chaotification of the plaintext message."

SEARCHING FOR INFORMATION

In 1973, Louis Kruh, seeking an interesting program for the 1974 annual convention of the ACA which he had volunteered to organize, thought of the unsolved Chaocipher. If he could obtain information on Chaocipher from Byrne's family it would make a fascinating program. Thus began a search for Byrne's heirs. After several months of research a letter about his hunt which appeared in the *Irish Echo*, resulted in a letter from John Byrne, son of John F, Byrne, with his telephone number. When contacted, John Byrne said he had helped his father with the cipher and was totally familiar with it. He said that perfect accuracy was essential when enciphering a message because one error would distort the rest of the cipher. But, unfortunately, he did not want to reveal how Chaocipher worked.

Kruh, not easily discouraged, continued to call or write to Byrne periodically for the next 15 years. In 1978, Byrne said his mother had died and that he was sorting out his father's papers. He also expressed concern that if Chaocipher had commercial application it shouldn't be taken from the family which was why he was reluctant to explain how it worked. In a call near the end of 1989, Byrne agreed that he had to make a decision on what to do with Chaocipher, i.e., determine whether it had commercial value and proceed accordingly or reveal its methodology in an article. Kruh suggested they meet to discuss how Byrne might make that determination and that to assist he would ask Dr. Cipher A. Deavours to attend the meeting. He explained that Deavours was a professor of computer science, co-author with Kruh of *Machine Cryptography and Modern Cryptanalysis* [2], an accomplished cryptanalyst, and an internationally known consultant in computer security matters.

Byrne agreed to the meeting and a few weeks later all parties met in Kruh's office in Manhattan. Byrne discussed his involvement with the cipher and his ambivalent feelings about revealing its secret before knowing if it had any value for commercial use. Deavours explained how Chaocipher could be analyzed for use as a computer security algorithm, how it could be protected if developed commercially, and some ways to market it. Byrne asked many questions and was genuinely interested but still hesitant when the meeting ended. But a few days later he called and said he was ready to demonstrate the Chaocipher for our evaluation.

CHAOCIPHER BECOMES A COMPUTER ALGORITHM

Saturday afternoon, March 10, 1990, in Deavours' home in New Jersey, Byrne showed Deavours and Kruh how the Chaocipher worked. They were quickly impressed with its simplicity, ease of operation and security. Deavours immediately suggested how the cipher could be expanded and adapted for implementation in computer systems and the next morning he wrote a computer program to simulate the original Chaocipher. This program was then used to decipher portions of Exhibit 1 taken from Silent Years in order to verify its accuracy.

Attempts to cryptanalyze the cipher appear to substantiate the elder Byrne's claim that "... any person on earth using a device similar to my own home-made contraption, could produce a cipher message which would be indecipherable by any other person except the one to whom the message is directed." Certainly, Chaocipher was a very secure cryptographic system for its time and even the original system would not be easy to solve today using a computer. And, an important shortcoming of the cipher – make one error during encipherment and the message is garbled beyond repair – is eliminated by its implementation on a computer.

Byrne says, "Without daring to suggest that my father would have been in the least surprised, I wish he could have experienced the speed, ease, and accuracy of computerized encryption. For me this is a stunning change indeed, considering my memories of tense and tedious hours striving for perfect accuracy while working with my father and his 'contraption' at the dining room table."

A CHALLENGE

In a letter, September 7, 1922, William F. Friedman, responding to a previous question of Byrne's about the type of material he needed to solve the Chaocipher, said, "... a series of fifty messages of approximately twenty-five words each might be sufficient ..."

Of course, in 1922, Friedman was asking for enough text to mount a pencil and paper attack on the cipher. Given the availability of computers today the need for adequate material should be greatly reduced. Accordingly, below are three messages of approximately 25 words each as an "in depth" specimen of Chaocipher (as opposed to the original exhibits in *Silent Years*, each of which was enciphered in a single continuous key). The plaintext, which is typical English prose, is taken from Stewart C. Easton's book, *Rudolf Steiner: Herald of a New Epoch* [3].

Thirty-seven years ago, John F. Byrne challenged "experts" to solve an enciphered message and collect a $5,000 reward. That message was never solved and now Byrne's son and his two associates, who have provided additional information about Chaocipher, challenge the readers of *Cryptologia* to solve the following messages within three months after publication. There is no monetary reward but all solvers who describe their method will receive a photograph of the Chaocipher equipment plus recognition in a future article.

The reader may consider this as Exhibit 5, continuing where Byrne left off in his book *Silent Years*.

```
MESSAGE 1:

MHRMC DEGBC WWQFN SDDQC UJHYH YNPYT QCXVR CSFUX QZZLH VVWIV SBFDF
QGEXV NXEAN SDEJF RVKYA RVRED QDEKZ QENIL QXVEV DXBZP URCQY QSHYY
VPLXP IKMGX Q

MESSAGE 2:

ENWSC EAQGI VIDEM WUMSN ZMNTV UFDLB JKKMR HHSNB KTJBH VPTWH FMQQJ
PGRWF FVJMD HFUZO XEOZT MKZSA MJYRL SQSXU ZYEKR JBFRE SGGFX FEGXL
PWTWL ZAVIM TBDTQ BLVRZ VEMMT LXITZ

MESSAGE 3:

JZHAS QNRTK TTLZD YOWLN VDMWN YHSMG XJMGZ QTHRI WTIFL XYHTK BOXUY
EANJU DXNVO GFZHM JEGRD GGPUG SXVBA CBEPK WHVSB IJGOH KVKAI BBNKF
HFFLS FMIIN TTXJX UHWQA PTSNB TBBNK FUCBP IONQS MVEHU XTLMR RA
```

REFERENCES

1. Byrne, John F. 1953. *Silent Years*. New York: Farrar, Straus & Young.

2. Deavours, C. A. and L. Kruh. 1985. *Machine Cryptography and Modern Cryptanalysis*. Norwood MA: Artech House, Inc.

3. Easton, Stewart C. 1980. *Rudolf Steiner: Herald of a New Epoch*. Spring Valley NY: Anthroposophic Press.

4. Mellen, Greg. 1979. J. F. Byrne and the Chaocipher, Work in Progress. *Cryptologia*. 3(3): 136-154.

BIOGRAPHICAL SKETCHES

John Byrne is an independent design consultant specializing in exhibits and visual merchandising. He is a member of I.E.A. (International Exhibitors Association) and N.A.M.E. (National Association for Museum Exhibition). After living in Ireland, San Francisco and New York, he now travels to clients from his home in Vermont where he and his wife remodeled an old slaughter-house in their "magic circle" ... a clearing in the woods beside a river. He professes almost complete ignorance of any cipher except the one he grew up with, his father's Chaocipher.

Cipher A. Deavours teaches cryptography courses at Kean College of New Jersey. He also does extensive consulting in the field of Electronic Funds Transfer.

Louis Kruh is an indefatigable researcher and collector of cryptologic memorabilia, particularly cipher devices. His interest in cryptology spans more than 40 years and he responds knowledgeably and promptly to all queries regarding literature and equipment. He is co-author with Dr. C. A. Deavours, of *Machine Cryptography and Modern Cryptanalysis*, (Artech House, Inc. 1985). Now that he has retired as Director of Advertising for New York Telephone, he is starting a new career as an attorney.

THE KEY-VOWEL CIPHER OF CHARLES L. DODGSON

Stanley H. Lipson and Francine Abeles

ADDRESS: Department of Mathematics, Kean College of New Jersey, Morris Avenue, Union NJ 07083 USA.

ABSTRACT: In his diary entry of 23 February 1858, Charles L. Dodgson (Lewis Carroll) described the first of two ciphers he invented within three days of each other. Analyzing this cipher, we show that it is a subset of a Vigenère cipher with the addition of systematically placed nulls. Discussing the Key-Vowel cipher and the later one, the Matrix cipher, we focus on the security of the key-word. Two exercises for the reader are also provided.

KEYWORDS: Dodgson, Carroll, cryptography, cipher, nulls, Vigenère, algorithm.

BACKGROUND

In two earlier papers [1,6] we have reported on Dodgson's work on ciphers. In the first paper we analyzed ciphers invented in 1868; in the second, a cipher from 1858. The subject of this paper is an earlier 1858 cipher that Dodgson described in his diary entry of February 23 [4]:

Invented a system of cipher, which I think looks promising, as it may be carried entirely in the head.

1. Agree on some word as the key: for every vowel in the key you write a letter representing some letter in the message, and for every consonant you write any letter at random.

2. The rule of the vowels is as follows: Suppose M the letter to be represented: imagine it written

$$\overset{a}{K}\overset{e}{L}\overset{i}{M}\overset{o}{N}\overset{u}{O}.$$

Then if "a" is the key - vowel, write K, and so on.

3. Begin with some unmeaning letters, the number regulated by the first vowel of the key [a = 1, e = 2, i = 3, o = 4, u = 5] and end with some, regulated by the last key vowel.

EXAMPLE

Let the key be $\underline{i}\ \underline{m}\ \underline{a}\ \underline{g}\ \underline{i}\ \underline{n}\ \underline{e}$. As there are 4 vowels every 4 letters of the message will use it once, producing 7 letters of cypher. Let the message be $\underline{\text{come directly}}$. We begin with 3 unmeaning letters, e.g. L A T – then for each letter of the key we proceed as follows:

for (i) symbol for $\underline{c}$ = C
(m) unmeaning – = H
(a) symbol for $\underline{o}$ = M $[\overset{a}{m}, \overset{e}{n}, \overset{i}{o}]$
(g) unmeaning – = E
(i) symbol for $\underline{m}$ = M
(n) unmeaning – = S
(e) symbol for $\underline{e}$ = D

this gives us LATCHMEMSD.

To retranslate this suppose it written

$$\text{LAT}\ \overset{i}{C}\ \overset{m}{H}\ \overset{a}{M}\ \overset{g}{E}\ \overset{i}{M}\ \overset{n}{S}\ \overset{e}{D}$$

Discard the first 3 letters, and every other one with a consonant over it –

As $\overset{i}{c}$ has an $\underline{i}$ over it, it = c. For $\overset{a}{M}$ we write $\overset{a}{M}\ ^{\underline{ei}}$ and run the alphabet on until till we get under $\underline{i}$ – this gives o. The next $\overset{i}{M}$ having $\underline{i}$ over it = m and $\overset{e}{D}\ ^{\underline{i}}$ by the same rule gives e.

ANALYSIS

If it is not immediately apparent, the Key-Vowel Cipher can be considered a subset of a Vigenère Cipher by replacing the key - vowels (a e, i, o, and u) respectively with Y, Z, A, B, and C and using just five of the twenty-six available alphabets. Which alphabets are used, and the precise sequence in which they are used, depends solely on the pattern of vowels within the key word. Consider the following tableau:

	A	B	C	D	E	F	G	H	I	J	K	L	M	N	O	P	Q	R	S	T	U	V	W	X	Y	Z
A	Y	Z	A	B	C	D	E	F	G	H	I	J	K	L	M	N	O	P	Q	R	S	T	U	V	W	X
E	Z	A	B	C	D	E	F	G	H	I	J	K	L	M	N	O	P	Q	R	S	T	U	V	W	X	Y
I	A	B	C	D	E	F	G	H	I	J	K	L	M	N	O	P	Q	R	S	T	U	V	W	X	Y	Z
O	B	C	D	E	F	G	H	I	J	K	L	M	N	O	P	Q	R	S	T	U	V	W	X	Y	Z	A
U	C	D	E	F	G	H	I	J	K	L	M	N	O	P	Q	R	S	T	U	V	W	X	Y	Z	A	B

Using the vowels in the keyword IMAGINE to index the rows, and the letters of the plaintext COMEDIRECTLY to index the columns, we locate the letters of the ciphertext in the table:

I	M	A	G	I	N	E	I	M	A	G	I	N	E	I	M	A	G	I	N	E
C		O		M		E	D		I		R		E	C		T		L		Y
C	_	M	_	M	_	D	D	_	G	_	R	_	D	C	_	R	_	L	_	X

Dodgson filled in the consonants with random letters:

C H M E M S D D F G E R X D C O R D L M X,

then added three random nulls, L A T, at the beginning since i–3 is the first vowel, and two, D N, at the end since e–2 is the last vowel in the key - word. Thus the complete encipherment is:

L A T C H M E M S D D F G E R X D C O R D L M X D N.

Observe that if a key word contains even multiple instances of a single vowel, the resulting ciphertext would be the equivalent of a Caesar type cipher with nulls embedded at regular intervals. Dodgson's message enciphered with SEVENTEEN, for example yields:

S	E	V	E	N	T	E	E	N	S	E	V	E	N	T	E	E	N	S	E	V	E	N	T	E	E	N
	C		O			M	E			D		I			R	E			C		T			L	Y	
_	B	_	N	_	_	L	D	_	_	C	_	H	_	_	Q	D	_	_	B	_	S	_	_	K	X	_

Furthermore, if the key word contained only the vowel 'i', as does MISSISSIPPI, the resulting ciphertext would contain the message in clear hidden amongst the nulls.

Commenting on the Key-Vowel Cipher Dodgson concluded his diary entry:

> The great advantage which I think this has is, that no one, even if he knows the system, can read a piece of cypher done by it, without also knowing the key - word. And this the correspondents may change as often as they like – or even use a whole sentence as key.

For his day, Dodgson may very well have been close to being correct. The introduction of nulls into the message, even in such a regular fashion, was probably too much for Victorian cryptanalysts to handle. These meaningless extra characters obscured the normal word divisions that Charles Babbage had exploited so successfully in the solution of polyalphabetic systems.[5, p.205] Étienne Bazeries, who is credited as being one of the greatest natural cryptanalysts, is known for reconstructing the "Grand Chiffre de 1691," a code with about 600 entries including nulls.[5, pp.245-246] However, the introduction of six nulls at the beginning and end of a polyalphabetic cipher in 1892 gave him such trouble that when he finally did solve it, he came to regard that solution as one of his best.[5, pp.245-246]

Dodgson abandoned and never published the Key-Vowel Cipher: indeed his diary entry three days later described the Matrix Cipher as being "far better".[3] On what basis? The advantages he claimed for the Matrix Cipher repeated the language he used with his earlier effort, with one important difference. For the Matrix Cipher, Dodgson added the claim that even if the plaintext is known, it is not possible to recover the key-word. If Dodgson had found reason to believe that the key-word in the Key-Vowel Cipher was vulnerable, he made no mention of it. Nevertheless his focus on it in the second cipher is sufficient motivation to examine the security of the key-word in the first cipher.

Consider Dodgson's example, and suppose that both the plaintext and the cipher text are available:

C O M E D I R E C T L Y
L A T C H M E M S D D F G E R X D C O R D L M X D N.

The general strategy is to slide the plaintext along the ciphertext looking for possible matches while watching for the inevitable repetitions of pattern to reveal the length of the key. Expanding the plaintext over the five alphabets:

```
A - a m k c b g p c a
E - b n l d c h q d b
I - C O M E D I R E C...
O - d p n f e j s f d
U - e q o g f k t g e
```

The first letter of the plaintext C could match with the second or fourth letters of the ciphertext. If it matches with the A, its key-vowel is A indicating one null, 'L', at the beginning. If it enciphers as the fourth letter, C, its key-vowel must be I, which indicates three nulls 'L A T' at the beginning. Since the second letter O of the plaintext can only be enciphered as M, N, O, P, or Q, it must correspond to either the 6th or 8th letters of the ciphertext indicating a key-vowel of A in either case. If, however, it corresponds to the 8th letter, there is no appropriate ciphertext for the next letter M until the 19th, and that would leave ten consecutive consonants. E, D, and I could map anywhere within 'D D F G', but if I does not correspond to G, once again there will be a gap with too many consonants in a row. Likewise, R must encipher to the 15th letter which yields:

```
Known              O   M           I   R
                   a   i           a   i
          L A T C H M E M S D D F G E R X D C O R D L M X D N.
                                E
Possible    C   C             E D D
```

At this point it is tempting to try a key-word length of 7 which forces:

```
      C   O   M   E D   I   R
      i   a   i   e i   a   i
L A T C H M E M S D D F G E R X D C O R D L M X D N.
```

The rest follows quickly from the pseudo-key 'I _ A _ I _ E'. A check of the dictionary reveals at least six words that will do as equally well as 'I M A G I N E'. This method works quite well and generally reveals a solution within two lengths of the key.

Recovery of the plaintext from the ciphertext alone is another matter, but modern tools provide sufficient entries. Surprisingly enough, statistical methods are sufficiently sensitive to reveal the periodicity in spite of the presence of a large percentage of nulls. The cipher may be solved simply by applying a kappa test on decimations to learn the length of the key-word.[3, pp.16-17] Next, a test of the distribution at each successive position will reveal the location of the nulls and permit them to be stripped off. At this point, the cipher has been reduced to a Vigenère and the solution is straightforward.

By now the reader may have anticipated the potential for a disasterous flaw to enter the system. If multiple copies of a message are sent to different individuals, possibly by different clerks, but using the same keyword, as might be the case if the cipher were used in the field, the ciphertexts would differ only in the randomly assigned nulls. A comparison of any two such texts would render the solution trivial.

Whatever reason Dodgson had for switching to a "far better" system, several patterns are evident in his exposition that were to characterize his approach both to ciphers and to mathematics in general for most of his life. First of all, he saw an underlying regularity and mathematical structure in the system:

"As there are 4 vowels [in IMAGINE] every 4 letters of the message will use it once, producing 7 letters of cypher." He exploited this cyclic feature to the utmost in producing the Matrix Cipher. The second characteristic was his insistence that the system be "carried entirely in the head." Third, and possibly the most significant was the procedural manner in which he presented the cipher. Much of Dodgson's work is characterized by a prescription of steps to be carried out. This algorithmic approach was not unique to Dodgson, of course, and cipher systems are particularly well described by a set of instructions.

In his approach both to cryptography and to mathematics in general, Charles Dodgson, like that other nineteenth century cryptographer, Charles Babbage, seems to have anticipated developments that were to evolve over the next century.

REFERENCES

1. Abeles, F. and S. H. Lipson. Some Victorian Periodic Polyalphabetic Ciphers. *Cryptologia.* 14(2): 128-134.

2. Candela, R. 1938. *The Military Cipher of Commandant Bazeries.* New York: Cardanus Press.

3. Deavours, C. 1987. *Cryptanalytic Programs for the IBM PC.* Laguna Hills CA: Aegean Park Press.

4. Dodgson, C. L. Diaries, 23 February 1858.

5. Kahn, D. 1967. *The Codebreakers.* New York: Macmillan.

6. Lipson, S. H. and F. Abeles. The Matrix Cipher of C. L. Dodgson. *Cryptologia.* 14(1): 28-36.

BIOGRAPHICAL SKETCH

Both Francine Abeles and Stanley Lipson hold doctorate degrees in mathematics from Columbia University and have recently earned masters degrees in computer science from Stevens Institute of Technology. Both teach mathematics and computer science at Kean College of New Jersey.

Dr. Abeles is the editor of *The Mathematical Pamphlets of Charles L. Dodgson* (in progress.)

EXERCISES

1. Recover the key in the following ciphertext having the plaintext:

LEWIS CARROLL IS CHARLES DODGSON

```
T P F S M X G E W J K Z T I E G A S M T S C Q B L M V G J J U

N C I P S B L T W L F J Q T N F K O E S O H P U X O O F W T C
```

2. Recover both the plaintext and the key.

```
DHQDR QXECC OPAAV CAERW YJPXO JEFTM HUKYE BMPLH AYHYI WENTD EDBJX

TCTFR RWOLO XMMTT HEKTE LGCIN PJLSH KPERV TLEBE XBXTD AJDBL QMONV

EVJVI CVSAE NLAXO NITHI EVXIR GIMAX TCDVT XCERE BMWBU ITTMM OTIUD

JDERW NOMQT QLOOM LJCDS EGPRT OYMEV KMIDW EBVHS HOUFV FIKHI NBCIC

EDSBN UPTEJ NTUIT QKRIE ETBNS RQSUI RPGWP IMRID SYKJI QXNGN LQCOY

LBENT OVIHU FXGHL SFWKT BHATY IJKLS METIN CIEDA TNLMM EBJIT HWHOT

DIDNS FVARE EUTKS PYUFX FJTCI UVCID EXYIN KCTLH YRYOS KOENV SJJQI

JHTIQ VGYLE NSTON RADBE UHVEA AJIOL WFTHQ EEVDP OWERU IBBAO LLDIO

CDHBI ODTYM IEBFN YLSPM IHCIT HKOOK FINUT YNHEP PRVFR KCESG EDKYN

GLCEQ OADTF AFALC ALLSR TUGEG PUKGE PHRCF ERLCN HETAR GUCUE CKOFX

NQEGU OGLLW SH
```

THE CYLINDER-CIPHER

David W. Gaddy

ADDRESS: The Riverside, Box 46, Tappahannock VA 22560-9765 USA.

ABSTRACT: Questions Thomas Jefferson's attributed invention of the "cylinder cipher," notes similar devices before and after his time, and introduces a "mystery device" of this nature in the National Cryptologic Museum, Fort George G. Meade MD.

KEYWORDS: CYLINDER CIPHER, Thomas Jefferson, Étienne Bazeries, Parker Hitt, Joseph Mauborgne, wheel cipher, Charles Babbage, Baron Fridric Gripenstierna, Francis Bacon, "Star" cipher, William F. Friedman, Riverbank Laboratories, M-94, CSP 488, CSP 493, French language cipher

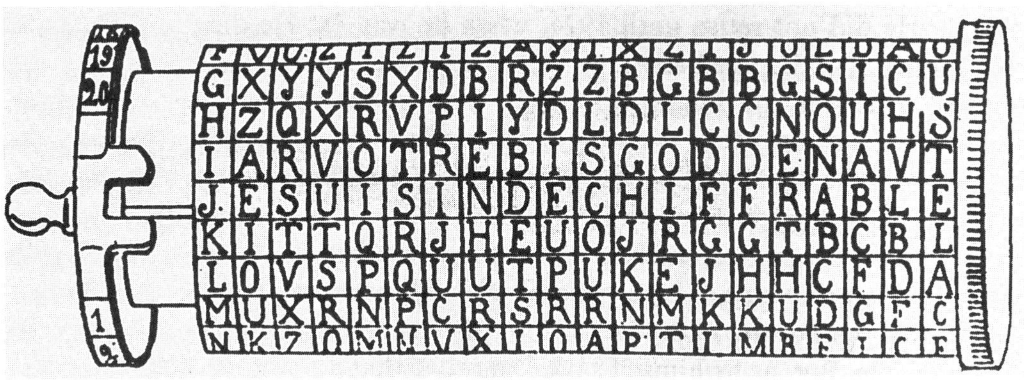

Figure 1. The Bazeries device:. The plain text along its middle line boasts "I am undecipherable" in French.

Friedman and Riverbank Laboratories called it "multiplex," or "the Star cipher;" Bazeries, the cylinder - orcylindrical-cipher; Thomas Jefferson used the term "wheel cipher." All three described the same type of cipher device: on a shaft a given number of disks (each individually numbered) are arranged in a prescribed order and placed side by side. Each disk or wheel (or, to use Bazeries' term, *rondelle*) bears on its outer edge scrambled alphabetical letters. Aligning any one row to the desired plain text, another row is copied as cipher text. The recipient arranges the line of cipher text on a row of his device (with disks set

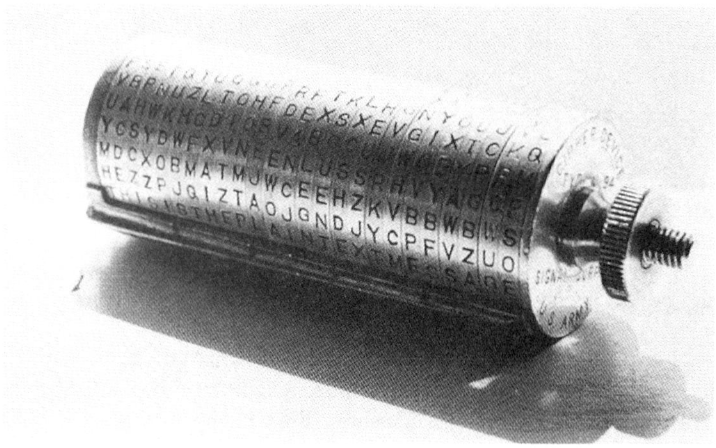

Figure 2. U.S. Army Cipher Device M-94.

in the same order as his correspondent) and turns the whole to find and read off the plain text. French army Major Étienne Bazeries described it in his 1901 book, *Les Chiffres Secrets Dévoilés*, published after a decade of frustration because French authorities declined to adopt the system, which was believed to have originated with the author (Figure 1). Two decades later a similar device, developed in the United States by Captain Parker Hitt of the U.S. Army, familiar with Bazeries' device, and enhanced by Major Joseph O. Mauborgne, was introduced by the Signal Corps in 1921 as its M-94 (CSP 488 in the Navy six years later; CSP 493 in the Coast Guard by 1939) (Figure 2). Virtually on the date of its introduction into military service, a scholar at the Library of Congress chanced upon a description of such a device in the papers of Thomas Jefferson. Marvelous! was the reaction: Jefferson, the intellectual giant, had anticipated the twentieth century inventions. Given his reputation, attribution came easy, and it was flattering to American ego that their exceptional fellow countryman had accomplished yet another feat. This is the accepted, romantic story surrounding the M-94, unwitting copy of Jefferson's "wheel cipher".[1]

[1] See, as a basic reference, David Kahn, *The Codebreakers* (New York: The Macmillan Company, 1967), pp. 181-186. The eminent American cryptologist, William F. Friedman, accepted the Jefferson story, as reflected in Friedman's "Lectures" of the mid- to late 1950s. See National Security Agency, Center for Cryptologic History, [Donald K. Snyder, ed.,] *The Friedman Legacy: A Tribute to William and Elizebeth Friedman* , 1992, pp. 192-193, at 193: "Jefferson anticipated Bazeries by over a century, and the Hitt-Mauborgne combination by almost a century and a half." (Bazeries had discussed his idea in an earlier (1891) paper.) In his seminal work, *U.S. Diplomatic Codes and Ciphers, 1775-1938* (Chicago: Precedent Publishing Inc, 1979), pp. 170-173, 178-179, Ralph E. Weber found no reason to challenge the Jeffersonian origin. The "official" army history of strip cipher devices (SRH-366) credited Jefferson as the originator (p. 20), doubtless reflecting Friedman's view. (The present article intentionally skirts the relationship of strips to disks and the evolution of the M-94.)

Figure 3. Unidentified Cylinder-Cipher in the U.S. National Cryptologic Museum.

In the early 1980s, the National Security Agency acquired for its museum collection an obviously old, but otherwise unidentified, device of this general type. Its size made it a "desk model," as opposed to a hand-held device (Figure 3). Accented letters showed its application to the French language. Its evident age and source (it was acquired from West Virginia) made speculation of Jeffersonian association tantalizing and reopened curiosity about antecedents of the cylinder-cipher. The device obviously is incomplete: A brass frame, comprised of a base and two uprights, holds a wooden shaft on which are placed 35 wooden disks (probably boxwood) numbered in red ink - from spacing and numbering, evidently five are missing. The disks had been placed arbitrarily, some reading upside down with respect to others (i.e., front and back are not obvious as the device is viewed).[2] Remains of a cloth covering - perhaps linen - can still be seen on the shaft, possibly placed there to keep the disks from slipping too freely while being adjusted. Other pieces are missing: evidently a horizontal brass bar held the two uprights against the shaft and perhaps served as an aid to viewing; the disks have 42 holes or perforations near the outer edge and were evidently

[2] An elderly lady recalled playing with the device as a child. Further, missing screws from the base have been crudely replaced with wood screws.

locked in place by a thin rod passing through an upright and through the aligned holes. On the edge of each disk, 42 characters (letters, including accented vowels; numerals; punctuation) are hand-printed (not stamped or incised) in black ink. (The accumulated grime and friction of handling and age blur or obliterate many of the inscriptions; some have been lost in the separation of chips of wood from the edge.) Each disk is numbered on the face in red ink (numbers 35, 36, and presumed 38, 39, and 40 are missing.) Informal examination of the device by specialists at the Smithsonian in 1991 concluded that it was of European origin, possibly French, made, in all probability, during the second quarter of the 19th Century. (Jefferson died in 1826.)

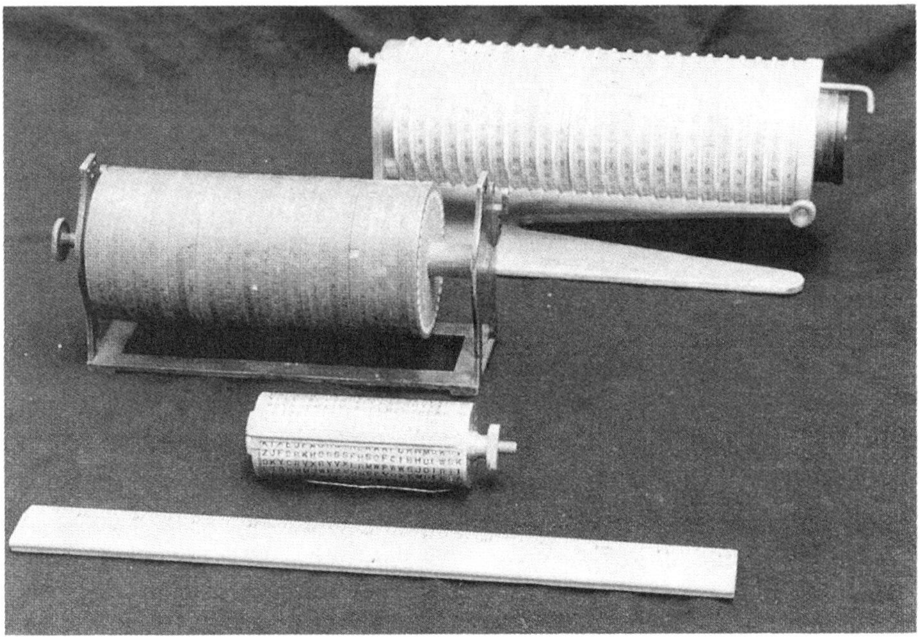

Figure 4. M-136 (1933), "Mystery Device," M-94 (1921).

But what is it? Nicely executed, it is no toy, but who used it, and for what purpose of communication? Does its companion lie in some museum of French or European cryptographic relics, or, is it described in a manuscript? How did it come to West Virginia? From such questions came, first, a reexamination of the Jefferson "invention" and devices of this design.

A recent Jefferson study by Silvio Bedini of the Smithsonian[3] repeats the

[3]Silvio A. Bedini, *Thomas Jefferson, Statesman of Science* (New York: Macmillan Publishing Company, 1990).

attribution of the seminal idea to Jefferson and quotes one of two copies of the manuscript in which the device is described. While accepting the idea that Jefferson invented it, Bedini was nevertheless careful and cautious with qualifying remarks which could support the notion that Jefferson was not inventing but adopting an idea seen elsewhere - not an unusual practice on his part.[4]

Jefferson's device, that of Bazeries, and the M-94 have a remarkable similarity in size and make-up - they can easily be held in one hand and manipulated with the other. The M-94 uses 25 disks bearing the 26 letters of the alphabet, mixed. Bazeries used 20 disks of 25 letters. In the surviving description of Jefferson's version,[5] the dating of which is speculative, but seems to have been around the turn of the nineteenth century,[6] he does not specify the number of disks, but simple arithmetic converts his "cylinder of wood" 6 or 8 inches long, sliced in disks of about one-sixth of an inch thickness, into 36 or 48 disks (not allowing for kerf waste).[7]

In a 1980 article,[8] cryptographic collector-scholar Louis Kruh traced these relationships (as well as the relationship between disks and strips to produce the same effect) and suggested that, in concept, they stemmed from ancient letter locks of similar design, not unlike today's bicycle combination locks that use digits instead of letters. Mr. Kruh also called attention to an overlooked statement by British mathematician-scientist Charles Babbage, published in 1854,

[4]Bedini (ibid.) also pictures between pages 462-463 in Figure 29 a Vigenère variant - a conventional, keyed alphabetical tableau, but with the addition of the ampersand at the end of the alphabet, and a provision for enciphering digits - possibly Jefferson's innovations. This President Jefferson provided to Lewis and Clark for their 1804 expedition of discovery in the west. If no earlier version of the Vigenére were known, Jefferson may as well have been credited with inventing that cipher.

[5]Bedini, 504-505, Document 7: "Project of a Cipher."

[6]Bedini, 236, concluded c1792-1793, but other scholars suggest a later date. Weber, *op.cit.*, 170, suggested "sometime prior to March 22, 1802."

[7]Mr. Bedini had a replica made after the Jeffersonian description, using 36 disks with the 26 letters Jefferson prescribed stamped into the wood with dies. The Bedini replica is executed in a different, and evidently softer, wood than the boxwood of the "mystery" device in Figure 4. The disks of the latter have a hard, thin outer layer that would resist, if not flake off under, the pressure of a die. Bedini's speculation that Jefferson intended to impress his letters with a die was based upon the discovery that Jefferson had purchased a set of "punches" perhaps for that very purpose . . . or to mark his garden tools. While impressed letters would have avoided the tendency over time to rub off, as has been the case with the "India ink" of the mystery device, whether or not Jefferson used one technique or the other remains open. There is no indication that Bazeries intended other than inscription, raising the question of how early stamped or incised lettering might have been employed for such purposes (Sumerians and Babylonians excluded). See Bedini, 236-243, with speculation concerning the 1792 purchase of an unspecified "set of punches" from a Philadelphia type-founder at 237.

[8]Louis Kruh, "The Genesis of the Jefferson/Bazeries Cipher Device." *Cryptologia* , October 1981 5(4): 193-208. (An earlier version appeared in *The Army Communicator* , Winter 1980 5(1): 48-54, under the title, "The Evolution of Communications Security Devices.") Cryptanalysis of this type of device has appeared in Greg Mellen and Lloyd Greenwood, "The Cryptology of Multiplex Systems," *Cryptologia* , January 1977 1(1): 4-16 and April 1977 1(2): 150-165, reprinted in Cipher A. Deavours, et al., eds., *Cryptology - Yesterday, Today, and Tomorrow* (Norwood, MA: Artech House, 1987.)

that cipher systems of this general type were not new, and that "the best form is, perhaps, rings of box-wood placed side by side on a cylinder, and having the twenty-six letters on the circumference of each," a concise description of the Jefferson-Bazeries "invention."

The theory of independent invention continues to be an interesting subject, but surely logic compels that that be a last-choice explanation. Jefferson did not say or even imply that he had invented something new, any more than in the case of the Vigenère-variant tableau he supplied to the Lewis-Clark expedition, which could as well (barring evidence to the contrary) be attributed to the fertile brain of Jefferson. It would seem more reasonable to assume that, in both cases, Jefferson was aware, through his wide reading, his acquaintances, or his direct observation, of useful ideas, which he adopted directly or with innovations of his own.

In fact, Jefferson's use of the term, "wheel cipher," may well point us to an earlier source: Francis Bacon's 1605 *The Advancement of Learning* (in the 1640 English translation by Gilbert Wats) recounts various forms of cipher, among which are the "Wheele-Cyphars," not further elaborated. (Bacon's 1623 *De Augmentis Scientarium* renders the Latin as "Ciphare Rotae").[9]

Over a century and a half later a somewhat comparable device was put forward in 1786 as the invention of a Swedish officer, Baron Fridric Gripenstierna. He offered to King Gustav III: "a wooden construction in the form of a cylinder with a number of alphabetic rings."[10]

As an alternative theory, then, suppose that, behind the scenes, unknown to the outside world, certain devices were used by cryptographers serving the courts of Europe - devices known only within that small corps of initiates, learned about through espionage or other means, but rarely documented or falling into the hands of outsiders. It is reasonable, then, to believe that - through Bacon, the manuscript describing the Swedish device, the Jefferson manuscript, the Babbage allusion ... and now this "mystery device" - we have been afforded those rare glimpses, hinting at, but not clearly giving us, a sense of connectivity behind the scenes: That Jefferson had seen an earlier version of the device he described. That Brazeries, too, was putting forward a device not entirely of his own invention. Hitt and Mauborgne relate to Bazeries. Somewhere in that continuum behind the scenes we place this "mystery" device.

---

[9]Quoted in Penn Leary, *The. Second. Cryptographic. Shake-speare.* The Enlarged Second Edition. Omaha, Nebraska: Westchester House Publishers, 1990, pp. 123 and 125-6.

[10]Josef Garlinski, *The Enigma War.* New York: Charles Scribner's Sons, 1979. Some additional detail is given in Boris Hagelin, "The Story of the Hagelin-Cryptos." Zug, Switzerland: Crypto-AG, 1981, pp. 5-7. A conjectural illustration of a "desk model" operated by two people is shown by Garlinski following p. 60, as in Hagelin, p. 7.

Finally (for the moment), we can add another ingredient to this mystery of behind-the- scenes continuum, another "desk model," as similar to the "mystery device" as the M-94 was to the Bazeries or Jefferson: the U.S. Army Signal Corps M-136 of 1933 (Figure 4). Friedman was experimenting with a method whereby the advantages of the M-94 could be expanded by being able, not only to re-order the disks themselves on the shaft, but to change the letters appearing on the edge of the disk. "Cipher device, type M-136" was the result: It employed 25 disks, each bearing an alphabetical strip of 26 mixed letters, the strips held in place by a spring-and-clip arrangement. The resultant device (evidently only one of which was constructed before the idea was abandoned) bears a striking similarity to the "mystery device" in size and appearance, even to the inclusion of a rod running the length of the shaft through the disks to hold them in place. [11] No information at this writing suggests that Friedman was aware of our "mystery device," or was thinking in terms other than modifying the M-94.

These musings have in common three points: doubt that Jefferson was the inventor or originator of the device associated with him thus far; curiosity as to the identity of the "mystery device;" and a renewed conviction that, given the nature of state cryptography, one must be cautious in ascribing "firsts" in that field as isolated and fragmentary examples come to light. The "mystery device" is on display at NSA's National Cryptologic Museum as the oldest such device known to its curator. Perhaps a reader can solve the mystery.

BIOGRAPHICAL SKETCH

After a forty-one year career in cryptology, Mr. Gaddy retired in 1994 as chief historian of the National Security Agency, having established the Center for Cryptologic History and the National Cryptographic Museum, both at Fort Meade, Maryland. He is a student of American Civil War history, with emphasis on the Signal Corps, cryptography, and secret service activities of the Confederate States of America.

[11] The only known model is in the collection of the National Cryptologic Museum, Center for Cryptologic History, at NSA. SRH-366 (pp. 43-44) implies that only one was made (by National Electric Supply Company, Washington DC, at a cost of $84) and that no instructions were ever written.

A WORLD WAR II GERMAN ARMY FIELD CIPHER AND HOW WE BROKE IT

Charles David

ADDRESS: 75 N. Strathmore Street, Valley Stream NY 11581 USA.

ABSTRACT: This article offers a first-hand account of the breaking of a German medium-grade cipher during Word War II and personal recollections of the process and history.

KEYWORDS: Vint Hill Farms Station, Weymouth St. London, Signal Security Detachment "D", Doppelkastenschlüssel, cryptanalytic procedures.

In 1942 the U. S. Army Signal Corps was looking for college graduates to train in radar work, which was then new and promising. I enlisted and after six months of pre-radar courses in calculus, physics, chemistry, and radio theory at Rutgers University, I found myself in Camp Crowder, Missouri, the Signal Corps basic training center.

Once there, a classification sergeant informed me that Camp Murphy, the radar facility in Florida, was overcrowded, and that I'd be assigned to some other training. Checking my academic record and Army Classification Test score, he suddenly asked me if I had ever heard of "cryptography".

As he thumbed through his manual, my memory reverted to a cryptanalysis course that my fiancée had taken at Brooklyn College, given by Professor Jack Wolfe of the math department. We sat in a sunny meadow in Prospect Park, and I helped with the frequency counts. My reverie was broken by the information that I was to be sent to Vint Hill Farms Station, near Warrenton, Virginia—close enough to my New York home and my wife-to-be to please me no end.

Vint Hill was an unusual army camp—in a bucolic setting, with evergreen trees surrounding the barracks, and woods all around. It was a hush-hush place, and we were constantly warned to keep it so. One of its two parts was devoted to cryptanalysis studies, and the other was a working radio facility.

The students were men of strong academic backgrounds and achievements. Of course, there was a large supporting cast of administrative cadre who tormented us with the usual army routine—drills, hikes, KP, inspections and the like. However, there was communion among the "crypt" people due to the learning atmosphere, and the intellectual interest engendered by these new and uncommon studies.

Our classes were taught by sergeants. Once in a while an officer would appear from Arlington Hall, but Vint Hill was an enlisted men's camp and school. We religiously followed the texts of William Friedman, and considered him our mentor. Men left after unpredictable time periods to join active units in both main theaters of the war. Most seemed to end up in our own signal centers, and were involved with security matters.

After seven months I was called in for an interview, and asked if I knew any German. I had studied the language for two years in high school, and was able to read and translate a newspaper handed to me by an officer. He said that would do just fine, and I was assigned to a special class being formed.

The new class of about 30 men was taught by a very bright Sergeant Dineen. He had just returned from Arlington Hall where he had been briefed on two German systems—one a lower-grade code, and the other a medium-grade field cipher which the British had started to break. We learned about the cryptographic nature of the systems, and then something of the analysis procedures. I knew then that working on them would be intriguing beyond compare. I found out I was right when I got a chance to work on the cipher as a member of Signal Security Detachment D, an element of General Omar Bradley's 12th Army Group.

In April, 1944, I and other embryonic codebreakers debarked from a large troop ship in the bombed out port of Liverpool, England. Our group was brought to London and billeted near Marble Arch. The next day we were brought to an apartment building in Weymouth Street that had been given to the Signal Corps for its intelligence work. A number of comrades and I were assigned to a breaking section working on the German army medium grade cipher. We had been familiarized with it during our last month at Vint Hill and told that the British had made good progress with the breaking. The system was used by all levels of the German army, from army groups on down, seemingly where the highest-level means (Enigma) was not necessary or available. The encryption method was a clever variation of the Playfair that rendered breaking very difficult. We called it **NI**, short for Non Indicator, as the early intercepts showed no indicator. To the enemy it was Doppelkastenschlussel (Two-Box Cipher).

At Weymouth Street it became apparent that all the operational work was performed by us enlisted men. This contrasted with the British Army, where our equivalents were officers. Our men and officers were mostly professional people and scholars. On the whole they were older than the average G.I.

Our first weeks were spent on practice problems, and perfecting our skills. We also took a class in military German. Although all of the men had had some past school contact with the language, this taught us likely army vocabulary. Oddly, however, many of the people who succeeded best in codebreaking had limited facility with German. The language experts, both trained and German-born, tended to see more than really was present in cipher text, and neglected the more important principles as frequency and combination which led to gradual, but surer results.

Soon we were thrown in with the earlier arrivals, and we began to work on current intercepts. Before the 1944 invasion, the enemy signal units on the Atlantic Wall practiced their cryptography and radio transmission constantly. Often they sent personal messages, newspaper articles, nursery rhymes, and the like. As they practiced, so could we, and thus our skills were improved.

When the Normandy invasion began, the breaking proceeded apace. We gleaned both tactical and strategic information which was coordinated with other forms of signal intelligence, such as direction finding and traffic analysis. As our armies pressed inland, it was decided to form a mobile unit to accompany the advancing allied armies. This was necessary for two reasons. Firstly, our radio intercept companies had to be near the front in order to read enemy traffic more accurately. Secondly, they could send intercepts to us quickly by messenger.

I was chosen for this unit along with a mixture of the newer and the more experienced men. The unit was called Signal Security Detachment D, afterwards always S.S.D.D., and was commanded by Lieutenant Colonel Charles Allen. We were attached to General Omar Bradley's 12th Army Group, but we always stayed by ourselves in the field.

S.S.D.D. consisted of various sections representing all phases of signal intelligence:

1. Cipher-breaking team. This was my group. It was headed by Sergeant Howard Arnold, a fine cryptanalyst from Providence, Rhode Island who later became president of a large department store in his home city. Besides continuing his own breaking activity, Howard supervised the group in a laid-back and thoughtful manner.

2. Codebreaking section, headed by Sergeant James Wallace of North Dakota, a skilled and assiduous codebreaker. After the war Jim became a newspaper columnist in Brainerd, Minnesota. The enemy used code for lower-grade messages and for smaller units. He and his men were very successful.

3. Traffic analysis was the largest group. It studied message headings, radio frequencies, code signs of sending and receiving units, volumes of traffic, and more, to derive indications of enemy placements and intentions. Led by Sergeant George Bauer, it was consistently productive.

4. Direction-finding team, headed by my tentmate Sergeant Leonard Netzorg, successful Yale Law School graduate. It located and identified enemy units by triangulating bearings. Len later became an eminent attorney in Portland, Oregon, whose career and views were the subject of a recent magazine article.

5. Emending Section. These German language experts removed mistakes from interpreted deciphered messages. Much of what was broken was so garbled that it required people very conversant in the language to make it coherent. Sergeant Hyman Sobel, a Harvard instructor, was in charge. The group's members were amazed at how we cryptanalysts broke traffic that was so full of errors, and we were surprised at their ability to make it understandable.

6. Intelligence Coordination Section. This group collated the information, evaluated it, and passed reports based on it to the 12th Army Group headquarters. It was headed by a very bright Master Sergeant Samuel Lieberman, whose ability was respected by all of us.

There was also a small photographic unit, a signal center, and the usual administrative personnel, motor transport, mess people, etc.

S.S.D.D. landed in Normandy. We started by living in the field, and working under tents. We bivouacked in the countryside near such towns as Coutances, Laval, Chartres, and Meaux, and in the tiny Lorraine village of Mangiennes. Then came three months in the city of Luxembourg, where we were quartered in a school not far from the headquarters of Generals Bradley and Dwight Eisenhower in the Hotel Alfa. After the Battle of the Bulge in December 1944, we went through parts of Belgium, entered Germany at Aachen, and after four months of movement, ended the war in Forchheim, north of Nuremberg.

As noted previously, all the operational work was performed by enlisted men. Many of the officers were sent by the Signal Corps to supervise subsidiary functions, as transport and billeting, and had no knowledge of our work. Occasionally, men from the intercept units drifted in and out of our unit, seemingly to absorb some notion of what was being done. Of course we had our own intelligence officers, and two in particular deserve special note for their hands-on attitude and productive work.

The officer supervising my cipher breaking section was the very erudite Captain Benjamin Schwartz. He was a family man, over 45, who had volunteered to contribute his expertise to the war effort. Schwartz was skilled in Sanskrit, other ancient Indic languages, and several modern tongues and had headed departments for those exotic languages in the New York Public Library and the Library of Congress. He instructed us in cryptanalysis, watched over our daily efforts and assumed a rather paternal role toward his men. He was a person who cared, and that meant a great deal.

Captain Howard Mendel supervised the traffic analysis team. He worked at it incessantly, and was an intelligent and effective participant throughout. He was wounded in Luxembourg when a stray airplane bullet went through his side, but much to the relief of his men, their Howie soon recovered and came back.

Soon after we landed, an unexpected development struck our cipher-breaking section. Many of the approximately 16 cryptanalysts—no matter how adept they had been in training classes—seemed unable to cope with a real life situation in which a solution might not exist. A substantial portion of the enemy traffic could not be solved, and this uncertainty created a psychological barrier that stopped men before they started. As a consequence, many faltered, and produced few results.

Four of our men did most of our breaking. They were George Hussey of Bronxville, New York, and Eric Porter of El Segundo, California who worked marvelously together, Howard Arnold, and this writer, Charles David.

SSD - Cipher or Code Breakers - Emenders - Intelligence control. September 1944 in Mangiennes FRANCE.

Though we were the signal intelligence arm of the 12th Army Group, S. S. D. D. traveled by itself in the field. We were an isolated outfit, known only to the intelligence people. Our security and secrecy were superb, even within the outfit.

People in one section knew very little about the doings in other sections. On a few occasions, senior officers from headquarters stopped by to express appreciation for our results, and would mention a specific case or two that were outstanding.

Our daily intelligence production was notable from the start, but reached a crescendo during the three months in Luxembourg. Our deciphered intercepts, traffic analysis, and direction finding indicated unusual enemy activity before the Battle of the Bulge began on 16 December 1944. I judge that people at army group headquarters simply did not evaluate these data properly. It probably seemed impossible to them for the Germans to stage a major armored offensive over unfavorable terrain, especially when the bulk of their forces was tied up on the Russian front. After the Ardennes offensive was stopped a month later, the cipher section broke the radio traffic to the German units retreating from the Bulge. It revealed the detailed instructions to each unit as to exactly where it should be virtually every minute of the day. With the skies now clear, the allied air forces made very effective use of this information.

In January, S.S.D.D. moved again through different parts of Belgium and entered Germany, remaining there until the end of the war. We stayed near Cologne, crossed the Rhine at Remagen, camped at Bad Wildungen, and were in Forchheim, Bavaria on V.E.Day. We then moved to the town of Russelsheim, where everyone wondered what our destination would be in the war with Japan. But that soon ended, and gradually we worked our separate ways home.

Many of us had a strong affection for S. S. D. D. Our fellow soldiers were often quite accomplished and thoughtful. There were professors, attorneys, classical musicians, writers, and businessmen—all on their way to join the hopeful post-war world.

A short personal epilogue: Several years later, I visited Captain Schwartz in Washington, D. C. He had stayed in the army, and had become a colonel in Army Signal Intelligence. We were happy to see each other and had much to discuss. He thanked me for my wartime efforts, and then offered an immediate commission to re-enlist and come to work with him. I'm afraid that my destiny was as a civilian, but it was gratifying to have the praise and appreciation of the boss.

NI DESCRIPTION AND BREAKING

The medium-grade field cipher was called "Doppelkastenschluessel"—double box cipher by the Germans. Because the first messages when intercepted by the British bore no indicator, the Allies called it the Non Indicator or **NI** system.

The **NI** was used from army groups to all lower units down to company level and carried both tactical and strategic information. It ranked just below the

Enigma, the high grade electro-mechanical machine cipher. The Enigma could not be distributed to the multitude of lower field units, and thus the **NI** served as a more practical hand system.

During the North African campaign the British overran a German signal center and discovered the nature of the system.

System Rules

The **NI** cipher combines a transposition, or more precisely, a seriation of the plaintext, and then two Playfair-like encipherments. The apparatus of the system consists of two 5 × 5 random alphabet squares (which omit **J**) such as those in Figure 1 below.

Box 1	Box 2
K X N Z Y	G S A O R
E M O B P	V F H Z W
L Q F V I	B N Y C Q
R A W G U	U I E M X
H S C D T	K L D P T

Figure 1. Enciphering Boxes.

A plaintext bigram, say **EU**, is enciphered by finding the first letter in Box 1, and the second letter in Box 2. In this case **E** is on row 2 of Box 1 and **U** is on row 3 of Box 2. The two letters are considered as diagonal corners of a rectangle, and the intermediate cipher bigram is those two letters which complete the rectangle, namely **VR**.

Box 1	Box 2
K X N Z Y	G S A O R
E M O B P	V F H Z W
L Q F V I	B N Y C Q
R A W G U	U I E M X
H S C D T	K L D P T

Figure 2. First Encipherment.

The bigram **VR** is now itself enciphered just as the original plaintext bigram was, and the resultant diagonals form the final cipher letters,— **QZ**.

Box 1	Box 2
K X N Z Y	G S A O R
E M O B P	V F H Z W
L Q F V I	B N Y C Q
R A W G U	U I E M X
H S C D T	K L D P T

Figure 3. Second Encipherment.

In this way EU becomes QZ in cipher.

If the two letters of a bigram to be encrypted lie along the same line in both boxes, we take the letters immediately to the left as cipher values. These we called liners. For example, **DL** = **KC**. The rectangular cases we called diagonals.

Box 1	Box 2
K X N Z Y	G S A O R
E M O B P	V F H Z W
L Q F V I	B N Y C Q
R A W G U	U I E M X
H S C D T	K L D P T

Figure 4. Linear Encipherment.

For cryptanalytic purposes it is important to see that four combinations of diagonals (**D**) and liners (**L**) exist:

	D–D	D–L	L–D	L–L
Plaintext	en	ea	gi	yr
Intermediate	FL	HK	UW	OZ
Ciphertext	NC	TT	XP	HM

The seriation of the plaintext consists of writing in blocks of two lines of 21 letters each. For example, using the first line of Heine's poem "Die Lorelei":

Ich weiss nicht was soll es bedeuten
dass ich so traurig bin.

and replacing the common **ch** combination with **q**:

```
i q w e i s s n i q t w a s s o l l e s b
e d e u t e n d a s s i q s o t r a u r i

g b
i n
```

The *vertical* bigrams (**I-E**, **Q-D**, **W-E**, etc.) are the pairs actually enciphered:

```
G G Y Q Q D C M A R N E R Q W M B G Q T B
i q w e i s s n i q t w a s s o l l e s b
e d e u t e n d a s s i q s o t r a u r i
U Z Y Z S Z I Q I Q I Y A R U F S T Z U N

X B
g b
i n
P O
```

The cipher text is then read off horizontally in groups of five letters for transmission.

GGYQQ DCMAR NERQW MBGQT BUZYZ SZIQI WIYAR

UFSTZ UNXBP O

German Communication Protocols

Each German division had its own set of cipher boxes. It was assigned six different boxes for each day. These were paired in different combinations for each day's eight three-hour periods. In effect, there were eight keys per day.

The German encipherers followed certain conventions, though they grew lax as the war progressed. They placed an **X** before and after all numbers, proper names, place names, sentences, within abbreviations, and at the end of plaintexts having an odd number of letters. As a result, plaintext **X** had an abnormally high frequency and tended to mask the identification of the usual high frequency German letters.

The word *zwei* ("two") was changed to *zwo*, and *ch* plaintext was changed to *q*, distorting the frequency of *q* also. The use of occasional "quatsch" (German for nonsense) sequences was encouraged to distort combinations and frequencies.

All in all, the Germans seemed to feel that the **NI** system was impervious to timely cryptanalysis. They had readied a different system on general security grounds, but kept the **NI** right to the end.

Some Cipher Security Considerations

The seriation (vertical reading of the plaintext bigrams) ensured that the enciphered bigrams were not letter combinations of normal German text. The most frequent ciphertext bigrams certainly did not reflect the most frequent German combinations (*en*, *er*, *ei*, *ie*, *in*, etc.). What the ciphertext bigrams did reflect was the individual frequencies of disconnected German letters taken two at a time. The most frequent **NI** cipher bigram had a good possibility of representing plaintext *ee*, for example, but this was by no means always the case.

Even knowing the mechanics of the system, we found that the reconstruction of the cipher boxes was extremely difficult because of the double encipherment feature. No doubt the German signal experts counted heavily on this.

By using different boxes for each division, and by changing the boxes every three hours of every day, the Germans made the cipher uncomfortably close to a one-time pad system—unbreakable in theory and practice. Nevertheless we broke it, and generally in useful time.

AIDS TO CRYPTANALYSIS: THE INITIAL BREAKS

Each message was preceded by three-letter call signs of the sending and receiving units and the time of day. Our traffic-analysis teams were quite adept at identifying them. This plus the assigned radio frequency used aided in sorting the messages by unit—and therefore by key.

In cryptanalyzing the **NI** system, we used very large sheets of paper, colored pencils, and good erasers. These sheets and colors cannot be fairly reproduced in a journal. I will describe our analytic tools, show how we used them, and then leave it to the initiative of the reader to complete the analysis to reach the final plaintext.

Usually we needed a satisfactory number of letters in the same key to make inroads into the cipher. We were delighted if many more letters arrived for a three hour period, but often not enough traffic was intercepted to afford us an entry.

To illustrate the solution of the cipher, I have composed a sample problem of 18 lines of ciphertext supposedly intercepted from units of one division during a three-hour period. Also listed are the times of transmission and the call signs of the sending and receiving radio stations. Each pair of lines is given a line-identification letter (Figure 5).

The messages have been manipulated to allow showing a number of techniques in a relatively short space. This sample is much less difficult than the messages we encountered, although the language is typical of the German radio traffic of the time.

Knowing how the system worked, we were able immediately to rewrite the intercepted messages in 2 lines of 21 letters each. Once the messages were in this form, we made a bigram frequency—one of our most useful tools. The count was made in a 25 × 25 matrix on graph paper.

Each line of the problem was given an identifying letter at the left, as can be seen in Figure 6. When a bigram was entered on the frequency count that line-identification letter was entered in the appropriate square instead of a tally mark allowing us to locate the occurrences rapidly (Figure 6).

With the frequency count as a guide, we attacked the messages using a technique we called anagramming. This started with an assumption—an educated guess—followed by a good deal of trial and error. The intent was to substitute these plaintext assumptions throughout the problem, check surrounding values, and search for clues to expand our guesses.

We could not immediately start reconstructing the boxes even when our assumptions were corroborated because of the double-encipherment feature. The

```
1205
MXY   A - V D L V V K X R K C X V R B X A B A S K D
URT       D F X E G K M N Z I W B T A W P F Y O K E

      B - G K S Q A V I S K S N U T E K O C K C G F
          V U A Z N R I M Z R W I F G K K K L T T D

      C - W P N M Y Y K G C C D Y M W O K F I S M Q
          T D Q Y C X K D Y P O P I S Z F L S O A E

      D - U Q D O U N K I N D H G O Y P H M K N Y B
          I Q M Q Y D K B F A G Q D M D A A I O Y F

      E - D O E R F B P V P U
          L K D I O R E M Q O
-----------
1215
REX   F - M U M V I B V O
IXY       P Q K O I W D U
-----------
1240
ZBT   G - K L O A Z W S E Z U C K S P K O T X C Y U
QLS       K S L F Q M F D X D P Z Q X N M O V K D S

      H - C X S Z O K W
          K C A C M K P
-----------
1305
RLS   I - K N D C V M K S A N I L Y S M C A P W B Q
BNQ       K S U P O K Z K A Q I K C L S K W S X C G

      J - Y K C S I U T O S G M D O G K K C K O O H
          W L D M Z Z P K A S Y A R Z K Q M K V Z B

      K - N C F Y K F
          L U T T Z A
-----------
1340
DEF   L - K U G Y X K W S K R M K A U O V H K K N C
GHI       K K B W W Z H M M Z U K E B K B G K F I P

      M - V C G D S O L C K O K I W A S V K F S U A
          B Q Q O V S R P L K Q I B F R M K S X I N

      N - S N O O V A
          A Z M K R P
```

Figure 5. Transcribed messages for a single period.

```
1350
POR   O - TCKU
CLV       NUKO
-----------
1415
QZB   P - VDLVV KBUOUS HWVDYEBQOG
LTS       DFXEG KZWZKAZ X DFPDOAZQ

      Q - KFUKY LKUGOK I XWZLOXFKC
          KPDKEWI DVKTMM X CS NRXBP

      R - XPBTMKFOTLSCD ANIMPSPZ
          NQFBDKS ZQPAAB R I I PQAXT
```

Figure 5. (Continued) Transcribed messages for a single period.

middle bigram is unknown, and a workable number of plaintext values must be ascertained before a start can be made in forming the boxes.

In the frequency count, we find that **KK** ciphertext stands out with the highest count by far. We can start by assuming that it represents *ee*. We replace **KK** with *ee* throughout the text, and see where it leads us.

The Germans often sent very short messages, which reflected basic military necessities, and which were therefore constantly repeated. The most common 6-, 8-, 10-, and 12-letter messages were known to us. Curiously, we were almost never given information about the military situation, that might yield more specific probable words for our attacks. I don't know why. Perhaps it was security, perhaps ignorance, perhaps lack of time.

Note Line O in our sample problem:

```
               e
Line O -  T C K U
          N U K O
              e
```

To the experienced eye, this a dead giveaway. The Germans were constantly asking "What is your position?" or *wie lage*. The phrase might be preceded by a "please" or "request" or an addressee, but time was critical, and the eight-letter message was common.

We thus have assurance that our **K K** = *e e* assumption is probably correct, and we have:

```
          w i e l
Line O -  T C K U
          N U K O
          a g e x
```

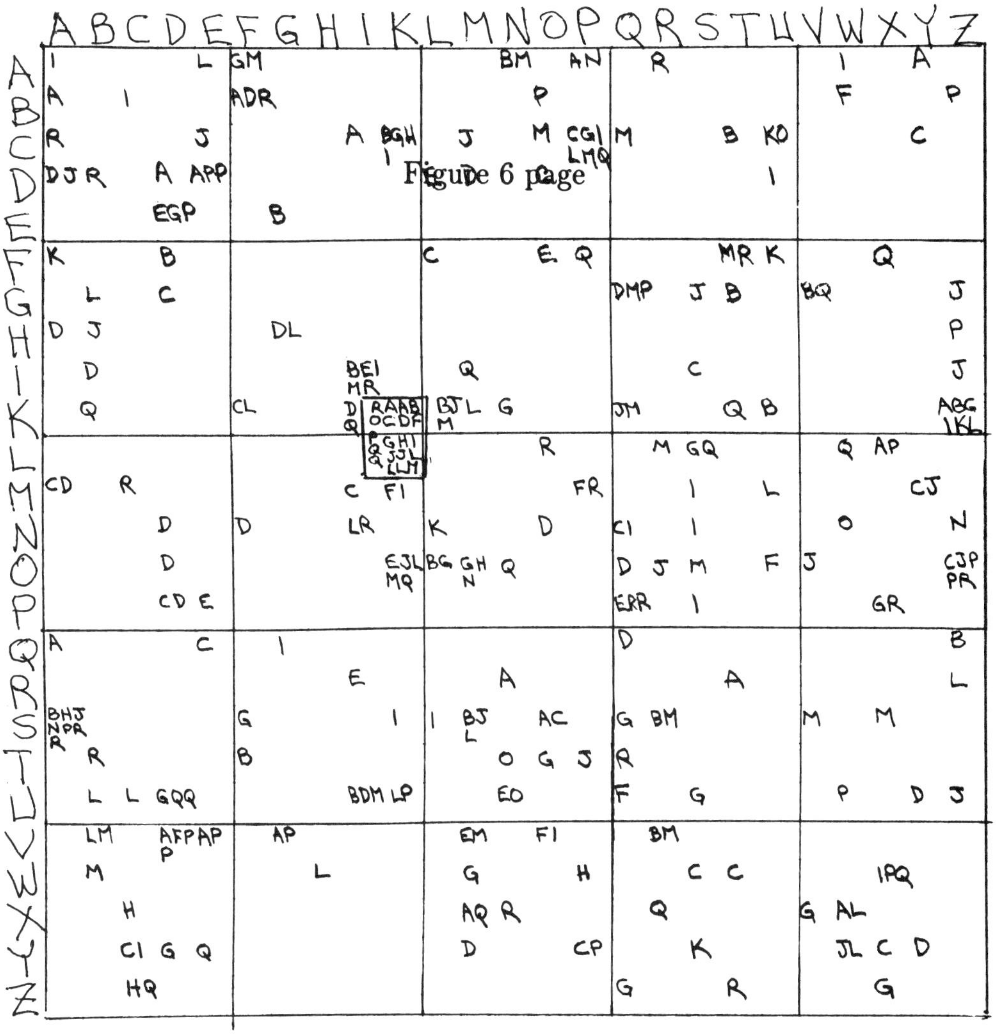

Figure 6. Line location and frequency count of cipher bigrams.

Our initial assumption has grown: We now have plaintext pairs *w a*, *w g*, and *l x*.

The Germans used many stereotyped openings. With experience, we cryptanalysts could recognize them. For instance, messages frequently began with *an* ("to") followed by the person or unit receiving the signal.

We are fortunate in having three of our practice messages begin with **KK** ciphertext. In German military text, about 95 percent of the time a message starts with *e*, the opening word will be one of four: *erbitte* ("request"), *eigene* ("our, our own"), *ein* ("a") and *eins* ("one"). Let us see how this works in our text.

If we try *erbitte* in Line I, the final *e* falls on ciphertext **KZ**, which has a high frequency. (Frequencies are always our strong ally.)

```
           e  r  b  i  t  t  e
Line I  -  K  N  D  C  V  M  K  S  A  . . .
           K  S  U  P  O  K  Z  K  A  . . . .
           e
```

K Z also occurs in line L in position 6. This is a perfect opening for *eigene*:

```
           e  i  g  e  n  e                 e
Line L  -  K  U  G  Y  X  K  W  S  K  R  M  K  A  . . .
           K  K  B  W  W  Z  H  M  M  Z  U  K  E  . . .
           e                                e
```

Although these are only the upper values of bigrams, when they are substituted throughout the text, we gain opportunities for assumptions for additional anagramming. For example, bigram **CP** in Line I has a rather high frequency, and it is a good assumption that the bottom plaintext letter is *e*.

We now examine other lines, with our assumed plaintext entered, and see what can be done:

```
                     e     e   n       n         e
Line A  -  V D L V V K X R K C X V R B X A B A S K D
           D F X E G K M N Z I W B T A N P F Y O K E
                     e
```

Units in the field constantly sent reports to various headquarters. German operations officer in a unit's staff (G-3 in American nomenclature) had the designator 1a. Messages to higher headquarters often began "*an roem eins Anton*" ("to roman numeral one Anton"—"Anton" being the German phonetic equivalent of our "Able"). Line A has several values that fit this possibility:

```
           a n x r o e m x e i n s x a n t o n   e
Line A  -  V D L V V K X R K C X V R B X A B A S K D
           D F X E G K M N Z I W B T A N P F Y O K E
                     e                           e
```

Often this opening was followed by "enemy something or other" as "enemy tanks, or planes, or troops". ("Enemy" is "Feind" (noun) or "Feindliche" (adjective) in German.) That seems a likely entry here because of the **KK** (*e e*) second from the end on top, and around the bottom where "Feindliqe" connects with it. Now we have six more probable equivalents to substitute throughout.

```
         a n x r o e m x e i n s x a n t o n f e i
Line A - V D L V V K X R K C X V R B X A B A S K D
         D F X E G K M N Z I W B T A N P F Y O K E
         n d l i q e                           e
```

Let us examine one more short message to see if anything can be made of it:

```
             t t e   a
Line F - M U M V I V B O
         P Q K O I W D U
```

That first word just has to be *bitte* ("please"). But please what? A common request was for a situation report, and a frequent 16-letter message was therefore *bitte lage meldung* ("situation report, please"). Checking the frequency chart also shows higher counts for a few of the cipher bigrams that accord well with their plaintext equivalents (e.g., **I I** = *e d* has a count of 5 and **V D** = *a n* a count of 3). We always looked for frequency indications that tend to confirm what are still only assumptions. Thus we have eight more probable bigrams:

```
         b i t t e l a g
Line F - M U M V I B V O
         P Q K O I W D U
         e m e l d u n g
```

It was a good break if we found two messages with the same beginning. It was especially helpful if the beginning was longer than 21 letters and thus continues on the second line. Such a case may be seen in Lines A and P. We have already anagrammed some of Line A and it seems as though it may share the first 27 letters with Line P. Note the six-bigram repeated block at the beginning of Lines A and P and the repetition of vertical bigrams at positions 1 and 14 of Line P.

```
         a n x r o e m x e i n s x a n t o n f e i
Line A - V D L V V K X R K C X V R B X A B A S K D
         D F X E G K M N Z I W B T A N P F Y O K E
         n d l i q e                           e

         a n x r o e m x e i n s x a n t o n f e i
Line P - V D L V V K B U O U S H W V D Y E B Q O G
         D F X E G K Z W Z K A Z X D F P D O A Z Q
         n d l i q e                 n d
```

Thus it seems we have another 15 upper letters to substitute in the problem.

One of the greatest helps to the cryptanalyst is to know about what the enemy is likely to talk. There were plenty of the German equivalents of "enemy", "aircraft", "troops", "armor", and so on. But by far the most common words were the cardinal numbers: *eins*, *zwo*, *drei*, etc. Supply reports, casualty figures, radio frequencies, times of day, location grids and much more all required numbers.

Often when no other openings were visible, numbers could be spotted in the middle of messages—after some judicious assumptions based on frequency and combinations. The 10 digits were frequently used in groups (e.g., 4 digits for the time of day plus *uhre*, "hours"), and were separated one from another by X.

Security violations and operator mistakes greatly helped us. Of course, we were always looking for that rare happening—the same message sent in plain and ciphertext—but that was rare.

Let us examine Line G of the problem:

```
          e       e i e
Line G -  KLOAZWSEZUCKSPKOTXCYU
          KSLFQMFDXDPZQXNMOVKDS
          e       e
```

We already have some letters of Line G, one of them being an initial e. Now look in Figure 5 at the header of the message:

ZBT
QLS
1240

The alert cryptanalyst now rubs his hands in glee, because he realizes that the careless German code clerk has included the time of day at the beginning of the message. (This occurred once in a long while.) We now have the following:

```
          einsxzwoxvierxnulluhr
Line G -  KLOAZWSEZUCKSPKOTXCYU
          KSLFQMFDXDPZQXNMOVKDS
          e       e
```

There is a slight problem here because the code clerk accidentally omitted the X after *null*, but that is seen when *uhre* fits into the bottom e of cipher **K K**.

The above examples should give a good idea of how the initial plaintext wedges were made. Again, I caution that the sample problem is contrived— it was never this easy. Very few of these entry points, if any, appeared within the same three-hour period. It took a very patient cryptanalyst to find correct plaintext in the average set of messages.

	A		B		C		D		E		F		G		H	
A		G								B						
	T								A							
B									P	M						
C										S						
									B							
D								D				M				
							A				O					
E		Y		P				I		K		K				
	W		L				I		K		Z					
F		Q								B						
	A								P							
G								Q		N				O		
							Q		P				U			
H																
I		Z		G						C				C		
	W		L						P				U			
K																

Figure 7. Portion of ENCIPHER CHART

This involved daring assumptions, and detailed tracking of surrounding bigrams to find promising combinations. Trial and error, persistence, and some sixth sense could lead to progress. We stuck with a three-hour period as long as it seemed to show promise—a large volume, a good frequency count, likely plaintexts. Conversely, we abandoned a period for a new one—usually on the basis of instinct and experience—when we were getting nowhere.

Once a sufficient number of bigrams was recovered—one cannot quantify it exactly—we set about trying to reconstruct the cipher squares, a process we called "boxing". This could be the most daunting task of all. As the squares were completely random, we got no help from any pattern or keyword.

The first step was to make "encipher" and "decipher" charts. These were large sheets with 25 × 25 matrices, each cell with four quadrants (Figure 7). The charts were over a foot square, and were preprinted for us in horizontal sets of two which made it easier for us to fill in and work from both at the same time.

In the encipher chart, the bigram coordinates of the matrix represented the plaintext, with the cipher equivalents written in the appropriate square. The first letter of the cipher bigram was written in the upper right quadrant of the square, and the second letter in the lower left quadrant, thus serving as a reminder of which letter belonged in which box.

The decipher chart was the reverse. The coordinates were the cipher bigrams, with the plaintext bigrams written in the upper right and lower left quadrants of the inner squares. The upper left and lower right quadrants were reserved for the unknown intermediate bigrams of the cipher which might be discovered as we progressed with pencil and paper. These charts now served as the software for the computer in the cryptanalyst' mind.

With graph paper and pencils, the problem then was to somehow string these letters together so as to recover the original cipher boxes. For a start we used numbers to represent the middle letters, and then by using other related values tried to "hook" some of the letters together. Except in the unlikely event of a very great amount of anagrammed plaintext, this became an eye-twisting exercise in frustration for several reasons.

First, we had no idea of the middle bigram. Second, as I explained earlier, the final cipher bigram can be the result of four possible enciphering combinations: diagonal-diagonal, liner-diagonal, liner-diagonal, and liner-liner. Third, we had no idea of which letters in each box were on the same line, or in the same column, or opposite which line in the other box.

Fortunately, the inherent properties of the boxes and the enciphering method offered ways to help reconstruct the boxes. I will use the cipher boxes of Figure 1 to illustrate the three most useful phenomena.

1. Reversibles — If the middle bigram of the encipherment is a repeated letter, the result of the second must be the reverse of the original plaintext bigram:

   ed = HH = DE

   This can only result from a double diagonal, which is a clue we really need in our boxing effort.

2. Reciprocal — Assume a plaintext bigram enciphered via two diagonals. If the reverse of its cipher bigram happens to be a plaintext bigram, its own cipher equivalent is the reverse of the original plaintext bigram:

(1)	n e	=	A W	=	X M
(2)	m x	=	W A	=	E N

   Thus if we find two of our anagrammed bigrams in this forward-reverse relationship we know that they can only result from a diagonal-diagonal encipherment. We encircle these in red on the encipherment and decipherment charts. Eliminating the possibility of liners is invaluable in rebuilding the boxes.

3. Appendixing — "Appendixing" aided the analyst in determining which letters were on the same line in a box. If any letters are on the same line (e.g., **K**, **X**, **Z**, **N**, or **Y** in Box 1 of Figure 1), and are combined with one particular letter in Box 2—say the letter **E** in the fourth row—then the first letter of the middle bigram must be the same for all the bigrams:

k e	=	A R	=	X X
x e	=	A A	=	E X
n e	=	A W	=	X M
z e	=	A G	=	U X
y e	=	A U	=	X R

When the middle bigram is re-enciphered, since the first letter (in this case **A**) is always the same, the final bigram can consist only of one of five values in each box, instead of the usual twenty-five: (in this case **U**, **I**, **E**, **M**, or **X** in Box 2, and **X**, **M**, **Q**, **A**, or **S** of Box 1).

So if we find a pattern of repeats going down a column of our charts, there is a greater than normal chance that the corresponding outside letters on the left are on the same line in their box. This can be fortified if another column shows a similar pattern. None of this is certain, but it bolsters the odds.

From this point, the cryptanalyst resorts to trial-and-error boxing. If the reader has experience with a particularly difficult Playfair cipher, where it appears that some letters must be all at once on the same row, in the same column, and on the diagonal with others, he will have some appreciation of what we faced.

Common Boxes

As mentioned earlier, each German unit was supplied with a group of six boxes for the day. Since two boxes were necessary for each of eight 3-hour periods, some of the boxes were used more than once. For instance, Boxes 1 through 6 might be used as follows:

Time of Day	Box Combination
0001-0300	1 & 2
0301-0600	3 & 4
0601-0900	4 & 1
0901-1200	5 & 6
1201-1500	3 & 5
1501-1800	2 & 6
1801-2100	5 & 2
2101-2400	6 & 3

The difficult and crucial breaking was of the first workable period that appeared during the day. This was not necessarily the 0001-0300 slot, but the first period with enough text to attack. We called that the original period. Which it would be was not predictable: no period was regularly the busiest. After that a search was made for other periods that used one of the two boxes now known. If one was discovered, its period was a rather easy one to solve.

Locating a common box in the double encipherment was not simple. A period had to be anagrammed and boxing started before there was a basis for comparison. However, once a common box was spotted, an experienced analyst could reconstruct the second one quickly.

I would like to discuss the major reasons for the **NI** cipher-breaking being so difficult and daunting.

1. Volume: The enemy changed its boxes every three hours of every day. The quantity of cipher text available in a three-hour period was often insufficient to work with, or just marginally sufficient and might not bear fruit.
2. Peculiarities: Even with adequate volume, a particular text might have a make-up that frustrated continuity, even after sensible beginnings. Worst of all was an aberrant frequency count, which would result in interminable false starts. Though this is common with limited message volume, many longer texts displayed abnormal frequency indications. Despite this, we had the feeling that given time, we could break into most problems, which brings up the next obstacle.
3. Time strictures: Messages enciphered in **NI** dealt with both tactical and strategic matters. However, after a day or two, the intelligence became stale, and we had to go on to new and fresh material. Working under this kind of pressure proved impossible for some of the men.
4. Garbles: This was the greatest obstacle. The German cryptographers made mistakes, their radio men transmitted wrongly, and our interceptors made errors in picking up what was often a weak signal. At times we had the same messages from three or four intercept companies and the variation could be most marked. If breaking assumptions are based on wrong cipher values, the analyst is dead in the water from the start. Sometimes we displayed an almost sixth sense about incorrect letters, but garbles greatly hindered us.

A Simplification—Single Encipherment

With the invasion of Normandy in June 1944, much of the enemy traffic began to be enciphered only singly, the second, double, encipherment being omitted. The German communication experts perhaps realized that the double encipherment took too long and was too prone to error at both ends to use in combat situations.

To illustrate its use, here are the opening words of "Die Lorelei" singly enciphered in the boxes of Figure 1:

Y	Y	I	V	Q	D	L	A	Y	N	L	U	X	L	P	W	Q	Y	V	T	F
i	q	w	e	i	s	s	n	i	q	t	w	a	s	s	o	l	l	e	s	b
e	d	e	u	t	e	n	d	a	s	s	i	q	s	o	t	r	a	u	r	i
U	S	A	R	T	A	Q	C	Y	X	Y	A	Q	X	X	C	K	K	R	X	G

Ct: YYIVQ DLAYN LULXP WQYVT FUSAR TAQCY XYAQX XCKKR XG

It will be seen that our task was much simplified. The diabolical middle bigram no longer exists, and the analyst can use a reciprocal process of simultaneous anagramming and boxing. Some salient helping points are as follows:

1. When enciphered, a plaintext letter must result in a different cipher letter in the same box. Thus, for example, bigram **ab** cannot possibly show a cipher **B** on top, or a cipher **A** on bottom. This helps prevent wrong anagramming.
2. Single plaintext values allow a start to be made in boxing. If we know that plaintext **u-** is cipher **XY**, then we also know that **X** in Box 2 is on the same line as **e** in Box 1, and that **Y** in Box 1 is in the same column as **e** in Box 1, or in the same line adjacent to **e**. Single value are far less useful in double encipherment.
3. Plaintext assumptions lead more quickly to boxing contradictions than they do in double encipherment, letting us discover anagramming errors much earlier.

Unfortunately, while these aids are valid, they were rather moot in practice. Often the reason was the lack of sufficient traffic in any three- hour period. Add to this the numerous garbles, and breaking the cipher was still difficult—especially in a usable time frame. Nevertheless a large volume of the single encipherment was broken in time to be of use. Due to security necessities and our isolation, we were largely in the dark as to the use of our information, but were assured by officers from General Bradley's headquarters that it was of constant and great value.

To sum up, I found that breaking the **NI** cipher system was a complex and often mind-bending process. It would be interesting to learn if it were the last of the "pencil-and-paper" systems used by a major army in a major conflict. It was tempting to think of it as "interesting" or "intriguing", but in wartime that seemed incongruous. Perhaps now, many years later, we can study it with a greater degree of academic equanimity.

BIOGRAPHICAL SKETCH

Charles David was was born in Brooklyn NY in 1922 to a well-known songwriter father and a school teacher mother. He graduated from Erasmus Hall High School in June 1937 and from Brooklyn College in 1941. Interrupting graduate studies at Columbia University he enlisted in the Army Signal Corps in 1942 and was discharged in January 1946. He immediately married his fiancé, Helen Robertson, a high school biology teacher, and entered the business world as a sales representative. Mr. David and his wife are both retired now and enjoy extensive and exotic travels.

AN INTRODUCTION TO AN HISTORIC COMPUTER DOCUMENT THE 1946 PENDERGASS REPORT CRYPTANALYSIS AND THE DIGITAL COMPUTER

Colin Burke

ADDRESS: 6319 Red Haven Road, Columbia MD 21045 USA.

ABSTRACT: In 1946 James T. Pendergrass of the American Navy's code-breaking agency, OP20G, attended the University of Pennsylvania's seminar on electronic computers. That experience convinced him that the navy should turn from the many special-purpose cryptanalytic machines it was building and develop a general-purpose electronic computer. His argument was persuasive and the navy went on to sponsor the Atlas computer. Built by Engineering Research Associates and delivered to the navy in 1950, Atlas became one of the first modern computers in operation. Pendergrass' 1946 report was classified top-secret for decades because it showed how the most significant problems, such as the Ultra-Enigma, could be conquered by a general-purpose computer.

KEYWORDS: Computer, cryptanalysis, OP20G, Pendergrass, Wenger, Atlas, Able, Abner, Monogram, Moore School, Enigma, Hagelin, Vannevar Bush, von Neumann, ENIAC, ERA.

The code-breakers of the American Navy's OP20G had struggled since the early 1930s to obtain the funds needed to design and build high-speed electronic analytic machines.[11, 2] Unfortunately, their mid-1930s project, which was led by Vannevar Bush of MIT, did not produce a truly successful device. The developmental funds and the institutional supports provided by the navy were insufficient for the challenging move into digital electronics. The relatively special-purpose, optical- electronic Comparator that Bush constructed for the navy in 1938 was unreliable and OP20G began World War II with only a handful of IBM electromechanical tabulators and the hope that massive funding of electronics research would finally allow the creation of the more powerful Rapid Analytic machines Bush had suggested to the navy's S. C. Hooper and Joseph Wenger in 1936.[21] They hoped that such machines, distant cousins of the programmed digital electronic computer, would allow OP20G to create and apply advanced mathematical and statistical cryptanalytic techniques.

A stable, long-term program to develop the Rapid Machines might have led OP20G to be among the very first to conceive of and build a modern electronic computer, but the chance for such a program had been lost in the 1930s. It could not be reborn during the war. The demands created by Pearl Harbor and the Battle of the Atlantic turned OP20G away from speculative projects. Although the outbreak of World War II and a new faith in cryptanalysis led to the allocation of millions of dollars for machine development and the creation of a special group for machine design and construction, there were too many pressures to allow OP20G to do more than build a series of very limited and special-purpose electromechanical and electronic devices during the war. Most did not even attain the flexibility of Bush's original mid-1930s design and few provided significant help to advanced statistical and mathematical cryptanalysis.

The need to immediately read the ever-changing codes and ciphers of the Axis powers caused the OP20G research group under Joseph Wenger and Howard Engstrom to take the cryptanalytical and technological paths of least resistance. In the early 1940s, least resistance usually meant brute-force cryptanalytic methods and the creation of machines that were either physical representations of an enemy's encryption devices or fixed physical embodiments of decryption processes. The men in OP20G's research team knew that such a strategy could lead to expensive machines that suddenly became useless when enemy cryptographers made slight changes in their cipher systems. They realized, however, that the creation of more general-purpose machines based on then blue-sky cryptanalytic and electronic techniques was impractical if not irresponsible. When the only sound cryptanalytic alternative was to attack systems through statistical methods, more general-purpose computing devices were constructed. But, their architecture had to be sparse and their components based upon as much off-the-shelf technology as possible.[9]

The crises of World War II were too grave and too frequent to allow creative urges to have priority over practical solutions. Reliable electromechanics and analogue electronics thus triumphed over digital technology at OP20G. Wile easy to construct and relatively fast special-purpose devices were chosen over more general-purpose designs, the young engineers that OP20G plucked from the leading electronic computer projects in American industry and academia lobbied for the exploration of technological frontiers and for more general-purpose machines. But they never recommended the option of constructing one machine to perform all cryptanalytic functions. A "universal" cryptanalytic machine, let alone a high-speed device to perform all logical and mathematical routines, was something at the farthest edge of imagination in the early 1940s. Even Alan Turing, Bletchley Park's mathematical genius, left his philosophical defense of the possibility of a universal mathematical device as a set of formula in an abstract article. [1,13,27]

Immediate problems, not long-term goals, shaped the accomplishments of the Rapid Machine group during the war. OP20G's teams at the National Cash Register Company, Eastman Kodak and OP20G's Nebraska Avenue Annex in Washington DC, created a host of amazing machines that frequently used components that were far beyond the state-of-the-art in electronics and electromechanics.[12] But all were compromises between engineering visions and practical needs. The machines created between 1942 and 1945 were limited in technology as well as in their range of application. Although most used some electronics, many were based on analogue rather than digital logic and components. A few were unashamed reversions to much older electromechanical technology. Such choices were rational in the context of the war. Analogue methods and electrical and electromechanical parts were reliable and inexpensive, ran faster than electronics in some applications and demanded much less development time than the still experimental digital electronics. For example, in the attack against the German's Enigma systems, it proved quite efficient to use a special-purpose design and to base the American Bombe on electromechanics. The early 1942 dream of a fully electronic and perhaps relatively general-purpose digital Bombe, or even one centered on electronic versions of the wheels of the Enigma, devolved into a variation on England's combination of spinning shafts and electric commutators.[14, 15] Electronic digital counters played a very secondary role as the hard-rubber and brass wheels of the Bombe proved to be a combination of speed and reliability that surpassed the power of the tubes and circuits of the early 1940s.[16, 8, 26]

Even when OP20G's engineering wizards were told to create devices to speed the application of general cryptanalytic methods, such as the Index of Coincidence, or to go far beyond the power of tabulating machines to perform searches for repetitions and correlates of code-patterns, they had to follow that least resistant path. For example, digraph analysis machines were hurriedly constructed using the very primitive technologies of mechanical teletype readers and electromechanical industrial counters.[9, 10: p. 23] And, while a new version of Bush's 1930s Comparator was constructed, it barely achieved the power of its mid-1930s design. It merely tallied 'coincidences' of varying lengths on electronic counters as off-set messages punched on special paper-tapes flew by an optical reading- station.[9]

Other war-time applications of Bush's original ideas were even more limited. The hand operated, table-top IC machine was based on analogue electronics. Its operators had to slide two photographic plates over each other until optical-cells sensed enough light passing through the off-set messages to indicate a desired level of coincidence.[9] The later and more sophisticated optical-electronic Copperheads, some of which were built for code-pattern look-ups, were also essentially analogue devices. They did not even count or tally, they just stopped when the messages on the two high-speed tapes had enough coincident code spots. The operator of the machines had to rewind the films by hand to locate the position of the code overlap.[9]

Other machines built by OP20G's Naval Computing Machine Laboratory contained more advanced electronics and logic. But even the complex extensions of the Bombe remained special-purpose as did the electronic digital machines built to model the encryption devices of the Japanese.[7, 5] OP20G was not alone in achieving less than its young engineers might have desired. The American Army's cryptologic machine builders were also limited by the state of digital electronic technology and the need to respond to emergencies. For the same reasons, England's engineers did not build a universal machine.[22, 20]

In 1945, when there were fewer crises, the engineers and mathematicians at OP20G urged the higher-ups to allow them to go beyond what they had accomplished during the critical years of the war. The crew at OP20G knew of the similar and in some ways more advanced work in England, work which led to the famous electronic pre-computer, the Colossus.[22] But they wanted to do more than catch-up with their British counterparts. They had their own dreams of extending the power of digital electronics, optical memory devices and magnetic data recording.

The new mathematical cryptanalysts at OP20G also wanted to continue and extend the World War II automation efforts but they feared that peace would mean the end of the Rapid Machines project. Luckily, the sacrifices and accomplishments of the war were appreciated by the navy's leaders. The ongoing contributions of Ultra and Magic and the role of cryptanalysis at critical times, such as the battle of Midway, had finally won recognition for the code-breakers. Unlike the situation at the end of World War I, when funding was suddenly ended, OP20G was promised several million dollars for post war technological development. Immediately, great plans were laid to produce the type of machines that would have been built during World War II if there had been time to develop digital electronics and to overcome the barrier that frustrated all computer builders of the period, the lack of high-speed memory.[19, 27]

By the end of 1945, OP20G had its Program Monogram computing wish-list ready. No one envisioned a universal machine but there were great hopes that the special devices of World War could be replaced by much more general mathematical engines. A new, much faster, multi- purpose digital Comparator was to be designed and built. It was to be one that would be able to perform the functions that had been assigned to separate machines during the war. Along with the search for the new Comparator was a quest for the best memory medium. Microfilm and the unproven magnetic recording were to be explored for use on the machine that later became known as Goldberg. There was even a great deal of thought given to the possibility of constructing a Super-bombe, America's version of the British Colossus.[19]

There was no thought given to the creation of a general- purpose, electronic and program driven computer, however. Monogram was not a computer project, it was an effort to develop what we would now call powerful special-purpose machines. OP20G's team included some of the best electronics engineers, mathematicians and logicians in the nation but even in 1945 they did not think a general-purpose computer was a realistic option for 1940s code- breaking. Cryptanalytic functions, they thought, had to be hard-wired into a machine. Thus, Monogram's original list did not include a 'programmed' computer.

Then, OP20G's technical leader, Howard Engstrom, decided to send one of his men to see what the University of Pennsylvania was offering at its Moore School's 1946 summer course on computers. The Moore School's computer group had just completed the ENIAC, the astounding yet special-purpose digital electronic version of the giant analogue differential analyzers. Unlike the engineers at the cryptanalytic agencies, the Moore School's team had been allowed enough time to experiment because Army Ordnance's computing crises were not of the same order as those of the cryptologic agencies. It could balance future usefulness against meeting production schedules.[24]

Although the huge ENIAC was a special-purpose machine, it was fully electronic and its construction led to something more significant, the design of universal devices that later became known as the EDVAC and IAS computers. By 1945, the Moore School's men had sketched out a machine that would perform any mathematical routine through what we now call 'software'. When the young Philadelphia engineers, led by John Mauchly and Presper Eckert, were joined by the famous and powerful John von Neumann, their idea for a universal engine became more precise and gained recognition in the scientific community.

Although torn by policy and personal conflicts, the original ENIAC group was so famous that engineers and mathematicians from the most advanced computer projects flocked to Philadelphia in the summer of 1946 to gain insights into the latest computer developments. Von Neumann's design for a more sophisticated universal machine, which he planned to build at the Institute for Advanced Study, was revealed as were the ideas of Eckert and Mauchly who were beginning to sketch the outlines of their own new computer, the UNIVAC. They were joined by a host of others who described the advances in digital electronic hardware and logic achieved at other American computer centers since the onset of the war.[4, 24]

Because of the continued security restrictions on OP20G's work, it could not reveal its computing secrets. But, Engstrom accepted an invitation to send a representative to the summer meeting. He was to listen and learn. At the same time, Engstrom and Wenger began to establish a long-term relationship with John von Neumann. The contacts with von Neumann and the Philadelphia group were important to OP20G because although they were aware of the general nature of the work at all the American and British computer centers, OP20G's men had not yet realized the implications and potentials of the design of the proposed EDVAC and IAS computers. They, like most others, did not fully appreciate the critical importance of the idea of a general-purpose programmed computer.

James T. Pendergrass, a young academic physicist who had been drafted into naval cryptanalysis, was selected to represent OP20G at the summer conference and workshops. He was impressed by the size of the group and the credentials of the men who attended the near month long meeting. He was even more impressed with what he heard. He very quickly became an advocate for the construction and use of a general-purpose digital computer for code-breaking, one based upon John von Neumann's scheme.

During his weeks in Philadelphia, Pendergrass convinced himself that most if not all cryptanalysis could be performed through digital methods. He came to believe that general-purpose digital computers would soon be fast enough to compete with the special-purpose and analogue devices being constructed under Monogram. His next step was to persuade the navy that OP20G should have its own version of the proposed IAS universal mathematical and information machine.

With the blessing of the operational head of OP20G, Joseph Wenger, Pendergrass devoted much time to preparing a long and detailed report. His two part document was submitted to OP20G in late 1946. It remained Top Secret for decades. Only recently has one part been declassified and the censors continue to black-out and delete significant portions of that text.[1][3]

There is a good reason for the continued caution about the Pendergrass Re-

[1] In the appended copy of the report the censor's deletions are indicated by the use of '[DELETED]'. Because the copy provided by the government contained some words and partial sentences that were unreadable, the symbol [UNREADABLE] will be found in a few places in the retyped copy. To allow computerized publication, some alterations were made to Pendergrass' report. Certain brackets used in the original text were deleted and superscript plus and minus signs were placed after variable names. The agony of interpreting and retyping the original text was borne by Mary Claire Monroe who coaxed Word Perfect into doing things that seemed impossible.

port for it contains secret information. Pendergrass had to demonstrate to the navy that a programmed computer could perform all the critically important cryptanalytic tests that were embedded in the many special-purpose devices, both analogue and digital, that were in OP20G's operational center in Building Four in Washington, D.C. Those tests included the secrets of Ultra and Magic. His report had to go further, however. He had to show that cryptanalytic procedures of a more advanced nature, which were to be built into the planned special-purpose Project Monogram machines, could be done with a computer.

The nature of his audience shaped the content and style of Pendergrass' report. The men at OP20G did not need a tutorial on basic electronic technology. They were leaders in the field and thought they would be able to convince the navy's Bureau of Ships, the agency that supervised all of OP20G's materials acquisitions, of the viability of electronic machines. What OP20G's cryptanalysts did need from Pendergrass was a demonstration that the yet to be born 'programming', digital methods and the non-existent general-purpose computer were reasonable cryptanalytic options. It was critically important to build a convincing argument that a computer could attack the new automatic enciphering devices that had led to the first Rapid Machine of the 1930s. Pendergrass and his allies had to show that the general-purpose machine could attack the Purple, Enigma and the more worrisome machines like those used in the German Fish system.

Thus, Pendergrass went into great detail about the logic of a general-purpose computer and the programming of cryptanalytic tests. His report outlined the architecture of the new computer, gave a version of John von Neumann's programming language and hinted that a more practical computer language would soon emerge. More importantly, the report showed how the most difficult World War II cryptanalytic problems could be programmed. Because time was always critical to the code-breakers, his reports also contained estimates of how long it would take the programs to solve problems familiar to the crew at 'G'.

Because upgraded versions of Bush's old designs were central to the plans for Monogram, Pendergrass had to prove that his proposed machine could replace the Comparator. Thus, one section of the report was devoted to the description of the steps necessary to compile the statistics needed for the ubiquitous Index of Coincidence analysis.

Then, Pendergrass went on to OP20G's truly secret techniques and to the problems that had been the focus of its attention during the war. Unfortunately, only one of the perhaps three World War II special cryptanalytic procedures Pendergrass explored in the first report have survived the censor's scissors and black pen. He demonstrated the solution to the now famous target of the American

'Bombes', the four wheel Naval Enigma machine. And, in Appendix II, Pendergrass programmed a solution to the Enigma 'Grenade' problem. It was one of the less time-consuming tasks assigned to the Bombes. A Grenade demanded a shorter run than other anti-Enigma procedures but followed the general logic of the Turing-Welchman attack on the Enigma.[26] Their critical method was crib-based and analogue and very fast. So, Pendergrass had to prove that his all-purpose digital machine could at least match the Bombe's efficiency.

Pendergrass wanted to show that his device would be useful to everyone at OP20G. In Appendix III he demonstrated how the digital computer could be used as an ultra high-speed version of the Hagelin ciphering machine. With a very different mechanism and logic than the Enigma, the Hagelin devices were used by several nations, including the United States, during and after World War II.[10] To aid the attack on Hagelin systems used by an enemy, and perhaps to illustrate how to speed the decryption of friendly messages, OP20G had constructed an electrical analog of the Hagelin during the war. Pendergrass showed how his all-purpose electronic digital machine could take its place.[18]

The particular systems discussed in the missing Appendices of the first report are unknown but they perhaps included an example of how to program a computer for an attack against Japanese encryption machines and a demonstration of how to solve the more sophisticated machine ciphers such as those produced by the American ECM. The other part of his report, which remains classified[3], perhaps contained proof of the ability of a universal digital computer to replace the dozens of tabulating machines and special-purpose Rapid Machine devices that were arrayed against code problems. Pendergrass perhaps demonstrated how to strip additives and, although he realized that there was no mass memory technology that could keep pace with the speed of electronics in 1945, showed how the computer could search for code groups.[10]

The Pendergrass report hit its mark, at least with the operational head of OP20G. As a result, Joseph Wenger began to lobby the very reluctant Bureau of Ships to allow yet another 'computer' project. The job of convincing the Bureau was not easy. 'G' already had the more special-purpose Goldberg and other machines with something like proven track records on the way to completion. The Bureau wondered if it wasn't wiser to let others, such as John von Neumann, carry the long and costly burden of the first stages of computer research.

The Bureau's men must also have wondered if the new computer would be cost-effective. The electromechanical Bombes which worked so well against the Germans cost $45,000 each and the other World War II machines were built for similar amounts. The ENIAC, however, had cost hundreds of thousands of dollars and the projections for the EDVAC and IAS computers were even higher.[12]

Even Pendergrass' optimistic estimates of the electronic computer's run times must have raised some concern. The machine he proposed would take seven minutes to run the regular Enigma problem. The inexpensive and reliable conglomeration of spinning wheels, the Bombe, took twenty minutes plus machine set-up time. A perhaps three fold advantage at a probable ten fold cost didn't seem a wise choice. An outside observer, even one who realized the other advantages of a general-purpose machine, might well have thought that OP20G was as or more interested in 'engineering' than cryptanalysis. [17]

Wenger won his battle but it took some time to secure the go-ahead from the Bureau of Ships and to piece together the funds to study and then build OP20G's general- purpose Atlas computer. Its design went through several alterations as the engineering team learned more about computer logic. The army's cryptanalysts were also inspired by the Pendergrass report and soon began their quest which led to the Abner computer. [22, 23]

Neither service found it easy to fund or build a machine, however. Using the private company established by the men who had been the core of OP20G's World War II computer program, Engineering Research Associates, the navy had a working computer at the end of 1950, some three and one-half years after letting the initial contracts. The army had an even more difficult time arranging for the construction of its device which was finally delivered in the spring of 1952.[25, 23]

The price of the Atlas was much greater than expected, more than twenty times that of a Bombe. Just the assembled hardware cost some $950,000. That was much, much more than the navy paid for the many fast special- purpose machines ERA and other contractors continued to build throughout the 1940s and 1950s.[22 p. 8]

The special-purpose machines continued to fill many cryptanalytic needs but the great advances in computer technology in the 1950s vindicated Pendergrass and the others who had been willing to take a very great leap of faith in 1946. Although it was more than a decade before the general-purpose computer, even when augmented with special-purpose add-ons, could match the speed of dedicated devices for all cryptanalytic problems, the code-breakers held to their commitment and continued to be major sponsors of the development of universal computers and their underlying technologies. [22]

REFERENCES

1. Bashe, Charles J. *et al.* 1986. *IBM's Early Computers*. Cambridge MA: MIT Press, 59.

2. Bureau of Engineering, November 18, 1931. To CNO, 'Automatic Machines for OP20G.'

3. Campaigne, H. and J. T. Pendergrass, December 12, 1946. Second Report on Cryptanalytic Use of High-Speed Digital Computing Machines. OP20G, Washington DC.

4. Campbell-Kelly, Martin and Michael R. Williams (eds). 1985. *The Moore School Lectures: Theory and Techniques for the Design of Electronic Digital Computers*. Cambridge MA: MIT Press.

5. Chief of Naval Operations, U. S. Naval Communications, 1945. Technical Paper TS-6, "Rattler".

6. Chief of Naval Operations, U. S. Naval Communications, May, 1945. Communications Intelligence Technical Paper-41, "Copperhead I Punch and Scanner".

7. Chief of Naval Operations, U. S. Naval Communications, March, 1946. "Duenna Operations Manual". Washington, D.C.

8. Chief of Naval Operations, U. S. Naval Communications, May, 1946. Communications Technical Paper TS-43, "The Bombe". Washington DC.

9. Chief of Naval Operations, U. S. Naval Communications, June, 1946. "Machine Comparisons". Washington, D.C.

10. Deavours, Cipher A. and Louis Kruh. 1985. *Machine Cryptography and Modern Cryptanalysis*. Norwood MA: Artech House.

11. Hooper, S. C., DNC, September 26, 1930. To OP20G, 'Automatic Computing', NARA RG457, SRH355, "Naval Security Group History to World War II," p. 79.

12. Meader, Ralph. USNR, January 21, 1949. To Capt. J. N. Wenger, USN, 'Report of 14 Day Training Duty.'

13. OP20G, November 3, 1941. 'Report of Meeting to Discuss New RAM Comparator.'

14. OP20G, May, 1942. 'Memo to Station X, Future E Policy.'

15. OP20G, August 5, 1942. Wenger to Ely, 'Electronic Wheel.'

16. OP20G, September 15, 1942. To Joseph Desch, 'Report on Proposed Bombe Design.'

17. OP20G, February 21, 1944. W. Wright to OP20G, 'Report on Meeting with ASA.'

18. OP20G, September 7, 1944. OP20G to NCML, 'Electrical Hagelin.'

19. OP20G, December 20, 1945. 'Project Monogram'.

20. Randell, Brian. 1980. The Colossus. In N. Metropolis *et al.* (eds) *A History of Computing in the Twentieth Century*. New York: Academic Press. pp. 47-92.

21. Safford, L. R. to S. C Hooper, November 19, 1940. 'Comparator'. NARA RG457, SRH355. p. 404.

22. Snyder, Samuel S. 1964. *History of NSA General-Purpose Electronic Digital Computers*. Washington: Department of Defense.

23. Snyder, Samuel S. 1980. ABNER: The ASA Computer. Part 1: Design. *NSA Technical Journal*. 25(2): 49-67.

24. Stern, Nancy. 1981. *From ENIAC to UNIVAC: An Appraisal of the Eckert-Mauchly Computers*. Bedford MA: Digital Press.

25. Tomash, Erwin. 1980. The Start of an ERA: Engineering Research Associates, Inc., 1946-1955. In, N. Metropolis *t al.* (eds) *A History of Computing in the Twentieth Century*. New York: Academic Press. pp. 485-495.

26. Welchman, Gordon. 1982. *The Hut Six Story: Breaking the Enigma Codes*. New York: McGraw-Hill.

27. Williams, Michael R. 1985. *A History of Computing Technology*. New York: Prentice-Hall.æ

BIOGRAPHICAL SKETCH

Colin Burke is an historian who became interested in the history of computers after many years researching the course of American politics and education. He resides in Columbia MD and is now completing a history of computer development in the American cryptanalytic community from 1930 to 1950.

ENIGMA BEFORE ULTRA POLISH WORK AND THE FRENCH CONTRIBUTION

Gilbert Bloch[1]
Translated by C. A. Deavours[2]

ADDRESS: (1) 7 Rue du Cher, 75029 Paris FRANCE and (2) Department of Mathematics, Kean College of New Jersey, Union NJ 07083 USA.

INTRODUCTION FROM THE TRANSLATOR

The Enigma story seems to be an unending series of revelations and corrections to previous revelations. It is then with genuine happiness that *Cryptologia* begins, in the current issue, to publish a series of chapters from the French work *Enigma Avant Ultra (Enigma Before Ultra)*. The author of this work, M. Gilbert Bloch, has done a real service to those of us interested in this subject. Mr. Bloch has examined all of the publicly available documents and put together what is the most comprehensive documentation on the subject of Enigma which has yet to appear. He has been able, in many cases, to clarify and correct dates and to establish the sequence of events. Where there are discrepancies and mysteries remaining, this too has been pointed out for the reader's benefit. In particular, the crucial period between May and September 1940 has received the author's attention. It now seems likely that the explanation of how the British *got by* may yet prove more interesting than the story told by the late Gordon Welchman in his remarkable book, *The Hut Six Story.*

I am not a professional translator as will, no doubt, be evident. Although Mr. Bloch will examine every bit of material published before its appearance, the final guilt for any mistranslations or missed meanings is with me. In making the translation, I have tried to stay somewhere between the *just give me the facts* school of translation where the author's personality and style disappears and is replaced by that of the translator, and the *don't change a comma* school, which, in this case, would mean quarter of a page sentences with accompanying clauses and parenthetical remarks. As the old maxim goes "What is not clear is not French" — but, then again, it's not English either. *Au boulot!...*

Avis à Nos Lecteurs Francophones

Il y a disponible, en quantité strictemente limitée, des exemplaires de cette oeuvre en version originale. On peut en obtenir une copie (gratuite) du traducteur (ne pas la demander de l'auteur, svp).

PRESENTING "ENIGMA BEFORE ULTRA"

Books and articles on what could be called *The Enigma Saga* are many; it is even tempting to say, taking into account the irrelevance of part of these publications, *too many* ... The production of a new study and its partial reproduction (not contemplated initially!) in *Cryptologia* necessitate therefore some explanation, if not an excuse.

The fact that the author of this new study is a Frenchman (and consequently, the original text written in French) does not constitute a recommendation. Rather on the contrary ... French literature on Enigma has until recently been coarse and poor [1]. One could therefore expect that the main purpose of a French study would be the filling, in French knowledge, of gaps that do not exist anymore in Anglo-Saxon minds; but, on some important points, new documentation, explanations and interpretations are brought to light.

It is only fair, before listing the areas where the reader can expect to find new developments, to warn on the *limitations* of the study.

1. *Enigma before Ultra* (*Enigma avant Ultra*), as indicated by its title, embraces only the 1930-1940 period.

2. The study is not a technical one.

3. The extraordinary spying process through which the French obtained secret German documents is not examined in detail [2].

Turning now to the more positive aspects, *Enigma before Ultra* will, among other things, provide the reader with improved views in the following fields:

1. For the first time, a precise and comprehensive time-table of the events is given, particularly for the 1931-1933 period, where the real course had been until now difficult to retrace. The interrelations between France and Poland are shown and dated: a coherent picture emerges from this process.

2. The knowledge of the exact time-table of events shows clearly how the Polish mathematician M. Rejewski was able to organize his work during the 4 decisive ending months of 1932.

3. As could be expected from a French study, particular attention has been given to the French contribution to the Polish success. One could have feared at that stage a burst of French *chauvinisme*. This is not the case, and it is made perfectly clear that the French contribution was limited to obtaining secret German documents and transmitting them to the Poles, who receive full credit for their work. But, what is shown beyond any doubt is that

   a) The documents transmitted by the French *conditioned* the Polish success (and not only facilitated it, as it has till now been widely believed).

   b) Besides the four documents (explicitly identified, and with their dates of transmission duly indicated) that were put at Rejewski's disposal, the Poles received, from 1931 till the end of 1938, a lot of other Enigma documents. These were not transmitted to the Polish mathematicians by their superiors, but were used for monitoring the quality of their work, and perhaps warned the Poles of procedural changes.

4. Hypotheses liable to explain how the British re-entered the 22nd May 1940 Enigma Luftwaffe net (after having lost all their prior means of access with the dropping, the 1st May, of the duplication of the message key by the Germans) are examined in detail.

In such a field as the Enigma story, nothing will every be considered as *final*; but, on many aspects, the new study hopes to clarify things, and to allow future research to make further advances.

Cryptologia will not furnish a *complete* translation of *Enigma avant Ultra*. The introduction and the two first chapters (devoted respectively to the description of the Enigma machine, to a review of its procedures of use — and the evolution through time of these procedures) are omitted, as *Cryptologia* readers should be familiar with such aspects. Also omitted will be an Annex 1, devoted to the problem of Marine Enigma: more specialized studies on this subject already exist, and better ones are in preparation.

The introduction of *Enigma avant Ultra* includes acknowledgments that it would be unfair to omit. The author owes large debts of gratitude to Sir F. H. Hinsley (Cambridge), Colonel T. Lisicki (London), J. Garlinski (London), P. Calvocoressi (Bath), R. Erskine (Belfast), and on *the other side of the hill*, to Pr. J. Rohwer (Stuttgart). Enigma instructions have been obligingly furnished by the *Militärgeschichtliches Forschungsamt* (Freiburg in Breisgau) and other German archives.

The study would never have been written without the constant encouragement of Colonel P. Paillole.

NOTES

1. It must nevertheless be remembered that the book that started the unveiling of the Enigma mystery was published in France in 1973. *Enigma, ou la plus grande énigme de la guerre 1939-45* by General Gustave Bertrand (Plon ed.) was not the first publication on Enigma: a Polish book on the subject *Bitwa o tajemnice* (*Battle for Secrets*) was issued as early as 1967, under the signature of Colonel W. Kozaczuk, but went practically unnoticed... Without Bertrand's book, it seems doubtful that Group Captain F. W. Winterbotham would have been granted the authorisation to publish in 1974 *The Ultra Secret*. Bertrand's book, despite the crucial role played by the author in the first phase of the Enigma story, is unfortunately, both from the historical and technical points of view, not very satisfactory: it is furthermore marred by errors on dates (some of these errors are simply misprints!) that have prevented many historians (Garlinski and Stengers are notable exceptions!) to draw a correct time-table of the events. Winterbotham simply did not take notice of Bertrand's book, and what he wrote in his second chapter on *the Birth of Ultra* for the period 1930-39 was nonsensical.

2. These aspects of the story are fully covered in the recent book of Colonel Paillole, *Notre espion chez Hitler*, (Laffont ed. 1986.) An English translation is due to appear shortly.

CHAPTER III

POLISH WORK AND THE FRENCH CONTRIBUTION

A. THE POLES AND THE INTRODUCTION OF THE ENIGMA

The reconstituted Polish State of 1919, sandwiched between two equally hostile neighbors, thought cryptology could bring some factor of security to its precarious existence. The Polish cipher bureau, always very active, included excellent specialists. Initially, the decryption of the radio messages of the Reichwehr and Kriegsmarine was accomplished without great difficulty.

This favorable state of affairs was not to last. The naval messages became unreadable beginning in February 1926. The same thing occurred with the army traffic in July 1928.

The German ciphered messages, which were still being intercepted by listening stations, resisted all the established methods of cryptanalysis.

The Poles rapidly arrived at the conclusion that the German armed forces were using a new method of encipherment *by machine*. This presented the Poles with a triple problem, or rather, three problems whose solutions depended upon one another.

1. To determine the type of machine being used
2. To obtain or reconstruct the machine
3. To develop methods of decipherment.

The study of these problems fell, logically, in the *Services*, i.e., the Second Department of the General Staff of the Army at Warsaw.

This Army Department, as well as its directors, had undergone changes during the period under review. For our purposes, it is sufficient to note that the division charged with interception, radio intelligence, and cryptology directed until January 15, 1930 by Major Pokorny was led thereafter by Major (later Lieutenant Colonel) Gwido Langer. In the middle of 1931, this division became the *Biuro Szyfrow* (Cipher Bureau) which was in turn divided into sections covering each geographic zone. The most important of the sections was BS4 which was responsible for Germany. This section was placed under the direction of Captain (later Major) Maksymilian Ciezki.

From 1928, the work on the Enigma was confined to a team of three officers: Ciezki, Michalowski, and Czajsner.

B. DETERMINATION OF THE MACHINE TYPE

The attention of the Poles seems to have been drawn to the commercial Enigma machine and its possible use by the German Army during 1927-28. A curious story about the interception by customs officials at the railroad station in Warsaw of a package containing an Enigma and its subsequent examination over the weekend by Polish technicians is found, in several versions and with conflicting dates, in the works of Lewin, Garlinski, and Rejewski [1]. At the indicated dates, the machine examined could only have been a commercial model of the Enigma. Our interest in this incident is only to indicate at what time the Poles became interested in the Enigma. The result was a purchase (perfectly legal but done under a cover address) of an Enigma machine by the Poles.

It was logical for the Poles to believe that there was some connection between the commercial Enigma which they had just acquired and the new model of the machine used by the German Army. They began by checking, with the aid of their commercial model, whether or not the undecipherable messages *could have* been generated by a similar machine. The answer to this question was in the affirmative.

The Poles were not devoid of sources of information. Their intelligence service soon began investigations in Germany, and the Poles rapidly acquired confirmation that the machine introduced into the German armed forces was a modified Enigma (in fact, significantly modified with the introduction of the Type I model on June 1, 1930).

C. RECONSTRUCTION OF THE MACHINE

1. Differences Between the Commercial Enigma and the Military Model

The commercial Enigma was a machine having 3 movable rotors and a *reversing drum* but it did not have a *plugboard* (*steckerverbindung*).

The military Type I Enigma model of 1930 differed from the commercial model in 4 ways:

a. The three movable rotors had completely different wirings from those of the commercial machine;

b. The reversing drum (*Umkehrwalze*) was of a new type and not settable;

c. A plugboard with changeable connecting plugs (the steckers) was added to the machine;

d. The wiring connecting the keyboard to the entry/exit drum (via the plugboard) was different from the corresponding wiring on the commercial machine which connected the keyboard directly with the entry text drum.

2. 1928-1931. The Initial Deceptive Attempt.

The first task of the Polish team was to identify with precision the differences between the commercial Enigma and the military version. Up to the end of 1931, only slight progress was made in this direction.

It is certain that the Poles, who had at their disposal at least one copy of the commercial Enigma, received from their intelligence service some information about the military Enigma [2]. It was thus that the *Stöpstellstellung*, the early version of the *steckers* was identified and described. But, the information remained imprecise and fragmentary. At the end of 1931, the Poles were still very far from having a clear and precise notion of the general structure of the machine and, therefore, could not envision that they might ever be able to reconstruct the internal wirings. In fact, research on the machine was abandoned.

One all important point had, however, been noted. Systematic studies of the encrypted German messages had shown that the first 6 letters of these texts had some particular significance. Very probably, these first 6 letters formed some sort of *key*.

3. The French Contribution (1931-1932)

In October 1931 [3], an employee of the *Chiffrierstelle* of the *Reichswehrministerium* (Cipher Bureau of the German Ministry of Defense) [4] proposed to deliver documents to the agents of the French *Service de Renseignement*, (S.R.F.).

Since October 30, 1930, the S.R.F. included a *Service Crypto* (called *Section D*) whose role was to obtain information concerning foreign codes and ciphers. (Section D had no decrypting or analytic functions and was limited to transmitting any information gained to the experts of the *Service de Chiffre*). Captain (later General) Bertrand, head of the crypto service of the S.R.F. (and at this date about the only person in its service), was put in charge as *case officer* of individual contacts with new agents.

It was thus that Captain Bertrand received from *Asche* or H.E. (the pseudonyms of Hans Thilo Schmidt) [5], on the 8th of November 1931, the first set of secret documents of which the most important were:

a. *Gebrauchsanleitung für die Chiffriermaschine Enigma*
(Instructions for Using the Enigma Cryptograph). Document H.Dv. g 13 of the Army (H=Heer) and L.Dv g 13 of the Luftwaffe.

b. *Schlüsselanleitung zur Chiffriermaschine Enigma*
(Instructions for Ciphering on the Enigma Cryptograph). Document H.Dv g 14 of the Army and L.Dv g 14 of the Luftwaffe.

These two documents, in fact, the instructions given to machine operators, furnished a detailed description (with photos) and the operating instructions for the military Enigma I (Eins), put into service in 1930. Nevertheless, nothing about the internal wirings of the machine was included [6].

Bertrand transmitted his acquisitions to the S.C.F. which declared immediately that the Enigma was impossible to solve, the documents useless, and removed itself from further discussion of the question. This unfortunate state resulted, perhaps, from the character of Bertrand himself – a character which hardly facilitated relations with the other services.

Bertrand, to his credit, was not convinced by these peremptory affirmations, but he was practically alone and his limited cryptologic competence prevented him from personally attacking the problem. He then had the idea, and was able to obtain from his superiors the necessary authorization, to transmit the documents to the *services* of France's allies and to suggest to them a common attack on the problem.

London was contacted first: the Government Code and Cypher School – the modest name of the British cryptologic service, better known later under the name of *Bletchley Park* – received the documents, carefully filed them, and gave no follow up offer of cooperation [7].

From the 7th to the 11th of December 1931 [8] Bertrand himself visited Warsaw and recontacted Major Langer, the very competent head of the *Biuro Szyfrow*, who understood immediately that the documents furnished would allow resumption of the interrupted Enigma studies. Bertrand and Langer, provided with codenames *BOLEK* and *LUC* respectively, promised each other to share all information. This meant that in exchange for these documents (and those to follow), the Poles promised Bertrand to keep him informed of the results which they obtained.

Major Langer, Captain Ciezki, and his team renewed their work with enthusiasm. The documents brought by Bertrand indicated clearly the points where the military Enigma differed from the commercial models: different wirings, the new type of reversing drum, addition of the plugboard and its plugs. The Poles also learned the frequency and procedures for changing the machine's settings as well as the significance of the first six ciphertext letters of each message. In other words, the Poles knew, from this time, that in each message the first and the fourth letters, the second and fifth letters, and the third and sixth letters resulted from the encipherment of the *same* cleartext letter.

Enthusiasm waned rapidly. All efforts to reconstruct the inner wirings of the machine failed: the magic mountain which had been opened by Bertrand's documents had ended in a blind alley – and this in spite of the receipt of new documents from *Asche*. The visits of Captain Bertrand to Warsaw on 9-11 May and 17-21 September 1932 [8], and, probably, other transmissions of documents through the Polish Services in Paris, placed in Major Langer's hands monthly schedules giving the daily settings of the Enigma. However, in the absence of a complete reconstruction of the machine (the difficulty of which has been discussed in Chapter II), all these documents were, apparently, useless.

4. Mathematicians to the Rescue

Being in despair, Langer and Ciezki decided to seek the intervention of other persons. For a long time, they had had the idea that the level of mathematical competence necessary for cryptanalysis surpassed that of the Army cryptologic personnel available and would require professional mathematicians. Beginning in 1929, they had organized a course of study in cryptology at the University of Poznan [9]. This course was taken by about 20 students already skilled in higher mathematics. The professors were officers in civil guise: Major Pokorny, then responsible for the interception service and cryptology (the predecessor of Langer), Captain Ciezki, future chief of the German section of the Cipher Bureau and a radio specialist, A. Palluth, employed as a high level technician by the Cipher Bureau.

Among the students, three distinguished themselves particularly: Marian Rejewski, Henryk Zygalski and Jerzy Rozycki. Rejewski, who had a degree in Mathematics (1st Class), spent the academic year 1929-30 at the University of Göttingen in Germany in order to pursue a special course of actuarial mathematics and returned in the summer of 1930 to take a post as assistant at the University of Poznan. At Poznan, he recontacted Rozycki and Zygalski who both worked several hours per day in an annex that the Cipher Bureau had set up in the town. They were occupied with the study of cryptologic problems related to the double transposition cipher. Rejewski rejoined his two fellow workers. The problem of Enigma was not considered.

At the beginning of 1932, the three mathematicians were warned of the impending closing of the annex. On the 1st of September 1932, they were officially *enlisted* into the Cipher Bureau at Warsaw. There, for more than a month, they studied the decryption of a German Naval code which they mastered. Suddenly, about the middle of October, Major Ciezki assigned Rejewski to work, alone and in great secrecy, on a new problem.

5. The Extraordinary Success of Mr. Rejewski (October 1932 - January 1933 [10])

What Major Ciezki had asked of Rejewski was to attempt the complete reconstruction of the German military Enigma. Rejewski was very young – born in 1905 he was then 27 years old – and had never even heard mention of the Enigma machine. But proving himself to be a worker of prodigious talent, he succeeded in *10 weeks* (from mid-October to the end of December

1932) in establishing a mathematical theory of the workings of the Enigma, in reconstructing the internal wirings, and in devising the first cryptanalytic methods to decipher the Enigma messages.

a. *The First Phase* (approximately October 15 - November 15, 1932).
Rejewski initially had at his disposal the following information and tools:

i. The two basic German documents on Enigma – *GebrauchsAnleitung* and *Schlüsselanleitung* – obtained from Asche and brought by Bertrand on December 7, 1931;

ii. A brand new model of the commercial Enigma;

iii. The continuous flow of German ciphertext messages captured by the Polish interception stations.

The two Asche documents allowed Rejewski to become rapidly familiar with the structure of the machine and the procedures used to operate it.

In studying the implications of the *repetition of the message key* – the first six letters – Rejewski realized that in 81% of the cases (21 out of 26) the typing of the six letters on the machine's keyboard caused only a movement of the righthand rotor. The rest of the machine could then be considered as a single fixed ensemble. Moreover, the first and the fourth letters, the second and the fifth letters, and the third and the sixth letters each corresponded to the same cleartext letter.

Rejewski then deduced the mathematical consequences implied by his two observations. Namely, that it was possible to express the encipherment of these six letters by equations representing the six successive permutations involved. Further, if given a sufficient number of messages from the same day (which would therefore use the same settings of the machine), one could draw out certain characteristics of the permutations of that day. The system of six equations was therefore *theoretically* capable of a solution which would, in turn, yield the wiring of the rightmost rotor.

Since the Germans changed the order of the rotors periodically, each of the three movable rotors would, from time to time, find itself placed on the extreme right. This allowed the three rotor wirings to eventually be reconstructed. In a similar manner, the reversing drum (Umkehrwalze) wirings could also be found.

The method which had just been described was *theoretically* applicable even if the machine settings (order of rotors, *Ringstellung*, *Grundstellung*, and plugboard setting) were completely unknown. Nevertheless, in this case, carrying out the *actual* calculations involved was practically impossible because no powerful calculating devices such as today's computers were available. Theoretically, one could dream of a solution but it implied use of a very large number of messages and a succession of luck too extraordinary to be believable [12].

Once again, the work seemingly landed in a blind alley.

b. *The Second Phase* (approximately November 15 - December 31, 1932).

In October, Major Ciezki had only furnished Rejewski with the two basic documents on Enigma. No doubt he wished to determine the ability of the young mathematician to use them. Experience having decided the question (beyond all possible expectations) Ciezki gave Rejewski, probably around November 15, two new documents also acquired from Asche, and transmitted by Bertrand. These documents were two monthly schedules giving the daily settings for the Enigma. Rejewski had at his disposal the ciphertexts of the messages intercepted during these two months. He knew from the latest documents the order of the rotors, the setting of the rings, the rotor starting positions (for enciphering the message key), and the plugboard connections. All this information allowed Rejewski to simplify his equations enough to solve them. By luck, the two months covered in the schedules fell in different trimesters of the year [13]. Because the rotor order was altered each trimester, two different rotors occupied the rightmost position. The solution of the wiring of these two rotors allowed the solution of the wiring of the third rotor. With this done, the wirings of the reversing drum (*Umkehrwalze*) could be determined without difficulty.

Rejewski immediately threw himself into his calculations. Initially, no valid results were obtained. In seeking the cause for this failure, Rejewski had the idea that his assumption about the wiring from the plugboard to the entry/exit drum had been wrong. In the commercial Enigma, the wiring was *straight through* from the keyboard to the 26 contacts of the entry/exit drum. This placed the wires in the same order as the letters on the keyboard of the standard German typewriter: the letter Q connected to contact 1, W to contact 2, E to contact 3, etc.

Rejewski had thought initially that the military Enigma was wired similarly. The keys being wired according to a straight alphabetic correspondence (key A joined to plugboard connection A, B to B, etc.) and the plugboard letters joined in the order of the entry keyboard letters with the contacts of the entry/exit drum.

Rejewski told himself that if the order had been modified, the methodical Germans would not have changed to a random wiring. He then tried the wiring where the letters were arranged in regular alphabetic order (A connected to the first contact, B to the second, C to the third, etc.). This inspiration was correct and, as if by a miracle, the internal wirings began to emerge from his calculations.

By the end of 1932, the combination of mathematical genius displayed by Rejewski and the trove of secret Asche documents sent to the Poles by Bertrand, as well as a good dose of luck (Rejewski's *intuition* concerning the entry wiring), had led to the complete reconstruction of the German military Enigma.

This reconstruction was not only carried out *on paper.* An eminent technician, the engineer Antoni Palluth, one of the directors of the AVA radio equipment fabrication factory in Vasovie and also a consultant for the Cipher Bureau had, following Rejewski's progress, modified a commercial Enigma into the military model of the machine. This perfectly operational prototype served first to verify Rejewski's results. Second, this machine was used as an industrial prototype for the AVA manufacture of replicas of the military Enigma destined for the Polish cryptologic services.

D. THE FIRST METHODS OF DECIPHERMENT

The knowledge that the first six letters of each message was the twice enciphered key to that specific message furnished Rejewski with the means of reconstructing the internal wirings of the Enigma. It also led to his decipherment of the messages coming from the machine.

A detailed description of the decrypting methods developed by Rejewski while at the same time reconstructing the Enigma machine will not be attempted here. Suffice it to say that these methods allowed, from some characteristics of the 6 letter groups corresponding to the twice repeated enciphered *message keys, one to determine* the settings of the machine *compatible* with these characteristics, to test these settings and finally to find out the exact message key used. After this had been done, the decipherment could be carried out under the same conditions as the German clerks to whom the messages were addressed had. In other words, using the same machine settings, each message could be deciphered by typing the ciphertext on the keyboard of the machine.

During his work in establishing the first methods of decipherment, Rejewski could exploit certain errors committed by the German operators. In particular, the use of particular groupings of letters such as AAA, BBB, SSS, etc. as message keys. The use of such letter groups was forbidden to the operators at the end of 1932, which was, of course, too late.

From the beginning of 1933, the decipherment methods perfected by Rejewski showed themselves to be entirely adequate.

The Polish success was complete.

E. THE ROLE AND IMPORTANCE OF THE FRENCH CONTRIBUTION

The Poles and *only* the Poles attempted the reconstruction and cryptanalysis of the German military Enigma. Alone, they carried out the totality of intellectual, cryptologic, and mathematical investments necessary. And they alone arrived at the solution. Chapter IV will show that they were the only ones, between 1933 and the beginning of 1939, to retain mastery over the Enigma and its decipherment. For these extraordinary exploits, the Poles must receive all of the credit and retain all of the glory.

It does not diminish the value of their accomplishment to recall that there was a French contribution and to try to evaluate its importance.

The French contribution to the Polish success was *both* mediocre and fundamental:

a. *Mediocre*, because that contribution was limited to the transmission by Bertrand of the documents obtained from Asche;

b. *Fundamental*, because the documents supplied by Bertrand played a decisive role. The writings of Rejewski and a careful study of the timetable of events leads to the following inescapable conclusions: the first two documents, *Gebrauchsanleitung* and *Schlüsselanleitung*, constituted the indispensable base upon which Rejewski's genius built his reasoning and equations; the two monthly schedules of daily machine settings allowed the theory thus established to be actually carried out. Without these four documents, success could not have been attained.

It would not be necessary to emphasize the role of the French contribution had that role been correctly described and appreciated in the available documentation. But, if the existence of a French contribution is acknowledged at all, most works on the Enigma appear to agree that the documents furnished by the French intelligence service merely *facilitated* the tasks of the Polish mathematicians and, that, in their absence, the same result would have been obtained with only an additional delay. These affirmations do not correspond to reality. The documents furnished by Bertrand not only *facilitated* the Polish success – they made it possible. One may consider that the methodical and obstinate efforts of the Poles have a *value* superior to the *stroke of luck* which was offered to the S.R.F by Hans Thilo Schmidt. But, without the Asche documents furnished to the Poles, it is nearly certain that the Enigma would have kept its secrets a lot longer [14].

There remains a secondary problem to discuss in regard to the French contribution. At first glance, this problem may appear quite strange. Rejewski has stated publicly, and there is every reason to believe him, that he only received four secret German documents (the two basic ones and the two monthly sets of machine settings). Bertrand has indicated that, thanks to Asche, a very large number of documents (notably, the monthly machine setting schedules from December 1931 to December 1932) were obtained, and it would seem implicit that this bulk of information was immediately repeated to the Poles, i.e., Langer and Ciezki. (However, there do not exist accompanying letters or memoranda explicitly proving that this transmission took place.)

Could this situation be explained by the fact that Rejewski was only allowed to see a fraction of the material received by his superiors? On second thought, such a situation would not appear surprising. The *need to know* principle is the rule in intelligence work. Rejewski, a completely new employee of the *Biuro Szyfrow*, had proved his worth when confronted with two documents (*GebrauchsAnleitung* and *Schlüsselanleitung*) which were, in turn, followed by two other documents (the two monthly schedules of machine settings). Those documents were sufficient to lead to the success of his enterprise and there was no cause to reveal the existence of the other *papers* (which were no longer indispensable to his work but would serve to verify it). Finally, it is worth noting that, up to 1939, Rejewski ignored that the base of his work had been furnished by the S.R.F. Only Langer and Ciezki knew of the liaison with Bertrand.

NOTES FOR CHAPTER III

1. R. Lewin – *Ultra Goes To War.* See note on page 30, paperback edition. J. Garlinski – *Intercept*: p 2-3. M. Rejewski – *How the Polish Mathematicians Broke Enigma* – English translation of an article which appeared in 1980 (posthumously) in a Polish mathematical magazine; this translation is to be found in the appendix of the book *Enigma*, by W. Kozaczuk, p. 246.

2. T. Lisicki *Die Funkaufklärung und Ihre Rolle im zweiten Weltkrieg*, p. 67.

3. And not 1932 as indicated by Bertrand (*Enigma*, 23). The mixture of dates in Bertrand's book which sometimes place the events of 1931 in 1932 did not help the Enigma historians. The chronology now established beyond any doubt.

Most historians of Enigma (excluding Garlinski and Stengers) have been deceived by the date 1932 which has, in turn, led to inextricable difficulties in reconstructing the correct timetable of events. A more serious fact is that Rejewski himself in his last articles also adopted the date 1932. Existing documents including accounts by Bertrand himself leave no doubt as to the actual date.

4. The *Chiffrierstelle* was not the only German service dealing with cryptology. E. Höttenhain in *Die Funkaufklärung und ihre Rolle im zweiten Weltkrieg*, p. 100 citation 8, shows clearly the waste and disorder resulting from the lack of coordination.

5. The extraordinary story of Hans Thilo Schmidt, whose importance extends beyond Enigma material, is recorded by Paul Paillole in his work *Notre espion chez Hitler*, Editions Laffont, 1985.

6. The wartime destruction in the Polish, French, and German archives was such that no copy of the original 1930 documents exists any longer in continental Europe. The only place where, perhaps, these documents might still be found is in the archives of the British

Government Code and Cypher School (GC&CS), which received copies of them from Bertrand in 1931. These archives are inaccessible.

The original 1930 documents did not carry the reference "L" indicating *Luftwaffe* since this organization did not exist officially until 1935. The documents Dv. g 13 and 14 were the object of numerous editions which incorporated the successive modifications of the Enigma machine and its method of use. These modifications do not seem, at least before May 1940, to have changed any essential aspect or spirit of the documents. Thanks to the amiability of the *Militärgeschichtliches Forschungsamt* (the historical service of the German Army located in Freibourg in Brisgau), the author of this study possesses copies of the following:

a. The *Gebrauchsanleitung* of January 12, 1937 (the edition is of January 1940 but specifies that the text is unchanged); this is a document of 16 pages of which 11 are text and the remaining are photographs of the machine pointing out its different parts.

b. The *Schlüsselanleitung* of January 13, 1940. 14 pages.

7. Cf. Hinsley, *British Intelligence in the Second World War*, Vol. 1, p.488. Bertrand did not go to London on this occasion. The documents were given to the representative of the I.S. in Paris, Commander Wilfred Dunderdale, who carried them to England. (He left on the 23rd of November for England. Source: P. Paillole, *Notre espion chez Hitler*, p. 39).

8. The dates of Bertrand's visits to Warsaw must be connected with those of the Asche meetings, each of which resulted in the transmission of documents. In 1931 and 1932, Bertrand met Asche four times

- November 7-8 1931 at Verviers
- December 19-20 1931 at Verviers
- May 7-8 1932 at Verviers
- October 29-30 at Liège.

Besides this, REX (aka Lemoine), the intermediary of the S.R.F., met Asche in Berlin August 1-2, 1932 and received from him documents which arrived in Paris August 10 by diplomatic pouch.

9. The University of Poznan was excellent. All of the students there spoke both Polish *and* German.

10. Rejewski's complete and detailed explanation of the mathematical theory of the Enigma, the corresponding system of permutations and their solution is beyond the scope of this study. The interested reader can consult the writings of Rejewski himself, writings from (1984) are now available in English. These writings consist of appendices D and E from the book *Enigma*, by W. Kozaczuk. Appendix B of the same book, a liberal transcription of an interview with Rejewski recorded July 24, 1978, furnishes an extraordinarily vivid illustration of Rejewski's career and work.

Another work of Rejewski cited (with numerous extracts) but not furnished *in extenso* in the work of Kozaczuk must be quoted. This article has been, fortunately, fully translated and published in English in *Cryptologia*, 6, No. 1, Jan. 1982, pp. 75-83. The *Remarks on Appendix 1 to "British Intelligence in the Second World War"* consists of Rejewski's comments of the text published in Volume I of Hinsley and furnishes precise corrections.

11. And it appears that the computer time required may, even today, be prohibitive.

12. Cf. Rejewski – *How the Polish Mathematicians Broke Enigma*. Appendix D of the book *Enigma*, W. Kozaczuk, p.258. Rejewski explicitly indicates that, without further information, it was impossible to solve the permutation equations.

13. The fact that the two monthly schedules belonged to different trimesters is perfectly clear. By contrast, the months covered by the two schedules and the date and circumstances of their possession by the Poles present several problems whose elements are given below:

a. Bertrand (*Enigma*, p. 32) indicates having received from Asche (and then transmitted to the Poles) *les tableau mensuels de configurations (comportant une configuration par jour) pour Décembre 1931, les années 1932, 1933, et 1934 (1er semestre).*

b. Rejewski states in his works, and there is every reason to believe him, that only *two* monthly schedules of machine settings were put at his disposition covering two months belonging to different trimesters. However, he has varied the dates. In his last article (posthumous, 1980) he indicates the months of September and October 1932 which appear not only plausible but probable (see below). But, he has also indicated that these documents were put at his disposal December 9, 1932 (cf. Kozaczuk p.256). This date, apparently based on the erroneous date (December 7, 1932 instead of December 7, 1931) of the first visit of Bertrand to Warsaw, is evidently false. It would have left Rejewski only three weeks to finish all of his calculations before December 31.

In the first paper that Rejewski wrote, in France during 1942 under conditions hardly favorable to composing memoirs, he quotes the months of October and December 1931 – a double impossibility. Bertrand has said that he did not have any table of keys at his disposal until December 1931. The months of October and December belong to the same trimester. (It is strange that, even in the conditions of 1942, Rejewski had forgotten a technical detail of such importance.)

c. The date of the Asche meetings, i.e., the dates of Bertrand's visits to Warsaw, do not allow one to settle, *a priori*, this question.

The monthly schedules could have only been transmitted by Asche during his meetings with Bertrand (or, in one case, with another agent of the S.R.F., Lemoine). These meetings took place November 7-8 and December 19-20 in 1931 and May 7-8 and October 29-30 in 1932. The meeting with Lemoine was on August 1-2, 1932.

Bertrand visited Warsaw from the 7th to the 11th of December 1931, from the 9th to the 11th of May, and from the 17th to the 21st of September 1932. (All of the preceding dates were furnished to the author by Colonel Paillole.)

Assuming, logically, but, as will appear, wrongly, that the monthly schedules of keys were not available in Germany until a few days before the first of the month in which they were to be used, one is led to think that the schedules for September-October 1932 could have been furnished by Asche only during the meeting of October 29-30. This would be too late to have been provided by Bertrand to Warsaw at the time of his visit of September 17-21, 1932. There would nevertheless remain other possibilities: the schedules in question could have been transmitted to Langer by an intermediary representative of the Polish *services* in Paris (the contacts here were constant). It would be equally possible that Langer himself came to Paris at the beginning of November 1932 to take delivery of these documents. Whatever hypothesis is adopted, the schedules provided by Asche to Bertrand during October 29-30 could have found themselves in Warsaw about November 15 without difficulty. This fits perfectly into the timetable of events.

The book of P. Paillole, *Notre espion chez Hitler* (p.53), provides the solution to the problem. The monthly schedules for September and October 1932 (the October schedule in provisional form) were provided by Asche to Lemoine in Berlin at the beginning of August 1932. This is early enough to have been given to the Poles by Bertrand at the time of his stay in Warsaw, September 17-21.

The date (approximate) of November 15, 1932 when Rejewski received the two monthly schedules of keys suggests they referred to the months September and October 1932. It was logical to furnish Rejewski with the *latest* schedules available since these schedules would correspond to the most recent ciphertexts intercepted by the Polish listening stations and Rejewski already had these ciphertexts.

14. The relative misestimation of the French contribution is understandable. The only French work concerning Enigma was that of Bertrand, published in 1973. This work had the distinction of being the first one to describe what had happened. But, the style of this book, the understandable tendency for the author to discuss more his personal role in the events rather than to provide a complete, clear and technical documentation of the facts, arouses suspicion. Historians can not be blamed for having preferred to utilize two short reports (unpublished but easily accessible) drawn up by Colonel Mayer in London during 1974 in response to the books of Bertrand and Winterbotham. Colonel Mayer was the former head of the Polish Intelligence Services. Mayer has written, "At the stage in which the research of the *Enigma* was at that time in the Polish cryptological section, these documents [those sent by Bertrand] appeared not to be indispensable for the final solution of the problem. But, unquestionably, they facilitated it." In writing this sentence Colonel Mayer was certainly sincere, but, even if he had closely supervised the work of Langer, Ciezki, and Rejewski, this supervision would not entail familiarity with the difficulties encountered, the methods used, nor the solutions found.

The works of Rejewski, partially in English, have been available only a little while. Their interpretation is sometimes made difficult by the fact that Rejewski had ended up being persuaded that the first visit of Bertrand to Warsaw was on December 7, 1932 instead of *1931*.

Of course, recourse to the original documents is impossible since the French and Polish archives were destroyed or lost during the war.

The book of P. Paillole, *Notre espion chez Hitler*, published in September 1985 describes in detail the extraordinary role played by Hans Thilo Schmidt, alias Asche, and, finally, permits one to appreciate exactly the magnitude of the Asche documents and their importance to the Poles.

BIOGRAPHICAL SKETCHES

Gilbert Bloch is retired and lives in Paris. He is member of the French Military Reserves. His hobbies include cryptology and parachute jumping.

C. A. Deavours teaches a unique sequence of courses in cryptology at Kean College of New Jersey. The courses address both technical and historical issues in cryptology.

CRYPTOLOGIA October 1987 Volume XI Number 4

ENIGMA BEFORE ULTRA THE POLISH SUCCESS AND CHECK (1933-1939)

Gilbert Bloch[1]
Translated by C. A. Deavours[2]

ADDRESS: (1) 7 Rue du Cher, 75029 Paris FRANCE and (2) Department of Mathematics, Kean College of New Jersey, Union NJ 07083 USA.

We continue the translation of selected sections of Gilbert Bloch's book, *Enigma Avant Ultra (Enigma Before Ultra).* See the previous issue of this journal for the first part of the publication, Chapter III — Polish Work and the French Contribution

CHAPTER IV

A. OVERVIEW

At the beginning of 1933, the Poles had attained their objectives:

a. They had completely reconstructed the German military Enigma. By this time, they possessed a replica of the machine — a prototype, of course, but fully operational;

b. They had developed methods which, exploiting the German error of the twice enciphered message key, permitted the determination of the machine settings and the individual message keys, and, therefore, the decipherment of intercepted messages on an Enigma replica with as much ease as that of the German addressees.

From 1933 until the end of 1938, the Poles kept their advantage. The *Biuro Szyfrow* read the majority of German radio Enigma messages transmitted by the Army, the Luftwaffe, and the S.S. [1] All through this period, the Germans changed certain workings of the machine, increased the frequency of key changes, and altered ciphering procedures. The Poles always, and with surprisingly short delays, kept up with these changes and continued their decipherments. To better aid them in their tasks, the Poles — particularly the trio of mathematicians Rejewski, Rozycki, and Zygalski — devised numerous methods, invented hand and mechanical *devices*, and, finally, created electro-mechanical machines which, from characteristic data of the intercepted messages, could find the daily settings used for the encipherment. During all of this period, the Poles alone were able to accomplish these things. [2]

On the 15th of December 1938, following new German modifications, the Polish lost the ability to decipher messages even though they had correctly diagnosed the new changes made, and reconstructed the new workings. In order to successfully confront these changes, the Poles would have needed to augment considerably (in fact, tenfold) their available materiel. For them, this was impossible.

On the 24th of July 1939 — one month before the beginning of the Second World War — the Poles revealed, during a meeting with French and British experts, their results. They gave to each of their astonished allies a copy of their German Enigma replica (complete with 5 rotors!), a detailed description of all decipherment methods, and schematics of all the devices they had developed to decipher the machine in which the Germans placed — and kept — complete confidence.

Until this date, the Poles had kept absolute secrecy even towards Bertrand who, from 1933 to 1939, continued to make numerous trips to Warsaw (compensated for by an equal number of trips of Langer to Paris) in order to transmit documents. This absence of reciprocity, this lack of *tit for tat*, should not surprise. Examples abound which show that relations between the *Services* are never free of hidden motives and that *an eye for an eye* is not always the rule. [3]

The Polish *gift* of 24 July 1939, which would have never been possible without the French aid received, was to play in its turn an essential role in the future British success.

B. THE CONSTRUCTION OF THE ENIGMA REPLICAS

Possession of only one copy of the Enigma machine was, even in 1933, not sufficient, and it was clear that in the future many more machines would be needed. Consequently, the Poles launched the industrial manufacture of Enigma replicas. This task was the responsibility of A. Palluth, one of the associate directors of the AVA company in Warsaw who specialized in the fabrication of radio equipment. He had been closely connected with the work of the *Biuro Szyfrow* and it was certainly he, following Rejewski's progress in reconstructing the schematics of the Enigma, who translated these schematics into actual pieces of equipment.

The AVA factory produced the first Enigma copies in 1934. In 1939, the Poles had 17 Enigmas of which two were given to the Allies (one to the French and one to the British). Taking into account the considerable number of Enigma *rotors* used in the mechanical devices invented by the Poles (*cyclometers* and *bombes*) to aid them in the tasks of decipherment, the number of Enigmas (or parts of Enigmas) produced by the factory between 1934 and 1939 was much greater (by at least 4 or 5 times) than one would think from the number existing in 1939.

C. THE POLISH SUCCESS (FROM 1933 TO THE END OF 1938)

To achieve mastery of the Enigma, the Polish aimed at numerous methods "from the simplest to the most complicated, from manual to mechanical, from the cheapest to the most costly." [4]

1. Manual methods

As has already been mentioned (c.f. Chapter III, Polish Work and the French Contribution), the methods used by the Poles consisted, essentially, of listing first characteristics from the first six letters of the encrypted part of the intercepted German messages — the portion corresponding to the encipherment of the *message key* (one group of three letters repeated twice). From these groups, one could then determine the settings of the machine which were *compatible* with these characteristics and, if the number of messages were sufficient, these *compatible* settings could be reduced to a small enough number that could be *tested* on the Enigma replicas. The tests permitted one to determine *the* actual setting used for the day under consideration (the *daily key*) and, by another process, the specific key for each message. It *sufficed* then to read the messages using the Enigma replicas.

2. The First Mechanical Methods

In order to make their work easier and faster, the Poles invented an apparatus, the *cyclometer* [5] composed of elements (rotors) from two Enigma machines joined together. This instrument allowed one to setup mechanically a catalogue of *compatible settings*, at least as long as the Germans made use of a daily master *Grundstellung*. The cyclometer began service in 1937.

3. The First Enigma Modifications — 1937

From 1930 to 1937, The German military Enigma hardly changed.

Of course, the method of use evolved: increase in the frequency — ending in a daily alteration of all elements of the settings — increase in the number of *steckers* used. [6] This evolution was not of such a nature as to seriously impede the Polish work.

It was the 1st of November 1937 when the Germans brought to their machine the first important modification. The original reversing drum (*Umkehrwalze*) was replaced by a new one (*Umkehrwalze B*). The Polish reconstruction of the wirings of the new reversing drum was very rapid and decipherments continued as in the past.

The facility with which the Poles adapted to this change would seem to pose a problem. It is certain that, once having determined the nature of the modification (the change of the *Umkehrwalze*), the Poles (by their mathematical theory of the machine, by the permutation equations which resulted from this theory, and by their methods of solving these equations) found themselves perfectly well able to determine the wirings of the new drum. [7]

But how was it possible for the Poles to establish an exact *diagnosis*, that is, how could they find the exact nature of the unexpected change? The Polish answer to this question is apparently clear: the Germans, always strongly methodical, sent by Enigma to all users a text reminding them that, beginning a 00:00 hours on 1 November, they must mount in their machines — formerly called the *Wehrmacht Enigma*, the new reversing drum. [8] The Poles, thanks to their decipherments, were therefore forewarned and mastered the new machine as well as the old one.

In January 1938, a *test* of two weeks by the Polish Cipher Bureau, showed that 75% of the intercepted messages had been decrypted. [9]

4. Procedural Modifications from September 1938 and the Polish *Bombe*

The 15th of September 1938, the Germans abandoned the daily master *Grundstellung*, used until then, for the encipherment of the specific message key for *all* messages of that day. From now on, a particular *Grundstellung* for each message had to be randomly chosen by the operator (and indicated in the clear in the first part of the message).

Here again, the Poles were very quickly informed of the nature of the change, and, equally quickly, found the means of parrying it.

a. A new electro-mechanical machine, the *bomba* incorporating elements (rotors) of six Enigmas was invented. Six such bombes, each corresponding to one possible rotor order of the Enigma, were constructed. These bombes became operational in November 1938.

b. The mathematician H. Zygalski devised a manual system based on the utilization of a series of perforated sheets. Each perforation corresponded to one configuration of the machine compatible with one characteristic of a message key. The juxtaposition, on a light table, of a certain number of sheets corresponding to the characteristics of the keys made apparent the cases in which these characteristics were juxtaposed by permitting light from the table to shine through perforations in the pile of sheets. These perforations corresponded to *compatible settings*. Once the sheets were established (by an initially long process, necessitating complicated calculations and mathematical checks, but done only once), the selections could be rapidly made in *a few minutes*.

D. THE POLISH CHECK (DECEMBER 1938-JULY 1939)

On December 15, 1938, the Germans added to the first set of I, II, and III (which had six possible rotor orders) two more rotors, IV and V. The machine itself was not modified and always used only three rotors, but the number of possible rotor orders was increased from 6 to 60.

This time also, the Poles are said to have had immediate knowledge of the nature of the change: they had known for several months — from intelligence — that in case of extreme international tension, the German operators would receive two new rotors (already stockpiled!) baptized *mobilization rotors* (*Mobilisierungswalzen*), and that the simple ordering of the three existing rotors would become in this case a selection of three rotors out of five, followed by an ordering of the three rotors selected in the machine. Moreover, once again, the Poles are said to have intercepted the Enigma message giving the order to put the new procedure into practice.

The Poles were even able to reconstruct the wirings of the two new rotors thanks to the S.S. During September 1937, a new network of Enigma users appeared transmitting S.S. and *Sicherheitsdienst* messages. Before being enciphered on the Enigma, these messages were enciphered manually using a method that the Poles easily mastered. For some reason, after the 15th of December 1938, the S.S. and the Sicherheitsdienst used, like everyone else, the entire set of five rotors but kept the old keying procedure instead of the *Grundstellung* modification of September 15, 1938. Under these conditions, the experienced Polish mathematicians were able to reconstruct the internal wirings of rotors IV and V, and continued to read the Enigma messages of the S.S. and the Sicherheitsdienst (until July 1939).

On the other hand, the Enigma messages of the other networks had become unreadable to the Poles in spite of the reconstruction of the two new rotors. This occurred because of a *lack of technical resources* and not because of a lack of cryptanalytic mastery. It is easy to see why: in order to read the Enigma using a set of three rotors the Poles had *six Bombes*. The addition of two more rotors multiplied the number of rotor orders by a factor of 10 (i.e., from 6 to 60). In order to expand the number of Bombes (i.e., to build 54 additional Bombes), the Poles would have had to quickly produce $54 \cdot 6 = 324$ more sets of Enigma rotors (since each Bombe used six Enigma rotor sets). (The Poles had only 17 Enigmas in service.) This task largely surpassed the production capacity of the AVA factory. In spite of their intellectual mastery, the Polish were circumvented by the lack of materiel. Also, the increase in the daily number of plugs (*Steckers*) used (beginning January 1, 1939) limited the effectiveness of the Polish Bombes.

Even if the lack of materiel could have been overcome, the Polish could not have hoped in so short a time to increase their cryptanalytic personnel for dealing with the work which had increased tenfold.

Zygalski's perforated sheets method suffered the same problems. The number of sheets required was also to be increased tenfold. The Poles put themselves to the task, but the magnitude of the necessary calculations made the process very slow, and only very partial results were obtained before the declaration of war.

It is Bletchley Park that, thanks to a prodigious concentration — both quantitative and qualitative — of staff and materiel, mastered the new problems. But the fact of having taken advantage of the Polish experience and having had access to the Polish work represented an enormous advantage for the English.

E. WAS THERE A FRENCH CONTRIBUTION TO THE POLISH SUCCESS OF 1933-38?

The French contribution to the Polish success of 1932 is universally recognized even if its real role may appear misestimated.

For the period 1933-38, there is no mention, either in Polish documents or other sources and available references, of a French contribution. Does this silence correspond to reality?

a. The Polish mathematicians, Rejewski, Rozycki, and Zygalski have always said that they had used no secret German documents other than the four used by Rejewski during the last trimester of 1932. There is every reason to believe them. But, this does not imply that their superiors, Langer and Ciezki, did not have possession of other documents which, while not indispensable to the mathematicians' work, which went along magnificently, permitted control of the quality of their results.

b. Bertrand states in his book that Asche delivered during the period 1931-34 a number of important documents relative to Enigma besides the basic ones, the *Gebrauchsanleitung* and the *Schlässelanleitung*. [10] It is practically certain that these documents were transmitted to the Poles.

c. Bertrand mentions (*Enigma*, p. 33) that in 1934 Asche left the *Chiffrierstelle*; the reader is then led to the conclusion that beginning at this date, Asche was no longer in a position to provide cryptologic information on the Enigma.

   The reality is quite different. Use of new — and irrefutable — documents allowed Colonel Paillole, in his book *Notre espion chez Hitler*, to show that Hans Thilo Schmidt alias Asche did not leave the *Chiffrierstelle* in 1934 but only his role there changed. Formerly, he was in charge of the liaison with the *Forschungsamt*, a department reporting to the Air Ministry (i.e., Marshall Goering) and one of the most important German cryptologic agencies. It was not until September 28, 1938 that Asche was administratively attached to the *Forschungsamt* (at an important level since he was director of the Templin station).

Up to this date, at least, he was marvelously placed to secure cryptologic documents and he made considerable use of this advantage.

Hans Thilo Schmidt had, in 1931-32, met six times with the agents of the French *Services*. Bertrand was present at four of these interviews (c.f., note 8, Chapter III and Appendix II of this work).

From 1933 to 1939, the records show 14 meetings between Asche and Bertrand (and there were other meetings at which Bertrand was not present). Each one of these encounters resulted in a transmission of documents.

d. Bertrand went to Poland three times during 1931-32 (c.f., Chapter III). From 1933 to 1939, there were 10 trips to Warsaw (and one to Czechoslovakia), [see Appendix II of this work].

   Langer went to Paris (on dates not precisely known) as many times as Bertrand went to Warsaw. Bertrand never went to Warsaw, and Langer never returned from Paris, empty-handed.

e. One part of the documents delivered by Asche between 1934 and 1940 contained general (but important) information on German armed forces and politics. These documents were *handled* by the German department of the French Services. But the mere fact that Bertrand (whose prerogative was limited to the cryptologic field) had continued to meet Asche between 1934 and 1938 shows unequivocally the duration of the deliveries of the cryptologic information. Bertrand himself gives some examples, not relative to Enigma.

   Colonel Paillole (op. cit.) is able to show that, during the course of this period, Asche continued to deliver Enigma documents and, in particular, a considerable number of monthly key setting schedules (see Appendix II of this work). At each interview, Asche furnished two of these schedules!

   These documents were surely transmitted to the Poles and, beginning in 1937-38, at least partly to the English.

f. It is therefore absolutely *certain* that Enigma documents continued to be transmitted after 1934. What were these documents and what role did they play?

   i. Appendix II gives a list (non-exhaustive) of the monthly key setting schedules furnished by Asche and their date of delivery.

      It seems certain that in the Polish case, no key setting schedules (apart from the two schedules which play a decisive role for Rejewski in the reconstruction of the machine) were ever delivered to the mathematicians who, by their own methods, determined the *daily keys*. Langer and Ciezki would have then used these documents to check on the work of the mathematicians. (Taking into account the *work rules* of a cryptologic service, this hypothesis is not unreasonable. [11])

      In the case of the British, this problem is studied in Chapter V.

ii. Aside from the transmission of the monthly key setting schedules, no explicit mention has been found as to whether or not Asche transmitted information about the procedural and technical modifications made by the Germans to their system between 1934 and 1938. These modifications, which have been detailed in Chapter II, are numerous. It cannot be excluded that such information had indeed been transmitted and would have been of the sort mentioned to the Poles. (The reasons that the Polish have given themselves to explain their knowledge could be *cover stories* — told in all good faith.)

g. There is a possible objection: if Enigma documents were transmitted after 1934, then the absence of mention of this fact, understandable in Polish reports and studies, in Bertrand's book is strange (to say the least).

This argument, while valuable, is not decisive. Bertrand wrote his book in 1973. He knew that from 1933 to 1939, the Poles whom he had pestered with questions about the course of their work had beat around the bush up until July 24, 1939. He could speak of what he had furnished from 1931 to 1934. Psychologically, it was perhaps difficult for him to admit that he had continued to go on helping the Poles for five years without receiving anything in exchange.

NOTES FOR CHAPTER IV

1. Enigma messages sent by the German Navy were decrypted only rarely. The lack of a sufficient number of intercepted messages is the reason.

2. The British seem to have begun study of the Enigma beginning in 1936. During the Spanish Civil War (beginning in 1937) the English deciphered German, Italian, and Spanish Nationalist messages enciphered on commercial models of the Enigma (slightly modified, but without the plugboard). In July 1939, at the moment of the Polish *gift*, the English had not yet reconstructed the German military Enigma.

3. Political factors can explain secrecy observed by the Poles. Political relations between Poland and France from 1933 to 1939 were far from always being excellent and information received from the Poles, partly due to their decryptments, made the Poles doubt the willingness of the French to aid Poland in case of attack, and also of the capacity of the French to keep secret the fact that the Enigma had been decrypted. As for the English, they gave their guarantee to Poland only March 31, 1939.

Besides, Bertrand was not a cryptologist and did not share the *fraternity* which joins specialists. It would not have been easy to explain to him the mathematical and technical complexities of the Polish decipherments.

4. T. Lisicki in *Die Funkaufklärung und ihre Rolle in 2. Weltkrieg*, p. 78.

5. So called because of peculiarities revealed in the encipherment of *message keys* leading to the determination of *cycles* of letters.

6. For this subject, see the part of Chapter II dedicated to setting changes.

7. Welchman (*The Hut Six Story*, p. 113-115) tells that following the interception of a message using a new *trial Umkehrwalze*, Bletchley Park reconstructed the wirings in a few hours (less than one night).

8. Source: interview with Colonel Lisicki.

9. Source: Memoranda of Colonel Mayer.

The percentages refer, without doubt, to messages sent by the Army and the Air Force. The alterations in machines and procedures used by the Navy rendered its messages practically inaccessible.

10. G. Bertrand — *Enigma*, p. 32

> "...the monthly key setting schedules (one key per day): Heeres-M., for December 1931, the years 1932, 1933 and 1934 (1st semester).
>
> ...10 documents concerning the Enigma cipher machine (1930), that became Enigma I in 1931 (when Enigma II was put into service) and, finally, Enigma *type Wehrmacht* (1937), when its use was extended to all of the Wehrmacht, among which two were of decisive value:
>
> ...a piece of information showing, for a given setting of the machine, the ciphered letters obtained by pressing each letter of the keyboard.
>
> ...a ciphertext, in several parts, with the corresponding cleartext and daily key used."

11. The following figures permit one to appreciate the importance that can be assigned to the delivery of the monthly Enigma key setting schedules. The combination of data furnished in the works of Bertrand and Paillole shows that, for the period extending from December 1931 to September 1938 inclusive (a total of 82 months), Hans Thilo Schmidt gave the French Intelligence Service 54 monthly schedules (at least!). Therefore, 66% of the settings of the Enigma used during this time interval were known.

The Poles received the greater part, if not all, of these schedules. In a letter sent to the author on January 15, 1986, Colonel Lisicki admitted the possible existence in the *Biuro Szyfrow* of a Section (whose existence would have been hidden from the Polish mathematicians) which decrypted the Enigma messages using the monthly key settings sent by Bertrand. For its work, such a section, which could not have begun operating until 1934 (the year of the first delivery of the Enigma copies by the AVA factory), would have been able to determine using the methods perfected by the mathematicians the particular key for each message.

BIOGRAPHICAL SKETCHES

Gilbert Bloch is retired and lives in Paris. He is member of the French Military Reserves. His hobbies include cryptology and parachute jumping.

C. A. Deavours teaches a unique sequence of courses in cryptology at Kean College of New Jersey. The courses address both technical and historical issues in cryptology.

ENIGMA AVANT ULTRA
ENIGMA BEFORE ULTRA

Gilbert Bloch[1]

Translated by C. A. Deavours[2]

ADDRESS: (1) 7 Rue du Cher, 75029 Paris FRANCE and (2) Department of Mathematics Kean College of New Jersey, Union NJ 07083 USA.

[We continue the translation of selected sections of Gilbert Bloch's book, *Enigma Avant Ultra (Enigma Before Ultra)*. See the July 1987 (Vol. XI, Number 1) and October 1987 (Vol. XI, Number 2) issues of this journal for the first parts of the publication, Chapter III — Polish Work and the French Contribution and Chapter IV — Polish Success and Check, respectively.]

CHAPTER V

A. THE ENGLISH, THE FRENCH, AND THE POLISH

The English do not seem to have shown much interest in the birth and childhood of the military Enigma; perhaps, because Germany, at the end of the 20's and beginning of the 30's, was not considered a potential source of trouble for them. When Bertrand tried, at the end of November 1931, to interest them in the documents which Asche had just provided, the English carefully filed the photocopies furnished (which consisted of a description of the Enigma I and its operating instructions, i.e., the *Gebrauchsanleitung* and the *Schlüsselanleitung*) and gave no follow up offer of collaboration.[1]

By 1936, the situation in Germany was not so reassuring, and the British Services had changed their tune; they asked the French if any new information was available. The English received the texts of intercepted messages as well as several goniometric fixes of the transmitting stations.[1] In 1938, probably at Bertrand's initiative, new contacts took place. The English then received the text, both clear and ciphered, of four Enigma messages [2], and a report on the radio network of the German Army. This time, they were quite interested and were provided with other documents as well, in particular, they were provided with monthly key settings for the Enigma.

By this date, there is no doubt that the English had begun their own studies of the Enigma and had come up with several of their own ideas on the subject.[3] They possessed two *commercial* Enigmas bought in 1928 at the request of the Admiralty. (The British cipher machine, "Type X," christened finally "TYPEX" and adopted in January 1935, was an improved Enigma.[4]) The documents furnished by Bertrand in 1931 described the military Enigma and its first operating instructions. But, unlike the Polish, the English were unable to reconstruct the internal wiring of the machine.

At the initiative of Bertrand, the Polish, French, and English experts held a meeting in Paris on the 9th and 10th of January 1939. The Polish revealed nothing of their success nor of their problems since the 15th of December 1938. The English avoided talking about their work and, as far as cryptology, Bertrand had nothing to offer. The contacts were useful, nothing more.

At the beginning of July, it was the Polish who summoned everyone together in Warsaw. During the 24th and 25th of July 1939, they revealed to their audience all of their work. They gave to France and to England two of their Enigma replicas (one for each country), the plans for their *Bombe*, copies of their perforated sheets, in brief, all they had accomplished in 10 years of effort. The Enigma destined for England was taken to London on the 16th of August by Bertrand and given to Colonel S. Menzies, the second in command of the SIS (Special Intelligence Service). [5] This machine had previously traveled from Warsaw to Paris by French diplomatic pouch. Langer himself, after a visit to Paris, had arrived in London several days before the machine and cemented more tightly relations with the British. [6]

The English were prompt to utilize the *Polish gift*. They had, for that, an organization and individuals which were to prove extraordinarily efficient.

B. THE GOVERNMENT CODE AND CIPHER SCHOOL

During the First World War, cryptanalytic work had be assigned, for specifically British reasons, to the Admiralty. The organization responsible for this work was known, both during and after the war, by the nondescript title of *Room 40*. Room 40 acquitted itself well in its tasks.

When England reorganized its intelligence services in 1922, cryptanalytic work (as well as the S.I.S.) was placed under the Ministry for Foreign Affairs. This guardianship was purely nominal; if the Foreign Office was among the *consumers* of the products of the *services*, it was perfectly incapable of telling them what they should produce or how.

The cryptanalytic service was modestly referred to as the *Government Code and Cipher School*, abbreviated GC&CS. In September of 1939, this service left London to move to its war headquarters which were about 80 kilometers northwest of the capital on a property known as Bletchley Park.

In 1938, the staff of GC&CS was numerically small but at the head of the organization were outstanding men. The director was Commander A. Denniston with the chief cryptologist being D. Knox. Both individuals were highly qualified specialists from the old Room 40 of the Admiralty.[7]

Denniston and Knox had grasped the necessity to cryptanalyse the Enigma machine along with other machines. They had, therefore, recruited a mathematical genius and Oxford professor, A. Turing, who was able to comprehend the entire gamut of possible electromechanical calculations (electronic means being not available at that time) and the practical methods of translating these visions into real and functioning cryptanalytic machinery. H. Keen, an engineer for the British Tabulating Machine Company (the English counterpart of the American IBM company), played the role that A. Palluth had played in Poland; i.e., he supervised the production of the machine which issued from the imagination of Turing. Turing had, perhaps, no need to know about the Polish work to conceive of his own version of a *Bombe*. This bombe, whose name was carried over from the Polish work, operated on principles different from those used by the Polish. These differences, as will be seen, made the British bombe far more powerful than its Polish namesake.

The beginning of the war coincided with the development of what came later on to be known as *Bletchley Park*. Among the new arrivals was another Cambridge mathematician, G. Welchman, whose role was to become considerable. From his mathematical acumen came proposals (immediately endorsed by Turing) of modifications to the first conceptions of Turing's bombe which greatly increased its power. Welchman also possessed managerial talents and laid some of the organizational foundations for the work at Bletchley Park.

C. FIRST WORK AT BLETCHLEY PARK

Examination of the Polish copy of the Enigma gave the British specialists a happy surprise. By making a few modifications to their TYPEX, this latter machine could be used to *simulate* the German military Enigma. [8] The Polish had had to set up an industrial production to obtain the AVA Enigmas; Bertrand had to order 40 Enigma replicas to be produced in France [9]; the English were instantly in possession of their own Enigma copies, and in quantity.

Blessed by this favorable occurrence, Bletchley Park began work in three directions:

1. *Design* (by a team of mathematicians led by another Cambridge Professor, J. F. R. Jeffreys) *of new perforated sheets* identical to those provided formerly by the Polish but taking into account the alterations introduced by the Germans at the end of 1938.

   In other words, the new sheets, consisting of 60 different sets, permitted the mastery of the new rotor orders whose number had been increased from 6 to 60.

2. *Research into new methods* for finding the machine settings and message keys. These new methods should not be based on the use of the *twice repeated message key*. The British were of the opinion that, sooner or later, the Germans would realize their mistake and, consequently, would change their procedures. The English did not want such an occurrence to take them unprepared.

3. *Development and construction of new Bombes.* As has been already pointed out, these bombes were based on more general principles than those of the Polish and would be more effective. While the Polish bombes determined the *compatible settings* using the twice enciphered keys from a set of messages, the British bombes could determine the *compatible settings* from a single message using certain words, parts of a phrase, or even entire stereotyped phrases used by the Germans. These stereotypic phrases were called *cribs* by the English. The *game* for the English consisted of constructing a *menu* for their *bombes* using these cribs. If the menu had been appropriately prepared, the bombe would determine from the portion of the message corresponding to the crib, the compatible settings which would produce the ciphertext. At that point, as with the Polish, it was sufficient to test the possible settings on the Enigma replicas to determine which setting was correct and then to establish the key for each message. The decipherment could then proceed in the same way as it was done by the German addressees. [10]

As can be seen, the British bombes bypassed the use of the repeated key altogether (cf. 2 above).

D. THE DIFFICULT BEGINNING (September 1939-April 1940)

1. P. C. Bruno [11]

The French also had their cryptanalytic service. Under the direction of Bertrand, "P. C. Bruno" regrouped at the Château de Vignolles (some 50 kilometers northeast of Paris) about 50 persons. Work on the Enigma messages was essentially the purview of *team Z* which was composed of 15 Polish from the *Biuro Szyfrow*, who, commanded by Langer, were able to escape capture during the invasion of their homeland by taking refuge in Romania from whence they were quickly transferred to France.[12]

The English proposal for a Franco-English cryptanalytic center was rejected by the French. However, P. C. Bruno had a British liaison officer, Captain MacFarlan, who possessed a good knowledge of cryptology; Bruno was connected to Bletchley Park by teletypewriter. This teletype liaison was both close and continuous.[13]

2. First Results

While waiting for the first fruits of labor from Bletchley Park, P. C. Bruno and Bletchley, during the first months of the war, collected Enigma messages and identified the various groups of users.[14] At the end of December 1939, Bletchley Park had completed the perforated sheets. A set of these sheets was immediately sent to P. C. Bruno. According to all sources, it was there that Polish team Z, in the presence of the British expert who had brought the sheets, *cracked* the first *daily setting* for the German Enigma during the war. This event occurred on 17 January 1940.[15] The setting found was for 28 October 1939 and referred to the *Green net* which consisted of communications among the 20 German military regions. Immediately afterwards, Bletchley Park found, for the same net, the daily setting for 25 October.

This was not an auspicious beginning in view of the content of the messages (administrative material) nor the delay – nearly three months – which separated the interception from the decipherment. Things rapidly improved. Between 17 January and the end of March 1940, the method based on the perforated sheets allowed the solution of about 50 daily settings from three user nets which were named after the pencil colors which Bletchley Park used to distinguish among them.

i. Blue- Traffic consisting of test messages sent by Enigma operators being trained by the *Luftwaffe*.

ii. Green- Traffic between the 20 German military regions.

iii. Red- General traffic from the *Luftwaffe* including operational and liaison operations between the Air Force and the Army. This traffic was quite important. The first daily setting was determined was that of 6 January 1940.

On 10 April 1940, a new *net* of messages appeared. These were the *Yellow* messages which concerned operations in Norway. The daily setting and the message keys were broken daily beginning on 15 April and, thereafter, until the middle of May (at which date the traffic disappeared). Channels for forwarding the information extracted from the decrypts were still embryonic and, therefore, its utilisation by the Allied staff could only be limited. Practically no operational use was made of this intelligence material.

The *Blue* traffic was without interest and the Green rarely decipherable (probably because of lack of messages). Efforts were therefore concentrated on the *Red* traffic which was of operational interest and for which there were plentiful messages. At the end of April, the results obtained were encouraging.[16]

3. Bletchley Park and P. C. Bruno

Officially, the British (Hinsley, Vol I, p. 493) state that between January and the 23rd of June 1940, (the date at which, due to the French defeat, Bertrand's organization was dismantled) Paris and the GC&CS worked in close cooperation.

Unofficial English authors, notably Lewin, as well as Polish sources are severely critical of the atmosphere, the rhythm, and the work methods of P. C. Bruno; the Polish team's expertise does not appear to have been fully utilized. It is also clear that P. C. Bruno, having taken to the road on the 10th of June in order to flee from the advancing Germans, found itself in such a state of disorder that useful work simply could not be performed.

Bertrand's book gives an impressive number of decrypted messages (4,789) which appear to be corroborated by Langer's notebooks. However, Bertrand's figures may include non-Enigma decrypts, but they do not furnish any information on the importance of the Enigma decrypts or on the delay in decryptment. Moreover, it is not indicated whether the daily settings were discovered at Bletchley Park or at Vignolles. The only (unpublished) account on this subject is to be found in the memoranda of Colonel Mayer (Head of the Polish S.R.). Colonel Mayer indicates that a division of labor was established and that Bletchley carried out the greater part of the work by determining 83% of the 126 daily settings found up until 16 June 1940. Taking into account the existing circumstances, it appears likely that the preponderance of work was, indeed, done by Bletchley.

NOTES FOR CHAPTER V

1. Sources: Hinsley, *British Intelligence in the Second World War*, Vol 1, p. 488. Paillole, *Notre espion chez Hitler*, p. 39.

According to Paillole, the documents were given to the representative of the Intelligence Service in Paris, Commander Wilfred Dunderdale. He left for London on 23 November 1931 and returned on the 26th with the British refusal for a joint study.

2. Source: Hinsley. Vol 1, p. 491. It was most certainly the Asche documents.

3. During the Spanish Civil War, the British, beginning in 1937, deciphered German, Italian, and Spanish Nationalist messages enciphered on *commercial* models of the Enigma (slightly modified, but without the plugboard).

4. The Typex was notably superior to the Enigma in three important respects:

A. The Typex was more convenient to use. Whereas the Enigma only enciphered letters (necessitating that numbers and figures in words be spelled out) the Typex enciphered letters and numbers;

B. The Typex incorporated a printing mechanism (ultimately, the Typex could be used to produce perforated paper tapes for direct radio transmission);

C. The Typex offered greater security because the rotor movement was more irregular.

For further information of this subject see L. Kruh and C. A. Deavours:

i. The Typex Cryptograph, *Cryptologia*: April 1983, pp. 145-163.

ii. *Machine Cryptography and Modern Cryptanalysis.* Artech Books, 1984.

5. The death of Admiral H. Sinclair made Menzies the head of the SIS on 28 November 1939.

6. Source: Paillole, op. cit. p. 167.

7. Denniston had to leave his post for reasons of health in 1942 and was replaced by his deputy Travis. Knox, one of the best cryptologists of that day, died in the beginning of 1943. The high degree of professionalism of the heads of the English and Polish teams contrasts with the situation on the French side where Bertrand seems to have possessed only a limited cryptologic knowledge and was unable to obtain competent aid. This fact possibly explains the situation which developed from 1931 to 1939 between Bertrand and the Polish. The Polish accepted the documents provided by Bertrand but probably never discussed with him the technical aspects of the problems which they faced and progressively mastered. Until July 1939, Bertrand was ignorant of the successes which began in 1933. This led him to believe that the success had only just occurred.

8. Source: G. Welchman, *The Hut Six Story*, p. 75 (note).

9. The deliveries had not yet begun at the time of the Armistice of 25 June 1940.

After the Armistice, Bertrand, in the course of dangerous trips to the occupied zone, gathered the parts necessary for the completion of four Enigma copies which were used at P. C. "Cadix" situated in the free zone near Uzès. One of these machines survived and is now at the Sikorski Museum in London.

10. On the British *Bombe*, see: G. Welchman – *The Hut Six Story*, p. 77-83, 295-309.

L. Kruh and C. A. Deavours – *Machine Cryptography and Modern Cryptanalysis.*

While a technical description of the British bombe will not be furnished here (besides, the author of the present work would not be able to do so), it is important to mention that in the original conception of Turing, the British bombe utilized the existence of *loops* to solve the problems. *Loops* were closed chains of letters. Welchman showed that by modifying this attack (by the introduction of a ***diagonal board*** into the machine) that the *loops* were superfluous. This increased the power of the machine dramatically.

11. The history of P. C. Bruno in 1939-40 has yet to be written. From what Bertrand says in his book, as well as from the few facts given in the writings of Rejewski which have been translated into English, one cannot reconstruct the activities that took place or judge the results obtained. One paper by Langer written in 1940 at Algiers (after the Armistice) could not be consulted.

12. Against the wishes of the British who were keen to have the Polish specialists in England. Perhaps, also against the wishes of the Polish themselves who had seen at the meeting of 24-25 July at Pyry that the English experts were better qualified than the French ones. Moreover, the Polish seemed to have quickly noticed that better possibilities of work existed at Bletchley Park.

When the Polish mathematicians Rejewski, Rozycki, and Zygalski arrived in Bucharest on 18 September 1939, they went first to the British embassy. Politely refused, they then went to the French embassy where they received a most hospitable reception. The French took immediate measures to see that they were transferred along with their military superiors (who had been interned by the Rumanians at Calimanesti) to France where they arrived the 28-30 September. At this date, France was still the main military adversary of Germany. The Polish brought with them two Enigma copies. P. C. Bruno possessed, therefore, three Enigmas (the third was the one offered on 25 July by the Polish). One of these Enigmas was dismantled in order to facilitate the production of the French replicas of the machine.

13. In fact, beginning in November 1939, after a visit from Langer at Bletchley, the teleprinter circuit which carried the Enigma information was no longer under the supervision of MacFarlan. The English and Polish cryptographers exchanged messages on this teleprinter circuit. These messages were, most certainly, enciphered before transmission on the Polish copies of the Enigma!

14. Interception was, for the English, the jurisdiction of "Service Y" and for France, the *Centres d'écoute spécialisés*. The problems associated with interception, the solutions which were found to these problems, and the efforts which were directed towards those solutions, would constitute a study in itself.

15. According to Polish sources, and, there is every reason to believe them. Curiously, the Bletchley archives seem to place this event at the end of December 1939. (See Hinsley, *British Intelligence in the Second World War*, Vol. 1, p. 493). According to Colonel Paillole, (op. cit. p. 80, and by a personal communication to the author) the *Journal de Marche* of Colonel Rivet, head of the 5ème Bureau SR-SCR mentions in an entry dated 3 January 1940 for the return from London of Captain MacFarlan, carrying new material for the decipherment of Enigma. This refers probably to the new perforated sheets. In order to explain the delay between the 3rd and the 17th of January, it may be supposed that, if Captain MacFarlan transported the heavy and voluminous package represented by the perforated sheets, effective utilization of the sheets could not have taken place until after the arrival at Bruno of the Bletchley specialist (Knox or Turing himself).

16. A table established by G. Welchman (*From Polish Bomba to British Bombe*, p. 105) on the basis of information appearing in the notebooks of Langer allows one to follow the increasing speed of the decipherments. For the Enigma intercepts of January 1940, 11 settings (for 10 different days) were determined with delays ranging from a maximum of 43 days to a minimum of 5 days (average delay, 21 days). For the February messages (13 settings found) the average delay was 13 days. This delay was reduced to 8 days during March (24 settings found). By April (25 settings found) the greater part of the decipherments were realized with a one day delay.

The *Journal de Marche* of Colonel Rivet mentions that, from 6 March 1940, the tactical exploitation of the decipherments appeared possible (Paillole, op. cit. p. 181).

BIOGRAPHICAL SKETCHES

Gilbert Bloch is retired and lives in Paris with cryptology and parachute jumping as hobbies. He is a member of the French Military Reserves. C. A. Deavours teaches courses in cryptology (both technical and historical issues) at Kean College of New Jersey.

THE TURING BOMBE: WAS IT ENOUGH?

C. A. Deavours[1] and Louis Kruh[2]

ADDRESS: (1) Department of Mathematics, Kean College of New Jersey, Union NJ 07083 USA and (2) 17 Alfred Road West, Merrick NY 11566 USA.

ABSTRACT: The effectiveness of Alan Turing's original bombe design is discussed in this article along with organizational methods of implementing it via a microcomputer. To illustrate the material presented, a three part German *Werhmacht* message is solved.

KEYWORDS: Turing bombe, Enigma, cryptanalysis.

It is well known that Polish methods of breaking the German Enigma cipher were based on the use of a twice enciphered rotor starting position. There were two such methods used by the Germans before May 1940 and the Poles overcame both methods in order to solve the Enigma ciphers. Beginning in 1939 when the Poles turned over the results of their work to the British, some method was sought which would not depend on the fact that the Germans used a twice enciphered key for each message since this keying method was likely to be changed in time (as it was during May 1940).

There were two such methods examined by the British. The first method was based on stepping through all rotor positions and computing, at each position, certain statistics which would indicate the correct cylinder positions had been attained. This method was a *ciphertext only* cryptographic attack and will not be discussed here. This first method was, apparently, never put into use, probably because of German foresight in keeping the individual message lengths to about 200 characters (the maximum allowed length varied from time to time).

The second method, and the one used by the British, was based on cribs (probable plaintext). In this second method, a crib of of suitable length was assumed to be in a certain position of a message and a device, the bombe, stepped through rotor positions trying to determine at each position whether the assumed plain/cipher correspondence was possible. Bletchley Park, where most of this codebreaking was done, had a separate section devoted entirely to obtaining cribs based on close study of previously deciphered texts. A fuller discussion of how the British bombe worked may be found in [1] and [2].

It was the Enigma plugboard that caused most of the difficulties in solution and the bombe was the machine that largely overcame these difficulties with the aid of available cribs.

The bombe as conceived by the British was the product of two ingenious individuals, Alan Turing and Gordon Welchman. Originally, the bombe was to be built according to principles devised by Turing but, reportedly, before the first bombe had been built, Welchman had an idea for an addition to the device which later came to be called the *diagonal board*. This board greatly augmented the power of the bombe. It is the purpose of this paper to inquire as to just how well the British would have done without Welchman's idea. To this end, a number of simulations have been prepared and used. These programs will be referred to in the following discussion and are discussed in the Appendix to this paper.

There are several reasons for interest in this subject. First, Turing claimed that his original idea for the bombe design came from Polish work. This cannot be wholly true since the Poles, who were themselves very able cryptanalysts, never seem to have come across the same idea. The author once discussed this issue at length with Mr. Welchman. The conclusion of that conversation was that Turing must have meant that the essential *trick* involved in the bombe, the use of plugboard invariance, was suggested to him by Polish work. At any rate, it would seem that someone other than Turing might have come up with his idea. This is not so of the diagonal board. How this board works is by no means obvious and even Turing, when first confronted by Welchman with the idea, took some time before he saw the wisdom in its use. It is unlikely that anyone other than Welchman would have come upon the concept of using a diagonal board.

Second, the British bombe relied heavily on the fact that the Enigma plugboard substitution was reciprocal, i.e., if E is mapped into Y then Y is mapped into E. The Germans were introducing an Enigma attachment called the "ühr box" towards the end of the war. The effect of this gadget was to make the plugboard encipherment non-reciprocal which would have, in turn, rendered the bombes ineffective. The British would most likely have returned to Turing's original conception which was based on *loops* and did not require, in its simplest form, the reciprocity of the plugboard.

Thirdly, the British TYPEX cryptograph, which was a sister to the German Enigma, used five rotors plus a reflector. The two entrance rotors did not move during encipherment and thus formed a sort of non-reciprocal plugboard. Could the Germans have solved this machine had they discovered Turing's idea?

To see how Turing's solution method worked, let us first look at a message enciphered on an Enigma with no plugboard. Suppose we have the following message beginning with suspected crib.

										1	1	1	1	1	1	1	1	1	1
0	1	2	3	4	5	6	7	8	9	0	1	2	3	4	5	6	7	8	9
F	O	R	F	I	R	S	T	D	I	V	I	S	I	O	N	S	T	O	P
Q	L	L	K	Q	G	D	E	M	D	R	G	I	V	A	S	F	G	L	T.

From this piece of text, we next draw up a *menu*. The menu consists of *loops* found in the above plaintext/ciphertext (pl/cr) pairs. In this case, the loops are

O⟶L⟶O	1	18		
S⟶D⟶I⟶S	6	9	12	
F⟶S⟶I⟶Q⟶F	16	12	4	0
R⟶V⟶I⟶G⟶R	10	13	11	5.

The first loop tells us that if we encipher O at position 1, we will get the letter L, which, when enciphered at position 18 gives O back again and thus closes the loop. The other loops are interpreted in a similar manner. Clearly a loop can start with any letter in its cycle. For our purposes however we only count a given loop once and ignore its rotations since these provide nothing new to us of computational value in this case.

If U represents the substitution produced by the machine at position 1, and V is the substitution produced by the machine at position 18, then we have

$$(\mathtt{o})\mathtt{U} = \mathtt{l} \qquad (\mathtt{l})\mathtt{V} = \mathtt{o} \qquad \text{or,} \qquad (\mathtt{o})\mathtt{UV} = \mathtt{o}.$$

That is, the letter "o" is a *fixed point* of the rotor path UV. Similarly, "l" is a fixed point for VU

Now, if we try every one of the 26^3 rotor positions for position 0, we can test for the presence of the four loops shown above. A positive result would indicate that we, most likely, have found the correct starting position for the coding cylinders. We do this calculation for every rotor order (e.g., 6 orders with 3 rotors). Actually, only 3 loops should be sufficient to exclude nearly all *false alarms* for a given rotor order. This is because the chance of a given letter enciphering to the same letter over any chosen path is about 1/26. Therefore the chance that an incorrect position will have three prescribed loops is about $1/26^3$ so that the expected number of false alarms in the trial of one rotor order is about

$$26^3(1/26^3) = 1.$$

There is, however, an important problem. We know that the fast rotor moves once for every character enciphered. Further, except for an infrequent anomaly when the medium and slow rotors move together, the middle or medium rotor moves once for every 26 letters enciphered. We have no way of knowing if or where the second rotor moved during the encipherment of the crib text. If, for instance, we decide that the middle rotor did not move at all during the crib encipherment then, taking pair 0 as cylinder position 0,0,0, then the position menu for the four loops given above would be (recall that the rotor order in machine is SLOW MEDIUM FAST):

1 18:	0,0,1	0,0,18		
6 9 12:	0,0,6	0,0,9	0,0,12	
16 12 4 0:	0,0,16	0,0,12	0,0,4	0,0,0
10 13 11 5:	0,0,10	0,0,13	0,0,11	0,0,5.

Suppose now that the medium rotor stepped between pl/cr pairs 11 and 12 (as it did). This means that at positions 12, 13, and 18 the medium rotor now occurs at a new position. In this case, the relevant positions in the above menu should be altered to 0,1,12 0,1,13 and 0,1,18 respectively. In the absence of further information, a number of *runs* through each rotor order might be necessary in order to determine the answer. Since the crib is 20 characters long, one might assume that 20 possible stepping positions for the medium rotor must be tried. This is not so since some stepping positions do not affect the loops because these positions are not used in forming them.

For the situation above, we see below assumed stepping positions which are equivalent for this set of loops. *Stepping position* means that the medium rotor steps between the indicated position and the next position.

EQUIVALENT STEPPING POSITIONS:

0	1	4	5	6	7	9	10	11	12	13	14	18
	2				8						15	19
	3										16	
											17	

Obviously, if all of these positions must be tried, those having the longest span (e.g., 14 15 16 17) should be tried first to save time.

Suppose now that an exhaustive search of rotor orders and positions has revealed that the required loops do exist when position 0,0,0 corresponds to coding cylinder positions JIV and the stepping position of the medium rotor was between pl/cr pairs 11 and 12. The solution is completed by determining the alphabet ring settings which will make the medium rotor step at the required location as well as the (adjusted) rotor starting positions.

	A	B	C	D	E	F	G	H	I	J	K	L	M	N	O	P	Q	R	S	T	U	V	W	X	Y	Z
Z:	Z	Y	X	W	V	U	T	S	R	Q	P	O	N	M	L	K	J	I	H	G	F	E	D	C	B	A
A:	A	Z	Y	X	W	V	U	T	S	R	Q	P	O	N	M	L	K	J	I	H	G	F	E	D	C	B
B:	B	A	Z	Y	X	W	V	U	T	S	R	Q	P	O	N	M	L	K	J	I	H	G	F	E	D	C
C:	C	B	A	Z	Y	X	W	V	U	T	S	R	Q	P	O	N	M	L	K	J	I	H	G	F	E	D
D:	D	C	B	A	Z	Y	X	W	V	U	T	S	R	Q	P	O	N	M	L	K	J	I	H	G	F	E
E:	E	D	C	B	A	Z	Y	X	W	V	U	T	S	R	Q	P	O	N	M	L	K	J	I	H	G	F
F:	F	E	D	C	B	A	Z	Y	X	W	V	U	T	S	R	Q	P	O	N	M	L	K	J	I	H	G
G:	G	F	E	D	C	B	A	Z	Y	X	W	V	U	T	S	R	Q	P	O	N	M	L	K	J	I	H
H:	H	G	F	E	D	C	B	A	Z	Y	X	W	V	U	T	S	R	Q	P	O	N	M	L	K	J	I
I:	I	H	G	F	E	D	C	B	A	Z	Y	X	W	V	U	T	S	R	Q	P	O	N	M	L	K	J
J:	J	I	H	G	F	E	D	C	B	A	Z	Y	X	W	V	U	T	S	R	Q	P	O	N	M	L	K
K:	K	J	I	H	G	F	E	D	C	B	A	Z	Y	X	W	V	U	T	S	R	Q	P	O	N	M	L
L:	L	K	J	I	H	G	F	E	D	C	B	A	Z	Y	X	W	V	U	T	S	R	Q	P	O	N	M
M:	M	L	K	J	I	H	G	F	E	D	C	B	A	Z	Y	X	W	V	U	T	S	R	Q	P	O	N
N:	N	M	L	K	J	I	H	G	F	E	D	C	B	A	Z	Y	X	W	V	U	T	S	R	Q	P	O
O:	O	N	M	L	K	J	I	H	G	F	E	D	C	B	A	Z	Y	X	W	V	U	T	S	R	Q	P
P:	P	O	N	M	L	K	J	I	H	G	F	E	D	C	B	A	Z	Y	X	W	V	U	T	S	R	Q
Q:	Q	P	O	N	M	L	K	J	I	H	G	F	E	D	C	B	A	Z	Y	X	W	V	U	T	S	R
R:	R	Q	P	O	N	M	L	K	J	I	H	G	F	E	D	C	B	A	Z	Y	X	W	V	U	T	S
S:	S	R	Q	P	O	N	M	L	K	J	I	H	G	F	E	D	C	B	A	Z	Y	X	W	V	U	T
T:	T	S	R	Q	P	O	N	M	L	K	J	I	H	G	F	E	D	C	B	A	Z	Y	X	W	V	U
U:	U	T	S	R	Q	P	O	N	M	L	K	J	I	H	G	F	E	D	C	B	A	Z	Y	X	W	V
V:	V	U	T	S	R	Q	P	O	N	M	L	K	J	I	H	G	F	E	D	C	B	A	Z	Y	X	W
W:	W	V	U	T	S	R	Q	P	O	N	M	L	K	J	I	H	G	F	E	D	C	B	A	Z	Y	X
X:	X	W	V	U	T	S	R	Q	P	O	N	M	L	K	J	I	H	G	F	E	D	C	B	A	Z	Y
Y:	Y	X	W	V	U	T	S	R	Q	P	O	N	M	L	K	J	I	H	G	F	E	D	C	B	A	Z

Beaufort Tableau.

ALPHABET RING SETTING TABLE
LEFTMOST COLUMN = LETTER READ FROM WINDOW
TOP ROW = CODING CYLINDER POSITION
INTERSECTION OF ROW AND COLUMN = ALPHABET RING SETTING

Enigma rotors 1, 2, and 3, when used in the fast rotor position, step the medium rotor when their alphabet rings reach locations Q, E, and V, respectively in the setting window of the machine (see [2]). From the above data, we conclude the following concerning the positions of the coding cylinders and alphabet ring settings in the crib. (The coding cylinder location is known from the rotor search while the ring positions are known from the stepping position of the medium rotor.)

	0	1	2	3	4	5	6	7	8	9	10	11	12	13	14	15	16	17	18	19
(1)	J	K	L	M	N	O	P	Q	R	S	T	U	V	W	X	Y	Z	A	B	C	D
(2)	U	V	W	X	Y	Z	A	B	C	D	E	F	G	H	I	J	K	L	M	N	O

Here, (1) is FAST ROTOR RING and (2) is FAST CYLINDER LOC.

Thus, the starting position of the fast rotor (as read from the viewing window of the machine) was J. At this time, the alphabet ring of this rotor was set so that the coding cylinder location was U. What is the ring setting?

The ring setting can be determined by consulting the ordinary Beaufort tableau.

We find the ring setting of the fast rotor to be P. The ring position for the medium rotor cannot be determined until the slow rotor moves. Hence, we should take as our solution at this point:

ROTOR ORDER = 1,2,3 RING SETTING = AAP ROTOR START = JIJ.

In the case of the plugboardless Enigma, the above method would not be the best one. First, a hand method, the *baton method* is available if we have the proper lookup tables. Second, if we had a machine capable of running through all of the rotor positions it would be much better to use a consecutive 3 or 4 letter crib in the solution in order to minimize the possibility of the medium rotor having stepped. For example, we might search for consecutive rotor positions such that the correspondence

PLAIN : F O R
CIPHER: Q L L

holds. This point deserves attention because, after World War II, the bombes were used on non-plugboard models of the Enigma in the manner above.

If the Poles had thought of the *loop* principle they could have built a bombe, which differed only slightly from their bomba, that would have been capable of solving Enigma ciphers based on cribs as long as not all letters were involved in the plugboard substitution. In order to see how this would have worked, consider the following sample crib:

O B E R K O M M A N D O D E R W E H R M A C H T
I F I X M W W X W V L T Z S L C L P E P J X K C

The relationship between the letters in the crib and the rotor positions is shown in Figure 1.

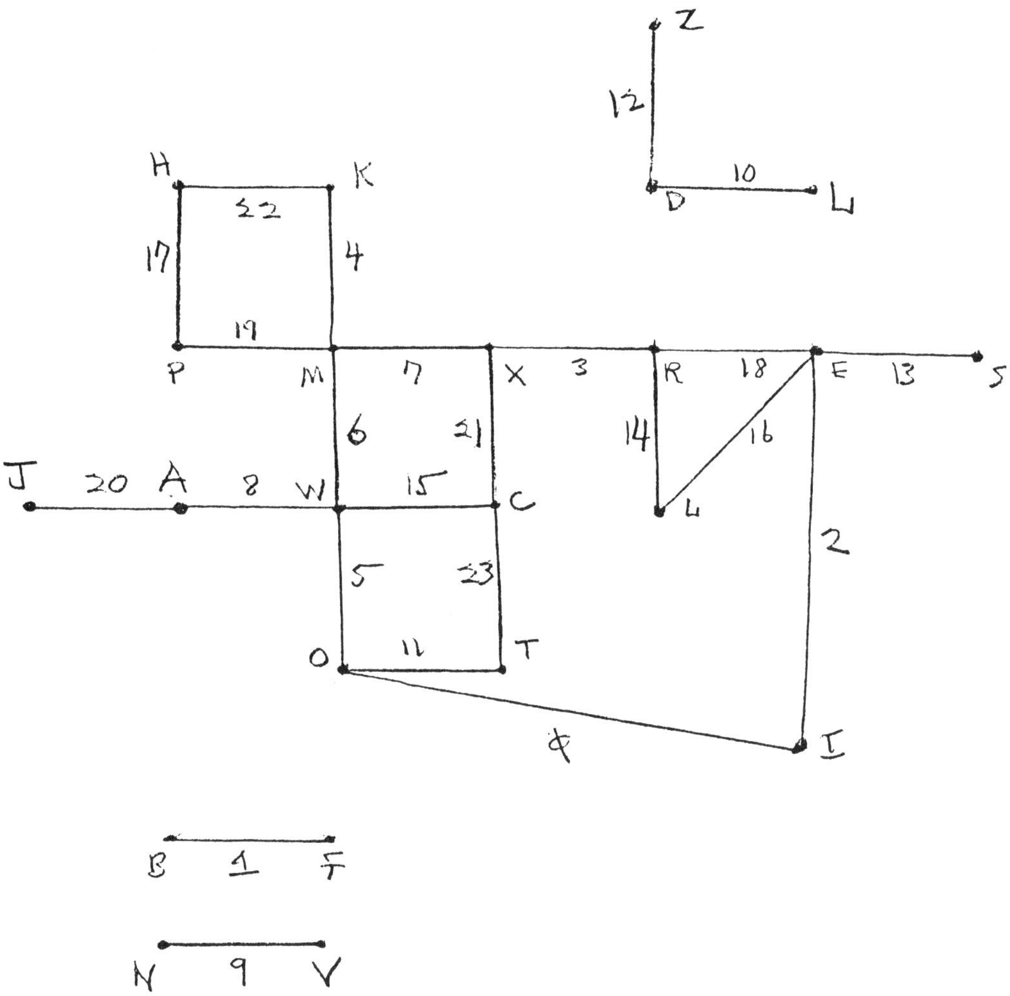

Figure 1. The relationship between the letters in the crib and the rotor positions.

The following loops are present.

(WCXM)
(ERL)
(TOWC)
(KMPH)
(IERXCWO)

Now, if the initial letter of a loop is unchanged by the plugboard, the loop still exists *regardless of the effect of the plugboard on any intermediate letters in the loop.* In order to see this fact, consider the loop (ERL). Suppose that U, V, and W are the three rotor positions involved and P represents the plugboard substitution. We have, noting that $P = P^{-1}$, $U = U^{-1}$, etc.,

$$(\mathtt{e})PUP^{-1} = \mathtt{r} \quad (\mathtt{r})PVP^{-1} = l \quad (\mathtt{l})PWP^{-1} = \mathtt{e}$$

or,

$$(\mathtt{e})PUP^{-1}PVP^{-1}PWP^{-1} = (\mathtt{e})PUVWP^{-1} = \mathtt{e}.$$

This last expression may be further written

$$(\mathtt{e})PUVW = (\mathtt{e})P \qquad \text{or,} \quad (\mathtt{e})UVW = \mathtt{e}$$

since $(\mathtt{e})P = \mathtt{e}$ if this letter is unchanged by the plugboard.

Amazingly enough, all plugboard references are missing from the final equation provided that the initial loop letter is unchanged by the plugboard.

This observation suggests a very simple form of bombe. Referring to Figure 1, we see that many points belong to two or more loops. We could, for instance, use as menus the following loops, all all of which have been rearranged to have the same initial letter.

(WCXM)	(TCWO)
(WCTO)	(TCXMWO)
(WOIERXC)	(TOIERXC)
(CXMW)	(XMWC)
(CTOW)	(XCWOIER)
(CWOIERX)	(XCTOWM)

etc.

If a letter such as W, C, T, or X is unchanged by the plugboard, then, we should find the correct cylinder setting and rotor order. Since there are three loops, there will be few, if any, false alarms. We now compare the solutions for position 0,0,0 looking for agreement between the different menus. Only two letters need be unchanged by the plugboard in order to reveal the correct solution. The fact that this form of the bombe is virtually identical to the ones which had already been built by the Poles is the most convincing evidence available that the Turing solution, based on loops, had not been thought of previously.

One can see from looking at Figure 1 that some letters do not fall into loops (e.g., Z, D, L, B, F,...). By the term **Turing bombe** we shall mean any solution procedure which uses only letters which are found in the crib loops. The diagonal board provided by Welchman permitted the use of all letters, whether in loops or not. Returning to the question of the Turing bombe, we may ask what modifications are needed when the plugboard is taken into consideration? All of the previous discussion regarding the ***canceling out*** of the plugboard substitutions in a loop goes through without change. The only difference is that the initial letter of a loop may have been effected by the plugboard. Thus, in the previous example, our final equation for the ERL loop would read

$$(\mathtt{e})PUVW = (\mathtt{e})P.$$

No matter what the plugboard substitution is, the effect of P on any letter can only be to produce another letter. There are only 26 possibilities for $(\mathtt{e})P$, for it can be only one of the letters A,B,C,... Z. All of these possibilities can be tested simultaneously by applying input to all 26 rotor input contacts and using *double ended* rotor sets where the reverse path of the current does not pass along the same rotor input paths. The current then flows through the resulting maze determined by the menu and indicates whether or not the rotor setup is correct or not. Once again, the reader is referred to [1] and [2] for further description.

For a loop like the one shown above, we can get candidates for the plugboard mapping by computing all of the fixed points, i.e., the solutions of

$$(\mathbf{x})UVW = \mathbf{x}.$$

Any such solution gives a possible plugboard mapping via

$$\mathbf{x} = (\mathtt{e})P, \quad \text{or, equivalently} \quad \mathtt{e} = (\mathbf{x})P.$$

The program **TBOMBE.BAS (TBOMBE.EXE)** emulates the solution process as it would have been carried out by the Turing bombe. First, the crib is examined and all loops are written out. Rotations of the same loop and loops derived from previously listed loops are deleted from the list. Next, the loops are rearranged to provide as many loops as possible which begin with the same letter. This arrangement greatly reduces the number of false alarms that the program finds. We shall call the loops arranged as indicated above *primary loops*. A *secondary loop* is any rotation of a primary loop.

To prepare the menu for a given rotor order, the program **LOOPFIND.BAS** is now run and the primary loops and their rotor positions relative to 0,0,0 are entered. The program will compute the secondary loops and add these to the menu and store the resultant menu on the disk. Note that something about the stepping position of the rotors must be assumed here in entering the rotor positions in the menu.

The main program **TBOMBE.BAS (TBOME.EXE)** is now run and the rotor order is entered while the menu is read from the disk. The solution process uses the following logic:

A. At each rotor position all fixed points for each loop are computed one at a time. If any loop fails to have at least one fixed point, then, no such loop exists and this rotor positioning is discarded.

B. If a loop has one or more fixed points, then the program tries to eliminate them using data from the other loops. If the elimination is successful, then the rotor position can be discarded.

C. If the elimination process fails, the cylinder positions and deduced plugboard connections are printed and the program proceeds to the next rotor position.

As an example, suppose the loops (and derived secondary loops) are as shown.

PRIMARY LOOPS:

(EV)
(EM)
(XAU)

SECONDARY LOOPS (ROTATIONS OF PRIMARY LOOPS):

(VE)
(ME)
(AUX)
(UXA).

Computing the fixed points of the (EV) loop we might find two such points, say, F and U. In terms of the plugboard mapping, we may have either $\mathtt{f} = (\mathtt{e})P$ or $\mathtt{u} = (\mathtt{e})P$. We now try to eliminate these two fixed points so as to rule out this rotor position. The other loops all either begin with the same initial letter, E, a fixed point letter, F or U, or some other letter.

If another loop begins with the same letter, such as loop 2, then we try the fixed points for the first loop at this position. On the other hand, if another loop begins with a fixed point letter, such as the 7th loop which begins with U, then we can test the assumption $\mathtt{e} = (\mathtt{f})P$, which is the same as $\mathtt{f} = (\mathtt{e})P$, by trying the letter E as a fixed point at this position. The letter U would be tested in a similar way. Loops that begin with neither letter are of no use in this testing process.

If we fail to rule out a loop, we move to the next loop and try again. If no loop can be ruled out then the surviving set of fixed points, which represents possible plugboard connections is printed.

We note that in the testing of the loops which begin with a fixed point of another loop, the reciprocity of the patchpanel is assumed. Thus, the Turing loop method does not preclude the use of this property of the plugboard as has been commonly assumed. In practice, however, this additional testing does not result in a greatly increased computational ability because the number of letters in the loops as well as the number of loops is usually small unless the crib is unusually long.

To illustrate the entire process as well as the program outputs, we will solve an actual set of three Enigma messages. These are seen in an actual Enigma message in Figure 2. These messages were kindly provided by David Kahn.

The messages in question are shown as they were received. In actuality we have one long message which has been divided into three parts. These messages were sent on 21 September 1939 to HEERSGRUPPENKOMMANDO 2. At this time only three Enigma rotors were in use and the 9 letter message key system in which the sender selected a groundsetting with which to encipher the rotor starting positions (twice) was the current message keying system.

The message preamble is

2109 -1750 - 3 TLE - FRX FRX - 1 TL - 172

The first two groups are the date and time (21 SEPT, 17:50 HRS). The next groups indicate that the message is in 3 parts. The groundsetting, FRX, is then repeated in the clear. Evidently, the same groundsetting has been used for all three parts. Finally, comes the indication that this is the first part of the message which is 172 letters in length. The term FERNSCHREIBEN indicates that these messages were received over land lines, i.e., teletype.

Fortunately, whoever deciphered the messages has left us some useful notations on the sheet. The first 6 letters of each message have been ruled out. These are the twice enciphered rotor starting positions. The third group of each message has also been ruled out. This is the indicator group for each message. A number of indicator groups were generally provided for each key schedule in order to make the process of sorting the cryptograms more difficult for interceptors.

The numbers 06 18 24 shown on the right border are the groundsetting FRX converted into numbers. The decipherer's machine had numbers, 01 – 26 on the rotor circumference instead of the letters A – Z. Inside the hood of the Enigma machine was a chart used to make this conversion, A=01, B=02, etc.

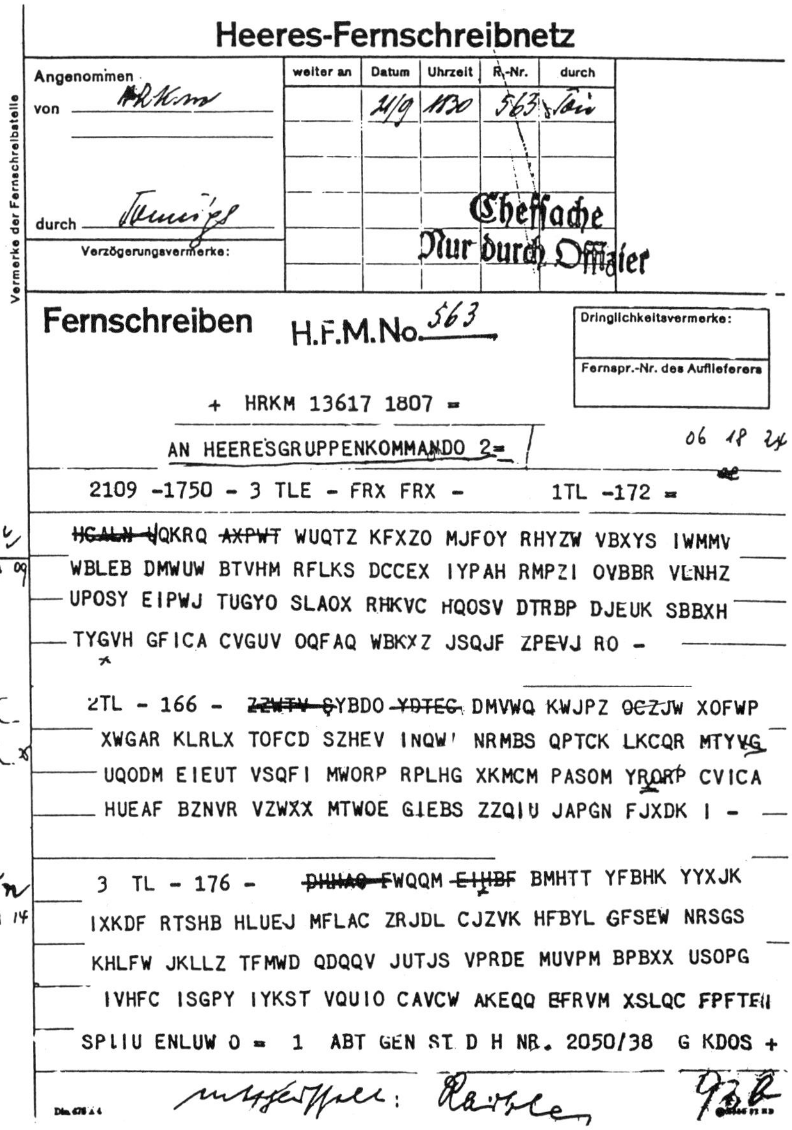

Heeres-Fernschreibnetz

Angenommen von	weiter an	Datum	Uhrzeit	R.-Nr.	durch
HRKM		21/9	1830	563	

durch

Verzögerungsvermerke:

Geheimsache Nur durch Offizier

Vermerke der Fernschreibstelle

Fernschreiben H.F.M.No. 563

Dringlichkeitsvermerke:

Fernspr.-Nr. des Auflieferers

\+ HRKM 13617 1807 =

AN HEERESGRUPPENKOMMANDO 2=

06 18 24

2109 -1750 - 3 TLE - FRX FRX - 1TL -172 =

agi 01 07 09

~~HCALN~~ ~~V~~QKRQ ~~AXPWT~~ WUQTZ KFXZO MJFOY RHYZW VBXYS IWMMV
WBLEB DMWUW BTVHM RFLKS DCCEX IYPAH RMPZI OVBBR VLNHZ
UPOSY EIPWJ TUGYO SLAOX RHKVC HQOSV DTRBP DJEUK SBBXH
TYGVH GFICA CVGUV OQFAQ WBKXZ JSQJF ZPEVJ RO -

ydx 25 04 24

2TL - 166 - ~~ZZWTV~~ ~~G~~YBDO ~~YDTEG~~ DMVWQ KWJPZ OCZJW XOFWP
XWGAR KLRLX TOFCD SZHEV INQW' NRMBS QPTCK LKCQR MTYVG
UQODM EIEUT VSQFI MWORP RPLHG XKMCM PASOM YRQRP CVICA
HUEAF BZNVR VZWXX MTWOE GIEBS ZZQIU JAPGN FJXDK I -

lun 12 21 14

3 TL - 176 - ~~DHHAQ~~ ~~F~~WQQM ~~EIHBF~~ BMHTT YFBHK YYXJK
IXKDF RTSHB HLUEJ MFLAC ZRJDL CJZVK HFBYL GFSEW NRSGS
KHLFW JKLLZ TFMWD QDQQV JUTJS VPRDE MUVPM BPBXX USOPG
IVHFC ISGPY IYKST VQUIO CAVCW AKEQQ EFRVM XSLQC FPFTEI
SPLIU ENLUW O = 1 ABT GEN ST D H NR. 2050/38 G KDOS +

Figure 2. Hardcopy of the three Enigma messages sent to HEERSGRUPPENKOMMANDO 2 on 21 September 1939.

Best of all, the deciphered rotor starting positions have been written for each each message on the left side of the paper. For part 1, these are AGI (01 07 09). One can see where the clerk has checked off the letters during the second decipherment of the key. This fact will save us some computation since we will know the turnover position for the middle rotor without having to guess it.

We now need a crib. This was provided by the fact that the plaintext of the message was also available. After some consideration the first 28 letters of the first message were selected for use. Having the plaintext does not make the solution process automatic since abbreviations, the use of X's as word and sentence dividers, and other complications often arise.

The crib is shown.

(1)	I	J	K	L	M	N	O	P	Q	R	S	T	U	V	W	X	Y	Z	A	B	C	D	E	F	G	H	I	J	K
												1	1	1	1	1	1	1	1	1	1	2	2	2	2	2	2	2	2
		0	1	2	3	4	5	6	7	8	9	0	1	2	3	4	5	6	7	8	9	0	1	2	3	4	5	6	7
(2)		A	U	F	B	E	F	E	H	L	D	E	S	O	B	E	R	S	T	E	N	B	E	F	E	H	L	S	H
(3)		Q	K	R	Q	W	U	Q	T	Z	K	F	X	Z	O	M	J	F	O	Y	R	H	Y	Z	W	V	B	X	Y

Here row (1) is FIRST ROTOR, row (2) is PLAIN, and row (3) is CIPHER.

The top line indicates the fast rotor position using a starting position of I. At this point, the menu loops are sought.

MENU LOOPS:

$E \xrightarrow{04} W \xrightarrow{23} E$

$E \xrightarrow{18} Y \xrightarrow{21} E$

$S \xrightarrow{11} X \xrightarrow{26} S$

$H \xrightarrow{07} T \xrightarrow{17} O \xrightarrow{13} B \xrightarrow{20} H$

$H \xrightarrow{20} B \xrightarrow{03} Q \xrightarrow{06} E \xrightarrow{18} Y \xrightarrow{27} H$

To prepare the bombe menu, the program **BOMBMENU.BAS** is run and the primary loop data is entered. The output from this program is on the following pages.

MENU GENERATOR:

ENTER NUMBER OF PRIMARY LOOPS TO BE ENTERED? 5

ENTER LENGTH OF LOOP 1 : ? 2
ENTER LENGTH OF LOOP 2 : ? 2
ENTER LENGTH OF LOOP 3 : ? 2
ENTER LENGTH OF LOOP 4 : ? 4
ENTER LENGTH OF LOOP 5 : ? 5

NUMBER OF PRIMARY LOOPS = 5
NUMBER OF SECONDARY LOOPS = 10
ENTER LETTER CYCLE FOR LOOP # 1 :? EW
ENTER POSITION # 1 ? 0,0,4

ENTER POSITION # 2 ? 0,1,23

ENTER LETTER CYCLE FOR LOOP # 2 :? EY
ENTER POSITION # 1 ? 0,1,18
ENTER POSITION # 2 ? 0,1,21
ENTER LETTER CYCLE FOR LOOP # 3 :? SX
ENTER POSITION # 1 ? 0,0,11
ENTER POSITION # 2 ? 0,1,0

ENTER LETTER CYCLE FOR LOOP # 4 :? HTOB
ENTER POSITION # 1 ? 0,0,7
ENTER POSITION # 2 ? 0,1,17
ENTER POSITION # 3 ? 0,1,13
ENTER POSITION # 4 ? 0,1,20

ENTER LETTER CYCLE FOR LOOP # 5 :? HBQEY
ENTER POSITION # 1 ? 0,1,20
ENTER POSITION # 2 ? 0,0,3
ENTER POSITION # 3 ? 0,0,6
ENTER POSITION # 4 ? 0,1,18
ENTER POSITION # 5 ? 0,1,1

BOMBE MENU

PRIMARY LOOPS:

1	E	2	0	0	4	0	1	23											
2	E	2	0	1	18	0	1	21											
3	S	2	0	0	11	0	1	0											
4	H	4	0	0	7	0	1	17	0	1	13	0	1	20					
5	H	5	0	1	20	0	0	3	0	0	6	0	1	18	0	1	1		

SECONDARY LOOPS:

6	W	2	0	1	23	0	0	4									
7	Y	2	0	1	21	0	1	18									
8	X	2	0	1	0	0	0	11									
9	T	4	0	1	17	0	1	13	0	1	20	0	0	7			
10	O	4	0	1	13	0	1	20	0	0	7	0	1	17			
11	B	4	0	1	20	0	0	7	0	1	17	0	1	13			
12	B	5	0	0	3	0	0	6	0	1	18	0	1	1	0	1	20
13	Q	5	0	0	6	0	1	18	0	1	1	0	1	20	0	0	3
14	E	5	0	1	18	0	1	1	0	1	20	0	0	3	0	0	6
15	Y	5	0	1	1	0	1	20	0	0	3	0	0	6	0	1	18

The menu is now stored on the disk. Three separate menus must be prepared (depending on the stepping position of the fast rotor). We will need menus for rotor orders 123 and 213 (stepping position V), orders 132 and 312 (position E), and orders 321 and 231 (position Q). The menu shown above is prepared for orders 123 and 213.

We can run the main program, **TBOMBE.BAS (TBOMBE.EXE)**. After specifying the rotor order and name of the bombe menu which is to be loaded from the disk the solution process commences. When rotor order 213 was tested the following output was printed.

CYLINDER POSITION=BKG

```
E E H H S W Y T O B B Q E Y X
Z Z H H O R Y C D L L Q Z Y P
        P Z   P J
        S     T P
              Z S
```

This program output shows the coding cylinder locations and the final fixed point data. We see that the letters H, Y, and Q are unchanged by the plugboard. Probably S and T are also unchanged. The plug connections which have been found so far are (EZ), (BL), and (XP). The connection (WR) can also be inferred since W must be mapped to either R or Z but the letter E is already mapped to Z.

Thusfar, we know the following key data:

ROTOR ORDER = 2,1,3.

PLUGBOARD: (EZ) (BL) (XP) (WR)

CODING CYLINDER STARTING LOCATIONS = BKF

ROTOR STARTING POSITIONS = AGI (from message sheet).

Using the Beaufort table, the ringsettings are quickly found to be ZWD.

We now have only to work out the remaining plug settings (if any). Using the above data and deciphering the cryptogram using the program **ENIGMA.BAS** we have

PLAIN : A I F B E R E H L D E S J U R G S W E N B E F E Q L S H

The decipherment is correct at 19 of the 28 positions. It is now comparatively easy to work out the remaining plug settings which are found to be (IU), (VM) and (JO).

The rotor starting positions for the three parts are found from deciphering the enciphered rotor positions to be AGI, YBE, and LUN, respectively. The messages can now be deciphered.

DECIPHERMENT (WITH GARBLES):

PART I:

```
AUF BEFEHLD DES OBERSTEN BEFEHLSHABERS SIND IM
FALLE XZXZTX UNWAHRSCHEINLICHEN FRANZOESISQEN
ANGRIFFS DIE WESTBEFESTIGUNGEN JEDERZAHLENMAESSGEN
UEBERLEGENHEIT ZUM TROTZ ZU HALTEN X
```

PART II:

```
FUEHRUNG UND TRUPPE MESSEN VON DIESER EHRENFPLIQT
DURQDRUNGEN SEIN X ABS X DEMGEMAESS BEHALT IQ MIR
DIE ERMAEQTIGUN ODER AUQ OON TEILEN AUSDRUECKLIQ
```

PART III:

```
PERSONLIQ VOR X ABS X AENDERUNG DERUNGERANWEISUNG
X OKH X GEN X ST X D X HXERSTE ABT X NR X DREI
DREI ZWO EINS DREI AQT G X KDOS X VOM JULI EIN X
WEUNDREI AQT BLEIBT VORBEHRLTEN X DER
OBERBEFEHLSHABER DES HEERES
```

The following conventions are to be noted.

X = PERIOD OR WORD DIVIDER
ABS = NEW PARAGRAPH
AQT = /
XZXZTX = ? MAP COORDINATES?
Q = CH.

The erratic use the X's, which has been noted in [3], must have been a tremendous problem in establishing cribs.

If Turing's loop method had been the sole principle used by the British bombes, could penetration of the Enigma ciphers have been successful? The answer rests mostly on the availability of cribs. Welchman states in [1] that the British bombes contained 10, later 12, rotor sets. This indicates that the cribs used by the British were about that length. This figure is disputed by persons who worked on and serviced the bombes, who recall 25-30 sets of rotors per machine. (A drawing in [4] shows a bombe with two rows of 12 rotor sets each.) In

[2], it is revealed that bombe personnel were given the following table for *false alarms*.

BOMBE SETUPS NECESSARY FOR ONE STOP PER ROTOR ORDER

NUMBER OF LOOPS	NUMBER OF NON-LOOP LETTERS
4	0
3	9
2	11
1	12
0	13

With a 12 letter crib, most loops are 2, 3, or perhaps 4 letters in length. Thus, the above table indicates that the cribs were about 12-18 letters in length. We also know from Welchman that Turing was basing his machine on the fact that 3 loops could be found in a crib. Turing must have known from available data that this was a practical possibility. The table below shows an experiment in which 20 cribs of lengths 12, 15, 18, and 21 letters were tested and the numbers of loops found were recorded.

	LENGTH OF CRIB			
# LOOPS	12	15	18	21
0	13	9	5	0
1	6	8	6	1
2	1	3	6	9
3			3	6
4				4
AV. # LOOPS	.40	.70	1.35	2.65

Unless a very large number of cribs are available it would seem that about 20 or so letters would be needed to have a good chance of securing the three loop situation. Further, since bombe operators were provided with a "4 loop" figure it appears that this case could occur, at least occasionally. This suggests once again that cribs of 20 or so letters must have been available.

Turing based his need for at least 3 loops on the following reasoning. Without the plugboard 3 loops are sufficient to produce only about 1 false alarm per rotor order. Because of the plugboard 26 more possibilities need to be investigated for the initial loop letter. Therefore 3 loops could produce about 26 false alarms per rotor order. This figure could be coped with but the actual situation was usually not so bad. A frequently occurring letter in a crib was quite likely to be in more than one loop. When this happens, the number of false alarms falls off. If, for instance, the first loop turns out to have 1 fixed point then this is the only letter which has to be tested in a second loop which begins with the same letter – the other 25 possibilities having already been ruled out.

From the above considerations, it would appear that, although the loop method requires a somewhat longer crib than if non-loop letters can be used, the more lengthy cribs appear to have been obtainable.

A more serious problem is the stepping position of the medium rotor. We do not know how this problem was handled. Even if short 12 letter cribs were used, there is a good chance (12/26) that the middle rotor will step somewhere in the crib. This would mean rerunning the test using different assumed stepping positions. Such a requirement could multiply greatly the time required to find the solution. Whether or not the diagonal board is used in the bombe, this obstacle is major. It would seem that, in many cases, the stepping position must have been known. One can conceive of how this might have happened from reading about the operator carelessness that was so prevalent, but, until more is known, we cannot say for sure. In any case, both methods would suffer from this problem.

In conclusion, it would seem that breaks in the Enigma system could have been made even if Welchman's diagonal board had not been conceived. The solutions could not have been accomplished with the same degree of work, but, the additional work necessary and the crib data required for the solution appear to have been available to the best of our current knowledge.

PROGRAM DESCRIPTION

The programs used in this article are described below. These programs are available from the author on an MS-DOS diskette. Most programs are in BASIC but machine language versions have been supplied when the execution time of the program was considered too slow. Both the programs and brief instructions and problems can be obtained from the author at the address given in the front of this journal. The price is $10 and covers the costs of preparation, shipping, and handling.

ENIGMA.BAS - Simulation of the 3 rotor Wehrmacht Enigma.

BOMBE.BAS (BOMBE.EXE) - This program runs through all rotor positions trying to match preassigned data. It is intended primarily for solution of plugboardless Enigma messages. The compiled version of the program, BOMBE.EXE, examines a rotor order in 30-45 seconds (4 MHZ).

LOOPFIND.BAS - Plaintext and matching ciphertext are entered and a list of primary loops is generated.

BOMBMENU.BAS - Primary loops and positions are entered and secondary loops are generated and stored on the disk. This program is used to prepare menus for the TBOMBE program.

TBOMBE.BAS (TBOMBE.EXE) - Simulation of the Turing bombe. TBOMBE.EXE requires 7-12 minutes to examine one rotor order.

REFERENCES

1. Welchman, Gordon. 1982. *The Hut Six Story.* New York: McGraw-Hill.

2. Deavours, Cipher A. and Louis Kruh. 1985. *Machine Cryptography and Modern Cryptanalysis.* Norwood MA: Artech House.

3. Calvocoressi, Peter. 1980. *Top Secret Ultra.* New York: Pantheon Books.

4. Johnson, Brian. 1978. *The Secret War.* London: Methuen.

BIOGRAPHICAL SKETCHES

C. A. Deavours teaches cryptography courses at Kean College of New Jersey. He also does extensive consulting in the field of Electronic Funds Transfers. His wife says that he has mastered the art of speaking in ciphertext and he continues a full schedule of teaching, consulting, and writing about cryptology at Kean College of New Jersey.

Louis Kruh is an indefatigable researcher and collector of cryptologic memorabilia, particularly cipher devices. His interest in cryptology spans more than 40 years and he responds knowledgeably and promptly to all queries regarding literature and equipment. He is co-author with Dr. C. A. Deavours, of *Machine Cryptography and Modern Cryptanalysis*, (Artech House, Inc. 1985). He is recently retired from his position as Director of Advertising for New York Telephone.

AN ENIGMA CHRONOLOGY

David Kahn

ADDRESS: 120 Wooleys Lane, Great Neck NY 11023 USA.

ABSTRACT: **A list of important events in the making and breaking of the German Enigma cipher machine of World War II is given in the order of their occurrences.**

KEYWORDS: **Enigma, chronology.**

In view of the importance of ULTRA to the history of intelligence, and in view of the errors in dating that appear in so many accounts, I thought it would be worthwhile to establish a list of important dates in the story of the Enigma cipher machine and its solution. Most deal with the naval machine.

This chronology comes mainly from my book, *Seizing the Enigma* (New York: Houghton Mifflin, 1991), which is based as much as possible on primary or on solid secondary sources, chiefly F. H. Hinsley et al's magisterial *History of British Intelligence in the Second World War* (Cambridge: Cambridge University Press, 1979-1988). For my book, the Naval Historical Branch in London made available to me a key primary source for much naval dating, but I promised not to cite it by name, a promise that I am honoring. Information about the times of naval solution comes from analyses of intercepts in the Public Record Office, London, and in the National Archives, Washington. To avoid burdening this list with a heavy biographical apparatus, I have given as sources the page numbers in Hinsley or in *Seizing the Enigma* that specify these dates or detail those incidents. Notes in the form 2:659 refer to Hinsley; those in the form 140 refer to Kahn, in which the notes to the text cite the source. "GC&CS" is the Government Code and Cypher School, the British codebreaking agency at Bletchley Park. Names in capital letters of colors, insects, and birds are GC&CS's covernames for Enigma keys.

An important element in the history of the solution of Enigma is that the Luftwaffe Enigma was solved more easily by GC&CS than the naval Enigma, though until 1 February 1942 both used the identical machine with identically wired rotors. One reason is that the air force (like the army) chose the three rotors inserted into its machines from a set of five rotors while the navy chose its three from a set of eight. In addition, air force key settings could often be recovered in the first few months of the war from the key indicators at the head of the cryptograms and later from blunders by the ciphering clerks in making up their part of the key. This gave GC&CS insight into Luftwaffe messages, providing enough cribs for a daily recovery of keys. The naval settings, on the other hand, were taken from books and permitted neither the indicators nor the blunders approach; without some early solutions, GC&CS could not at first get enough messages to yield future guesses of plaintext for cribs. Captures of the books were thus the necessary first step for solution of the naval Enigma. The Enigma keys used by the army were seldom solved because the army cipher clerks, more experienced than those of the Luftwaffe, seldom made blunders.

Solution of the Enigma consisted of guessing an intercepted cryptogram's plaintext (either from knowledge of the traffic previously passed between posts or from solutions of the same messages in other cryptosystems) and then seeing whether the bombes (which were in effect multiple Enigmas) found a setting of the Enigma rotors that would produce such a ciphertext from the assumed plaintext. If they did, that setting was the key for the day on that cipher net and permitted the reading of other messages on the net. The process had to be repeated for each net each day.

Not included here are minor matters, such as the dates of early improvements in Enigma construction or keying, of techniques the Poles used to overcome these changes, or of the solution of secondary keys, such as the Luftwaffe's practice key. (I consider keys minor if they did not contribute significantly to the fighting. Some people may of course differ with my judgment.) A full list of solution dates is given in Hinsley, 2:658-668, 3:1:483-487, 3:2:855-857. In some cases, Hinsley does not state for how long the solutions continued. The date of 24 May 1943 at 2:667 is a misprint for 24 May 1945, Hinsley told me.

Date		Enigma	The World
1918	Feb 23	Arthur Scherbius files for a patent for the Enigma cipher machine (Patentschrift Nr. 416291)	
	Apr 15	Arthur Scherbius offers the Enigma to the German navy (31)	
1918	Nov 11		World War I ends
1921	Feb 19		France and Poland sign accord for "concerted measures for the defense of their territory"
1926		German navy begins using Enigma (40-41)	
1928	Jul 15	German army begins using Enigma (51)	
1931	Nov 8	Hans-Thilo Schmidt betrays army Enigma instruction manual and keying methods to the French, who give them to the Poles (59)	
1932	Dec	With the help of keys betrayed by Schmidt and given by the French to the Poles, cryptanalyst Marian Rejewski reconstructs the wiring of the three Enigma rotors, enabling the Poles to begin solving German army Enigma messages (65-66)	
1933	Jan 30		Führer of the Nazi party, Adolf Hitler, named chancelor of Germany
1938	Dec 15	Two new rotors (making a toitl of five) go into service on army Enigma; Rejewski recovers wiring, but tenfold increase in number of keys overwhelms Poles, requiring them to ask aid from their allies (77)	
1939	Mar 30		Britain and France guarantee to help Poland if she is invaded by Germany as Czechoslovakia was
	Jul 24	Poles reveal success to British and French (79)	
	Sep 1		Germany invades Poland
	Sep 3		Britain and France declare war on Germany

Date		Enigma	The World
1940	Jan 6	British break into Luftwaffe Enigma using hand methods (1:108, 2:659)	
	Feb 12	British seize two of the three unknown rotors used in the naval Enigma from a crew member of the U-33 captured after the submarine was sunk (114; 3:2:957)	
	Apr 9		Germany invades Denmark and Norway
	May	First bombe installed at GC&CS (1:494, more likely than August installation date given at 1:184 because regular solution of RED began May 22)	
	May	First break into naval Enigma: using documents salvaged 26 April from the trawler *Polares* (303, 3:2:957), GC&CS solves April messages (117)	
	May 10		Germany opens massive offensive against France; Winston Churchill, a long-time supporter of communications intelligence, named prime minister
	May 22	GC&CS begins solution of Luftwaffe general purpose key (RED) continuing virtually uninterrupted to end of war (1:109, 144)	
	Jun 10		Last Norwegian troops surrender
	Jun 22		France surrenders
	Aug	The last unknown naval rotor obtained from a naval capture (3:2:957)	
	Aug 12		Adlertag (Eagle Day): Germany attempts to win air superiority over England as invasion preparation
	Sep 15		Battle of Britain D Day: it becomes clear that Germany's attempt has failed
	Oct 12		Hitler postpones invasion of Britain until spring
	Dec 10	SS (Schutzstaffel) general-purpose key (ORANGE I) first broken by GC&CS (2:668)	

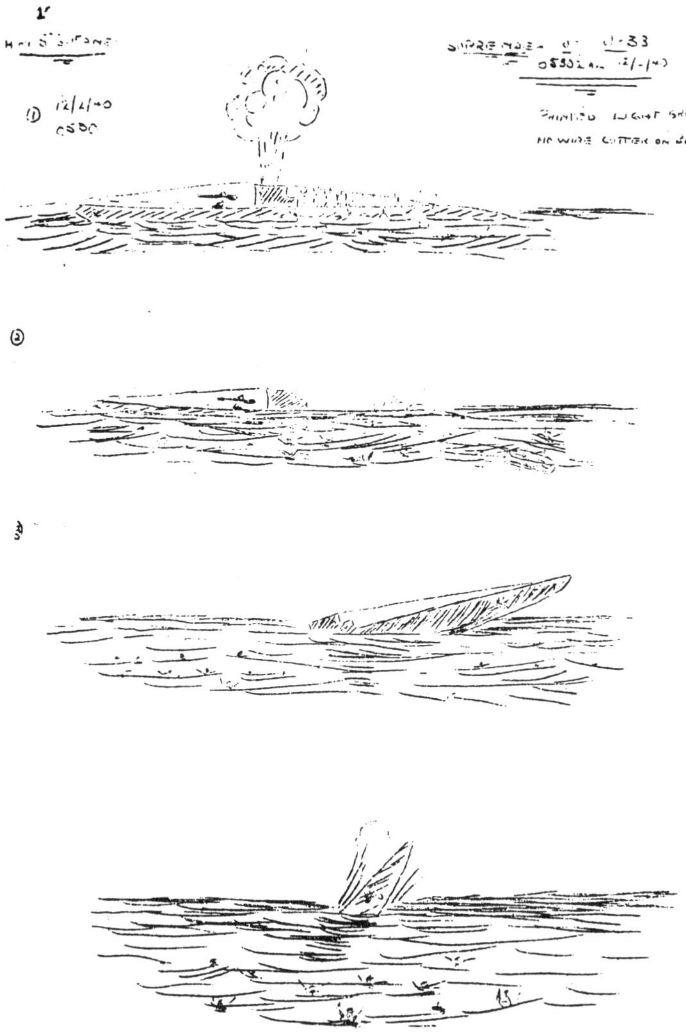

12 Feb 1940

A series of sketches – artist not indicated – showing the end of the U-33 after it surfaced in a vain attempt to escape the minesweeper *H. M. S. Gleaner* in Scotland's Firth of Clyde 12 February 1940. The sketches show (1) a shower of sparks from the coning tower; (2) crew members in the water as they abandon ship; (3) more in the water as the submarine starts to sink; (4) the U-boat in her final dive with survivors swimming away from her. Three rotors were recovered from a crew member who forgot to take them from his pocket and drop them in the ocean.

(Public Record Office, ADM 199/123, p. 54)

May 1940

The former trawler *Julius Pickenpack*, now a German attack vessel disguised as the Dutch ship *Polares*. From one of its canvas bags, thrown overboard on the approach of British warships but saved by a British sailor, the British obtained cryptographic documents useful in solving German naval Enigma traffic.

(Public Record Office, ADM 199/476, p. 243)

1 June 1941

The officers and crew of the U-110, in port, await an approaching inspecting officer while the skipper, Lieutenant Fritz-Julius Lemp, 28, approaches the gangplank to greet him. On patrol, the submarine was seized by the British and, before it sank, yielded an Enigma machine and keying documents. These enabled the British to read U-boat messages for June 1941. Lemp was lost during the action.

(David Kahn collection)

2 July 1941

Hinrich Gewald, 58, skipper of the *Lauenburg*, whose capture by the British led to the seizure of Enigma keying documents that enabled the British to read U-boat messages for July 1941.

(David Kahn collection)

2 July 1941

The *Lauenburg*, just before being boarded on 28 June 1941.

(David Kahn collection)

Date		Enigma	The World
1941	Jan 28	Luftwaffe operational key for Africa first solved by GC&CS; solutions end 31.12.42 with introduction the next day of separate Fliegerkorps keys (2:660)	
	Feb-Mar	2 U. S. Army and 2 U. S. Navy cryptanalysts visit GC&CS and learn Enigma cryptanalysis (235-237)	
	Feb 15		General Erwin Rommel named commander of German forces in North Africa
	Mar 12	Documents seized from patrol ship *Krebs* permit reading of some March and all April and May naval messages in the Home Waters key net, also used for U-boat messages (137)	
	Jun 1	GC&CS reads Home Waters (and U-boat) messages for a month with keys seized from weather ship *München* and U-110 (169)	
	Jun 27	GC&CS breaks German army Russian front key (VULTURE I); solutions last until spring 1944 (2:662)	
	Jun 22		Germany invades the Soviet Union
	Jul 2	GC&CS reads Home Waters (and U-boat) messages for a month with keys seized from weather ship *Lauenburg* (182)	
	mid-Aug	Home Waters key (DOLPHIN), then also used by U-boats, begins to be solved daily in 36 hours without seized keys on bases of knowledge gained in two preceding months; solutions continue to 7 May 1945 (1:338, 2:663-664)	
	Sep 6	Churchill visits GC&CS at Bletchley Park (184-85)	
	Oct 5	TRITON key net for U-boats separates U-boat keys from Home Waters key (205); solutions continue on both nets in times ranging (in January) from 12 to 120 hours	

Date		Enigma	The World
	Dec	GC&CS first solves Enigma of the Abwehr, the armed forces espionage service; solutions, called ISK (Intelligence Service [Dillwyn] Knox), continue to end of 1944 (2:668)	
	Dec 12		Germany declares war on the United States
1942	Jan 1	Keys for Luftwaffe's Fliegerkorps IX (WASP) and Fliegerkorps X (GADFLY) broken on day of first appearance and continuing to end of war; key for Fliegerkorps IV (HORNET) broken to December 1943 (2:660)	
	early	New Short Weather Key goes into service (2:750), ending GC&CS's ability to use meteorological messages as cribs to read naval Enigma (189-190)	
	Feb 1	4-rotor Enigma goes into service on TRITON; U-boat solution blackout begins (2:179)	
	Apr 3		Rommel begins offensive against British
	Apr 22	GC&CS breaks the Luftwaffe key for ground-air cooperation in Africa (Fliegerführer Afrika, or SCORPION); this ends February 1943	
	Jun 30		Rommel halts offensive at El Alamein
	fall	U. S. Navy liaison officer starts at GC&CS	
	Oct 23		British attack at El Alamein begins German rout
	Oct 30	New Short Weather Key captured from U-559 (220, 226)	
	Nov 8		Americans and British invade North Africa
	Dec 13	GC&CS, using new Short Weather Key to get cribs, breaks 4-rotor Enigma keys (SHARK); blackout ends (226, 2:667) but solutions irregular and at first slow (sometimes 8 days), by end of month much faster (around 12 hours)	

Date		Enigma	The World
1943	Jan and Feb	SHARK solutions stutter, but when working some are fast enough to be of operational value (229)	
	Mar 10-Jun 30	90 SHARK keys solved in 112 days (230)	
	Mar 14-20		Greatest convoy battle of the war, in which 21 ships are sunk, leading Allies to fear that the Atlantic lifeline to Britain could be cut
	May 24		Commander of U-boats Admiral Karl Dönitz withdraws his submarines from the North Atlantic, signifying a major Allied victory in the Battle of the Atlantic
	Aug	SHARK being read almost solid and fast (242)	
1944	Jun 6		Allies invade Normandy
	Nov	Individual U-boat keys end SHARK solutions, but so great is Allies' strength in ships and airpower that this no longer matters much (262)	
1945	May 8		Germany surrenders

THE SIEMENS CIPHER TELETYPE IN THE HISTORY OF TELECOMMUNICATIONS

WOLFGANG W. MACHE

ADDRESS: Zielstatt St. 133, D-8000 Munich 70, WEST GERMANY.

ABSTRACT: This paper spans a period from early pioneers in electrical telegraphy and telephony to the roots of the *Geheimschreiber*: the (electric) Siemens *Schnelltelegraphy* and the development of the first electrical teletypewriter "T25" by *Wernerwerk 1*, Berlin, in 1928. The paper expresses the opinion that the *Geheimschreiber* T52d/e was not cracked by Bletchley Park's deciphering machines during WWII.

KEYWORDS: Teletype, electric teletypewriter, *Geheimschreiber*, *Schlüsselzusatz*, Bletchley Park, *sägefisch*, Fish, Tunny, Sturgeon, COLOSSUS.

The following is a translation of a speech given at the 23rd Meeting of the Association of Former Navy Teletype Operators held at the Telecommunications School of the Federal Navy, Flensburg-Muerwik WEST GERMANY on 26 August 1984.

Much appreciation goes to the Siemens Museum (Munich) and Deutsches Museum (Munich) for supplying photographs.

Today when one hears the word *America* in discussions of countries that have produced great inventions, in communications technology what people associate with it are names like Samuel Finley Morse for the telegraph, Alexander Graham Bell for the telephone, and John William Mauchly for the computer. All this is a one-sided view, however, and does not do justice to any serious history of technology. It is true that the men I have just named have made widely known and immortal contributions, but in regard to the title of *the inventor* or *the first* we must differentiate.

It was from this point of view that I welcomed and supported the article by Dr. Jarotskiy of the Post Office in Moscow in remembrance of the 150th birthday of Johann Philipp Reis, born 7 January 1834 in the Barbarossa town of Gelnhausen, as a *pioneer in telephony* in No. 1/1984 of the technical journal *Electrical Communications*. The author made use of a quotation, you see, which I would like to share with you.

It runs: "A Critical history of technology would generally show that hardly any invention of the 18th and 19th centuries belongs to one individual or another."

Incidentally, this quotation comes from a gentleman who is known to all of you. He is the jurist, philosopher and utopian Karl H. Marx.

I think we must agree with Marx, because his comment also holds true to a remarkable extent in regard to the development of telegraph equipment.

The processes of the development of telephone and computer technology also bring up a multitude of names, which I will not enumerate here.

Instead, let us stay in the realm of teletype technology. Its rapid development was based on discoveries by scientists whose names, gentlemen, you know as well as I do. I remind you of Galvani, Oerstedt, Faraday, Henry, Ohm, and Ampere, to name only a few.

Control Number 1135

STAFF INFORMATION. SECRET!

The Siemens SFM T52d Cipher Teletype Machine

SECRET STAFF INFORMATION!

1. This is a state secret within the meaning of Section 88 of the Reichs Criminal Code.
2. To be forwarded only by handcarrying or addressed by name, in double envelope, and upon written acknowledgement of receipt.
3. Transmission by courier or by a trusted person if possible; with a declared valuation of more than 1000 Reichsmarks if mailed.
4. Reproduction of any kind or taking extracts is forbidden.
5. Storage is upon the responsibility of the recipient, to be in a safe or under exceptional conditions in a steel locker with a combination lock.
6. Violations of these rules will be most severely punished.

Printed Naval Regulation No. 35 IV

Cover sheet for the Siemens SFM T52d Cipher Teletype Machine

But how did it begin?

On the question of origin in the field of *pre-Philipp Reis* electrical telegraphy, I may emphasize its germ in Germany, namely the *electrochemical telegraphy* of the physician and member of the Royal Academy of Sciences at Munich, Samuel Thomas von Soemmerring, born in 1755 at Thorn.

After the invention of the electrical pile by the Italian physicist Volta in 1800, Soemmerring succeeded in setting up an impressive but slow parallel-wire telegraphic system over a distance of around 1000 meters, making use of the releasing of gas. Soemmerring demonstrated his method in Munich at the Academy in 1809, but he was unable to make any headway against the *optical telegraphy* (i.e., semaphore) which was dominant at that time, (e.g., the Chappe System, used by Napoleon Bonaparte at that time). However, his labors were of inspirational value for the later telegraph systems that Schilling von Canstatt and Moritz Jacobi developed in Petersburg. They determined the further progress of two-wire telegraphy. For *electrical telegraphy*, in Russia 21 October 1982 was the 150th anniversary of the invention of the electromagnetic telegraph, the day when the German-Russian Baron Paul Schilling von Canstatt successfully demonstrated his 6-pin apparatus for the Russian alphabet to the Czar and the Academy in Petersburg in 1832. After his death in 1837 his work was continued until 1874 by Jacobi, a Prussian master builder in Koenigsberg and from 1837 on a scientist in St. Petersburg.

Let us not trouble now to go into the numerous events in the evolution of telegraphy from 1833 to the start-stop teleprinter of the 1920's in detail. But emphasis should be given to the essentials. Some systems established milestones and others sooner or later were replaced. Thus, for example, the needle telegraphs did not achieve lasting prevalence, whereas the American painter Samuel Morse, in the years from 1837 to 1844, in creating so-called recording telegraphs and inventing the alphabet that is named after him made an inestimable contribution that is still influential at the present time. Morse gave credit publicly to the German Steinheil, who first proved to use the ground as the return loop, which saved the installation of a 2nd wire and eased the introduction of the Morse system in the US Continent in 1844.

Still more successful was the technology of the type printing telegraphs that came into use in 1854. A well known representative of this type of equipment is the so-called Hughes printer, a very remarkable device with a life in service of almost 8 decades.

I have been told that there are several members represented in your Association who were trained on this machine here in Muerwik at the Torpedo and Communications School before the beginning of the 1930's. If you should ever come to Frankfurt am Main some time, I recommend to you a visit to the Federal Postal Museum. There on display are two different models of the Hughes printer, in working condition. In 1900 there were already 3500 of his teletypes

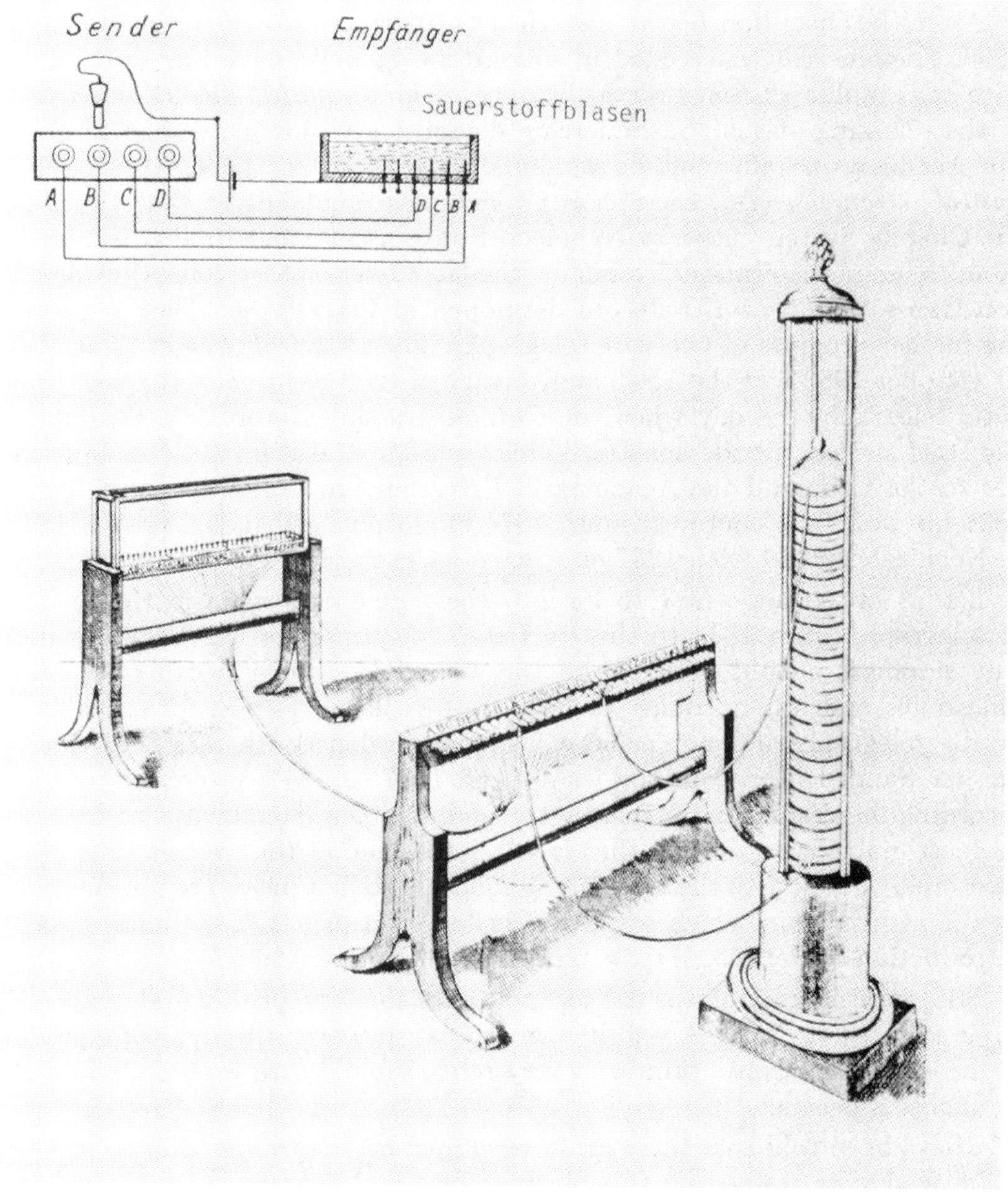

Figure 1. Soemmerring's electrochemical telegraph. Munich, 1809. Later improved by Schweigger for use over only two wires, 1811 (but not realized).

Figure 2. Schilling von Canstatt's 6-pin telegraph. Petersburg, 1832. Courtesy: Deutsches Museum, Munich.

in existence in the whole world when Edward Hughes dies a millionaire and one of the leading consultants in telegraphy. As early as 1895, Hughes printers were being developed further at Siemens, and by the beginning of the teletype generation of the 1920's they were being built. A parallel development in 1903, that of the Siemens automatic telegraphs with a speed of 2000 characters per minute, (i.e., 33 characters per second), was far in advance of its time. It was followed in 1924 at Siemens by the 56 Baud keyboard printing telegraph, which still came in two pieces (i.e., separate sender and receiver. See picture). That became the starting point for the electrical teletype. You see, there was more demand for a machine that would send, receive, be operated by only one person, and still be cheap as well. What was aimed at was a machine similar to the type bar carriage typewriter in common use at that time, which had been factory made since the second half of the 19th century. It was in this way that the start-stop teleprinter patents of the German-American Krum originated in the USA, financed by Morton and with the trademark MORKRUM, and in 1923 for Kleinschmidt, a German-American from Bremen. The two firms merged in 1924 to form the Morkrum-Kleinschmidt Corporation. Next came the development of the "Teletypewriter Model 14," which was imported under the name *Springschreiber* (i.e., start-stop teleprinter) in 1926 by Lorenz of Berlin. Lorenz copied the "Teletype 14," and Siemens & Halske successfully met the challenge from America too. Between 1927 and 1933 Wernerwerk of

Figure 3. Siemens high-speed telegraph. Berlin, 1903.
Courtesy: Siemens Museum, Munich.

Berlin likewise developed a series of different teletype machines in rapid succession. In the course of doing so, a transition was made from steps under patent to taking new directions.

In 1927, when the German licensee of the principle of the *lengthened stop signal* in the mechanical teletype, patented by Kleinschmidt, would not share his monopoly with other manufacturers, F. Geissler, a designer in our firm, had the idea of cleverly circumventing the Kleinschmidt patent through an *electrical* plan. There came into existence an electrical teletype machine, "T type 25", whose further development – available in 1932 – is familiar to all of you gentlemen. I refer to the T of "T type 36." Until now it remained unclear why it was this particular machine, which later served as the basic module for a *secret teletype machine*, that attracted the interest of the Navy in the first place. The Navy preferred the electrical machine over the mechanical machine because of its superior stability at sea.

This machine superseded the Hughes printer that had been in use up to then, and the *sounder*. The change from the piano-like keyboard to a completely new set of keys may have caused a few problems at the time. Perhaps one or another of you can confirm that.

Of the further developments that followed at Siemens, I would like to point out the following mass-production teletypes, all of them page printers:

- 1933: T type 37a (successor to the T type 26a of 1929),
- 1957: T 100, 50 Baud capacity,
- 1961: T 100, 100 Baud capacity,
- 1976: T 1000, already highly integrated.

The following figures point to the worldwide significance of these developments:

By 1984 Siemens had supplied over 1 million teletype machines and had become the unquestioned market leader among 13 manufacturers.

Our firm's share in the total number of 1.6 million telex machines in the world is 30%.

By the end of the second World War, 15,000 machines had been produced; in recent years, 1978 to 1983, it was 70,000 machines a year, just on the average.

From the point of view of types of technology, we get the following distribution figures:

- 55% electromechanical, including Model T100
- 40% electronic, first generation
- 5% electronic, second generation, microprocessor controlled.

After this short outline of the history of the development of telegraphy, I would like to take up a machine that you all know, gentlemen, and which even yet – almost 40 years after the end of the was – is surrounded by secrets. I mean the Siemens cipher printer "T type 52", models "a" through "f". On the construction and manner of operation of this machine I do not need to say much. You, gentlemen, have worked with this machine during your time of active service. Many of you have been at Siemens in Berlin, and have taken

Figure 4. Siemens keyboard printing telegraph. Berlin, 1924.

part in the course of instruction there. Instead of going into that, allow me to make a few remarks on the development of this machine and the enemy efforts at decipherment.

Figure 5. T type 25 without protective cover.

In 1969 a colleague in our firm, Ehrhard Rossberg, director of the Technical Division of Switching Technology of Siemens retired. The patent department marked the departure of this extraordinarily skillful engineer with a separate volume which contained more than 100 of his patent descriptions. Mr.

Figure 6. Electrical teletype, T type 36. Berlin, 1932 (without casing).

Rossberg, who was born in 1904, attended the Chemnitz Industrial Academy after a period of practical training, and came to Siemens & Halske in Berlin-Siemensstadt in 1926.

At Siemens, Mr. Rossberg was on the scene at the new starting point of the T type 25 electrical teletype and in 1928, when the Navy in Kiel selected it. A productive collaboration ensued shortly with additional, very successful colleagues such as Honorary Dr. of Engineering Herbert Wuesteney, Honorary Dr. of Engineering August Jipp, Eberhard Hettler, Fritz Geissler, and others.

The Navy, concerned because of the British decipherment successes in the first World War, made it a requirement that the electrical teletype should be provided with an enciphering device.

This idea was unusual, because at that time teletype text requiring protection were enciphered before transmission in a fussy and time-consuming manner. But what they had in mind now was the truly sensational development of what today is called an on-line machine. It is possible also to ascribe this to the design advantages of the electrical teletype, whose creation was provoked (i.e., by the Kleinschmidt patent problem), but was fortunate.

Figure 7. T25 with cipher device. Berlin, 1930/31.
(Laboratory model of the cipher printer).

On the way there, we come in 1930 to the first applications for patent with the title *Arrangement of Switching for the Transmission of Information in Cipher*.

A further application from the year 1931 contains the principle of irregular cam discs, i.e., cam discs of different sizes, set up chiefly on the basis of prime numbers. In the same year the patent application was filed in the USA by Jipp, Rossberg and Hettler, there US patent 1,912,983 was granted June 6, 1933.

Figure 8. Cipher printer, T52a/b, with power supply in the lid, before WW II.

It has been reported by several veterans that the Navy was the first service branch to be interested in using the cipher teletype machine, also called a *cipher printer* (*Geheimschreiber*), or *G-Schreiber* for short. The Navy wanted secure written communications, for example, from a ship at the pier to the high command of the fleet.

I owe it to a fortunate circumstance that in 1982 in Munich I was able to discover – or rather rescue – a photograph that shows the laboratory model from this first couple of developmental years. It still consisted of two separate machines, the T type 25 electrical teletype and the cipher device with 2 x 10 adjustable wheels. This *attachment* was later, in the commercial version of 1932, united with the T36, in *one unit*, the "T52".

Figure 9. Cipher printer, T52c, without casing.

Now we go to the introduction of the Siemens T type 52a cipher teleprinter.

In 1932 the Hungarian armed forces were the next to be supplied with machines and complete documentation.

Navy tests soon showed that the cipher printer interfered with the sensitive radio receivers. For this reason the model T type 52a/b with noise suppression circuitry was created in 1934.

A further step in the development of the cipher printer was the T type 52c, for which the Air Force gave specifications in 1936.

Figure 10. Cipher printer, T52e, without casing, a *Norway machine.*

In the meantime, the security authorities of the Third Reich imposed secrecy measures which still make themselves felt today in the considerable scarcity of information.

Nevertheless, after the signing of the German-Russian treaty on 28 September 1939 in Moscow, the Siemens SFM T52 cipher teletype machine was demonstrated and shown opened up to Soviet military delegation in Berlin-Siemensstadt.

Development continued during the second World War. Models T type 52d and T type 52e were created, and towards the end of the war Model T type 52f was in the works, at least in specification form.

In 1944 or '45 the technical documentation in the factory for the strictly secret T52 became lost. For one thing, the production facility in Berlin was badly hit by Allied bombs in October 1944. But what stands today as the main reason for unsuccessful endeavors to obtain authentic factory records and technical descriptions is the rigorous security regulations and their application before the occupation in 1945.

After the end of the war Dr. Wuesteney was brought to the British headquarters in Berlin-Gatow and given the task, together with Fricke, a Siemens mechanic who worked on T52 assembly, of reconstructing the circuit diagram of a T52. At the end of a week the British discontinued the project without result, although they are said to have gotten possession of no documents, but only machines, with Mr. Fricke, among others, serviced. The machines were later put to use on teletype connections between the British Sector in Berlin and the headquarters of the Army of the Rhine.

In Norway, too, a number of T52e's survived the end of the war, and were later put into use again there. Of these there are specimens in operating condition on display today in the Siemens Museum in Munich, in the Science Museum in London, and in the Technology Museum in Oslo. The exhibit in the Military Museum at Dieppe-Pourville on the other hand, a "d" (in phonetic alphabet Dora) machine (T52d), which I saw last month, is incomplete, poorly maintained, and hence not in working condition, although the T rls 43a telegraph relays, for example, look like new.

From the former armed forces' telecommunications equipment depot which was in in Elmshorn between 1948 and 1960, the remaining cipher printers, which had previously been demilitarized, i.e., partly dismantled, were brought to Trier and there at the *Electromechanical and Electronic Plant* were put in circulation, as the French security authorities had decided on utilization of the cipher printers. Trier supplied a total of 280 T52d's and e's, which is an estimated 25% of all T52's supposed to have been produced in Berlin up to 1944. The cipher printer was classified down to the present time, and may still be in use somewhere. Investigations become all the more difficult; informants prove to be a decided rarity.

This may be sufficient at this point as a presentation of the development of the cipher printer.

But now to the question of the extent to which the enemy was able to break into our machine's cipher texts. I can anticipate the answer: Up to the present time we have not been able to obtain any reliable information – but let us nevertheless attempt an approximation.

Soon after the first opening of the British secret records in 1975, 30 years after the second World War, the British journalist Brian Johnson wrote in his book *The Secret War* about technical facilities of the British decipherment center at Bletchley Park. According to that book, there were the so-called *bombes* for processing the Enigma material, and forerunners of the computer, designated *Colossus*, were available for deciphering the cipher printer texts. This statement went from one book to another, and was even broadcast by the BBC in both words and pictures.

Figure 11. Decipherment computer Colossus, Bletchley 1944.

Knowledge of our own and indications from England over the past years have confirmed us more and more in the belief that the above-named author's investigations were in behalf of the Siemens & Halske Geheimschreiber inadequate. Recently our belief has become a certainty. Because the third volume of the official British history of the operations of the intelligence service in the second World War has been published. In it, in Part I, the principal mission of the *Colossus* decipherment computer is described: Actually, it was used in decrypting German 5-character code messages. These messages however came not from the Siemens cipher printer, but rather from another teletype cipher printer, the model SZ40 and SZ42 cipher attachment from the firm of Lorenz – a machine, therefore, with which you may have had less contact, gentlemen.

In order to substantiate this statement further, at this point let a short review of the development of radio teletype technology be permitted:

As a result of the constantly increasing distances between the fronts, in 1940 a need grew up in the German armed forces for a serviceable radio teletype system that cipher equipment could be incorporated into.

To meet this need, in the course of 4 years a series of various devices was developed and in varying quantities was requested and operated by users, chiefly the Army. The first of these shortwave radio teletype sets with telegraphy channels went by the covername of *Saegefisch* (i.e. Sawfish) on the German side, by the way. The British intelligence service found out this designation from a decrypted Enigma message, and for their part used the codeword *fish* for it. Since about 1978, this covername has been incorrectly applied to the Siemens cipher printer, and indeed even when in reality what is being discussed is the German 5-character code traffic in general, or even the decipherment of Army communications done on the Lorenz cipher attachment.

It would be dishonest to the history of the field if it went unmentioned that the T52 designed by Siemens for wire communications lines was also used in its "b" and "c" models for the first radio teletype trials ([19]40/41). It soon turned out, however, that both models were suited only for limited use. This was because if it happened, for example, that as a result of atmospheric disturbances even a single letter of cipher text was suppressed, then the encipherment's synchronization was thrown off and subsequent text was decrypted incorrectly. The problem was overcome, however, with the Lorenz machines already mentioned.

Thus when the British systematically monitored the German experimental radio teletype link between Vienna and Athens in 1941, they thought they were on the trail of the Siemens cipher printer. We must even concede today that British mathematicians had then already succeeded, by awkward manual

Figure 12. SZ42, without cover.
(1984 pictures, Navy Telecommunications School, Flensburg).

procedures, in deciphering German teletype messages. In so doing, they realized that only automatic decrypting machines could give effective support to their own military operations. These were actually put into use in June 1942, and had to be constantly modified because of their severe rate of wear and breakdown.

The British were more successful with the Colossus computer, which was not put into use until February 1944, however. An example on this point:

Beginning on 6 January 1944 the radio teletype link belonging to the Commander in Chief in the west (Army Group D) between Paris and the OKW [Supreme Command of the Armed Forces] in Berlin (Stratssberg radio station) was intercepted, and starting in March, about 3 months before the beginning of the invasion, they were in a position to decrypt teletype enciphered on the SZ42 (*Schlüsselzusatz* 42).

But the "d" model of the Siemens cipher printer was also intended for radio teletype operation. Here is the proof: It is from among you, gentlemen, that valuable training materials from the Naval Communications School at Waren-Mueritz dating from 1944 have been made available to the Archive of the Naval Telecommunication School. They contain operating directions for the machine just mentioned. So far, however, it has remained unknown whether the Navy has at any time ever operated communications as we see them shown here. It would be very nice if this matter could be clarified with your kind assistance.

Figure 13. SFM T43, basic model of mechanical *page* printer, T type 37f.

With that I come to my conclusion:

I believe it has become clear that we are only at the beginning of the truth. The official British historiography has up to now not been prepared to make any statement regarding the extent of the decipherment of teletype texts. Might this perhaps be because they would have to admit to not having had unqualifiedly splendid success? I ask you to consider: Up to the present time we still have no information on how the attempts at decipherment turned out in so far as they relate to the T type 52 "d" and "e" cipher printer or the "SFM

T43" cipher teletype machine also developed by Siemens late in World War II. Wernerwerk T in Berlin built the "T43" on the basis of 2 types of teletype machine. These were the T type 37f mechanical page printer and the T type 34n mechanical tape printer, of which a hand full of machines started working in 1944 on a one-time-pad principle, e.g., between Paris and Berlin.

Figure 14. SFM T43, basic model of mechanical *tape* printer, T type 34n.

Thus for the present the question remains open as to the extent to which the teletype decipherment added to the significance of the Enigma decipherment or qualified its value. In the same way it must be kept for a later historical study to elucidate the degree to which the decipherment successes that have been presented influenced the strategy and tactics of the Allied invasion.

Gentlemen, and naturally ladies also, in so far as you were active in the communications service, the aim of my remarks has been, among other things, to present to you some facts about the T type 52 Siemens cipher printer which cannot have become known to you at any time, at least, during your period of active duty.

If I have succeeded in this, then it is an additional pleasure to me that I have been able to comply with your invitation to the Flensburg inlet, which I know well from former years. I would also like to express my obligation to the commandant of this school, *Kapitaen zur See* (rank of US Captain) Ehlert and *Kapitaenleutnant* (rank of US Lieutenant Commander) Timm for their support for this presentation: I wish all of you unforgettable days here in Muerwik.

WORDS USED IN FIGURES

		German	English
1.		*Sender*	transmitter
		Empfaenger	receiver
		Sauerstoffblasen	oxygen bubbles
3.		*Locher*	perforator
4.		*56Bd*	56 Bauds
5.	a.	*Tastenwerk*	keyboard unit
	b.	*Kontakteinrichtung*	contact device
	c.	*Sender*	transmitter
	d.	*Empfänger*	receiver
	e.	*Übersetzer*	converter
	f.	*Druckwerk*	printing mechanism
	g.	*Übersetzerrelais*	converter relay
	h.	*Locherschaltrelais*	perforator switching relay
	i.	*Auslöserelais*	release relay
	k.	*Regler*	control
	l.	*Weckerrelais*	relay controlling local bell circuit
12.		*Gleichlaufeinrichtung*	synchronizing device
		Funkfernschreibbetrieb	radio teletype operation
14.		*Abfühleinrichtung*	tape reading device
		Vorschubeinrichtung	feed device
		Stanzeinrichtung	punching device
		Gleichlaufschalter	synchronization switch

BIOGRAPHICAL SKETCH

Wolfgang W. Mache, member of VDE/ITG, derived his interest in WW II cryptology from Brian Johnson's book *The Secret War* (1978), whilst he wrote the *Lexikon der Text- und Datenkommunikation* (1980, Oldenbourg). He joined Siemens for telephone technician training, received a Dipl. Ing. degree in Electrical Engineering (1955) and then worked in Siemens' Telecom development Centers, Munich-Sendling for Power-line telephony and FDM/TDM-ARQ radio teleprinter systems. After time as Telecom Sales Manager in South America he is in marketing in the field of data transmission and switching.

NEW INFORMATION ON THE HISTORY OF THE SIEMENS AND HALSKE T52 CIPHER MACHINES

D. W. Davies

ADDRESS: Fair Winds, 15 Hawkewood Road, Sunbury-on-Thames, Middlesex TW16 6HL GREAT BRITAIN.

ABSTRACT: Two earlier papers in *Cryptologia* [1, 2] described the WWII German on-line cipher machine known as T52. The circuit which controls the irregular movement of the cam wheels appeared in two forms and it seemed likely that alterations had been made. The reason for these alterations can now be explained. The paper which follows this one is from the person who made the changes.

KEYWORDS: Cipher machine, Teletype, Geheimschreiber, World War II

BACKGROUND

The T52 Cipher Machines were Siemens teletype machines, modified by adding extra equipment which could encipher their transmissions and decipher what they received, on-line. They were used extensively by the German forces during World War II. Until near the end of the war, this secret traffic was not sent by radio, only by lines. Consequently it was not routinely broken by the Allied cryptanalysts.

The T52 machine is sometimes known in recent literature as a "Geheimschreiber", but there seems to be no clear evidence that it was called by this name. German documents of the time call it "Die Siemens - Schlusselfernschreibermaschine". These cipher machines should not be confused with the Lorenz cipher attachments, SZ40 and SZ42, which were the source of the "fish" traffic broken so effectively by the Colossus machines.

EARLIER REPORTS ON THE T52 MACHINE

Models T52 a/b, c, d and e were produced. I wrote a detailed account of the T52e [1]. Then [2] I traced the development of the T52 through its four models (a and b were logically similar, differing only in the suppression of electrical noise in model b).

All models featured 10 cam wheels which rotated step-by-step as Baudot characters were transmitted or received. The numbers of steps in a complete rotation of each of the cams were mutually prime so that, if all cams stepped each time, the maximum total cycle length would be achieved. The cam profiles produced a pseudo-random string of bits. The outputs of the 10 cam contacts were used both for modulo 2 addition (XOR) and for permuting bits in the 5-unit Baudot code. Details are given in the two papers quoted above.

An important feature of models d and e was that cam wheel movement was intermittent. Any given cam would move approximately three quarters of the time, its movement depending on a logical function derived from the outputs of the cams themselves. The outputs used for encipherment and for controlling cam movement came from different points on the cam periphery.

In the 1982 and 1983 papers there were unexplained features of the cam movement control circuits. New information has come to light which helps to explain the history of these machines. In this paper I will give more details of the circuit information that was included in the earlier papers. A companion paper by Ernst Selmer describes how the changes came to be incorporated in the cam movement control circuits.

THE TWO VERSIONS OF CAM MOVEMENT CONTROL CIRCUITS

My first observations of the wiring of these circuits showed that changes had been made. Many of the wires forming these circuits were attached to a multi-point switch called KTF or Klar-Text Function which, when set to the ON position, complicated the whole process by making control of cam wheel movement also dependent on one of the previously transmitted plaintext Baudot elements. (With KTF in operation, a single error in transmission would put sender and receiver permanently out of synchronization.)

Of the 10 wafers of the KTF switch, only 4 were in use but these and the others had loose ends of wire attached. I concluded that at some time all the wafers of the switch had been used and that a radical rewiring had taken place, but its purpose and history was obscure.

The machines I had studied were at the Siemens Museum in Munich and at the Science Museum in London, on loan from the Norwegian Technical Museum in Oslo. Both came from Norway and it soon became clear that most of the T52 machines still in existence had their origin in Norway.

Then new sources of information appeared. Circuits from the Oberkommando der Kriegsmarine dated March 1944 showed a different logic for controlling cam movement. Another source was Ing. Willi Reichert who, at the end of WWII bought T52 parts at scrap prices from a store at Elmshorn and rebuilt them into T52 machines in Trier. His circuits showed the same version as the Kreigsmarine. I understand that Reichert's machines were supplied to the French Government.

Finally, I had the opportunity to examine a T52 machine in the care of the ZFCh (German Cipher Bureau) in Bad Godesburg and found again the same circuit. It therefore seemed that all T52 machines with intermittent wheel movement, whether T52d or T52e, had the same movement control circuit and that only the Norwegian machines had the alternative version. My paper of October 1982 had been misleading because it showed in Figures 7 and 8 a circuit that resulted from a Norwegian modification of the machine, not the original as I had supposed.

The figures in my two earlier papers used different notations. to clarify the nature of the changes a new figure is given here comparing the circuits in the condition "without KTF". the wheels are labelled A to K. The figures show the cam outputs as changeover contacts, which are approximately balanced between the 0 and 1 states. The magnets are therefore operated approximately one quarter of the time. Each magnet, by means of an interposer, stops the stepping of its corresponding wheel. Note that contacts do not influence their own wheels, except in one case which is the C wheel in the modified circuit. Also in the modified circuit the B wheel might be expected to stop on about 3/8 of occasions.

WHENCE THE MODIFICATIONS?

At the Oslo Defence Museum I was told that T52 machines had been modified for use by the Norwegian Border Guards. This was all the information I had until a report from Eurocrypt '93 that Dr. Ernst S. Selmer had spoken about his own role in the "capture" of these machines and their modification. His part of the story is given in a companion paper.

He questions whether the modifications improved the machine cryptographically. One objective would be to obtain a very long cycle before the wheels were all back in a previous position and the whole cipher repeated. It is very hard to estimate the cycle lengths produced. In the case of KTF, the cycle structure depends on the text being transmitted, which rather reduces its importance.

The original wiring produced more difference between the behavior with and without KTF. I believe that KTF caused so much operational difficulty (and weakness due to restarting after a failure) that after a time it went out of use. The significant circuit is the one "without KTF".

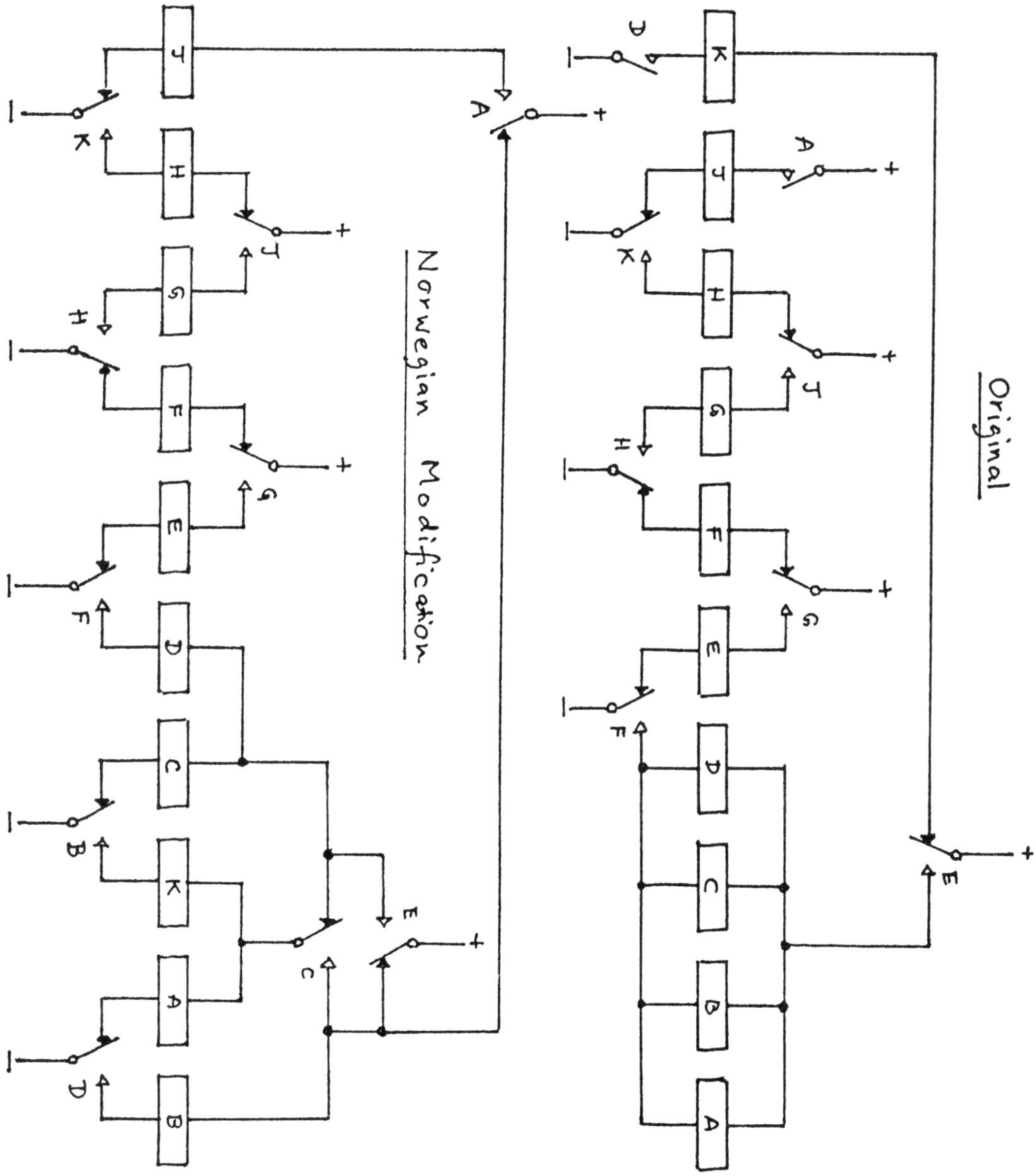
Norwegian Modification
Original

Comparing thc "without KTF" circuits it is noteworthy that in the original configuration cam wheels A, B, C and D always moved together. This ensured that any cycle would be a multiple of 73.71.69.67 (=23,961,009) steps. How this compares with the (less easily estimated) performance of the modified circuit is an interesting question. Perhaps this was considered a large enough cycle to exceed the life of one key period.

OTHER MODIFICATIONS OF THE T52?

Two other features of the machines studied might be evidence of modifications.

(I) The 10 binary outputs of the cam wheels go through switches or a plugboard which permute them. This permutation is the main cryptographic key. Then in models T52c and T52e the outputs go through the logical transformation of the H relays to operate the encoding relays SR. This logic is different in the two models. Compare October 1982, Figure 10 with July 1983, Figure 8.

I believe that this is a real difference between models T52c and T52e, since there is no evidence of extensive wiring changes and nothing else suggests that a change has been made.

(II) From the start of my work it appeared that the cam profiles were fixed - that they remained the same throughout the whole series of models and machines. These profiles are recorded in Figure 3 of the October 1982 paper. But when I came to examine the T52 machine at ZFCh the profiles were different.

The mechanical explanation was that the same cam parts had been used, but assembled differently. Each cam is fixed by three screws and can be assembled in one of three positions by rotation through 0, 1/3 or 2/3 of a turn. Also a cam can be flipped over, so that the sequence of output bits is reversed, making a total of six possible orientations. The ZFCh machine had different orientations from the Norwegian machines. It is not the kind of change that could easily be made in the field because there are consequential adjustments of contact lift and timing.

I found the modified cam orientations nowhere else. Either these were the wartime positions and all Norwegian machines had been modified or else it was the ZFCh machine that had been modified.

There is no conclusive answer. Ernst Selmer has no recollection of the change, which would have been a major task. Therefore I incline to the view that, for some reason, the ZFCh machine was modified and that in all the other machines we see the cam profiles in WWII.

REFERENCES

1. Davies, D. W. 1982. The Siemens and Halske T52e Cipher Machine. *Cryptologia*. 6(4): 289-308.

2. Davies, D. W. 1983. The Early Models of the Siemens and Halske T52 Cipher Machines. *Cryptologia*. 7(3): 235-253.

BIOGRAPHICAL SKETCH

Donald W. Davies helped to build one of the first electronic digital computers at the National Physical Laboratory. He was active in computer research and pioneered the concept of packet switching from 1965 onwards. Then his interests moved to computer and communications security. Since 1984 he has provided security consultancy to financial institutions, telecommunications providers and pay TV broadcasters, among others. He has received a number of honours, including Fellowship of the Royal Society.

THE NORWEGIAN MODIFICATION OF THE SIEMENS AND HALSKE T52e CIPHER MACHINES*

Ernst S. Selmer

ADDRESS: Idrettsvn. 20, N-1400 Ski NORWAY.

ABSTRACT: Two earlier papers in *Cryptologia* [1, 2] described the WWII German on-line cipher machine known as T52. The circuit which controls the irregular movement of the cam wheels appeared in two forms and it seemed likely that alterations had been made. The reason for these alterations are given in this paper, by the person who made the changes.

KEYWORDS: Cipher machine, Teletype, Geheimschreiber, World War II

In 1946, my friend Nils Stordahl became head of the Norwegian military cipher office, a position he held until 1983. One of his early initiatives was to engage me as a consultant.

My first big task, and my most fantastic cryptological experience, was to establish a (hopefully) safe communication system for the Norwegian equivalent of MI5 and Scotland Yard Special Branch ("Overvakningspolitiet" = "watch-over police"). We based it on the German Siemens teleprinter ("Fernschreiber" = "remote writer"), with an additional unit for encryption/decryption ("Geheimschreiber" = "secret writer"). There were many of them in Norway, and the Norwegian (public) Telephone and Telegraph Company had collected them to dismantle the cipher unit and use them as ordinary teleprinters. We had to prevent this, and could not let it pass official channels. I cooperated with another of Stordahl's men, a young electrical engineer named Asbjörn Mathisen. Some time in 1946, we were supplied with two large military trucks and 20 German prisoners of war. We commanded them in our school-Germanm, and drove to the Telecompany store just outside Oslo. The few attendants protested vigorously, but did not stand a chance against us and the Germans. Including a wooden case, each G-Schreiber weighed 180 kg, so the POW's were really needed. We got everything onto the trucks and drove it to a safe hiding place.

Mathisen used wires and torch bulbs to reconstruct the coupling diagram of the G-Schreibers, including 20 relays. I knew absolutely nothing about relays

*Extract from the Eurocrypt '93 Proceedings [5]

then, but had to find out how everything worked, from a maze of unsystematized wires.

The same reconstruction has been performed much later by other people. The Norwegian Technical Museum in Oslo had managed to get hold of 3 G-Schreibers. Two of them were given - for exchange purposes - to museums in London and Munich, and these were analyzed by Donald W. Davies [1, 2]. Some more historical information was supplied by Wolfgang Mache [4].

The cipher unit contained 10 large notched wheels, stepped by a pawl mechanism. The number of steps in a revolution varied from 47 to 73, all pairwise coprime. The cam pattern was the same for all machines. The cams activated two contacts for each wheel. Over 20 relays, one contact set controlled an irregular stepping of the cam wheels, while the other set performed the ciphering of each 5 bit teleprinter symbol, in two rounds: one bitwise addition key, and one permutation (only 32 out of the $5! = 120$ were used). Details can be found in [1, 2, 4].

I did not want to use the German outfit exactly as it was, so I made some modifications which Mathisen implemented. I could not change the notched wheels (with contact sets), but I fiddled with the above-mentioned relay control of the cam wheel stepping. I thought I made it more complicated, but had nobody to check whether the outfit had become cryptologically stronger.

The G-Schreibers came in successive versions 52 a/b, c, d, e. The first version was broken by the Swedish mathematician Arne Beurling (cf. Kahn [3]) from the traffic passing Sweden between Norway and Germany. Later versions were "occasionally, but not routinely" broken by the British cipher office at Bletchley. – All this has become known later; in my case, I just had to hope for the best.

In my later designs of cipher machines (for NATO), I always said like the waiter: "Not my table", about all questions regarding protocols, key distribution, initiating and closing routines, etc. But for the police cipher, I had to work this out all by myself. In 1948, we summoned 20 police officers from different stations to Oslo, where I drilled them in the use of the G-Schreiber for a 3 weeks course. And the system was used from 1949 until around 1960.

COMMENTS ADDED TO REPRINT

Andrew J. Clark, after attending the Eurocrypt '93 seminar, has established contact between me and Donald Davies, who wrote the above-mentioned two papers on the G-Schreiber in *Cryptologia*. In a third, companion paper, he now describes in more detail the differences between the original and the " Norwegian" versions of Model 52e, in other words, my modifications.

All my notes on the G-Schreiber, from the late '40s, were stored at the cipher office, and were unfortunately destroyed a long time ago. Of course, it is impossible for me to recollect the modifications I made, and it is very satisfactory that this has now been described by Donald Davies.

The only relevant archive information I have been able to trace is a note at the Norwegian Technical Museum, stating that they got their 3 machines from the Military High Command in 1961, so they were in all likelihood police machines which had been replaced by more modern equipment.

I have also been in contact with one of my police pupils from 1948. He remembered that they were instructed to use KTF (Klar-Text Funktion, cf. [1]) only for top secret messages.

REFERENCES

1. Davies, D. W. 1982. The Siemens and Halske T52e Cipher Machine. *Cryptologia.* 6(4): 289-308.

2. Davies, D. W. 1983. The Early Models of the Siemens and Halske T52 Cipher Machine. *Cryptologia.* 7(3): 235-253.

3. Kahn, D. 1967. *The Codebreakers.* New York: Macmillan.

4. Mache, W. 1986. Geheimschreiber. *Cryptologia.* 10(4): 230-247.

5. Selmer, E. 1994. From the Memoirs of a Norwegian Cryptologist. In *Advances in Cryptology - EUROCRYPT '93*, ed. T. Helleseth. New York: Springer-Verlag.

BIOGRAPHICAL SKETCH

Ernst S. Selmer received his Dr. philos. degree (Mathematics) from the University of Oslo in 1952. From 1957 to 1990 he was professor of pure mathematics at the University of Bergen, Norway. His main mathematical interest has been number theory, but he has also been strongly engaged in computer activities and cryptological work for the Norwegian government.

THE BREAKING OF THE JAPANESE ARMY ADMINISTRATIVE CODE

David Mead

ADDRESS: 1229 Glen Meadow Road, East Lansing MI 48823 USA.

ABSTRACT: This memoir relates the writer's experience as a Signal Corps cryptanalyst in World War II. The major focus is the effort of the Signal Intelligence Service at Arlington Hall to crack the Japanese Army Administrative Code. The technical operation of the code and the ultimate breakthrough into the indicator system are detailed. The process of decoding and translating messages is described, and the memoir concludes with comments on the influence of cryptanalysis on the prosecution of the war in the Far East.

KEYWORDS: Japanese Army Administrative Code, solution problems, 1943 breakthrough, decoding messages, impact of cryptanalysis.

Legion of Merit ceremony for Lt. David Mead at Arlington Hall, 1945.
(l) General Preston Corderman, Commanding Officer, Arlington Hall, and (r) Lt. David Mead.

In the summer of 1942, World War II occupied the mind of the American people and was requiring more and more of their energy. The Allies were preparing for the invasions of North Africa and of Guadalcanal. I was a graduate student at Ohio State University and had just completed the general exams for a doctorate in English; my dissertation, however, would have to wait a few years. Like many other young people, I decided to put aside my studies and to try to help in some small way in the war effort.

The armed services were expanding that summer at an unprecedented rate and had numerous openings for civilian specialists in language, mathematics and other fields. Although I had no specific knowledge of cryptology, I had one of the cryptanalyst's essential traits — an infatuation with solving puzzles, whether they were acrostics in the old *Saturday Review of Literature* or problems in identifying Middle English dialects in a medieval seminar. My years of literary study may also have given me the illusion that solving enemy codes would be a life of romantic adventure.

In any event, I wrote the Army Signal Corps, detailing my biography and aspirations to become involved in cryptography, as we then called cryptology. I soon received an offer to come to Washington as a research analyst in the Signal Intelligence Service. After a brief FBI investigation confirming that I was sound in character and innocent in politics, I reached Washington and was assigned to a four-week training course in the Munitions Building. The small class was supervised by Evelyn Akeley, an able teacher who immersed us in the cryptographic manuals of William F. Friedman, the eminent cryptanalyst who later gained fame for his part a few years before in solving the Purple diplomatic cipher. He was for many years technical chief of the Signal Intelligence Service. I met Friedman on a couple of occasions, but never really knew him. He was an impressive figure, a neat, urbane man who seemed in a nervous hurry to get on with whatever was engaging his restless spirit.

I still recall that August in Washington, partly because it was so oppressively hot that Akeley fed us salt tablets, but mainly because of the excitement I experienced solving Friedman's problems. Most of them were ciphers, using mixed alphabets with keywords, irregular form and other concealing devices. After working on them all day, I took them home at night, so that my meals and sleeping hours became as irregular as Friedman's alphabets. But I made satisfactory progress, although the ciphers which concluded the course were ingenious in construction, and their solution was painfully slow. Among other things, they taught me that patience is not merely helpful to the cryptanalyst, it is critical.

I was next assigned to B Branch, the intelligence section of S.I.S., where I worked in the cryptanalytic unit under the direction of Major Solomon Kullback, an expert mathematician and cryptanalyst. Kullback had been recruited by Friedman in the early years of the S.I.S. He was a blunt, forceful administrator, well respected for his organizational skills. S.I.S. soon outgrew its quarters in the Munitions Building. By the time I undertook my permanent assignment, S.I.S. had acquired a new home at Arlington Hall, a former school for girls in the Virginia suburb across the Potomac. The Hall, or "The Salt Mine," as we often called it, became the nerve center of American signal intelligence, the army counterpart of England's Bletchley Park, where British code operations were housed. When I arrived in Arlington Hall, it had a few hundred workers. Before I left in 1946 it had several thousand.

Kullback placed me in a section of about 15 cryptanalysts supervised by Frank Lewis, a small, wiry civilian with astonishing nervous energy. A former music teacher from Maryland, he was also gifted in mathematics. Lewis generously taught me his cryptanalytic skills. I also learned a great deal from Lieutenant Karl Elmquist, a linguist from Texas A&M. He was an innovator, who liked to try unusual approaches to cryptanalytic problems when conventional methods proved fruitless. I worked most often with Elmquist, but I had the advantage of knowing and working with other talented cryptanalysts. Among them was Eli Schuman, a fine mathematician. Like Elmquist, he was a good chess player. I often ate lunch with one or the other while we played chess on a portable board. They both beat me regularly. Other able cryptanalysts included Sam Snyder, a mathematician, and Delia Sinkov, wife of Lieutenant Colonel Abe Sinkov, cryptanalytic officer for U.S. Army Forces in the Far East. I was fortunate in having the opportunity to work at various times with them, as well as with other capable people in my code unit.

The major focus of our activity in the fall of 1942 was the Japanese Army Administrative Code, a machine-encoded system of four-digit groups layered over with additive to disguise the code text. We called the code the 7890 (seven-eight-nine-oh). I am not sure who coined the term, though I first heard it from Frank Lewis; nor do I recall that it had any special meaning beyond its convenience. The 7890 emanated mainly from Tokyo, although it was used wherever the Japanese had army units with radio communication facilities. The 7890 intercepts were sent to the Hall's IBM unit, where they were bound in thick sheafs of printouts with the messages arranged chronologically. These were the raw material for cryptanalysts to study. The volume of messages was so heavy that, even before the code was broken, the Signal Corps' skill in traffic analysis and radio direction finding sometimes furnished clues to Japanese troop positions, impending operations and other matters useful to military intelligence.

During the next few months I learned why Arlington Hall was called "The Salt

Mine." Through repeated trial and error we tried to find a way of penetrating the Japanese code. (We knew from contacts with liaison officers at the Hall that cryptanalysts in the Far East and in England were sharing our difficulties.) Our group occupied a spacious room where each of us had a desk, file cabinets and a long table where we could spread out messages. Both our work-station environment and even our methodology seem somewhat crude by modern standards, but in 1942 the electronic explosion had not yet occurred. Instead of computers we had piles of graph paper and pencils that could write in four colors. We were confident that the encoders were using an additive book of four-digit numbers and that somewhere in the messages there were indicators informing the recipient how to decode them. Of course the Japanese could increase their security by periodically changing the indicator system, the additive tables, or even the codebook.

One attack on the system was to search for code groups or paired digits that could serve as indicators pointing to messages that would overlap because they used the same additive numbers. If the encoders were using whole groups as indicators, this approach would be feasible. However, if they were using scattered pairs (as they were), the possibilities of success would be more remote. In this approach, as with others, we sometimes found interesting correspondences that encouraged us but that then, perhaps after lengthy analysis, proved meaningless. The cryptographer's random numbers often function in nonrandom ways that promise hope but end in disappointment.

The opposite strategy was to begin with encoded messages and work backward toward the indicator system. The Hall's IBM unit could provide us with message printouts and make a machine search for repeated groups that formed some kind of pattern. We tried various strategies, such as bunching messages from a single source or within a narrow time span. Identical groups in separate messages were clues for careful study. They were nearly always random coincidences, but occasionally the correspondence of a second group would encourage us to believe we had a true overlap. However, we did not have enough evidence to identify the indicator system.

Around the beginning of 1943 some of us stopped working in pairs or small groups, and I set out on my own. Maybe our nerves were beginning to show. Even before going to Arlington Hall, I had had my share of meetings with failure, but none so protracted as my experience with the 7890. (Fortunately my wife came to Arlington from Columbus, Ohio, and stayed for the duration. She was a welcome distraction from my cryptographic misfortunes. We luckily found an apartment near the post, and since she was an experienced social worker, she was soon happily employed in Washington.)

In February my life took another turn. Almost from the time I arrived at Arlington Hall, I felt somewhat uncomfortable not being in the Army Signal Corps. I worked daily among people in uniform, and some of them had become good friends. It is hard to define this longing to be part of the armed services in wartime, this feeling that somehow you belong there. It is a little like being homesick for a place you've never been. Anyhow, I wrote my draft board in Ohio and asked to be drafted, requesting also that I be inducted in Virginia. The board obliged; they did not care where I was inducted as long as they got credit for one more draftee. So in a few weeks I entered the army in Charlottesville as a private.

Northern Virginia can be cold in February, and at Camp Lee, where I was sent for processing, I found cold, drizzly weather and an outbreak of pharyngitis along with other draftees I was kept busy washing dishes and shoveling gravel while I awaited orders. Since Major Kullback had requested my assignment to his code unit, I anticipated a prompt release from Camp Lee and a rapid return to Arlington Hall. What I got was a bad case of pharyngitis and a slow trip to the hospital at Fort Myer. A few days later, however, I was fully recovered and back at my desk in the Hall.

My work was much as before except that some of my time was given to military exercises such as weapons training, drill sessions and special functions of the Second Signal Service Battalion, which intercepted messages for cryptanalysis. I enjoyed these activities. They were good for camaraderie as well as my health, both of mind and body. In March I received a corporal's rating (Technician 5th Grade), which was not very impressive at a post like Arlington Hall Station, but was progress.

The first signs of a breakthrough in the Japanese Army code came in mid-April. I had asked the IBM unit on numerous occasions for printouts of messages sorted on specific code groups in selected positions in the text. One of these revealed a pair of messages sent on the same date from Tokyo and apparently beginning with the same additive numbers. This could happen by chance, but more likely a Japanese operator carelessly or inadvertently sent almost identical messages without shifting to a different starting point in the additive table. The number of hits, or matching groups was much greater than random. Further IBM runs on these matching groups uncovered two more messages, neither sent from Tokyo, using the same table, but with different starting points. By close study of these overlaps I made a tentative identification of digits used in the indicator system.

Although I felt reasonably confident about these findings, I did not want to report them without strong confirmation. I requested a new IBM run based on the apparent indicator positions. Then unfortunately I had to delay the project for a brief time because my name was posted in the orderly room for KP duty. Privates and corporals were so few at Arlington Hall that they, and often sergeants, were expected to take their regular turn in the mess hall. Sometimes I was the subject of joking remarks that I was better at breaking dishes than breaking codes.

The new IBM printouts disclosed several overlaps that supported my conclusions about the indicator system. These were that eight indicator digits, which were scattered in the message heading as well as in the message text, represented the following:

BPPS RCRC

The B indicated which one of the ten additive books (1-0) was used to encode the message. PP revealed the page (01-00) in the 100-page book used to begin the message. The S was a sum check. For example, if BPP totaled to 23, S would be its last digit, or 3. The first RC marked the row and column on the additive page where the encoding began. The final RC denoted the row and column where the message ended.

I first showed these findings to Elmquist. He was an able cryptanalyst; moreover, I knew that if he found my solution to be flawed, he would react with understanding and useful criticism. More than once he had eased my depressed feelings with friendly encouragement. After studying my results briefly, Elmquist responded excitedly and hurried me off to report the news to Kullback. Unluckily, he was not available, as he was holding a staff meeting. To my dismay, Elmquist grabbed a sheet of paper and wrote in large block letters

WE'RE READING THE 7890!

Then he brashly shoved this premature announcement under the door. I never learned Kullback's immediate reaction to this interruption. Maybe I was afraid to ask. But he emerged from his office in about an hour and wanted an explanation of the note. After our brief account, he assigned Elmquist and Sam Snyder, an experienced cryptanalyst, to check my findings further and report back to him. They reviewed them carefully, and together we returned to assure a relieved Kullback that the Japanese Army Administrative Code had finally been broken.

As I walked the two miles home that day, I tried to sort out my feelings about the breakthrough. Relief, quiet elation, even a tinge of disbelief all combined in a kind of emotional tangle. I wondered, of course, what influence the breakthrough would have on the solution of the code. Most of all, I felt a bond, a sense of community, with the Allied forces who were enduring the hazards and hardships of the war in the Pacific.

The breakthrough apparently came on April 30, 1943. (Truthfully, I do not remember the precise date, but on May 10 I received a letter from the Commanding Officer at Arlington Hall Station, Colonel [later General] Preston Corderman, which began: "I wish to extend to you my commendation for your very important accomplishment of April 30....")

Kullback told me, along with some other military and civilian personnel, to report to the post that evening. It was a night to remember. Besides the Hall's headquarters officers, officials from the Military Intelligence Service and other Washington agencies attended the "celebration" of the long- awaited breakthrough. There was brisk discussion among them about it and its implications for their particular offices. I had little share in this, but sat at my desk observing the dignitaries and feeling somewhat out of place among them. Several came by to see the "evidence." On a table I had spread out several overlaps and circled the indicator digits with a red pencil. Most people looked at the exhibit like museum-goers scanning the Rosetta stone, not really understanding it, but realizing it was important.

The rapid decoding and translation of the Japanese messages was achieved mainly by a group of about 50 specially trained Signal Corps enlistees who had studied Japanese, mostly at Columbia, Harvard and Yale. They were supervised and given further intensive instruction in Japanese military language by Captain Edwin Reischauer at Arlington Hall. These talented students could read Japanese messages almost like natives. Reischauer later became a foremost advocate of U.S.-Japanese understanding, especially as our ambassador to Japan. The solved and translated messages of his group went to intelligence assessors and other military intelligence personnel for use in the prosecution of the war.

Reconstruction of the Japanese additive tables and codebook was aided not only by increasing traffic, but also by some curious practices of the Japanese themselves. Sometimes messages used stereotyped patterns or commonly repeated phrases. For example, Japanese syllables with the same pronunciation but different meanings were often identified after a code group meaning "open Chinese telegraph." Several groups later would come the group for "close Chinese telegraph." Decoders were adept at finding and interpreting such patterns on overlaps. Even more helpful was the carelessness of Japanese signal officers when they changed their additive books, usually every few months. Apparently they had complete faith in the security of their code and thought we were incapable of breaking it, because they regularly sent their new additive tables to their stations throughout the Pacific area by using the old tables to encode them.

As a result, we received their new additive books at the same time their cryptographers did. Probably most successes in military cryptanalysis benefit from the careless mistakes of the enemy. (I can recall a couple of rare cases in which the Japanese operator, apparently suffering from burnout or too much sake, omitted the additive and sent the raw code undisguised.)

By 1944 virtually every message between Japanese stations was being intercepted, and useful information such as the date, place of origin and destination of messages was being recorded on IBM equipment. On a high hill somewhere in America there ought to be a lasting monument to IBM and its creative personnel for their invaluable contribution to our success in breaking enemy codes in World War II. For all branches of the service, they supplied the materials, often ingeniously arranged and sorted, that aided the work of cryptanalysts. (I have sometimes wondered whether I could ever have broken into the Japanese Army Administrative Code without their help. It certainly would have taken a lot longer.)

The enormous volume of Japanese messages sent in a variety of codes and ciphers actually became a problem as well as an asset to the Military Intelligence Service. In describing "the importance of the 1943 breakthrough into the Japanese Army's signal traffic," the historian Ronald Lewin noted that at the direction of General George C. Marshall a "Black Book," summarizing intelligence information from the mass of intercepts, was sent daily to the White House. In a letter to President Franklin D. Roosevelt, Marshall wrote: "The recent discovery of the Japanese Army machine code has added a tremendous amount of such material.... The problem is how to avoid being buried under the mass of information, and I think the present arrangement satisfactorily meets that difficulty."[1]

A high proportion of Japanese Army intercepts did not qualify for Roosevelt's "Black Book" because they dealt with routine matters, such as local orders and reports. However, many messages had important operational value for Allied commanders. They provided information about the disposition and strength of Japanese forces, as well as troop movements and their strategic intentions. Other intercepts related to the enemy's shipping organization and the stocks of oil and other resources necessary for defense of its island garrisons. The sifting of all this material for useful intelligence was a prodigious task which required increasing numbers of intelligence assessors and other skilled personnel at Arlington Hall.

[1] *The American Magic: Codes, Ciphers and the Defeat of Japan* (New York: Farrar, Straus and Giroux, 1982) p.199. The entire letter is reproduced on p.201. I have frequently relied on Lewin's well-researched study for historical details about which I have little or no personal knowledge. The word "Magic" was a general term used to designate secret intelligence in both the European and Far East Theaters. The naval and military signal traffic of the Japanese was known as "Ultra."

"In July 1943, B Branch, the intelligence wing of the Army's Signal Security Agency [a later designation of the S.I.S.], numbered 1,804. By July 1944 it had expanded to 2,574, of whom 82 percent worked on the signal traffic of the Japanese Army."[2]

To illustrate the staggering volume of messages to be evaluated, we had a "show but don't tell" exhibit of overlaps which extended from the ceiling to the floor - overlapped messages eight feet deep using the same additive page. The display also illustrated how messages using additive could be read vertically before they were read horizontally.

After the breakthrough into the Japanese Army code, I found my later experiences at Arlington Hall somewhat anticlimactic. For a while I worked on overlaps to recover or confirm additive numbers. I was also assigned to instruct new personnel in my cryptographic section. By early 1944 I had been promoted to Technician Third Grade (which finally released me from KP chores, though I never found them very onerous). I spent the summer in Officers' Candidate School at Fort Monmouth, New Jersey, and was commissioned a second lieutenant in the Signal Corps in July 1944.

During the rest of my stay at Arlington Hall I had various supervisory assignments, but little opportunity for meaningful cryptanalysis. We also solved a few minor codes, but the Japanese Army Administrative Code dwarfed them in volume and importance. For several months I worked on a three-digit army code. Its solution was not especially difficult, but message traffic dwindled and was insufficient to warrant a serious effort to recover the additive and code books.

In the final stages of the war, when the Japanese empire was collapsing in the face of Allied assaults, my work in cryptanalysis virtually ended. In mid-1945, at the request of now-Lieutenant Colonel Kullback, I wrote a detailed analysis of the Japanese Army Administrative Code and its operation. I suppose it is somewhere in Signal Corps records. Later in the year, at a ceremony in Arlington Hall, I was awarded the Legion of Merit.

My final assignment was writing citations for military officers ending their wartime service. (The practice would prove useful later in my academic career.) In the early spring of 1946 I was promoted to first lieutenant, and in May I was discharged at the Pentagon. A week later, together with my family (which now included infant twin daughters), I returned to Columbus to write the doctoral dissertation (a study of 19th-century New England lecturers in the Middle West) that I had put aside four years earlier.

It would be difficult to assess the significance of the 1943 breakthrough and its

[2] Lewin, p. 198. The S.I.S. in 1942 became the Signal Security Service and was renamed the Signal Security Agency in 1943.

influence on the war. Such judgments are better left to scholars and historians whose knowledge of the war's events and personalities is much greater than mine. Several observations, however, can safely be made. The Japanese Army Administrative Code was one of a group of cracked enemy codes and ciphers, all of which had their special importance. Among them were diplomatic and navy ciphers and a water transport code, broken in early 1943, that traced the activities of the Japanese Army's own fleet of supply vessels. The war in the Pacific was mainly a maritime conflict, and the dramatic breakthrough into the main Japanese Navy code by Captain Joseph Rochefort in the spring of 1942 was of paramount importance.[3] Intelligence gained from the code was largely responsible for the American victory at Midway, and for many other successful attacks on Japanese-held islands. The solution of the main Japanese Army Air Force code in early 1944 added to the flood of information available to Admiral Chester Nimitz, General Douglas MacArthur and other Allied commanders.[4]

In the end, of course, the war in the Pacific was not won by cryptanalytic intelligence. It was won by superior strength on land, at sea and in the air, by astute military and civilian leaders, by a supportive civilian population, and by an industrial complex that could turn out five Liberty ships every day. Still, cryptography had its impact. In assessing the navy and army code breakthroughs of 1942 and 1943, Ronald Lewin wrote that, had they not been achieved, "the war in the Pacific might not have been lost, but victory would have been incalculably more expensive in time, effort and blood."[5]

In the years after the war I seldom thought about the Japanese Army Administrative Code or my experiences as a cryptanalyst. My civilian life was crowded with new interests and events. Moreover, I was so conscious of security that I avoided speaking to anyone about Arlington Hall. Occasionally there were unexpected reminders. For example, while waiting for a traffic light to change, I would be startled to see a half- remembered number, like 2042, on an automobile license plate. That one was the code group for "owari," meaning "end of mes-

[3]The mistreatment of Captain Rochefort and the struggle in the Office of Naval Communications in Washington to claim credit for his achievement are narrated in a number of war histories. A first-hand account is given in the memoirs of Rear Admiral Edwin T. Layton, "*And I Was There*"(New York: William Morrow, 1985). Layton, the Pacific fleet's Intelligence Officer, died in 1984, and his book was completed by Captain Roger Pineau and John Costello. Captain Rochefort, who died in 1976, was awarded the Distinguished Service Medal posthumously in 1985.

[4]Unlike Nimitz, Marshall and other leaders, MacArthur disliked agencies "whose responsibility did not run directly to himself," and was rarely willing to share credit for his striking exploits with military intelligence, though the intelligence gleaned from the Japanese Army Administrative Code and others must have been invaluable to him. In 1961 MacArthur was given an honorary Doctor of Laws degree at Michigan State University, where I was the English Department chairman. The citation which accompanied the degree was a well-deserved tribute to his military successes in the Far East. I enjoyed writing it.

[5]*The American Magic*, p. 18.

sage." I suppose that everyone who has been involved in cryptography is at times pursued by ghosts of the past. Now, like other people in their retirement years, I give more space in my life to reminiscence, and perhaps even nostalgia. When by chance I meet a veteran of the war in the Far East, I cannot help wondering whether, in some cryptic way, our lives may have touched a half-century ago.

BIOGRAPHICAL SKETCH

David Mead is a retired Michigan State University professor. In 1955-57 he directed an advisory group at the University of the Ryukyus, Okinawa, where he developed English language and literature programs. In 1964, as a Fulbright lecturer in the Philippines, he established the American Studies Association of the Philippines with his seminar students as its charter members. He is the author of numerous scholarly articles and books on American culture and co-author of the *Prentice-Hall Handbook for Writers*, a college text now in its eleventh edition. He was English advisory editor for Dodd, Mead & Company (1963-75) and editor of *The Centennial Review* (1966-81).

THE STORY OF THE HAGELIN CRYPTOS

Boris C. W. Hagelin

Edited by David Kahn

INTRODUCTION

"The Story of the Hagelin Cryptos" could never have been written if Russia after the revolution had become a democratic republic. The expected defeat of the Bolshevik minority and the return of a moderate government would have caused me to remain an employee of the Nobel Brothers Oil Production Company, and there would have been no Crypto AG - the present manufacturer of ciphering machines.

I was born on July 2, 1892, in Adshikent, a small summer resort in Azerbaijan. My father, a Swede who had been born in St. Petersburg in 1860, was at that time manager of the Nobel company's oilfields and refineries in Baku. Ha had joined the company in 1879 when it was founded by an older brother of the famous Alfred Nobel. My father was made a director of the company in 1899, and moved with his family to St. Petersburg. I went to a Russian school until 1904, when my father sent me to a Swedish school. After finishing this school I entered the Royal Technical University in Stockholm where I graduated in 1914 with a degree in mechanical engineering. It was taken for granted that my future would lie in Russia with the company, where my father held a leading position and had become the closest friend of Emanuel Nobel, head of this famous family.

My first job would have been the supervision of the construction of an electric power station on one of the Nobel oil fields in Baku. As I was a mechanical engineer, I had first to broaden my knowledge of electrical engineering. Nobels had ordered the equipment from the large firm ASEA in Vasteraos, and therefore it seemed to be a good idea for me to work in different departments of that firm for a year.

However, the Russian Revolution broke out, and it became obvious that the project would have to be delayed until normal conditions returned. I thus stayed on at ASEA in their foreign department since I spoke five languages.

In 1920 the Nobel family made an agreement with the Standard Oil Company of New Jersey for a future collaboration in Russia, and I was sent to the USA where I worked in their general engineering department during 1921. At the end of that year it became clear that stable moderate conditions in Russia could not be expected under the rule of the Bolsheviks, who had in the meantime confiscated all private enterprises, including the Nobel works.

I could have stayed with Standard Oil, but I did not feel at home in America and wanted to get back to Sweden. There was no future for me in the oil business but my ties with Emanuel Nobel remained unbroken. The Nobels managed to retain their non-Russian businesses, but had to live in exile, like my father.

Emanuel Nobel was very generous towards me and financed the establishment of a small engineering office in Stockholm. During my stay in the US I had acquired some inventions, which I developed and which made me financially independent. The decisive turning point in my life came, however, when Emanuel Nobel entrusted me with the supervision of a small company which he had begun to finance in 1921 – Aktiebolaget Cryptograph. This company was founded in 1915 with the objective to develop and manufacture ciphering machines invented by the Swedish engineer Arvid G. Damm. In 1925 I assumed the management of the company as well as the development of saleable products. This was a fascinating task although I did not have any knowledge of cryptography. Mr. Damm died in 1927. In 1932 A. B. Cryptograph was liquidated and replaced by A. B. Cryptoteknik.

A. B. Cryptoteknik manufactured only mechanical and electromechanical ciphering machines. After World War II the need for ciphered telegraph transmission became obvious. In order to be able to work without the interference of the Swedish government – ciphering machines were at that time considered war material – I decided to move to Zug, Switzerland. I first collaborated with the Swiss inventor Dr. E. Gretener, but later established a small independent laboratory. Crypto AG (Aktiengesellschaft) was incorporated on May 13, 1952, and had at first just one employee. My Swedish activities were transferred to Crypto AG , and since the name "Hagelin Cryptos" had become well known before World War II, the enterprise grew so fast that in 1966 a new manufacturing and administration building was built in the Steinhausen section of Zug.

The basis for the growth of Crypto AG was the mechanical machines. which were conceived and developed in Sweden – the original "Hagelin Cryptos." The first products of Crypto AG were the "Telecryptos". They were succeeded by the electronic ciphering equipment which, since the mid-1970s, has become predominant. The old mechanical machines are, in decreasing quantities, still being produced. Without their basis, however, the new developments which have culminated in a large number of different versions would never have evolved. The trade name "Hagelin Cryptos" has therefore been applied to all the products of Crypto AG . It would be presumptuous for me to claim the credit for the success of Crypto AG ; however, it was internationally acknowledged as the most important manufacturer of ciphering equipment before I retired from the company in 1970.

The art of invention – and I have mainly become known as an inventor – has for me been sudden ideas or the combination of existing designs for totally different purposes. But the hard detail work to obtain the finished products has been performed by many collaborators, for all of whom I retain a sincere thankfulness. In this context I feel that I must mention several names. First, C. A. Lindmark, who became indispensable to me when I developed my best known machine, the C-type, in 1934.

My son Boris Jr., who met a tragic death in 1970, had an inventive mind, and his contributions when the postwar CX-type was designed were very important. Oskar Stürzinger, my first employee when I began my activities in Switzerland, developed my original ideas for the Telecrypto machines. He also began to experiment with electronic ciphering devices, at first against my wishes, as this new technology was alien to me. To achieve success, however, it is not enough to have an excellent product but also superior management. Here, Sture Nyberg, who was first the manager and then the director of Crypto AG until his retirement in 1976, proved to be the ideal choice. He also contributed by analyzing the security of the machines which were produced.

In closing this introduction I must pay my humble respect to Emanuel Nobel and to my father, without whose financial and moral support during the long and not always easy years of my work with ciphering machines I would have remained an insignificant engineer.

– B. C. W. H.

I. THE BEGINNING

The use of cryptographic methods of various kinds stretches back some thousands of years. A few hundred years ago inventive minds began to make devices to facilitate the process of enciphering and deciphering messages. In 1891 the Frenchman Etienne Bazeries invented a device, which to the best of my knowledge was still used until World War II. The principle used by Bazeries had been propounded earlier by Thomas Jefferson for the construction of a cipher device, but his writings were not discovered until after World War I.

In Sweden too, interest existed very early in the use of mechanical devices for secure communication. My old friend Sven Wäsström, a Swedish cryptologist, has discovered in the national archives in Stockholm some very interesting documents on this subject. In 1786 Baron Fridric Gripenstierna received permission from King Gustav III to make a cipher device in accordance with Gripenstierna's drawings. In his letter to the King, Gripenstierna indicated that he had gotten the idea for the device from his father-in-law, the celebrated inventor Christopher Polheim. The machine itself, unfortunately, has been lost, but the invoice by the builder, which is dated August 26, 1786, shows that the device was actually built. A remarkably detailed description allowed Crypto AG to produce a replica of this machine.

Figure 1. Cipher machine of Gripenstierna

This device displays certain resemblances to the one developed much later by Bazeries but far surpassed the French device as far as security and method of operation are concerned.

Gripenstierna's device consisted of 57 disks, each of which carried two alphabets. These disks, mounted on an axle, could all be rotated independently of one other. A longitudinal slit with an index guide on each side of the device exposed two sequences of letters. The authorized person who was to perform the enciphering (at the right in Figure 1) adjusted the disks during enciphering so that the plaintext appeared row by row in "his" slit. The assistant (at the left in Figure 1) then copied down from the other slit the ciphertext, again row by row. During the decipherment the assistant adjusted the disks to show the ciphertext row by row in the slit on his side. The authorized person then copied down the plaintext row by row from his slit. Here we have a modern security concept which separates the ciphering process from the transmission of the cipher.

In the 19th century, more and more inventors concerned themselves with ciphering devices and machines. In the German patent office, which was founded in 1877, about 150 patents were issued before 1927 for cipher devices.

In this list the name of a Swede appeared in 1915: Arvid Damm. He was able to interest Swedish industrialists in his inventions and A. B. Cryptograph was founded in 1915 to exploit his inventions.

Figure 2. A-21 by Arvid G. Damm

His first fundamentally sound machine is shown in Figure 2. He used a revolvable drum, on which 26 alphabet strips could be fixed in any desired order. The alphabets were formed so that they could be considered to be a scrambled Vigenére square, with the alphabets in reversed order. Close to the alphabet drum a reference alphabet in the normal order (A...Z) was mounted.

When ciphering, the drum would be advanced one step for each operation, while the reference alphabet could take one of two positions. Through a slit at the top of the machine one could read off the letters to be ciphered from the reference alphabet to one of two drum alphabets. The positions of the reference alphabet were controlled by the low and high links of the chain, which advanced one step for each operation.

The length of the chain and also the position of low and high links could be varied within certain limits. These options as well as the possibility of changing the relative positions of the drum alphabets made it possible to obtain a large number of different cipher series. The use of reciprocal alphabets made it easy to use the machine, as the reading from the reference alphabet could be done both when enciphering and when deciphering. In emergency cases the machine could be opened and the chain disassembled into the individual links, thus destroying an important key element.

Figure 3. Office machine by Damm (cover removed)

Using the same principle, Damm built a number of office machines equipped with keyboards. When a key was pressed, a letter would be exposed below the alphabet strips. During encipherment, it would be a ciphertext letter; during decipherment, the plaintext letter (Figure 3).

Here too the chain was used as the controlling element. In operation the machine had a protective cover.

Damm also constructed another purely mechanical machine which printed both the plaintext and the ciphertext. Four of these machines were sold to Japan. Finally, Damm invented a system with so-called rotors, i.e., alphabet permutating wheels, an invention made at almost the same time by the American Edward Hebern, the Dutchman Hugo Koch, and the German Arthur Scherbius.

Figure 4. "Electrocrypto B-18" by A. G. Damm

A normal rotor at that time consisted of a disk that had on each of its two faces 26 metal contacts embedded around the circumference. These contacts were connected in pairs through the inside of the disk in a random fashion, one contact on one face connected to one contact on the other face. The best-known machine using rotors was the German Enigma, equipped with 3 or 4 rotors which were set to a new starting position for each message in accordance with the keying instructions. The machine was provided with a keyboard and a lamp field. Each lamp was arranged to light up one letter. To operate the machine only a flashlight battery was needed.

Damm's rotors were constructed somewhat differently from the Enigma's but produced a similar effect. They were made for machines which were intended for the ciphering of messages transmitted by radio telegraphy, because such messages could easily be intercepted.

Damm aimed to interest primarily the large telegraph companies that handled transatlantic telegraph traffic in particular. With such machines the radio traffic could have been just as secure as the cable traffic.

II. THE FIRST MACHINES

When in 1921 the capital of A. B. Cryptograph was exhausted, none of the stockholders would or could make further funds available. However, at this point Damm, the director of the company, succeeded in interesting Dr. Emanuel Nobel in the A. B. Cryptograph and its possibilities. Because of the Revolution the Nobels had lost their holdings in Russia: a large factory and the largest Russian oil firm, the "Nobel Brothers Oil Production Company".

With regard to the A. B. Cryptograph, Dr. Nobel consulted his colleague and close friend Karl Wilhelm Hagelin, my father, and they came to the conclusion that the possession of good cipher machines for written communications could play a decisive role in business affairs. Based on that conclusion Nobel decided to invest new funds in the A. B. Cryptograph, and my father also took a small part in the company. He was technically interested as he was an engineer himself.

Since Damm had a rather obstinate personality, Nobel and Hagelin needed a trustworthy person to monitor his activities. Because I was then living in Stockholm where I had opened my small engineering office, Nobel asked me to watch over A. B. Cryptograph.

When I joined the company, Damm's interests in cipher machines were concentrated in the field of radio telegraphy. When I first appeared at the company offices a complete layout was on display: a typewriter for the input of text, a cipher machine, and a typewriter equipped with electromagnets for the output of enciphered or deciphered text. Damm had succeeded in arousing the interest of the major radio telegraph companies in his machine and prototypes were to be built by one of these companies in Paris. After I joined the firm Damm moved to Paris, and I had to look after the business side of the firm and its technical development more than I had originally expected.

In the end Damm succeeded in winning the "big four" to his project: Marconi, Telefunken, TSF and Western Union. They financed the construction of four prototypes. But these did not appear to be reliable enough and were too slow, so the project failed.

In the meantime, however, Damm had developed a system with so-called simplified rotors, and some prototypes of these were built for wireless traffic. The machines consisted of three units: The keyboard, the ciphering portion with motor drive, and the output unit – either an electromechanical tape punch (a Creed Morse code punch), which was equipped with a magnetic control or an electric typewriter. Figure 4 shows an electric typewriter, keyboard and ciphering unit. This machine was Damm's Electrocrypto B-18.

The principle of the simplified rotor later became the salvation of the faltering business. But this came only after Damm's death in 1927.

In 1925 I was able to make my first positive contribution to the A. B. Cryptograph, and this proved to be of decisive importance for the future of the company. I happened to hear that the Swedish General Staff had received an Enigma machine for study and I rushed to visit the officer concerned with this matter. I explained to him that the A. B. Cryptograph already had ten years of experience in the field of cipher machines and that I would be able to offer something possibly superior to the Enigma.

The General Staff wanted our machine to be the same size as the Enigma and to operate in the same way. Time was short and I had only six months to produce the new machine. At the time we had only Damm's earlier constructions on hand, and they did not fulfill our needs. Nevertheless I promised to deliver. In those days I had no experience whatsoever in cryptography but I had a certain talent for tinkering: I believed I would be able to build a comparatively compact machine on the basis of Damm's "simplified rotors". This design made use of a 5×5 grid, whose principle is shown in Figure 5.

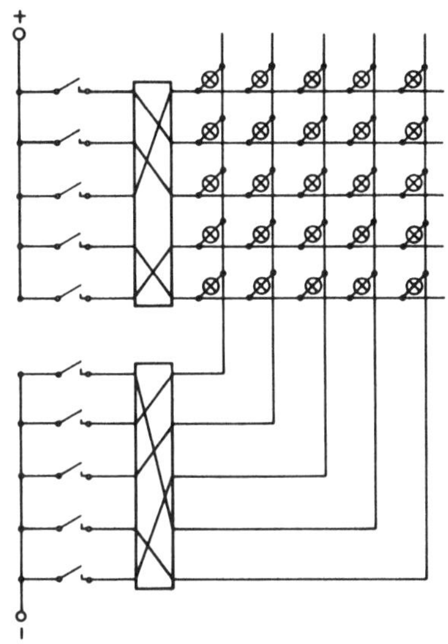

Figure 5. 5×5 matrix

To build the first model of the machine Nobel allowed me 500 kronor, then about $134.00(!) I succeeded in producing a prototype within the size limitations and in the time authorized, a somewhat primitive model but still adequate for the evaluation. This machine had a keyboard, 2 rotors whose stepping was controlled by two separate pairs of pinwheels (keying-wheels) with different divisions on their periphery, and a display of 25 electric lamps which served to indicate the output letters for encipherment or decipherment (Figure 6). This prototype was analyzed by a mathematician, and approved instead of the Enigma for use by the General Staff.

Figure 6. B. C. W. Hagelin's B-21 (prototype)

Because Damm used a 5×5 grid, it was necessary to eliminate one letter of low frequency or to use the same signal for two letters, e.g. for i and j. The keyboard was connected to two groups of five contacts each. By pressing a key, one contact was closed in each of the two groups, forming one of the 25 possible combinations. After the closing of two circuits through the two rotors, and a further circuit through a relay system, one of the 25 electric lamps lit up and indicated the resulting letter. The relay system was necessary because at that time there were as yet no devices like modern diodes, which would have simplified the construction considerably. In order to be able to vary the enciphered alphabets, so-called modificators were added in series with the rotors, i.e., interchangeable leads which could be plugged in as desired. This provided $5! = 120$ combinations for each group and for both groups together yielded 14,400 possibilities.

The designation "pinwheel" mentioned above needs to be explained since the pinwheel came to be used later in many other machines. It is a disk that carries on its periphery a number of axial holes or slots in which pins are located. These pins are so arranged that they can protrude on one side or the other of the pinwheel. One side is the "active" side, i.e., pins that protrude on this side exert a controlling effect. The pinwheels are stepped with each operation, each wheel moving one pin position during each operation. The cyclic lengths of the wheels have no common factors. Thus as several wheels work together, a very long period for the key is obtained.

Figure 7. B-21 with solenoid magnets (series production model)

The prototype had four pinwheels, with cyclic length of 17, 19, 21, and 23, which yielded a combined keying cyclic length of about 1.5×10^5, or 150,000 operations.

By using different combinations of pin settings theoretically $2^{17} \times 2^{19} \times 2^{21} \times 2^{23} = 2^{80}$ or approximately 10^{24} different period characteristics could be obtained. This meant that an enormously large number of variations was possible. The machine was given the name B-21.

Figure 7 shows the final model, in appearance quite similar to the Enigma. It used a power supply of 110 or 220 volts for a solenoid magnet which operated the stepping of the rotors. For the lamp display a flashlight battery was utilized.

For the convenient use of this machine in central offices with extensive operations, it was connected to a Remington electric typewriter instead of the lamp display. Typewriter keys were activated by small electromagnets and the lamp display was shut off.

Figure 8. B. C. W. Hagelin's B-211

III. THE B-211 MACHINE

After Emanuel Nobel died in 1932 his heirs did not want to make any more funds available for A. B. Cryptograph. I myself had been working without pay because the firm's income had been very meager. In spite of the limited possibilities, I was able to make several trips to other countries where I found buyers for the B-21. The decisive success, however, came from the interest of the French Army in our machine. Before the machines could be sold two requirements had to be fulfilled: the machines had to print the text and had to be portable. This called for the implementation of an electromechanical drive. I succeeded in solving this problem in a short time by replacing the lamp display with a special typewheel printing mechanism (Figure 8), which I had designed.

Since the machine, which we called the B-211, was to be built in France, we got as manufacturer the French subsidiary of the Telefon - A. B. L. M. Ericsson, in Colombes just northwest of Paris.

The B-211 was provided with shift keys like those on a typewriter. During encipherment only letters were printed as output, but during decipherment letters, numbers and symbols were printed. The machine also had a crank so it could be operated by hand in case the electric power failed. For the ciphering circuits a flashlight battery was used, as in the B-21 machine.

Before the outbreak of World War II about 500 B-211 machines were built and delivered. It was through the financial support of my father, who lived in Paris at the time, that this business could be carried through. It would not have been possible without his encouragement and funds, and it was not easy for him.

When World War II broke out, my father at the last moment succeeded in transferring the profits from France to Sweden. The amount was large enough to pay his expenses and to install and equip a modern workshop. It was our first workshop since the founding of the A. B. Cryptograph in 1915. The shop was dedicated on New Year's Eve of 1940 by the heirs of the A. B. Cryptograph and was renamed the A. B. Ingeniörsfirman Cryptoteknik.

After the war the French army bought 100 more B-211 machines.

I also might mention that I was obliged before the war to sell "two cipher machines" to the Russian Trade Commission in Stockholm (according to the purchase order). I sold them two B-211 machines which we had in stock in Stockholm. They were copied (with changes) in Russia and used during World War II. They were provided with a 5×6 grid. The Cyrillic alphabet had more than 30 letters but a few of them occurred very seldom, and could be omitted. Also the modificators were plugged in outside of the machine, probably under locked cover so that the operator would not know which inner settings were in use at any given time.

IV. THE C-MACHINES

In 1934, quite soon after the beginning of our business relationship with the French "Deuxième Bureau" I was asked if I could develop a compact pocket cipher machine that printed. As good luck would have it, I had already a few years before received from two Swedes an order to design a coin changing machine.

After I had built a functioning prototype, I got the idea that it should be possible to insert any amount in coins, to enter the invoice amount by a key, and finally to get back the change. This machine was equipped with a unique calculating mechanism but it was never built because we had not been reimbursed for our development effort. We agreed to cancel the debt and retain the rights for our designs.

Figure 9. B. C. W. Hagelin's C-35 (closed)

Trying to find a solution for the pocket device for the French, I got the idea of incorporating the calculating mechanism for the money changer into such a device. The "revolutionary invention" (in the opinion of distinguished cryptologists) consisted in a pure association of ideas that were transferred from one field of application to a completely different one.

The calculating mechanism of the money changer consisted of a drum with displaceable axial bars around its circumference which could be affected by operating the keys. Those bars that were affected in an operation were displaced to the left as the drum turned a typewheel meshing with the drum was turned the same number of steps as the number of bars moved to the left. I replaced the keys of the money changer with pinwheels, and the typewheel carried letters now instead of numbers. In this way I obtained a cipher machine. In order to estimate the dimensions desired, I cut out a wooden block to fit snugly into the pocket of a jacket. For the first models I used a drum with 25 bars and five pinwheels of the same construction as those in the B-21. These wheels were provided with pins around their circumferences, which could be placed axially in two different positions, either "active" or "inactive". "Active" pins worked upon the drum by means of a control arm.

The bars of the drum were provided with cams, i.e., raised areas on their edges, of such a design that the first bar would be moved by the control arm of the first pinwheel, the next two bars by the second wheel, the next four by the third, the next eight by the fourth, and the remaining ten bars finally by the fifth wheel (Figure 10).

Figure 10. B. C. W. Hagelin's C-35 (open) with Thai alphabet

Depending on the number and the arrangement of active pins in the pinwheel in question, the control levers were pushed against the periphery of the drum, or remained in their inactive positions. When the drum was revolved during an operation, the active levers pushed the corresponding bars (from 0 up to 25 bars) to the left and so engaged the gearwheel connected to the typewheel. The typewheel was therefore displaced a number of steps equal to the number of the active bars. The drum thus acted like a gearwheel with a variable number of teeth. As the numbers of pin divisions on the pinwheels had no common divisor, the length of the key period, i.e., the number of operations before the pinwheels returned to their common starting position, was very long.

For the first machines the divisions, or cyclic wheel lengths selected, were 17, 19, 21, 23, and 25, which resulted in a period length of 3,900,225 operations, a length which until then had never been achieved in any mechanical ciphering machine.

Moreover, since one could set all the pins theoretically in about 10^{29} different combinations, the number of possible variations was so high that these machines are fully adequate for certain purposes even today.

Figure 11. B. C. W. Hagelin's BC-543

Later I made various changes in direct response to the steady progress in cryptanalysis, which included the significant work of the Swedish cryptologist Yves Gyldén. The main improvement consisted of increasing the number of pinwheels from five to six, and to introducing moveable lugs on the drum-bars instead of fixed cams.

When the machine began to be produced in larger quantities, a bottom plate and a protective cover were added. The bottom plate was formed so that the machine could be used in the field strapped over the knee of the operator. If necessary the operator could even march with the machine strapped to his thigh. This new type of apparatus was given the designation C with one letter and a numeral suffix which originally designated the model year but after 1952 defined the specific model of the machine: e.g., CX-52.

The opportunity to establish a modern workshop with the royalties from our French sales occurred just before the outbreak of World War II. It thus became possible for us to make and sell a considerable number of machines during the war. Our first large customer was the Swedish defence forces, with whom I had maintained connections since 1925. From France we received an order for more than 5000 machines, which were to be manufactured by Ericsson at Colombes, but this order could not be completed.

Figure 12. B. C. W. Hagelin's BC-543 (open)

To promote the C machine I made several trips, first within Europe, but in 1937 to the United States, where the wish was expressed to have an electrically powered machine with a keyboard. In the summer of 1939 I took a prototype of the so-called BC machine to Washington. Unfortunately the construction needed improvement and I returned to Sweden that autumn after the outbreak of war with matters still unsettled.

Figure 13. USA version of the C-36: M-209

I made another trip to the US at the beginning of March 1940 on my own initiative without any request from Washington. I was able to leave, with two machines in my luggage, from Genoa on May 10, 1940, on the last ship from Europe before the Italians entered World War II. This trip was to lead to the largest sale of C machines ever made.

The American business began with an initial order for 50 machines which were shipped airfreight from Sweden to Washington. After extensive testing the machine was accepted. The Americans selected it for tactical use, as they did not have any comparable machine at the time. The C machine, designated by the U. S. Army the M-209, was manufactured by the Corona, New York, plant of the L. C. Smith typewriter company, with a daily output of up to 500 units.

I remained in America during the war and had a small shop in my home for servicing the BC machines that were used by an American organization during the war.

Figure 14. C-41 (German copy of the BC-543)

When it became possible for me to return to Sweden in 1944, more than 50,000 machines had been built. By the end of the war more than 140,000 units had been made in America.

During my four-year stay in the States, the workshop in Stockholm was kept busy with orders from several foreign countries. One delivery went by an extraordinary way to Japan. The machines were smuggled out by the Japanese military attaché in a night boat passage and were picked up by a U-boat off the coast of Sweden. But very few machines reached their intended destination in Japan.

It seems worth mentioning that the German authorities, who years before the war showed no interest in the demonstrated C machine, began, toward the end of the war, in the Wanderer works at Chemnitz to manufacture a copy of the BC machine for their own use (Figure 14).

When the Third Reich collapsed it had been able to manufacture only about

700 machines. It is even more interesting to note that in the 1950s, a postwar variation of the C machine was built in Germany under license. About 10,000 machines are said to have been produced. France also obtained the manufacturing rights for the C machine for the francophone community.

It should be stressed that the C machine was planned for tactical purposes, i.e., for use at the front. When, because of its simplicity and portability, the machine also proved to be interesting for use in the diplomatic service, improvements were needed to attain a higher degree of security.

I got the idea first of modifying the machine so that the rotation of the pinwheels would be irregular. Other improvements were introduced as well. The new type brought no large changes in the basic structure of the original C machine. This new machine was given the designation CX-52; later, after different pinwheel movement variations were made, new model numbers were used.

Before going into the details of the various improvements, it is important to emphasize that the requirements of our customers led in two opposite directions. One group required very long key periods, while the other wanted a movement of the pinwheels that was as irregular as possible. It required great mathematical competence to compile keys that would assure that sufficiently long periods were obtained. In order to satisfy our customers we also developed hybrid systems where usually one pinwheel was stepped regularly and the others irregularly.

The further development of the C machine, which even today, in the age of electronics, is being used, can now be followed in some detail:

a) The number of pinwheels was increased quite early to six. Instead of the original five wheels with divisions of 17, 19, 21, 23 and 25, 12 pinwheels became available with divisions of 25, 26, 29, 31, 34, 37, 38, 41, 42, 43, 46 and 47, of which six were to be used at one time, preferably those with no common factors. The wheels were inserted easily without using a tool.

b) The number of slide bars was increased to 32. In place of the fixed cams on the bars which caused the typewheel to turn, movable lugs were installed on each bar so that the movement arrangement for the typewheel could be changed as desired. For other ways of using the machine, some bars with fixed cams which control irregular forward movement of the key wheels were manufactured.

c) Even in the very first C machines, reciprocal alphabets were utilized. With a letter sequence on the typewheel running in the opposite direction to that on the indicating disk, the machine could be used both for enciphering and deciphering without any modifications. In later models a second typewheel was introduced, which was permanently connected to the indicating disk, and which printed the same letters which were set with the indicating disk. They were called "primary" and "secondary" typewheels, where the primary typewheel is connected to the indicating disk. There are also available indicating disks and typewheels, in which the letters can be rearranged at will, and without removing them from the assembly.

One model printed the plaintext in the characters of a non-European language (e.g., Arabic) and the ciphertext in Latin letters, and one text (Arabic) read from right to left and the other text from left to right. This arrangement was indispensable for Arab countries using the international telegraph service.

d) Between the indicating disk with the primary typewheel and the secondary typewheel was a disconnectable clutch. It could either retain the chosen relative positions between the primary and secondary typewheels during a series of messages or change each time at the start of the next letter in the text.

e) For the easy setting of the pins on the pinwheels a special tool was provided. This tool set in a single operation all the pins in one wheel.

f) A special device which was developed at the request of one customer was the mechanization of the one-time pad system, which had been a slow hand procedure.

Near the end of World War I the American Gilbert S. Vernam made the first attempt to mechanize this system for telegraph use. He used for the key tapes punched paper tapes with random sequences of teleprinter signals. The characters of the plaintext in the standard CCITT No. 2 code were combined by means of relays with the key tapes according to the equivalence truth table (compliments of the "exclusive-or"), associating plus and minus in the following way:

+ and + results in +,
+ and - results in -,
- and + results in -,
- and - results in +.

After World War II we learned that special devices had been built to generate such tapes. Consequently out firm also developed special electronic random generators for producing key tapes. In part VI below, as a curiosity,

I have described two of my designs for the automatic production of one-time pads.

g) While the pre-war C machine was supplemented with the keyboard BC-machine, for the modern C machine we developed a separate attachment, which was provided with keyboard and electrical operation. In smaller message centers the C machine alone was sufficient, while in large one the C machine was mounted on this device.

h) The PEB machine provided easier operation for telegraphic traffic. The device consisted of a tape punch and a tape reader and could be connected by a cable with a BC machine. During encipherment a punched tape with ciphertext was made on the PEB (possibly with the address in plain language placed at the beginning) and was dispatched directly over the punched tape transmitter of the teleprinter station. At the receiving end the cipher was punched on the tape which could then be deciphered automatically with the PEB (Figure 15).

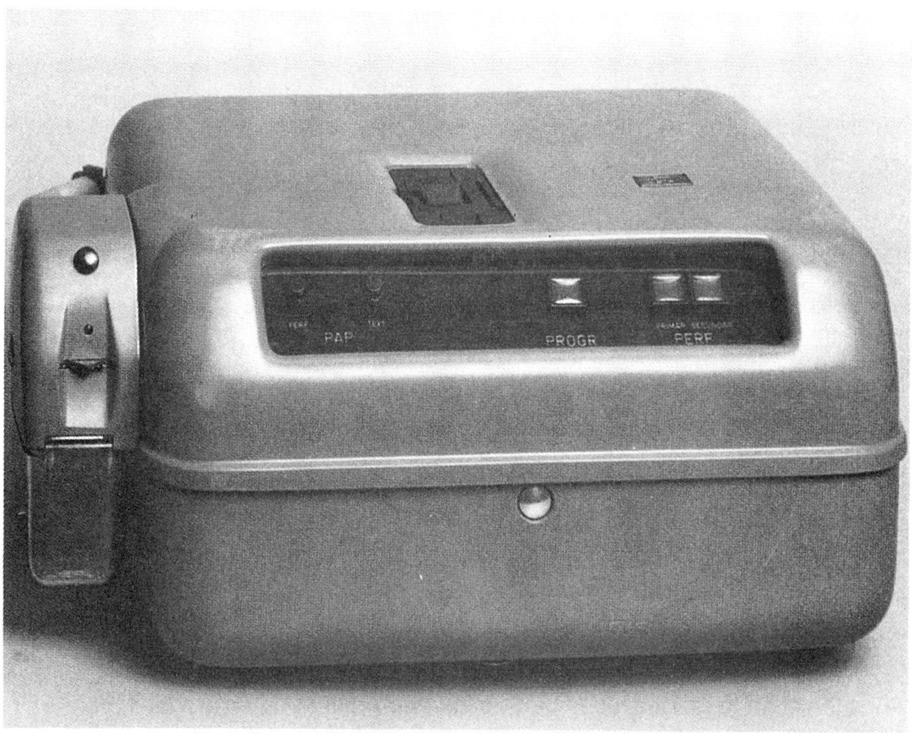

Figure 15. PEB-61

The various improvements caused the C machine to outgrow its pocket-size format. Now I wished to build a "pocket machine" in the true sense of the word. I was able to introduce a model which, because of its small size, could not print but nonetheless allowed an easy read-off of the text letter by letter. The cipher mechanism was quite different from that in the C machine but the device was fully compatible with a contemporary variation of the C machine, which worked with regular stepping of the pin wheels.

The French gendarmerie urged us to build this pocket machine and assured us of an order of at least 4000 units. Although this quantity was not actually reached, lots of these machines were sold, which came on the market under the designation CD-55 and CD-57 (Figure 16). About 12,000 have been delivered.

Figure 16. CD-57

The input and output elements of the device consist of a ring inscribed with an alphabet and rotatable disk inside the ring. A levergrip pressed by the left thumb displaces the alphabet disk in each operation. With the right hand the resulting text can be copied letter by letter. If the alphabets are not reciprocal the encipherment must be done from the ring to the disk and the decipherment from the disk to the ring. For convenience users normally use reciprocal alphabets and then can go from the fixed ring to the movable disk for the output. Even these machines can be equipped with a keytape reader.

Because the C machines have been sold in more than 50 countries and are still being sold, the characteristic components of these machines can be described in more detail. These parts are as follows:

1. Setting disk or keyboard plus typewheels. When the alphabet consists of 26 letters, these letters can be arranged in 26! or about 4.03×10^{26} different sequences. Moreover the indicating disk can be put into 26 different positions relative to the typewheel. There is also a mechanism that allows this relative position to change between the primary and secondary typewheels continuously during the processing of a message.
2. Thirty-two slide bars in the drum. Here 6×10^{24} possible arrangements of the lugs are available when all steps from 0 to 25, i.e., 26 steps, are included. These can be permuted in 720 different ways. Special cams on the drum bars can also be used to effect the movement of the pinwheels. These particular slide bars are equipped with cams which work as teeth.

   There are four possibilities of action by such teeth:

   a) The tooth moves the associated pinwheel when the slide bar has moved into active position.
   b) The tooth moves the wheel when the slide bar has remained in inactive position.
   c) The tooth moves the wheel no matter what position the slide bar is in.
   d) The tooth does not move the wheel no matter what the position of the slide bar (see more below on the performance of the CX and CXM). The possible actions just described make $4^6 = 4096$ possibilities for one slide bar or $(4^6)^{32} =$ about 4×10^{115} possibilities for the 32 slide bars.
3. Pinwheels. Here we have the largest number of possible variations. There are 12 different pinwheels to choose from, and 6 wheels are used at one time in the machine. The 12 different wheels, used 6 at a time, make 663,280 different combinations, giving period lengths between 10^8 and 2.7×10^9. The number of starting positions for the pinwheels is the same as the number of different combinations, but the number of possible different pin combinations reaches astronomical numbers. Assuming half of the pins on each wheel are active, which has been found to be most secure, we get for example with the wheels 29, 31, 37, 41, 43 and 47 about 10^{67} different pin combinations.

The C machines, as already mentioned, were originally designed for tactical use but gradually began to be used in the diplomatic services. Every cipher service has to set its own rules; nevertheless, the numbers given above showing the abundance of different possibilities for setting the machine may be of general interest.

There are three main varieties of the C machine:

The normal C machine (as well as the CD-57) when using regular movement of the pinwheels displays very long periods before the key repeats, but there are subperiods. For this reason messages with the same pin settings on the same wheels should not be too long, and the pin settings on the pinwheels for the important traffic should be changed frequently. In addition all other possible variations should be kept in mind and used.

The CX machine with irregular movement of the pinwheels brings the advantage that non linear movement sequences can be obtained. If certain mathematical conditions are met in the compilation of the material for the keys, then the periods occurring will be sufficiently long. In simple terms it can be said that the length of the period is determined by the number of pinwheels which control another pinwheel.

The CXM machine is a compromise between the C and CX machines. The movement scheme for the pinwheels here is as follows:

The wheel I (from the left) turns regularly one step in each operation. This wheel controls wheel II. Wheel III is controlled by wheels I and II, wheel IV by the three preceding wheels, wheel V by its preceding wheels, and wheel VI by its preceding wheels. All six wheels having the same number of divisions, the length of the period will be n^6, or with six wheels with 47 divisions a length of about 10^{10}. With this system only the first wheel has subperiods which affect the drum slide bars which are controlled by this wheel, while the remaining pinwheels get an irregular movement. With this machine somewhat simpler keying instructions can be used than with the C or CX machines without endangering security.

The three machines offer various advantages, and each user has to decide for himself which machine he wants. He has also to set up the instructions for the safe use of his machines, which is dependent on their deployment.

The numbers given in this section are hard to comprehend. Even the number 10^{15} corresponds approximately to the distance from the earth to the sun in millimeters. The numbers cited have significance only as far as they show that the possibilities for all machines of the C-type are practically inexhaustible. But these numbers are meaningless if the user does not carefully accept and exercise the instructions and does not make full use of the possible variations.

V. THE TELECRYPTO MACHINE

In 1948, as mentioned in the Introduction, I settled in Switzerland where I worked for a short time with the Swiss inventor Dr. Edgar Gretener on the development of machine for automatic enciphering and deciphering of teleprinter messages. These on-line machines were connected between the ETK teleprinter, designed by Dr. Gretener, and the line. Our cooperation did not last very long and I decided to build my own machine for use with all standard teleprinters. Dr. Gretener's machine is no longer in production. My preliminary work on a comparable machine was done in my workshop in Stockholm, but the manufacture took place at Crypto AG in Zug where some improvements were made to the original design.

Two types were built under my direction: Type T-52 and Type T-55. Both were installed between the teleprinter and the line, so that the teleprinter personnel worked only with plaintext while over the line only ciphertext was transmitted. Both types used as their key generator the C machine drum and six pinwheels. The T-55 was also equipped with two tape readers, one for the text tape and the other for a random key tape. It was possible to work either with the key generator or with the random key tape.

The T-52 machine (Figure 17) was produced in series in Zug in 1953 and 1954. It worked with six fixed pinwheels and a drum with 2×12 slide bars, using the principle of the C-36 machine.

The T-55 was designed in 1955 and 1956. It used six interchangeable pinwheels and a drum with 22 slide bars like the system in the C-52 series. The keying sequence was generated in both types mechanically, the information processing electrically by relays.

In the T-52 each telegraph signal resulted from a function step in which each telegraph signal was processed directly and passed forward. The keying signals produced were passed into the five keying signal relays, combined with the incoming telegraph signals according to the "exclusive-or" rule, and sent out. The T-52 could be set by hand to "Send", "Receive", or "Punched Tape Operation".

In contrast to the T-52, in the T-55 (Figure 18) the keying signals were not stored but were produced continuously as five keying signal elements during one cycle of the machine and combined according to the "exclusive-or" rule with each binary element of the telegraph signal. Therefore the slide bar drum had to run synchronously with the teleprinter signal, and consequently the five keying information elements always coincide with the text elements.

In addition the T-55 was equipped with a tape reader by means of which a superencipherment, i.e., mechanical keying generator plus random key-tape was

Figure 17. Telecrypto T-52 with and without hood

obtained. As an important improvement over the T-52 machine the T-55 offered automatic direction reversing.

Another important feature of the T-55 was the remote lock for the answer-back unit of the teleprinter. When the T-55 was not in use it remained on "plain language operation". Should this machine be called, then after operation the WRY-Key at the transmitting teleprinter a "Call" button at the T-55 was pushed which locked the sending loop of the remote station. By this action the incoming transmission could not be jammed by an answer-back sent during the transmission.

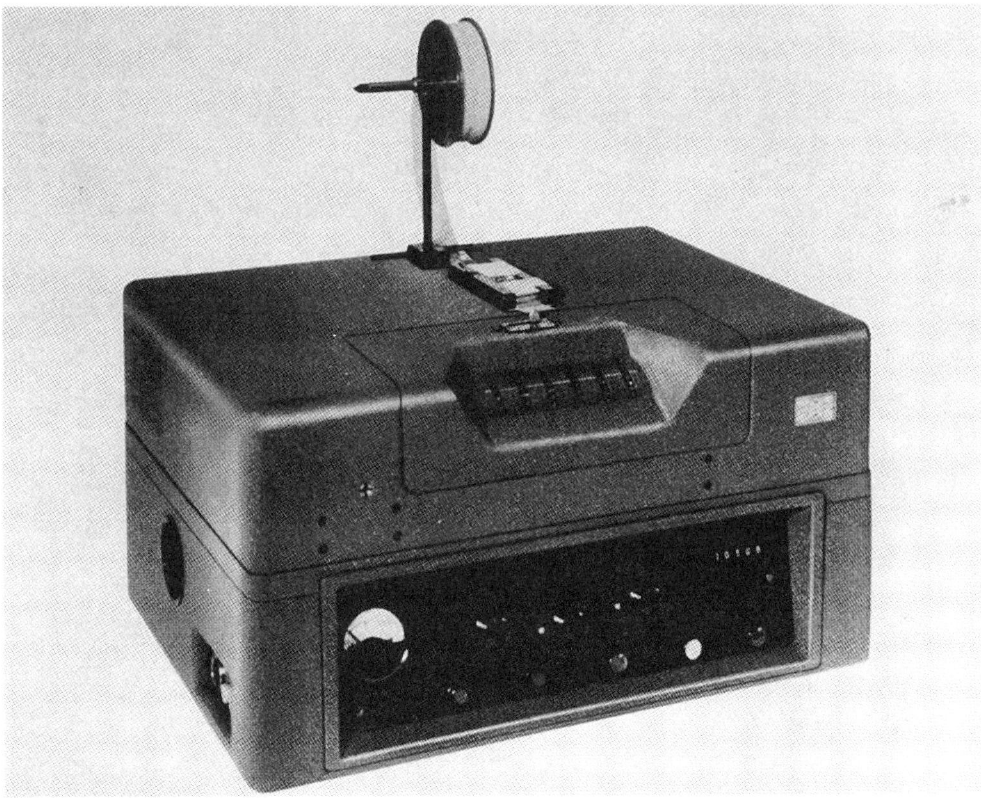

Figure 18. T-55 - office version

Later the tape could be deciphered in a local loop. The answer-back lock was released automatically when the lines were disconnected, and the station was the ready to operate in both ways again.

To produce the key tapes for these machines we constructed a three-part apparatus: an electronic random number generator, a punch, and a labeling machine. The punch, of American design, could do up to 120 characters per second. It produced two identical tapes at the same time, which were printed upon at regular intervals with consecutive identification numbers according to customers' specifications.

The use of random key tapes became very popular. But after some years the use of this system declined because of the problems of supplying and distributing key tapes. Key tape machines were then replaced by electronic telecrypto machines.

Both machines were single-shaft regenerators in terms of telegraphic technique. They delivered error-free telegraphic impulses, which left no possibilities whatever for electronic cryptanalysis by means of pulse analysis.

Figure 19. T-55 - field version

VI. MISCELLANEOUS MACHINES

Some machines are described which were built only as prototypes or in small numbers.

1. An Autokey Machine of the C-Type

Customers have always looked for cipher systems which are not only secure but also simple to operate. An example of such a system is the autokey cipher. The correspondents utilizing this system used code words to indicate the starting point in the keying sequence. It soon appeared that this system was not absolutely secure since probable words could help to analyze the ciphertext, i.e., to make decryption trials in order to discover where such words or sentences are located in the secret message. This system was thus discarded.

But the idea of integrating the text into the keying sequence was very attractive to me, and in 1948 I designed a machine that was built only as a prototype. The basic principle of this machine was to combine the text to be enciphered with a classical cipher system in such a way that the text component of the keying sequence would not be vulnerable to analysis. The keying sequence of a classical machine was thus changed into a pseudo-random keying sequence. I use the expression pseudo-random since the sequence of letters in plain language can not be considered randomly distributed. However, from a practical viewpoint the combination of a continuous keying sequence with this text to be enciphered could in practice be called randomly distributed.

For the classical cipher machine I used a normal prewar C machine with five pinwheels and a drum with 25 slide bars. This machine was modified so that in place of the toothed wheel drive, a drive using a latch drive for the regular movement of the pinwheels was used. A drum equipped with 26 slide bars was placed in front of the latches (Figure 20).

This auxiliary drum was connected via gears to the indicating disk, so that for each change of the indicating disk a corresponding auxiliary slide bar came into active position before the latches. The auxiliary slide bars were provided with fixed lifters. As an example the latches for pinwheels 1 and 3 of one slide bar could correspond to the letter A of the setting disk and could be found in the input position. They could then be brought into an active latch position by a movement of the drum, so that in the succeeding operation the pinwheels 1 and 3 would be moved one step forward. These slide bars were interchangeable, so they could be set into 26! different sequences in their drum.

The machine functioned as follows:

Before starting an encipherment the pinwheels were set into their specified positions. Then the letter to be enciphered was brought by a turn of the indicating disk into its input position. The slide bar on the auxiliary drum corresponding to the letter was advanced to the active position. Then the rotation of the main drum followed, which turned the setting of the indicating disk-typewheel unit the number of steps determined by the pin arrangement on the pinwheels and the distribution on the main drum. The resulting cipher letter was printed and the auxiliary drum directed against the latches to advance those pinwheels affected by the active latches one step forward at the end of the operation.

Figure 20. C-36 with Autokey (prototype)

In decipherment the active time for the movement of the latches was shifted to the beginning phase of the operation. After setting the cipher letter into the input position the turning movement was performed first, then the readjustment of the type wheel unit, and finally the printing of the deciphered plaintext letter.

However, I reluctantly discarded this device because transmission errors and garbles causing the keying to get out of step could lead to considerable delay in recovering the message. Today this objection is no longer valid since transmission technique, either by wire, radio, or even via satellite, has been improved immensely.

I regret that this simple and good machine came too early, and today has been outdated by its electronic successors.

2. The Ciphering Teleprinter

In 1953 and experimental ciphering teleprinter, Type TMX, was built.

It had four input possibilities: keyboard, text tape, keytape, and incoming signals, and three outputs: printout, tape punch and outgoing signals (Figure 21) I soon realized, however, that Crypto AG possessed neither the financial nor the technical means to compete with such a machine against the major teleprinter firms. We still have the prototype, but we dropped this large project and undertook something more suitable to our capabilities: the manufacture of a telecrypto machine which could be connected to normal teleprinters.

Figure 21. Front view of Cipher Teletype TMX-53 (prototype)

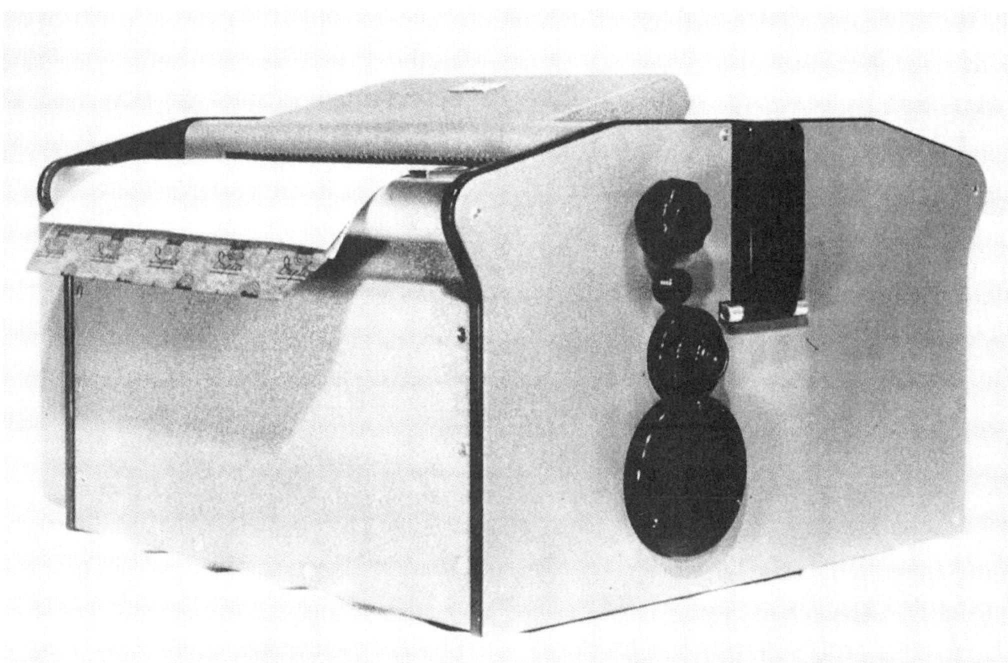

Figure 22. Rear view of Cipher Teletype TMX-53 (prototype)

3. "One-Time Pad" Machines

Before any electronic random generators existed for making keytapes, I made two attempts to build devices for printing one-time pads. Such pads at that time were made either by laboriously rolling dice or by haphazardly typing sequences of letters on a typewriter. It was found that with the latter method patterns developed which made such pads unusable.

The first device was rather primitive. It consisted of a shaft on which 50 typewheels in 10 groups were placed. Since these typewheels were not strictly uniform in weight, they gathered different momentums when the shaft was rotated. There was also a different friction between them and the axle-shaft. In each operation the shaft received a strong turn. The typewheels rotated at the same time and after the shaft was stopped they would run on at different speeds until locked. A line of ten 5-letter groups was then printed. By repeating the same operation many times a cipher table was formed. It may be mentioned that even today the same method is used for the winning numbers for lotteries.

The second design was even more unusual. typewheels 8×5 were used with each wheel coordinated to a mixing chamber which held 26 steel balls of which one ball was somewhat larger than the other 25 balls (Figure 23). For each operation the balls were mixed and run into a tube until the thick ball blocked the gate to the tube. A blocking arm obstructed the tube just below the thick ball. Then the number of free balls was determined in such a way that the associated typewheel received a corresponding turn. A sequence of 40 letters was printed on the paper, the balls were expelled from the tube back into the chamber, and the operation could restart.

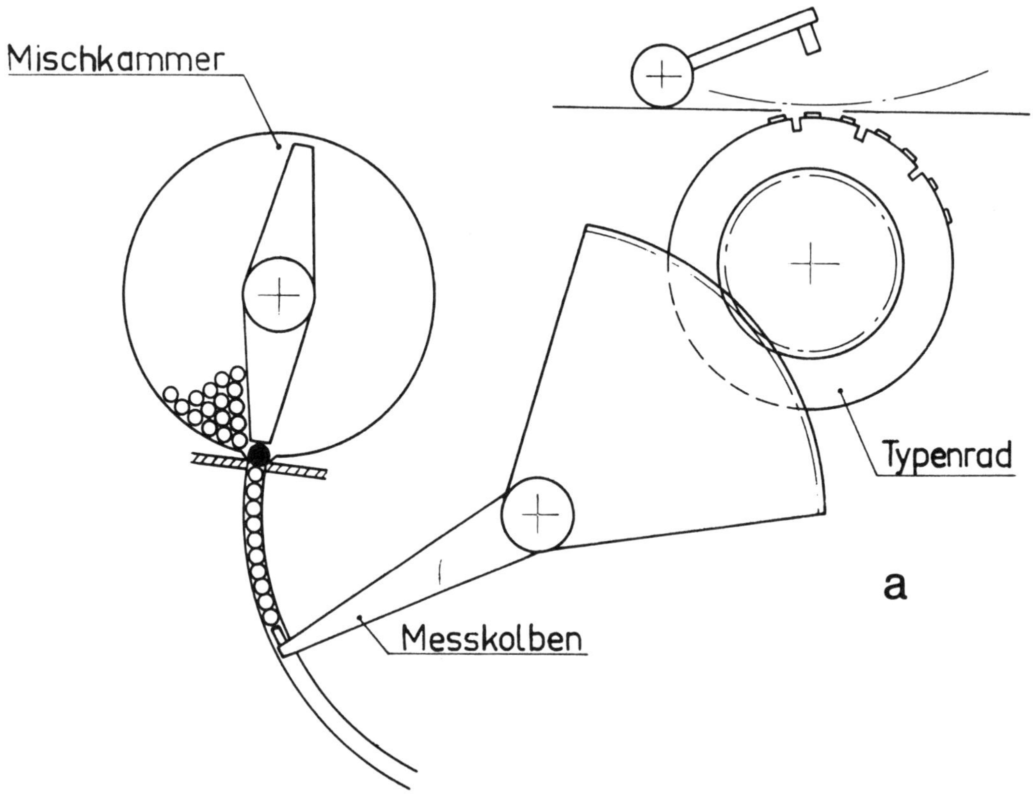

Figure 23. Random Number Generator with Printer CBI-53 (principle).
Mischkammer = Mixing Chamber; Messkolben = Blocking Arm; Typenrad = Typewheel

Each operation required about 7.5 seconds for the printing of one row and this speed corresponded to the capacity of a normal teleprinter. About 10 machines of this type were built, and some of them are still in use (Figure 24).

The chamber system as well as the typewheels could be changed for either 26 letters or 10 digits 0…9 so that either letter or number tables could be produced.

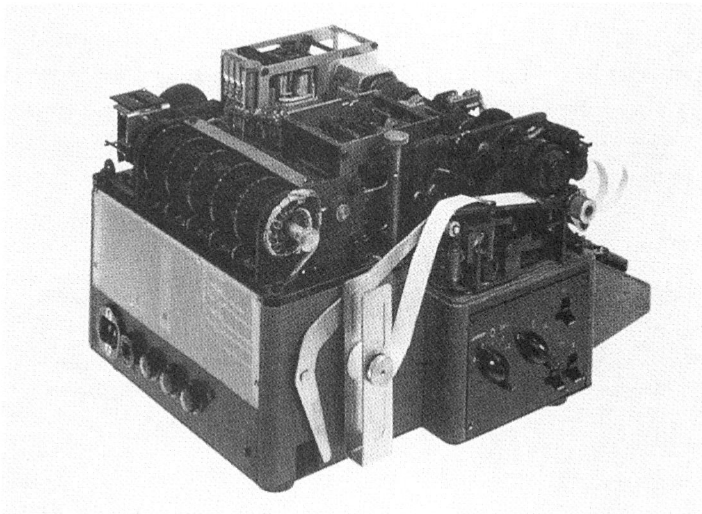

Figure 24. Random Number Generator with Printer CBI-53 (unit)

4. Rotor Machines

Even when mechanical and electromechanical machines became obsolete the rotor machine was still surrounded by a certain aura. We knew that such machines were being used by the great powers, and we decided to build a "super" version. Because all machines of this type had the same weakness – rotors with fixed wiring and simple motion, in 1952 we began the first attempts to produce a different machine with special rotors. One feature in our development was the use of feed back. While only 26 circuits are needed for the input (keyboard) and output (printer), the rotors have 41 circuits of which the surplus wires are looped back outside the rotors.

All circuits could be rearranged by the use of modificators, so that $26! \times 15!$ choices could be obtained (around 5.2×10^{38}).

The connections within each of the nine rotors could easily be changed in 2^{41} different ways. Furthermore, the rotors could perform every desired combination of movements like those used in the modern C machine. The program choice was enhanced greatly by the use of electrical switches. A complete block diagram is shown in Figure 25; the HX-63 machine itself in Figure 26.

A number of these rotor machines were supplied together with the tape punch device PEH-61 to a French government agency. The inconceivably large number of 10^{600} variations was theoretically possible with this machine. Due to the development of electronic machines, the manufacture of the HX-63 and the PEH-

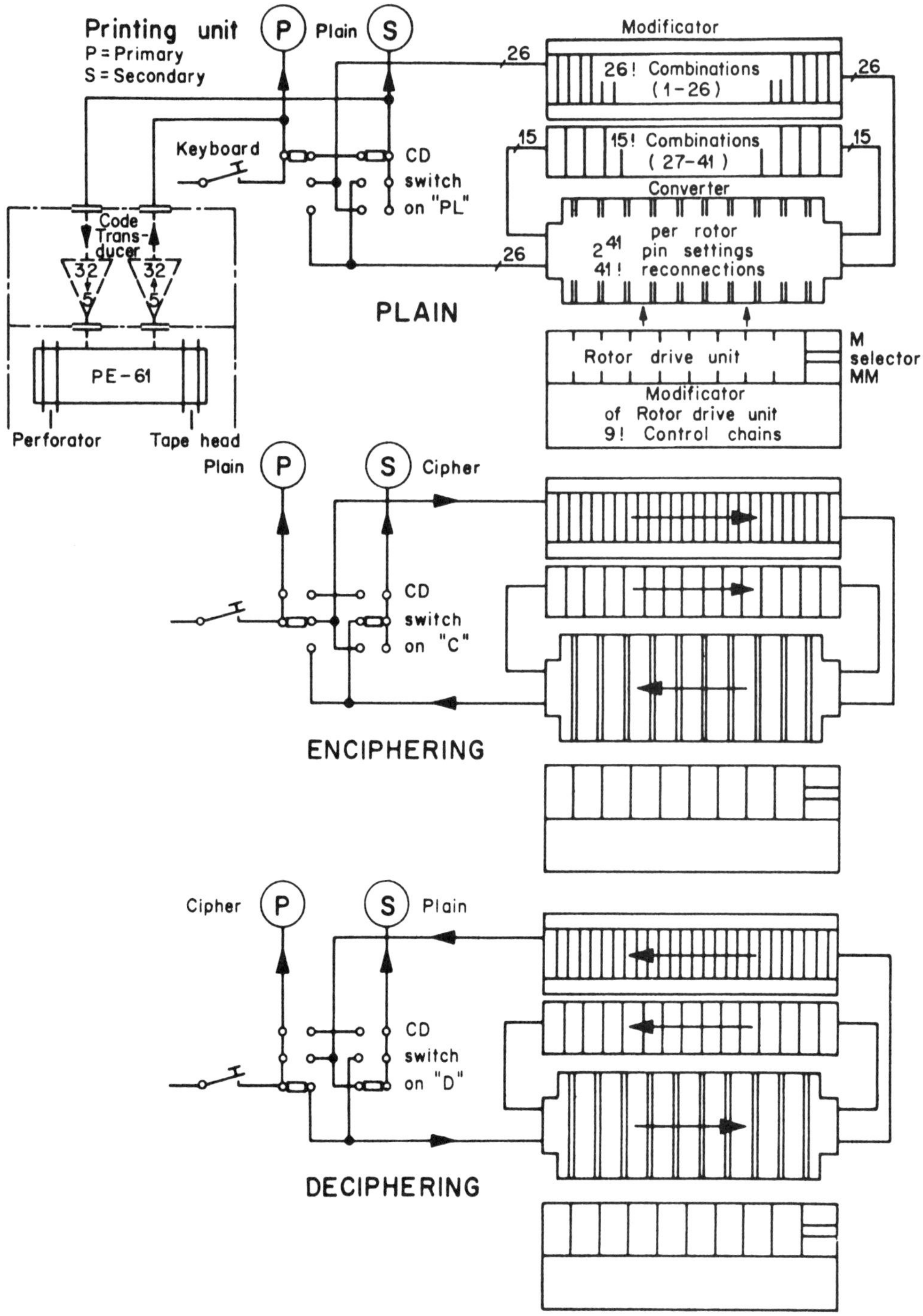

Figure 25. Block diagram of HX-63

61 machines was discontinued. We manufactured for another firm a sizeable number of high-speed printers and tape readers originally developed for this machine.

Figure 26. Rotor machine HX-63

5. The CDS-62 Machine

After 1955, I had the idea of improving the CD pocket machine further by the addition of a multi-alphabetical attachment. In 1950 the renowned cryptologist William F. Friedman had once suggested I redesign the C machine to enhance its applicability. It would use 16 different types of wheels which would engage in an irregular manner. This suggestion unfortunately did not work in practice. I used, however, a version of this suggestion as an improvement for the CD-57 machine.

In Sweden I made a first attempt by mounting on the turnable shaft of the CD machine a cylinder on which 32 alphabets were inscribed. These could be read off against a fixed reference alphabet. One could say that this was simply an improvement of A. G. Damm's first machine, but with a vastly longer period (Figure 27).

I changed the design and introduced interchangeable cards carrying random alphabets.

The alphabet ring of the CD-machine was removed, the turnable alphabet disk was replaced by a wheel with a control cable that could shift the positions of the strip which carried the reference alphabet from top to bottom over the 32 different alphabets of the card. On the back of the card were the corresponding reciprocal alphabets to be used in decipherment.

So that the machine could be used even if separate alphabet cards were not obtainable, there was imprinted under the removable cards a fixed Vigenére table whose alphabets were written in reverse sequence to the reference alphabet with rows 1 to 6 repeated.

Unfortunately, this course of development did not lead to manufacture of this device since at the time we were fully engaged with the development of electronic machines.

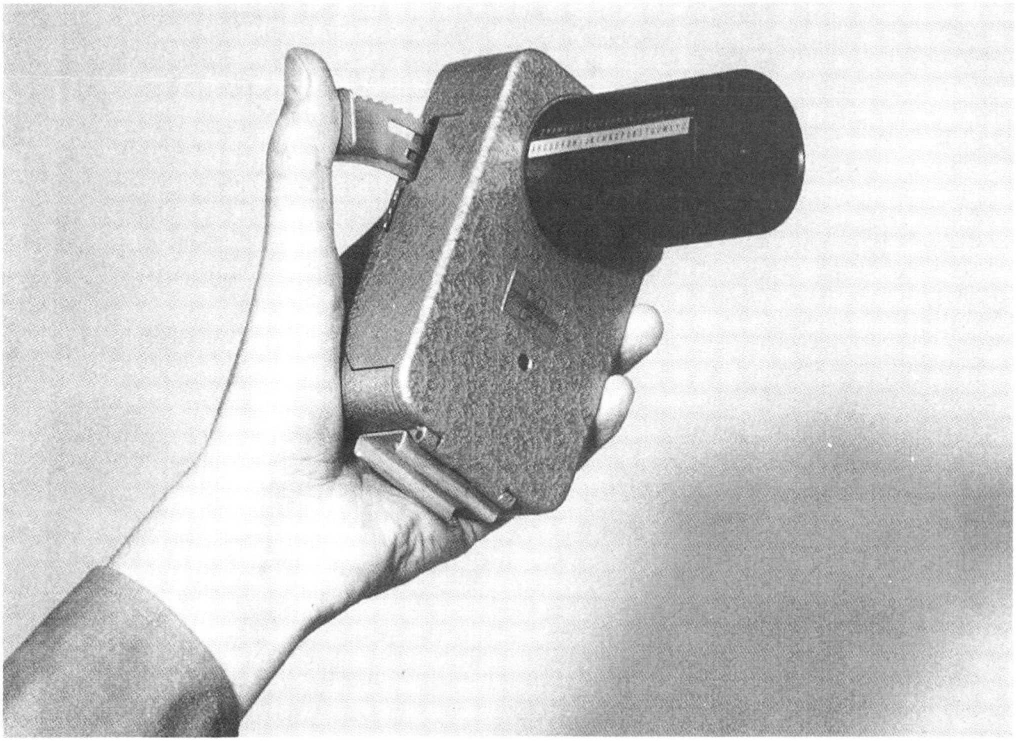

Figure 27. CDS-62: CD-57 with multi-alphabet attachment

CLOSING OBSERVATIONS

The literature of cryptography and cryptanalysis up to now has been primarily concerned with manual systems developed during the last five centuries. Although cipher machines have been used more and more during the past 50 years, the subject literature is scarcely informative. It is clear that the large countries with their unlimited material and technical resources have done much work in this specialized field. It is equally obvious that the results of such efforts are not published. Each nation has to look out for its own security and does not want to disclose any useful data to its adversaries.

The use of computers for the cryptanalysis of cipher machines was developed by the British during the Second World War, and has steadily been refined. It is now daily routine.

For cryptanalysis it is mandatory to know the characteristics of the machine. In addition the cryptanalyst gets considerable help from poor operational practices of the machine. Human errors can simply not be eliminated. On the other hand, the cryptanalyst today still has the chance of success through his own intuition as he depends on the "probable word", i.e., on stereotyped, often-used expressions. Direct betrayal or unintentional indiscretion naturally do not belong to the technique of cryptanalysis. This art depends mainly on statistical methods, which allows the recognition of repetitions of chosen machine settings which arise through overlaps of keying periods.

For this reason secure cipher communication requires not only good equipment but also cryptologically trained personnel to prepare the operating instructions for the machine, and very trustworthy, well trained operators who strictly observe these instructions.

The old rule is still true: The quality of a machine depends largely on its user.

THE LORENZ CIPHER MACHINE SZ42

Donald W. Davies

ADDRESS: Fair Winds, 15 Hawkewood Road, Sunbury-on-Thames, Middlesex TW16 6HL, GREAT BRITAIN.

ABSTRACT: The German cipher traffic called ***tunny*** which was broken at Bletchley Park using the Colossus machine was produced by an in-line "cipher-attachment" made by Lorenz, the Schlüssel Zusatz or SZ42. This paper is a description of the way that the SZ42 operates and its mechanism.

KEYWORDS: Cipher attachment, teletype, World War II, tunny, Colossus.

INTRODUCTION: THE SZ42 IN WORLD WAR II

The work of the Government Code and Cipher School at Bletchley Park in breaking the German ciphers of World War II is legendary. They worked on a large number of codes and ciphers and the wide range of their work is revealed in *Codebreakers*[5]. For many years the secrets of WWII cryptanalysis were kept but gradually the story of "Ultra" intelligence and the breaking of the Enigma cipher emerged. The Enigma machine was used to encipher written texts which were then transmitted by radio in Morse code. Breaking these signals provided quantities of information which, put together in an effective intelligence operation, was of the greatest value.

Bletchley Park broke another German cipher to great effect. *Tunny* was the code-breakers name for radio traffic which was not in Morse code but in the five-unit (Baudot) code of the teletype machine. At its source the traffic was generated from messages on punched paper tapes and enciphered (or deciphered) *on-line*, while it was being transmitted (or received). The machine which performed this cryptographic function was the Schlüssel Zusätz (Cipher Attachment) called SZ42, made by Lorenz. Because this non-Morse radio traffic served the army high command, the intelligence gained by Bletchley Park from tunny was extraordinarily valuable. At Bletchley Park the *Colossus* machines were used to break the tunny traffic

Plate 1. The Lorenz SZ42 "Cipher attachment".

There was another on-line cipher machine which was used in large numbers, the combined teleprinter and cipher machine called T52, made by Siemens and Halske. Its traffic was called *sturgeon* at Bletchley Park. I have written two papers and a note which detailed the construction and operation of the T52 series of cipher machines [1, 2, 3] and Wolfgang Mache has explored the history of these machines [8, 9]. The T52 seem to have been used for diplomatic messages and by the air force and navy. It is said that naval ships in port used a T52 connected by cable to the dockside. Much of the T52 traffic passed over cables, not radio, when it was not accessible to the British. I have been told that the T52 cipher was "never routinely broken". The term *fish* was used for all non-Morse traffic which included *sturgeon* and *tunny*. Individual tunny links had their own fishy names.

In this paper I describe the SZ42 machine, its mechanisms and method of operation. As an interesting mechanism it can be placed alongside the T52.

These are examples of the peak of development of mechanical cipher machines before electronics and the computer changed the whole nature of the technology. The SZ42 is especially attractive because most of its operation is mechanical, compared with the wider use of relay logic in the T52.

Plate 2. The code-wheels of the SZ42

Little seems to be known about the history of the SZ42. Mr. Augustin, who helped me examine the first machine I found, told me that its development started in 1938. Eric Hüttenhain, who was in charge of designing cipher breaking machines for the codebreakers of the OKW, was asked before the war began to evaluate the security of the SZ40 (predecessor of the SZ42) and the T52a/b models and found neither system was "up to the security standards then in force". He mentioned to me models a, b and c of the SZ42 and we can conjecture that these correspond in some way to the modifications of the SZ42 noted by Bletchley Park, which will be described later, but the exact correspondence is uncertain.

THE SZ42 EMERGES FROM THE SHADOWS

The first public reference to a non-Morse cipher broken by the British was in a television programme in the series "The Secret War" in 1976, later in a book of the same name[7]. The programme showed a T52 in action at the Siemens Institute in Munich and I was able to visit the Institute and examine this machine in 1980. But I already knew something of the cipher which was the target of Colossus and immediately saw that the T52 was not this machine.

A small clue appeared in a reprint in *Cryptologia* of the interrogation of two prisoners of war[4]. This report gave a garbled account of the T52 and ended with the words "Neither of the PW had ever seen nor heard of a GEHEIM-SCHREIBER made by LORENZ". I found that the most likely place to trace a Lorenz machine was the teleprinter section of Standard Electric Lorenz (SEL) at Pforzheim and wrote to them.

Plate 3. Holder for the code tables of initial wheel positions.

I had another piece of luck when visiting the Gesellschaft fur Mathematik und Datenverarbeiten, near Bonn. I mentioned my interest in German cipher machines and one of the people present told me his father had worked on them. That afternoon I met Eric Hüttenhain, who told me the name of the tunny machine and provided me with photographs of the SZ42 which matched my expectations. The name *Schlüssel Zusätz* means *cipher attachment* and signifies that the SZ machines did not include their own teleprinter (like the T52) but worked *in-line* between a standard teleprinter and the transmission medium, in this case radio equipment The earlier version was called the SZ40 and it seems possible that these numbers 40 and 42 refer to the dates of their introduction, since the first enciphered transmission of teleprinter traffic was detected in late 1940. But we do not know the difference between SZ40 and SZ42 and I believe all the variants described in this paper are of the SZ42 variety.

Plate 4. Telegraphic mechanism (above) and cryptographic unit (below).

From SEL at Pforzheim I heard, first that they had found a likely machine and then that they confirmed its identity from my SZ42 photographs. They sent me photographs of their machine, which were so clear that I could count the steps on the wheels and identify some of the mechanism.

I sought to confirm my identification of the origin of tunny traffic with the SZ42 from someone who played a significant part in the work of cryptanalysis. He replied with the short sentence "You have caught the right fish!"

It was 1981 before I could travel to examine the machine at Pforzheim and begin a close examination. My time with the machine was unfortunately very short. The part of the machine which receives and transmits the signals and powers the mechanism had been lost and only the cipher unit existed, but this was a great prize! The extreme historical importance of the device led the SEL people eventually to put it in the Deutsches Museum in Munich.

The Armed Forces Museum in the Akershus Fortress at Oslo has a magnificent Enigma collection and a T52, which was one of those I had used for my work. They told me they had found a new kind of machine and I identified it as an almost complete SZ42. The only part missing was the motor drive. I was able to explore the mechanism further but again with too limited time. Since I did not return to Oslo, my work on the SZ42 stopped there for 10 years. I knew that the UK Government Communications Headquarters (GCHQ) had obtained an SZ42 after the war, but it was still hidden.

As a result of the Secret War programme and publicity about Colossus it became wrongly accepted that the target of Colossus had been the T52. In correspondence with Tommy Flowers, the leader of the team that built Colossus, I raised this point and, to my great surprise, found that he also believed this. Then at a meeting arranged at the National Physical Laboratory I found that Professor Newman, who led the T52 breaking team also believed it. During a visit to Germany they had been shown rather briefly a machine said to be made by Siemens, probably a T52, and jumped to the wrong conclusion. The error has been repeated in public several times and seems likely to persist. Let me emphasise that it was the Lorenz SZ42 which produced the tunny traffic that was routinely broken using Colossus machines.

The end of this tale was the release of a complete SZ42 from GCHQ, to become part of the museum being assembled by the Bletchley Park Trust, in the place where the code-breaking was done. The opportunity to study it at length, arranged by Tony Sale on behalf of the Trust, allowed me to complete my work.

BASIC PRINCIPLES OF THE SZ42

From the front of the machine twelve wheels are visible (Plates 1 and 2). Each has a thumb wheel for setting its position and a number ring which indicates its current position. The wheels can be moved in steps and they click into position with a number showing. The wheel sizes, starting at the left, are 43, 47, 51, 53, 59, 37, 61, 41, 31, 29, 26 and 23. These numbers are all relatively prime, but include non-primes 26 and 51. The wheels are numbered from 1 to 12, starting on the left with wheel 1 having 43 positions.

Closer examination shows that each wheel has cam-segments on the edge furthest from the thumb wheel which can be pushed sideways, having two positions.

In their upright position these segments, which I shall call *camlets*, can engage a cam follower which cannot be seen, behind the wheel, and when pushed sideways they are inoperative. By setting the pattern of these camlets a binary sequence can be defined for each wheel and I shall adopt the convention here that the upright or operative position is a "1". In order to change the wheel pattern a cover must be unlocked and lifted. The wheel patterns were changed by an officer, monthly when SZ42 traffic was first encountered and daily later on. To send a message, the operator first set "random" wheel positions, which formed a *message key*. The method of transmitting the message key to enable the receiving end to start in synchronism has not been revealed but Plate 3 shows a device which could hold a sheet of paper with a code to represent wheel positions by letters A to Z.

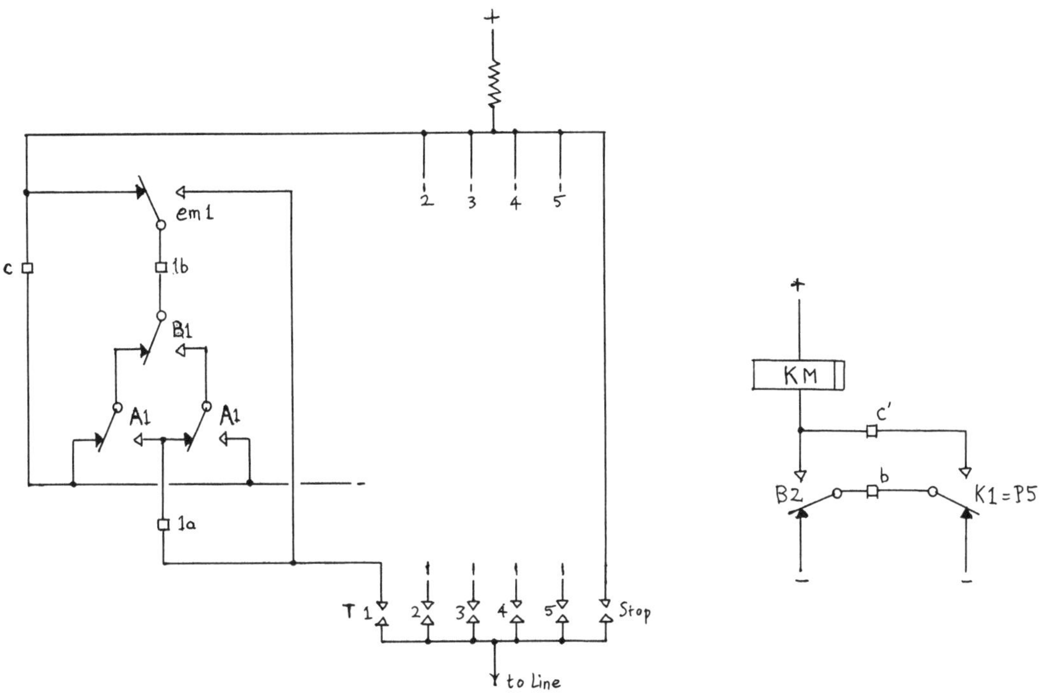

Figure 1. 1(a) Cipher circuit. 1(b) Limitations.

These details suggest the basic principle, closely related to the original form of the Vernam cipher. There are two groups of five wheels, numbered 1-5 and 8-12 on the machine, which correspond to the five unit code of the teleprinter. Each unit in the plaintext character is added, modulo 2, with the output of two wheels to produce the unit in the ciphertext character. Thus the units 1-5 of the character are modulo 2 added to the outputs of wheels 1-5 respectively and then to the output of wheels 8-12 respectively.

The circuit which performs the modulo 2 addition is shown in Figure 1(a). The contact em1 holds the state of the first element of the received telex character. Contacts A1 are operated by the camlets of wheel 1 and the contact B1 is operated by wheel 8. The result of the modulo 2 addition of the states of em1, A1 and B1 is applied to a cam-operated contact T1, which generates and times the first element of the transmitted telex character. Corresponding circuits for the other four elements control the outputs from the other four transmit cams. The rather unusual circuit reduces the number of connections into the cryptographic unit, shown here as a common c and 1a and 1b, which are replicated for each of the five elements of the coded character.

The tunny cipher is an additive stream cipher, in which a key stream, generated as the wheels step on, is modulo 2 added to the text. Encipherment and decipherment use the same key-stream, assuming that the wheels were set to the same starting positions. The key stream is the modulo 2 sum of the outputs of the two sets of wheels.

If the wheels all moved on by one step between each character, the cipher would have a cycle length equal to the product of the ten wheel lengths, yet be very weak. The SZ42 has two further wheels, numbers 6 and 7, which cause wheels 1-5 to move intermittently. The details of these wheels, and other features introduced from 1943 onwards, are described below in terms of the mechanisms which they employ.

The character code used was the Baudot or Murray code, now known as International Telegraph Alphabet No. 2. This telegraph code is commonly described in several different ways. To make our notation clear I shall use the convention that binary "1" represents polarity Z, or a current pulse or a hole in a paper tape. Thus A is coded as 11000. In this notation the start pulse is a "0" and the stop condition is a "1".

THE WHEEL DRIVEN MECHANISM

When the outer cover is removed, the SZ42 can be seen to comprise a main framework holding on top the teletype equipment and a cryptographic unit which contains the twelve wheels and their drive mechanism, which plugs into place underneath (Plates 4 and 5). The main unit has a drive motor with speed governor which connects by shafts and gears to three clutches.

Two light-duty clutches engage two telegraph receivers (Plate 6) each having a magnet actuated by the telegraph signal. They engage the clutch when the start pulse appears (i.e. the magnet releases) and then demultiplex the following five signal units and store their five bits mechanically. The mechanical memory actuates five contact sets labelled *emf* for the left hand set and *emo* for the right hand set. In the stop condition (magnet on) the clutch releases when the driven shaft has completed its revolution.

Plate 5. Cryptographic unit of the SZ42.

When either of the receive clutches engages, it operates a linkage which releases the third clutch, a heavy duty one, which engages spur gears that drive the crypto unit. It also turns a shaft which operates, via a camshaft, seven contact sets which define the timing for a transmitted signal. Thus the receive mechanism is slightly ahead of the transmission and the transmission is locked to the crypto mechanism. At the time that a signal is transmitted, the mechanical store holds the received character and contacts in the crypto unit have been set mechanically, by the camlet followers, so that circuits can be established via the transmit contacts and out onto the line. The task of the crypto unit is to move the wheels so that the camlets are in the best position at the time of transmission. The timing will be described in more detail below.

The transmit shaft engages with spur gears on the side of the cryptographic unit. The large wheel on the unit moves by 15 teeth for each cycle of operation. When the unit is replaced in the frame with the clutches disengaged, it must be set with the timing mark on the large wheel at the edge of the box. Two shafts are driven by this gear train, a camshaft which turns 180° per cycle, which we shall meet later, and a wheel drive shaft which turns 12° per cycle. This shaft turns the twelve coding wheels with their camlets, via jockey gears, at the appropriate rates for one step per cycle.

The wheels with larger numbers of positions (numbers 2, 3, 4, 5, and 7) move by one gear tooth per step of the wheel, having a 30 tooth wheel on the drive shaft. The others move by two gear teeth per step, having a 60 tooth wheel on the drive shaft. The jockey wheel sizes vary to suit.

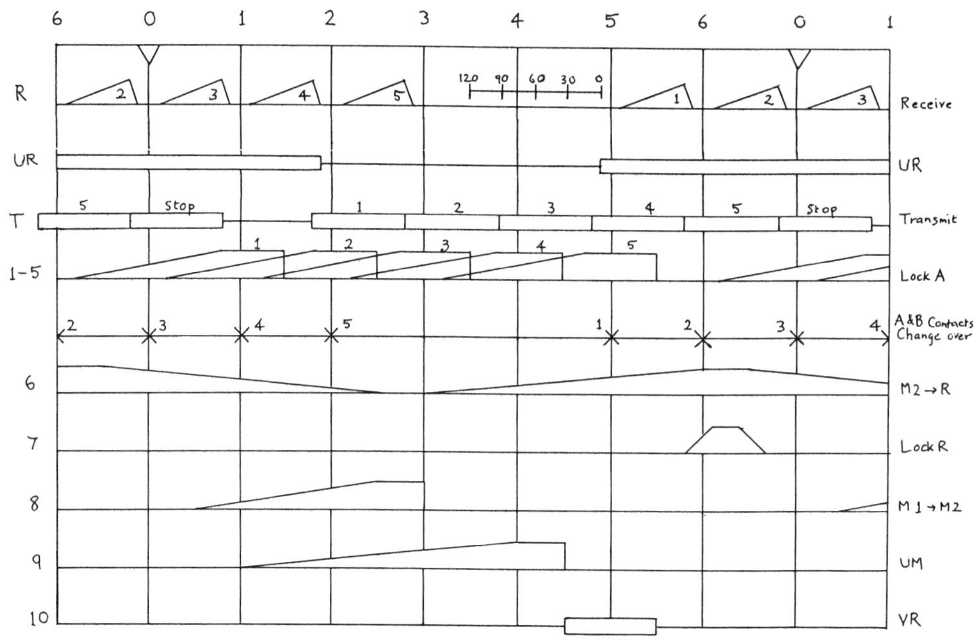

Figure 2. Cam timing of the SZ42.

By these means, each wheel is driven forward by one position at each cycle, moving continuously, not, as in the T52 cipher machine, by ratchet and pawl. A smoother motion and more durable mechanism are the result but it entails some complication in the rest of the mechanism.

Because of the rigid coupling of the gears, a provision must be made to enable the wheels to be set into their initial positions and to allow wheels 1-5 to stop moving. Each wheel is connected to its drive gear by a coupling which can slip a step at a time. Inside the wheel the drive gear is coupled to a wheel with rounded teeth into which a sprung detent is pressed, which is attached to the main part of the wheel. The detent passes from tooth to tooth as the wheel slips relative to the gear and holds the wheel accurately when the wheel is released.

Plate 6. Telegraph signal receiver mechanism.

THE RECEIVE-TRANSMIT SEQUENCE

Figure 2 is a timing diagram for the entire machine, which will later be used in the explanation of the mechanism of the cryptographic unit. The cams labelled 1-5 and 6 to 10 are part of this mechanism. The scale of the diagram is lined up with the main clutch, which operates at time 0 as shown by the triangle at the top.

Our present concern is with the receive mechanism shown at the top of the figure. The receive clutches (one each for receivers emf and emo) are simply friction drives held back by detents which is released by the receive magnets. This enables them to release without mechanical delay. The instant of release is somewhere in the scale 0 to 120 shown around time 4. Each receiver has an adjustable quadrant, visible in Plate 4, which holds the stationary cams in the place shown and thus compensates for delays and places the received pulses at the correct time for the operation of cams R1 to R5. These push the mechanical memories into place during the rising part of their motion. Note that this mechanism trips the main clutch between the reception of elements 2 and 3 of the signal. These five-bit memories hold their state until the next character reception.

The main clutch determines all the timing of the motions of the cipher wheels but also, in the main unit, drives a camshaft which times the transmission by cams T1 to T5 and "stop". The latter represents the timing of two contact sets, one producing the stop current in the transmitted character and the other, a changeover contact, concerned with control of the system. Note that T1 lies about halfway between R1 times in a continuous transmission, so the state of the contacts emf or emo is well stabilised. (The same will be found for the action of the wheel operated contacts to be described later). The main clutch will have some mechanical delay not shown here which, in effect, advances the timing shown here for the receive mechanism. This relative timing is slightly variable, for example the main clutch engages with teeth.

CHANGEOVER OF TRANSMISSION DIRECTION

The SZ42 is connected between a teleprinter which sends and receives cleartex and a radio equipment which accepts and delivers ciphertext. It can work in two ways, enciphering or deciphering, according to the source of incoming signals. The function of enciphering and deciphering is the same (except for one detail) but the source and destination of signals is different. This is the reason for the two receivers. Receiver emf is connected, via the fixed female connector, to the local teleprinter while emo is connected, via the male connector on the cable, to the radio equipment. When the main switch is off, both of these connections are disconnected.

The contact sets operated by the emf receiver have two extra contacts, K1 and UR. Contact K1 duplicates the output of R5 and takes part in a special feature called the P5 limitation, to be described later. Contact UR is cam operated with the timing shown in Figure 2. Then if the first character after a pause comes from the local plaintext source, the UR contact operates three relays labelled URI, URII and URIII which change the connections so that emf contacts are the source for encipherment, shown in Figure 1 as em, and the output from the T cams goes to the radio connection. On the other hand, if the first character comes from the radio side, UR is not operated, emo is the source (em) and the output goes to the local connection. The output also goes to emf in this case, which has to provide the K1 signal for the P5 limitation. The timing of emf in receiving the plaintext output is delayed about four units from the emo source.

When a continuous stream of characters is received, the clutches must release momentarily in order to keep synchronism, which is based on the start signals. It is likely that all transmissions were originated with a 7.5 unit cycle, that is a 1.5 unit stop element. In continuous transmissions the changeover relays UR would remain on or off, according to the direction of transmission. UR would release at the end of the main cycle following the last received character. A changeover therefore needs a distinct pause. Normally the SZ42 was used with long sequences of messages sent from punched tape readers but operator chat is also mentioned in histories of Bletchley Park and this is possible if the pause of about half a cycle is allowed before replying.

Figure 2 obscures one detail of timing, which is a difference in the lengths of the receive and transmit cycles. The transmit cycle, based on the main clutch and its seven unit scale, is shown correctly in the figure, but the drives to the receive mechanisms are via 26 to 24 tooth gears, making the receive cycle have a length of 6.46 units instead of 7. This ensures that the receive clutches meet their stops early. The effective gap between cam R5 and the earliest setting of the timing quadrant is closed up and cam 5 can be seen to be not quite released at the extreme setting. The reason for this gearing is not clear to me.

STOPPING MECHANISM OF A WHEELS

To identify the wheels functionally we shall give them names. Wheels 1-5, which can stop moving, were called at Bletchley Park by the Greek letter *Psi*. The name on the contacts of the crypto unit is *Spri*, the meaning of which is unclear. For convenience we shall call them A1 - A5. Wheels 8-12, which move regularly by one step per cycle, were named *Chi* wheels or *Spa* but we shall call them B1 - B5. The two remaining wheels which determine the stopping pattern of the A wheels were called *Motor* wheels or by the Greek letter *Mu*. We shall call wheels 6 and 7 by the names *M2* and *M1* respectively. So the wheel array consists of A1 - A5, M2, M1 and B1 - B5.

The back cover of the cryptographic unit can be removed, with its sets of contacts that are operated by the wheels. This reveals the mechanism which controls the stopping of the wheels, shown in Plate 7. Visible in this picture are twelve arms with small pads at their ends, which operate the contact sets. Ten of them are followers of the cams on the A and B wheels. The other two have special functions described later.

The six wheels on the left are capable of stopping, namely the A wheels and M2. For this purpose they have additional fixed square teeth on their periphery, as shown in Plate 2. Locking bars can rise under any of these wheels and hold them in fixed.

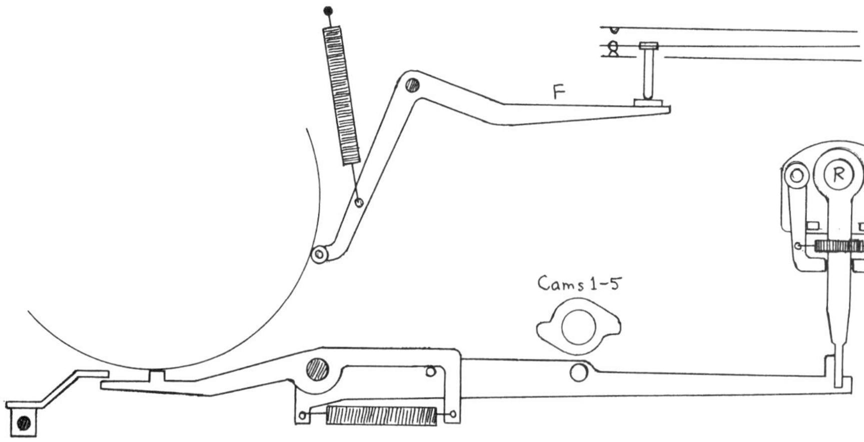

Figure 3. Section A. Shaft R controls stopping of wheels 1-5.

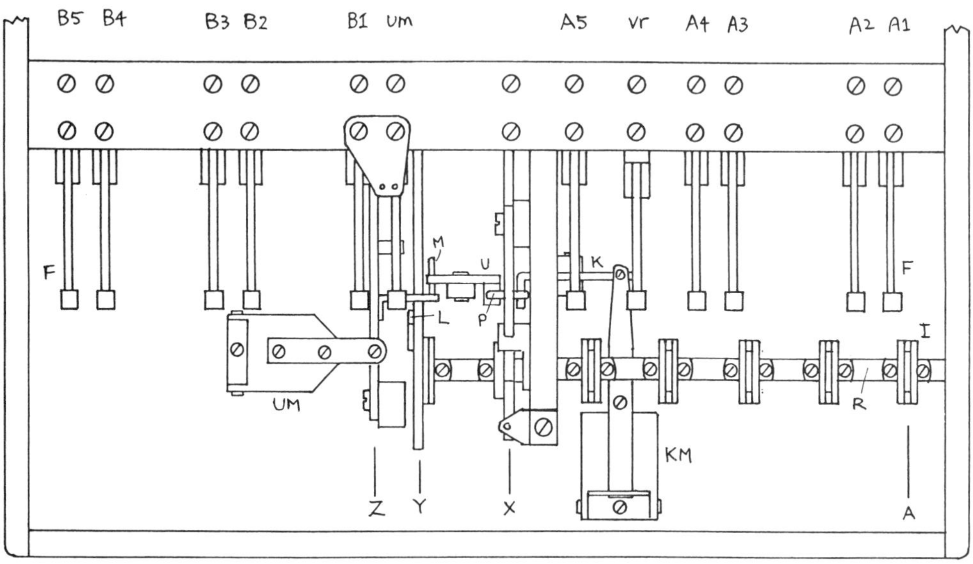

Figure 4. Plan view.

All the five A wheels stop or move together, being restrained by identical mechanisms shown in Figure 3. The locking bar under the wheel has a square lug which fits between the square teeth on the wheel and holds it firmly. It must do this because the springy detent mechanism can push the wheel either way as the gearing turns. The locking bar is hinged and driven by a cam, but is in two parts held together by a spring, allowing the front end to give way until the lug falls into place.

The figures which show the mechanisms have been simplified a little so that they show the principles but not all the construction details. Four separate mechanisms are shown in Figures 3, 5, 6 and 7. A plan view in Figure 4 shows the location of these mechanisms as sections A, X, Y and Z. Some of the bars which take part in these operations are cranked because they engage different wheels or cams. Fixed pivots are shown shaded and spring anchorages as black dots.

Plate 7. Part of the mechanism of the cryptographic unit.

The timing of events during a cycle of operation (the receipt and transmission of one telex character) can be followed using Figure 2, which shows the timing of various cams. Cams 1 to 10 are in the cryptographic unit and provide the power to drive its mechanisms, except cam 10, which simply operates a contact VR. This is one of the special contact sets, the other being operated by the UM mechanism in which cam 9 takes part. The other ten contact sets on the outputs of the A and B wheels change their states approximately where the crosses are shown in the timing chart. Cam profiles in this figure are conventional, showing the nature of the function rather than its precise movement. Cams 1 to 5 relate to the mechanism in Figure 3.

The downwards drive of any of cams 1 to 5 is a short movement at a time when the relevant wheel is not yet in place for locking. It has no permanent effect unless the locking bar is held in place by the interposer on the right of the figure. Each of the A wheels has an interposer assembly labelled I and all of these are fixed to the rocker shaft R. The five interposers on shaft R can be seen in the plan view of Figure 4 and in Plate 7.

When the A wheels are to be locked, shaft R moves clockwise at the very end of the previous cycle, pressing the interposers against the locking bars. The interposers can move relative to shaft R whenever they are restrained, because of the spring mechanism at their top ends. When the cam (numbered 1 to 5) briefly depresses the right end of the bar, the interposer falls into place and holds it. Shaft R will move back again if the next cycle does not require locking action, but the locking bar remains in place until the action of the cam (1 to 5).

Cams 1-5 operate in sequence, separated in time by one unit, which is the interval of the transmit pulses. Figure 2 shows the cam timing. Whenever a locking bar is depressed by its cam it will instantly be locked or unlocked by the interposer, depending on the state of shaft R. The cam profile is short and drops sharply. At this moment the locking bar is correctly placed for its wheel.

Figure 2 shows one cycle of operation (nominally 140 milliseconds), divided into seven units, between the triangular marks at time zero, which is the point at which the transmit clutch disengages or engages. The timing of the receive cams, as shown here, can vary a little because it is controlled by a separate clutch.

This stepped operation of the locking bars on the five A wheels means that the wheels do not stop in the same position. The reason for this feature is that locked wheels stop with their camlets in the best position for reliably operating their contacts, with the cam followers near to the middle of the camlets. Figure 2 shows by the numbered crosses the approximate timing of the camlet changeovers for each unit. The transmit pulse falls into the middle of the camlet and the wheel locks just before its transmit pulse is due. But the timing relative to the camlets is apparently not critical, so the need for the staggered timing could be questioned. Another factor, perhaps the most important one, is the drag of the wheel against the drive gearing. If all the wheels slipped over their point of maximum potential energy at the same time, the load on the motor would have large peaks. Staggering prevents this, as well as optimising the camlet timing. It is very hard to get the timing of all the hundreds of camlets correct. Our measurement showed some scatter of timing for the camlets on the B wheels and a much larger scatter on the A wheels, which have more positions and therefore smaller camlets. The timing error was as large as 15% of the total cycle time, but this leaves plenty of tolerance relative to the transmit pulse. The crosses in Figure 2 are approximately the mean changeover times, as measured on the more accurate B wheels.

When the wheels stop at the end of a cycle with the clutches out, a wheel's position depends on whether it is locked. A locked wheel will pull (either way) against its drive gear, but it will not be released until it is in the neutral position during the next cycle.

As well as the locking mechanism, Figure 3 also shows a cam follower F, visible in Figure 4 and Plate 7. This transfers the signal from the camlets to a contact spring assembly. The pivots of these cam followers are adjustable to get the changeover from one camlet to the next as close as possible to the optimum point shown by the crosses in Figure 2. Both the A and the B wheels have these cam followers and contact sets. Wheels A have the locking mechanism and contact assembly. Wheels B have only the contact assembly.

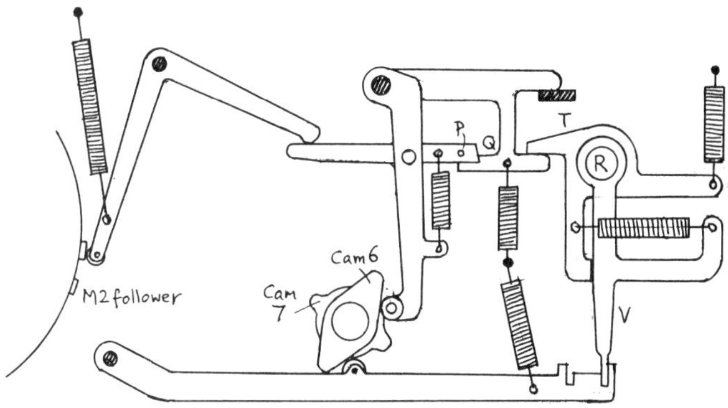

Figure 5. Section X. The cam follower of wheel M1 actuates shaft R.

ACTION OF THE M WHEELS

Figure 5 shows the mechanism which moves shaft R to initiate the locking of the A wheels. This is located at section X of Figure 4. Two cams are involved and their timing is shown in Figure 2. If a camlet on wheel M2 is operative, the cam follower does not contact the mechanism, so cam 6 pushes the three parts, latched together at Q, and these press under part T, which is hinged on Shaft R, tending to turn the vertical member V which is attached to shaft R. When the short pulse of cam 7 drops the bar underneath, at the end of the cycle, V moves left and the shaft turns. For the rest of the time, shaft R is locked in one of its two states. But if no camlet moves the follower, latch Q opens and the parts on shaft R are unaffected, falling back anticlockwise and allowing part V to move right when it is unlocked. Thus shaft R follows the movement of the cam follower in the previous cycle. A "1" on wheel M2 causes all the A wheels to be locked in the next cycle.

The pin P on the latch extends both ways and either end can be raised by the action of a magnet, as can be seen in Figure 4. Thus magnet KM presses on lever K, which is pivoted, and its other end rises under pin P. Also magnet UM, through a mechanism still to be described, can move lever U by a pin M moving down in a fork on its left end as shown in Figure 4. This raises its right end and lifts the latch. Either event prevents a "1" on M2 from stopping the A wheels.

The M2 wheel, which controls the stopping of all the A wheels, itself can be stopped by the camlets on wheel M1.

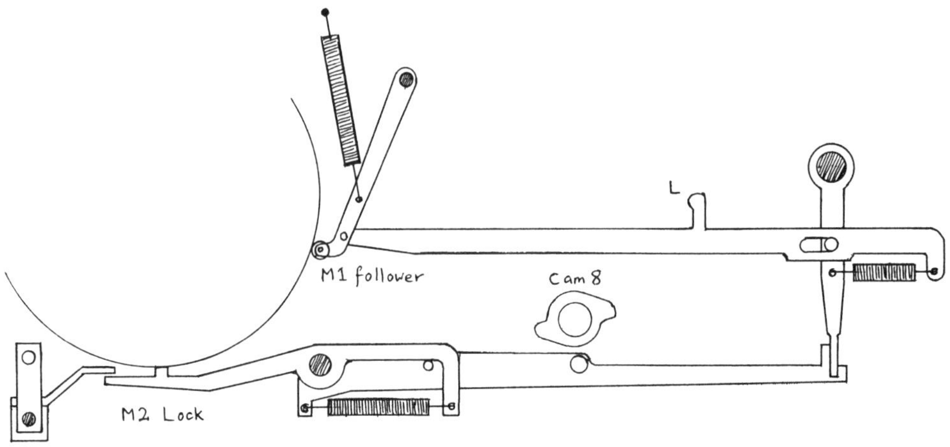

Figure 6. Section Y. The cam follower of wheel M1 controls stopping of wheel M2.

Figure 6 shows the operation of the locking bar of wheel M2. This mechanism is located at section Y on Figure 4. In our figure, the M2 wheel has been locked by the catch underneath it. Suppose now that a camlet on wheel M1 appears. A cam follower on wheel M1 moves the horizontal bar to the right which, through a spring, pulls the vertical interposer hinged loosely on shaft R to the right. When cam 8 depresses the right hand end of the locking bar, the interposer falls off its end and when the cam releases the bar this end rises. This allows the locking bar on M2 to drop away from the wheel. If there is a "0" on M1 the interposer moves into engagement and M2 is locked.

The moment of decision for locking M2 is between 2 and 3 on the timing diagram, just before cam 6 begins to read the output of M2 as a preliminary to setting the state of shaft R just after time 6. Shaft R must be in position during each of the operations of cams 1 to 5, so it is set while they are inactive.

The logic of M1 and M2 can be summarised thus; The motor wheel, M1 moves at every cycle. Where there is a "0" on its camlets it causes M2 to be locked. When there is a "1" on M2, wheels A are locked for the next cycle, unless pin P is raised by either of the magnets.

LIMITATIONS

The motor wheels, M1 and M2, generate an output in the form of shaft R's movement which causes the A wheels to stop moving, in a pattern which repeats with a long period. The operation of magnet KM (see Figure 4) allows wheels A to move, overriding the output of M2. This feature was called at Bletchley Park a *limitation.* The first version of limitation was introduced in February 1943, with the operation of KM dependent on the output of the B2 wheel during the previous character, so it was called the *Chi-2 limitation.* The dates given here for the introduction of the limitations are based on problems experienced at Bletchley Park as reported by Hinsley [6] and hints from people who worked there. But the dates given are not always consistent, therefore the history of these event is still uncertain. The sequence of introduction does seem to be clear.

Chi-2 was the first of three limitations, introduced in stages. As each appeared, the older ones remained and the signal reaching magnet KM was the modulo 2 sum of all the limitations in use.

The next limitation was brought into use briefly in March and April 1943, then fully on two links in December 1943 (one source says 11th December) and on all links by February 1944, see [6, pp. 481-482]. This was the P5 limitation in which KM was operated according to the 5th element of plaintext, measured two characters previously, added modulo 2 to the existing Chi-2 limitation. At the receiving end, the deciphered text was being used to control the deciphering process, so a momentary error on the transmission path would get the two machines out of step and they would remain that way until the operator at the receiving end saw the garble coming out and managed to alert the sender to start again. The P5 limitation caused operational problems to the Germans and it started to be dropped from September 1944. One account says it was dropped in January 1945, fully dropped, perhaps.

Figure 1(b) shows the circuit which operates magnet KM to impose the limitations according to the modulo 2 sum of the B2 output and the previous plaintext element P5.

Meanwhile the Psi-1 limitation derived from the A1 contact was introduced, in June 1944, on two links and in January 1945 on a third link. At the fullest development of limitations, before P5 was dropped, magnet KM was being operated by the modulo 2 sum of the B2 and A1 wheel outputs of the previous cycle and the plaintext element 5 of the cycle before that. In machines with all three limitations, A1 and B2 contacts are combined in a modulo 2 circuit to operate a relay in the cryptographic unit and this relay's contacts are used in place of B2 in the circuit shown here, which adds in the P5 output.

The probability that KM operated remained at 1/2 for all limitation versions. It cancelled half of the "1" outputs of wheel M2. To compensate for this, the patterns set on M2 had more ones, increasing the average number from 11 out of the 37 positions of M2 to 22. There was another rule that the Germans used for the wheel patterns. Stopping the A wheels reduced the number of 0-1 and 1-0 transitions in their outputs. To compensate for this, the patterns on the A wheels had more such transitions than a random binary pattern would have.

The Munich and Oslo machines have all three limitations but the Bletchley Park machine has only the B2 and P5 limitations. This confirms the sequence of their introduction and that the A1 limitation had not spread to all links. The P5 limitation can be controlled by a switch under a screwed-on cover, so that it could be inhibited to let operation continue if their was a high error rate. The B2 and A1 limitations, when they were fitted, could not be removed.

FUNCTION OF THE UM MAGNET

It is not difficult to determine the effect of the UM magnet but its operational purpose is less certain. Figure 7 shows its mechanism, which is at section Z of Figure 4. Magnet UM presses the long lever so that it catches on the cam follower at the small block B and is moved to the right by cam 9 at mid cycle. At first it operates the UM contacts but these fall off as the long lever moves right. At its rightmost position it locks onto the sprung lever and holds until UM releases.

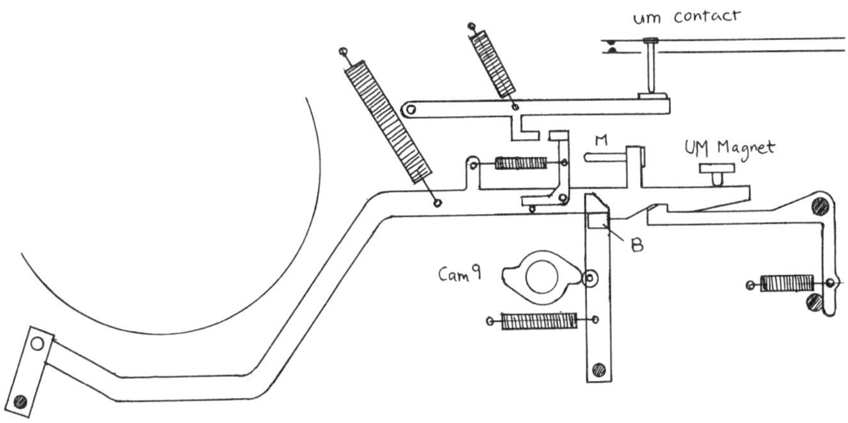

Figure 7. UM magnet operated. Section Z.

The long lever moves both right and down. Attached to it is an arm with a pin M which can be seen in Figure 4. The downwards movement of this pin in a fork of the small transverse lever U presses the other end up under pin P, so it inhibits the stopping mechanism for the A wheels. The rightwards movement as seen in Figure 7 makes this same arm hit a lug L which can be seen in Figure 6 and this removes the locking action on wheel M2. At the same time the long arm pulls a lever which turns a shaft in front of all the wheels and six fingers on this shaft press down on the locking bars of the A wheels and M2. These can be seen in Figures 3 and 6.

So UM has an immediate effect of removing locking and it also, as the camshaft turns, removes the mechanical memory (in the positions of the interposers in Figures 3, 5 and 6) that determine whether locking will happen in the next cycle.

The obvious purpose of UM is to unlock all wheels so that they can be set in position for the start of a transmission. In a photograph of an earlier model, which has a KM and is probably an early SZ42 ,the UM magnet is not in its present position but there is a larger magnet at the B end of the unit which pulls directly on the front shaft with its six fingers. This would be sufficient to allow wheel position setting.

Suppose that wheels had been positioned in this way but some of the mechanical memories shown in Figures 3, 5 and 6 were in the locking position. When transmission started again, the memory would cause a locking action either on M2 or an A wheel, which might not be matched in the machine at the other end of the line, resulting in a loss of synchronism. So the triple action of the UM mechanism can be explained.

It is unclear, at present, what causes the operation of UM, since there are no other controls than the ON/OFF switch. When the switch is ON, the machine is liable to spring into action to receive a message from the distant machine, therefore wheel setting must be done with it OFF. The switch disconnects the telegraph lines but does not stop the motor or the power supplies, therefore the OFF state probably operates UM to clear the mechanical memories. The locking bars are depressed first by the six fingers, enabling the wheels to be set. An operation of the main clutch is required to complete the clearing of the memories and put the locking bars out of action completely.

COMPARISON OF THE LORENZ SZ42 AND THE SIEMENS AND HALSKE T52 CIPHER MACHINES

Both machines are fine examples of the technology of electromechanical cryptography. As on-line cipher machines they probably represent the peak of its development. The T52 made some use of relays for processing data but the SZ42 is predominantly mechanical. Wheel motion uses ratchets in the T52 and spur gears in the SZ42. I will compare the later models in each range.

Both use a reversible combination of a "keystream" with the text. In the SZ42 this is simply the Vernam version of modulo 2 addition of bits but in the T52, five bits are used for Vernam addition with a character of the text and five other bits produce a permutation of the five signal elements in each character.

The binary patterns on the wheels are fixed in the T52 but they are the basic key in the SZ42. In the T52, all 10 wheel outputs are processed (linearly) and the basic key is a permutation of these wheel outputs. Because of the processing, there is no permanent link between any wheel and a signal element.

In both machines there is intermittent wheel movement. In the SZ42 this is rather simple because five of the wheels move together while the other five are intermittent but move or stop together. In the T52, each wheel movement is a function of outputs from other wheels, taken off the wheel profile by a cam follower at a different point from the one supplying the main signal.

To compare the strengths of the ciphers would be difficult. The SZ42 has the virtue of a changeable set of wheel profiles but the weakness of simple wheel movement. That it was routinely broken is significant, but the lack of attention to the T52 may have resulted from lack of access to its signals. The limitations of the data processing available at the time gave them both their brief period as usable cipher machines.

REFERENCES

1. Davies, D. W. 1982. The Siemens and Halske T52 cipher machine. *Cryptologia.* 6(4): 289-307.

2. Davies, D. W. 1983. The early models of the Siemens and Halske T52 cipher machine. *Cryptologia.* 7(3): 235-253.

3. Davies, D. W. 1994. New information on the history of the Siemens and Halske T52 cipher machines. *Cryptologia.* 18(2): 141-146.

4. Detailed interrogation report. 1979. Interrogation on German secret teletypewriters. *Cryptologia.* 3(4): 211-214.

5. Hinsely, F. H. and A. Stripp, eds. 1933. *Codebreakers.* Oxford: Oxford University Press. (This work is not to be confused with David Kahn's *The Codebreakers.*)

6. Hinsley, F. H., *et al.* 1984. *British intelligence in the second world war. Volume 3, Part 1.* London: H. M. S. O. Appendix 2, Geheimschreiber (Fish).

7. Johnson, B. 1978. *The Secret War.* London: BBC.

8. Mache, W. W. 1986. Geheimschreiber. *Cryptologia.* 10(4): 230-242.

9. Mache, W. W. 1989. The Siemens cipher teletype in the history of telecommunications,. *Cryptologia*. 13(2): 97-117.

BIOGRAPHICAL SKETCH

Donald W. Davies helped to build one of the first electronic digital computers at the National Physical Laboratory. He was active in computer research and pioneered the concept of packet switching from 1965 onwards. Then his interests moved to computer and communications security. Since 1984 he has provided security consultancy to financial institutions, telecommunications providers and pay TV broadcasters, among others. He has received a number of honours, including Fellowship of the Royal Society.

THE AUTOSCRITCHER

C. A. Deavours

ADDRESS: Mathematics Department, Kean College of New Jersey, Union NJ 07083 USA. Internet: deavours@luau.kean.edu

ABSTRACT: In the latter part of WWII, the Army Signal Security Agency built several machines known as Scritchers. These machines mechanized the solution of a variant of the German Enigma cipher machine having a rewirable reflecting rotor. This article discusses the probable method by which the machines worked.

KEYWORDS: Enigma, Scritcher, Autoscritcher, Bombe.

INTRODUCTION

In 1992, David Crawford and Philip Fox published an article, [1], in which they described two versions of a cryptanalytic device built by the Army Signal Security Agency. The first of these "scritchers" was a relay based device while the second machine was a faster electronic version of the first. However, both machines used copies of the five German Army Enigma rotors in their operation. The reader is referred to [1] for a detailed design description of both devices.

The scritchers were built and housed at Arlington Hall Station, as were other cryptanalytic machines built by the Army. The autoscritcher seems to have become operational in the 1944-45 period while the Superscritcher was in use from early 1946. Crawford, Fox, and others were a part of F Branch at Arlington Hall. This section was then under the command of Colonel Leo Rosen who had, as a young Signal Corps Lieutenant in the late '30s, suggested the use of stepping switches for mechanizing the solution of the Japanese Purple machine [2, p. 238]. The engineers working on the scritchers were never told exactly what cryptanalytic problem was being attacked by the system other than that the German Enigma cipher machine was the intended target of the analysis. The purpose of this article is to describe, in more detail, the method of cryptanalysis implemented by the scritchers as well as the primary problem that was attacked.

The cryptologic ingenuity of German designers is well known [2, Chapter III]. Among the "Enigma variations" produced was a version of the Enigma having a rewirable reflecting rotor [2, p. 138]. "Scritching" was a method of solving the ciphers generated by this variant of the Enigma family of cryptographs. The bombe devices constructed by the British and Americans could not deal with the ciphers produced by the rewirable reflector since those machines relied on the wiring of all rotors, including the reflector, being known before a problem could be attacked. The fact that the first device built was known as the "auto"-scritcher indicates that scritching was not new but that the process was being automated for the first time.

The method of scritching uses a "meet-in-the-middle" type of cryptanalytic attack and can be conceived as having evolved from the earlier Method of Batons used to attack commercial versions of the Enigma. Like most machine analysis methods of that time, a break in the system was initiated using a plaintext crib. It seems at this point that the amount of matching plaintext and ciphertext required was large, perhaps in the range of 150-200 letters. This requirement in turn indicates that the rewirable Enigmas were probably used in communications circuits where the regular version of the machine was also employed. This fact would have resulted in lengthy plaintext cribs becoming available when a message was encrypted and transmitted using both versions of the machine and solved for the simpler system.

To understand how the scritching technique may have evolved, consider the following Enigma crib which might constitute the beginning of a cryptogram (German plaintext with "X" for word division).

```
CIPHER: Y Q P J Z N V K I R O W E W S Y X Y O Z C Y Y I Z X E
 PLAIN: V O N A U F K L X A B T X D R E I A N D R E I X D I V
```

The equation governing the machine is

$$pPC^iRC^{-i}C^jSC^{-j}C^kT^{-k}UC^iTC^{-i}C^jS^{-1}C^{-j}C^kR^{-1}C^{-k}P^{-1} = c$$

where p denotes a plaintext letter, c, the corresponding cipher letter, R, S, T are the fast, medium, and slow rotors respectively, positioned at i, j, and k, U is the reflecting rotor and P is the plugboard permutation.

The operator C^i denotes, as usual, a circular shift of i letters forward in the A-Z alphabet.

This equation can be written as

$$pPC^iRC^{-i}C^jSC^{-j}V = cPC^iRC^{-i}C^jSC^{-j} \qquad (1)$$

where $V = C^kTC^{-k}UC^kT^{-1}C^{-k}$ is the permutation induced by the slow and reflecting rotors in combination. We shall call V the "crossover permutation" or "crossover wiring". It should be observed that V is idempotent (self-reciprocal) since $V * V = I$. In words, equation [1] asserts that if we encipher both p and c first using R and then S to obtain p' and c' respectively, then the results are related the equation

$$p'V = c'.$$

As long as the slow rotor, T, does not move, V remains constant.

PLUGBOARDLESS SCRITCHING

Suppose that, in the previous crib, we knew which rotors were the fast and medium rotors, and, in addition, the plugboard was left unplugged ($P = I$), then we could encipher both plain and ciphertext starting at the correct rotor positions and stepping the medium rotor at the appropriate point to obtain the p' and c' strings. In this particular case, we would have

```
CIPHER: Y Q P J Z N V K I R O W E W S Y X Y O Z C Y Y I Z X E
    c': D P A D U P Z K X J G U R A L V F B T H X E G G G Q U
    p': U R I U X R C E Q V S D P I W J M Y H T Q K S S S X D
 PLAIN: V O N A U F K L X A B T X D R E I A N D R E I X D I V
```

By reading the c' and p' lines we see parts of the (reciprocal) V permutation, e.g. DU, PR, AI, UX, etc. In fact, the repetition of some of the 13 pairs constituting the V mapping is evidence that the correct rotors and starting positions and turnover point have been chosen. If we had selected the wrong rotors and/or starting positions, we would have obtained something like the following sample.

```
CIPHER: Y Q P J Z N V K I R O W E W S Y X Y O Z C Y Y I Z X E
    c': A D N S T U B N P F Y L B R X Q I P K J E J R H H U E
    p': Q X O W H T V D O E G F K H J O N J P X W W Y U J G S
 PLAIN: V O N A U F K L X A B T X D R E I A N D R E I X D I V
```

In this second case, the p' and c' lines contain numerous contradictory pairings such as DX/XJ, YJ/YR, XS/XU, EE. This observation can be developed into a method of cryptanalysis.

To determine the identities of the fast and medium rotors and their starting positions, we could take our plaintext crib and encipher both strings using assumed identities for the fast and medium rotors as well as assumed starting positions for these rotors. The resultant enciphered strings constitute "deductions" about the crossover permutation, V. If we had an N letter crib, we would then have N enciphered pairs of letters. If any of the letter pairs contain contradictions, as was seen above, then, the rotors used or the starting positions must be invalid. When the true rotor order and starting positions are tested, we would see no contradictions in the enciphered pairs and, moreover, we would probably see some of the deduced pairs repeated if the crib is long enough. This latter phenomenon of a repeated deduction is termed a "hit". Of course, we can also expect some number of false alarms, settings that satisfy all our selection criteria but are nevertheless incorrect, in this process. It is also necessary that the stepping position of the medium rotor be known - something which is not too difficult to determine in this case.

Using the principles discussed in reference [1], we can show how an ingenious machine using stepping switches, rotors, ' wiring, and associated control circuitry may be constructed to perform the above process. A simplified diagram of the device is shown in Figure 1. In the illustration, the stepping switches and rotors have only 4 positions instead of the 26 positions that would be necessary.

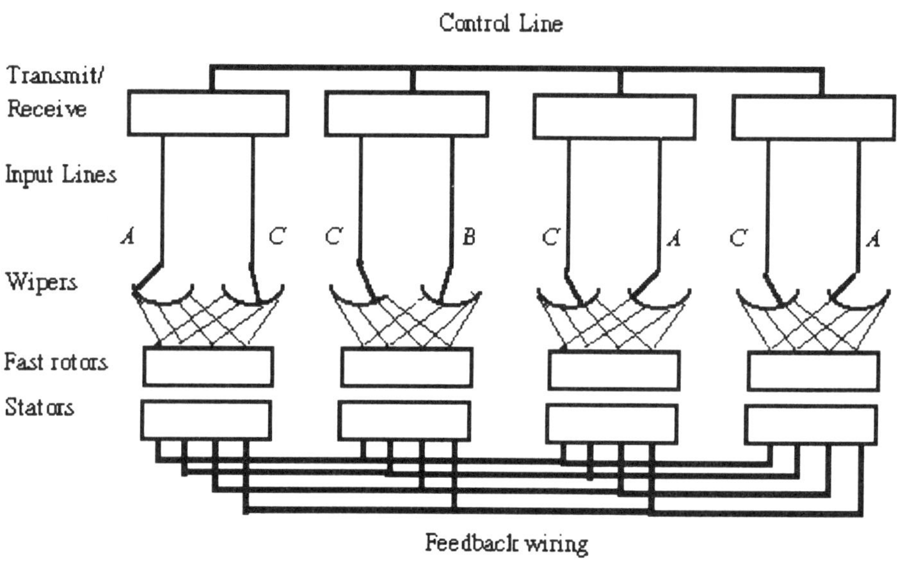

Figure 1. Hypothetical Scritcher with 4 Plain/Cipher pairs.

In general, we would have N two-rotor sets. Each rotor set contains the same two rotors in the same order but set to different positions. The input to each rotor set is via two input lines determined by the settings of the two stepping switches shown. The wiring from the 26 position stepping switches to the 26 rotor contacts is such that the "A" positions on both switches are wired to the "A" rotor entry position, the "B" positions are wired to the "B" entry points and so on. Electrical signals input along the two input lines selected by the stepping switches then pass through the two rotors representing the fast and medium rotors in the Enigma machine. Upon exiting the second rotor, the two output signals are fed back into each of the other rotor sets and traverse these rotors in reverse order.

Thus, if we were to send pulses through the two input lines selected for the first set of rotors, then, the two signals exiting from that rotor set constitute a crossover wiring deduction. This deduction is tested for consistency by feeding it into every other rotor set. If there is another rotor set that would have "deduced" the same crossover wiring, then, this fact can be determined by the presence of current in both of input lines of that rotor set. In other words, we "pulse" each of the rotor sets one at a time and look for the presence of current in the two input lines of all of the other rotor sets. A hit is scored when we detect current simultaneously in both input lines of any other rotor set.

If none of the other rotor sets indicate the presence of current in either of their input lines, then, there is no contradiction in the deduced crossover wiring (but no confirmation either). In the case that current is sensed in only one of the two input lines of another rotor set, then, a contradiction has been deduced.

It should now be clear how to use this device to determine the correct rotor order and positioning. First, the two rotors to be tested are chosen and inserted into the machine. Next, the pairs of wipers for each of the N sets are set to correspond to the known plain/cipher pairs of the crib. The positions of the rotors must be set according to same arbitrary starting point taking into consideration the stepping position of the medium rotor (as shown below). Finally, we need to run the rotors through all possible positions and to test at each position for hits and/or contradictions.

```
   FAST
          1 1 1 1 1 1 1 1 1 1 2 2 2 2 2 2
   POS:  0 1 2 3 4 5 6 7 8 9 0 1 2 3 4 5 6 7 8 9 0 1 2 3 |4 5 0
CIPHER:  Y Q P J Z N V K I R O W E W S Y X Y O Z C Y Y I |Z X E
 PLAIN:  V O N A U F K L X A B T X D R E I A N D R E I X |D I V
   MED:  0 0 0 0 0 0 0 0 0 0 0 0 0 0 0 0 0 0 0 0 0 0 0 0 |1 1 1
   POS                  (Stepping position for medium rotor)→ |
```

Instead of rotating both rotors while keeping the stepping switches locked at their starting positions we may also keep the medium rotor fixed (stator) and rotate only the fast rotor. In this case, the stepping switches must be passed through all 26 possible positions with each pair of switch wipers maintaining the same distance apart as in their initial setting. This latter procedure was the stepping method used by the Autoscritcher and the Superscritcher.

When the second rotor is fixed in a static position, the solution is obtained is a relative one. For example, suppose the correct starting positions of the fast and medium rotor were "H" and "X" respectively. (This means that the fast rotor is actually at position "Y" when the first letter is enciphered.) Further, assume that the first three matching plain/cipher pairs of the crib were VY, OQ, and NP. If we fix the medium rotor at position "A" then the solution will be found when the fast and medium rotors are at positions "R" and "A" respectively. This is because the distance from "Y" to "H" is 9 letters. Therefore, if the medium rotor is fixed at position "A" the relative displacement between the two rotors must be the same at the solution point. The distance between "R" and "A" is also seen to be 9 letters. This will also cause the stepping switch wiper positions at the solution point to be shifted 9 letters forward. The wiper positions at the solution will be EH, XZ, and WY. The reason for this phenomena can easily be seen mathematically.

If

$$pPC^iRC^{-i}C^jSC^{-j} = p' \tag{2}$$

then,

$$pPC^iRC^{-i}C^jS = p'C^j,$$

which can be written,

$$pC^jC^{-j}C^iRC^{-i}C^jS = p'C^j$$

or,

$$pC^jC^{(i-j)}RC^{-(i-j)}S = p'C^j. \tag{3}$$

In Equation 2, rotors R and S are at positions i and j respectively. In the equivalent Equation 3 S is at rotor position 0, i.e. "A", while rotor R has been shifted backwards j positions. Note that p and p' are both shifted forward by j positions. These two equations also show that hits remain hits and contradictions remain contradictions when the medium rotor is fixed during the search process although the deduced crossover wiring pairs are all shifted forward as are the original plain/cipher pairs as represented by the setting of the stepping switch wipers.

THE AUTOSCRITCHER

When the principles of the forgoing hypothetical machine are understood, the operation of the Autoscritcher and its more electronic counterpart, the Superscritcher, can be understood easily.

Extending the previous crib into a full text and enciphering it with the same settings but now using plugboard connections as well, we have

	0	1	2	3	4	5	6	7	8	9	10	11	12	13	14	15	16	17	18	19	20	21	22	23	24	25
0	R	Q	Z	J	I	N	D	U	I	Y	I	C	E	C	W	R	X	R	O	P	F	R	R	I	P	X
	V	O	N	A	U	F	K	L	X	A	B	T	X	D	R	E	I	A	N	D	R	E	I	X	D	I
1	E	H	F	Y	G	Q	H	H	U	Z	T	E	A	C	X	L	A	M	R	G	W	M	D	D	L	O
	V	X	S	C	H	W	A	C	H	E	X	F	E	I	N	D	L	X	X	K	R	A	F	T	E	X
2	W	G	M	R	H	P	U	R	Y	T	N	U	R	S	T	L	Y	Y	A	D	K	N	M	Q	H	A
	S	U	D	L	X	H	O	H	E	X	Z	W	O	X	V	I	E	R	X	A	C	H	T	X	A	N
3	F	P	F	T	P	J	G	W	L	L	B	P	E	N	N	T	X	X	V	Y	P	V	G	V	R	X
	Z	W	O	X	V	I	E	R	X	A	C	H	T	X	F	E	S	T	G	E	S	T	E	L	L	T
4	U	P	Y	D	R	M	Y	O	V	S	U	T	Y	M	T	L	S	F	R	W	V	B	Q	F	L	U
	X	X	M	I	N	E	N	S	P	E	R	R	E	X	N	O	R	D	A	U	S	G	A	N	G	X
5	J	M	E	N	H	R	S	C	H	P	W															
	E	I	C	H	W	A	L	D	E	X	X.															

The fast and medium rotor positions have been numbered relative to the arbitrary points 00 (i.e., "AA"). The assumed stepping position of the medium rotor is also indicated.

We cannot proceed as previously in this case because the unknown plugboard mapping prevents us from enciphering our crib correctly. The crib contains a number of plain/cipher pairs in which the same two letters are paired. Some of these are shown in boldface. Extracting four of the most frequent such pairs and noting the positions of their occurrences we have

XT	EY	AR	NF
10-1	8-2	17-0	5-0
9-2	16-2	18-4	14-3
3-3	19-3	5-5	23-4
17-3	12-4		
25-4			

These four sets of data are now treated as four "groups" by the Autoscritcher. In each of the four cases we have a letter pair such as XT which has been enciphered by the plugboard. The Autoscritcher attempts to find the plugboard connections for each pair of letters in a given group using much the same techniques as before to check for hits and consistency between the groups. To begin, the rotors are all set up according to the relative positions shown above. The modifications from our previous method are easy to see. For each group we try all possible combinations of plugboard combinations. This means that the wipers must try all 26 × 26 possible positions for each pair of letters. In contrast to our previous attack, all wiper positions are the same at all times for the rotors in a given group since the same plugboard mapping must work simultaneously for all rotor pairs in the group. When a hit is attained for a rotor position and wiper positions in group 1, then, we attack group 2 the same way by trying all possible plug connections. If we fail to find any usable wiper positions at some group i, then, we go back to group $i-1$ and continue from where we left off searching. If no viable solutions are found after all 26 settings of the fast rotor are tried, then, the rotor set is changed and the search continues.

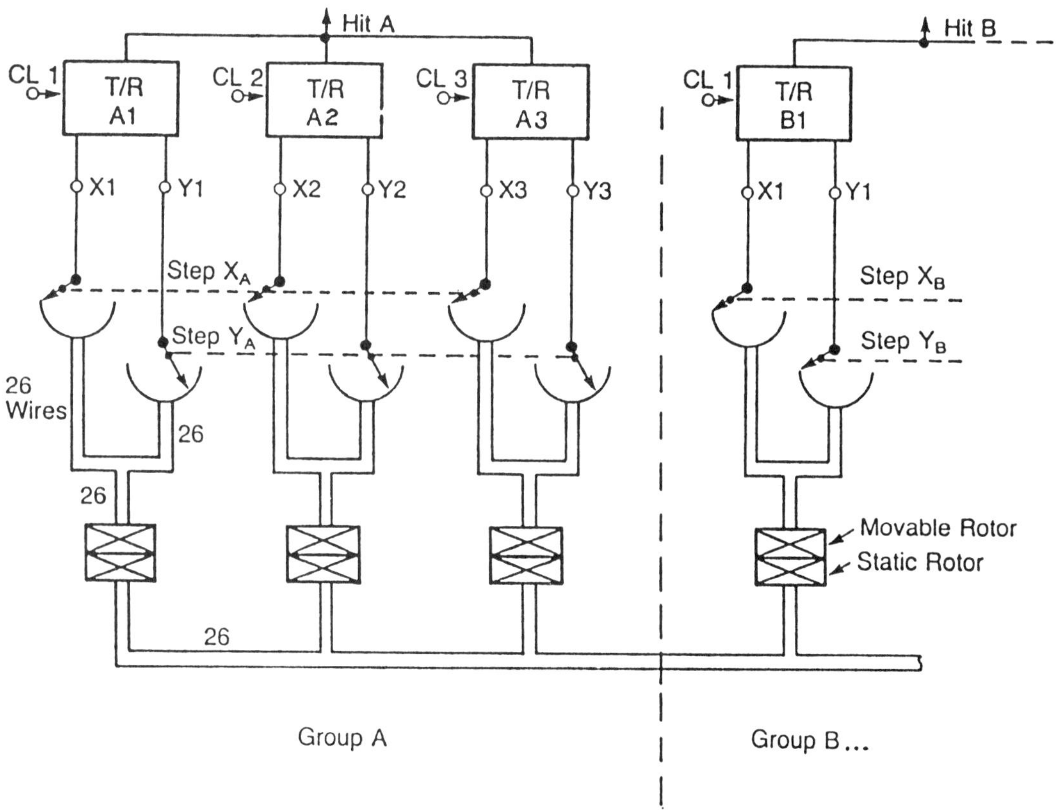

Figure 2. Condensed Scritcher system diagram. Figure reproduced with permission from [1], Copyright 1992 IEEE.

SEARCH LOGIC

The description given here of the search logic used in the Autoscritcher differs in important ways from that described in reference [1] which was, admittedly, based on memories recalled from decades before. As described in the referenced paper, the machine seeks a hit in the first group. When a hit is found, the first group search halts at its current positions and the search continues with the second group with the search surging back and forward between groups until a hit is recorded in all groups. The scritcher diagram accompanying the article is shown in Figure 2. There is no interplay between the groups and the 26 wires that connect all of the rotor sets in Figure 2 are for "convenience only". A typical set up involved 5 rotor sets with the number of rotors in each set going from about 7 down to 3. Thus, five groups with 7, 6, 5, 4, and 3 rotors might be considered normal.

With no interconnection between the groups, the Autoscritcher could never have delivered a solution. Not only is there a serious problem with false alarms, but, the true solution would never emerge. To see why this is so, consider the fact that when we have a group of N rotors at the correct settings, the probability that two randomly selected crossover deductions will coincide is about 1 in 13. Since there are $N(N-1)/2$ such comparisons, the expected number of hits is given by $N(N-1)/26$. So, the expected number of hits in the above situation is given by the following.

N	Expected Hits
7	1.6
6	1.2
5	.8
4	.5
3	.2

Clearly, it is only remotely possible for all 5 groups to score hits simultaneously. For proper functioning of the scritcher, it would seem necessary that hits are based on comparisons with all previous groups and not just within a single group. But, even this modification, while it would assure the correct solution would be found with high probability in the above case, would leave too many false alarms with which to deal. The logic used in the machine must also have rejected cases where only one input line of a previous or current rotor set sensed current. Without this additional proviso, the correct solution is literally buried among hundreds of false alarms.

There are other types of restrictions that could have further reduced the probability of false alarms. For example, if all the plain/cipher pairs used are chosen so that no letter occurs more than once in the complete set of pairs, then we can assert that the plugboard deductions must all contain distinct letters. In reality, I doubt that this condition, which would check to see that no two wipers between the groups were ever positioned at the same setting, was implemented.

The article further notes that all possible 26×26 wiper settings for a group were explored in seeking a solution. This exhaustive attack would be unnecessary since, if a wiper setting, say X/Y is valid, then so is Y/X. For each setting of one of the two stepping switches, the other switch only has to try the next setting up to the last setting. This means that, for one switch at setting A, the second switch must try settings B-Z. When the first switch is at setting B, the second switch must try C-Z, etc. The number of combinations that need to be tried in this case is $25 + 24 + 23 + ... + 1 = 325$ instead of $26 \times 26 = 676$.

SAMPLE SOLUTION

In order to study the scritching process more thoroughly, a simulation of the machine was written in BASIC. (This program is offered as freeware to readers.) When the four sets of group data for the above cryptogram were entered and the program was run, a solution was obtained with the fast rotor being Enigma rotor III and the medium rotor being Enigma rotor II. The solution was attained at a fast rotor setting of "R" and the (fixed) setting of "A" for the medium rotor. The wiper positions at the solution were EA, LY, HF, and UM. The deduced crossover wiring connections were as follows:

GROUP	CROSSOVER DEDUCTIONS:				
XT	JG	IF	KB	TM	BK
EY	YW	HP	FI	AO	
AR	IF	JG	RL		
NF	YW	MT	CQ		

We can see that 8 of the 13 distinct mapping pairs of the crossover permutation, V, have been deduced in the solution process. By examining the above list, it can be seen just how the solution was attained. A hit, "KB" was scored in the XT group. A hit in Group 1 is necessary to get the solution process started. The EY group had no hit within itself, but, it did have one in comparison with the previous XT group deductions, namely, "IF". The solution process then continued with two hits in the AR group ("IF" and "JG" from the XT group. Finally, the hit "MT" was scored between the NF group and the XT group. Since there were no contradictions encountered, the program stopped at these settings. A number of false alarms were also encountered due to the paucity of the data used.

The reader may wonder if the "relative" solution achieved about can be translated into "absolute" form. In this case, the answer is affirmative. If the four solution wiper settings are backed up, the correct value is easily seen because only 6 plugs were used in this cryptogram leaving 14 letters unchanged by the plugboard.

SOLUTION	EA	LY	HF	UM
-1	DZ	KX	GE	TL
-2	CY	JW	FD	SK
-3	BX	IV	EC	RJ
-4	AW	HU	DB	QI
-5	ZV	GT	CA	PH
-6	YU	FS	BZ	OG
-7	XT	ER	AY	NF

At the last position the correspondence between these values and the starting plain/cipher pairs of the crib is apparent. We have

P/C Pairs	Soln - 7
XT	XT
EY	ER
AR	AY
NF	NF.

This last comparison shows that the letters X, T, E, A, N, and F are probably unchanged by the plugboard while another of the plugboard mappings is (Y, R). Since the original plain/cipher pairs have been shifted 7 letters forward after plugboard encipherment, the starting positions of the fast and medium rotors must be shifted forward 7 positions from RA to YH indicating that the fast rotor started at position X and the medium rotor at position H. The author does not wish to leave the impression that determining the absolute setting from a relative one is always so easy.

We see that the scritching solution process provides the identities of the fast and slow rotors along with the their relative displacements from one another as well as a set of plugboard and crossover wiring values. Generally, more than enough data is obtained in order to deduce the remaining values through trial and error.

CIPHERTEXT ONLY

A remaining question of interest is whether or not the ideas involved in scritching can be extended to provide a solution when no cribs are available. The logic involved in picking out a correct solution and rejecting incorrect solutions is considerably more complex (and thereby all the more interesting) than in the probable plaintext case. In view of this observation, it would probably be necessary to use a general purpose digital computer instead of a special purpose machine for the analysis.

If we know only ciphertext, then, we could extract the settings of identical letters and these would form our groups. When a correct rotor setting and the correct wiper position for a letter is attained the situation differs from our known crib case because a single wiper position in the second stepping switch will not correspond to all of the unknown plaintext letters. Analysis becomes more complicated, particularly the problem of rejecting false alarms.

FREEWARE

The Autoscritcher simulation, sample problems, and other related material is available from the author. Send a formatted DOS disk along with a self addressed return mailer to the author by the address given in the title to this article. (US residents only, please)

REFERENCES

1. Crawford, David and Philip Fox. 1992. The Autoscritcher and the Superscritcher: Aids to Cryptanalysis of the German Enigma Cipher Machine, 1944-1946. *IEEE Annals of the History of Computing*. 14(3): 9-22.

2. Deavours, Cipher and Louis Kruh. 1985. *Machine Cryptography and Modern Cryptanalysis*. Norwood MA: Artech House.

BIOGRAPHICAL SKETCH

C. A. Deavours is a Professor of Mathematics and Computer Science at Kean College of New Jersey. Aside from teaching and consulting, most of his time is spent directing the Mathematics Department computer network facilities.

The Artech House Telecommunications Library

Vinton G. Cerf, Series Editor

Access Networks: Technology and V5 Interfacing, Alex Gillespie

Advanced High-Frequency Radio Communications, Eric E. Johnson, Robert I. Desourdis, Jr., et al.

Advanced Technology for Road Transport: IVHS and ATT, Ian Catling, editor

Advances in Computer Systems Security, Vol. 3, Rein Turn, editor

Advances in Telecommunications Networks, William S. Lee and Derrick C. Brown

Advances in Transport Network Technologies: Photonics Networks, ATM, and SDH, Ken-ichi Sato

An Introduction to International Telecommunications Law, Charles H. Kennedy and M. Veronica Pastor

Asynchronous Transfer Mode Networks: Performance Issues, Second Edition, Raif O. Onvural

ATM Switches, Edwin R. Coover

ATM Switching Systems, Thomas M. Chen and Stephen S. Liu

Broadband: Business Services, Technologies, and Strategic Impact, David Wright

Broadband Network Analysis and Design, Daniel Minoli

Broadband Telecommunications Technology, Byeong Lee, Minho Kang and Jonghee Lee

Cellular Mobile Systems Engineering, Saleh Faruque

Cellular Radio: Analog and Digital Systems, Asha Mehrotra

Cellular Radio: Performance Engineering, Asha Mehrotra

Cellular Radio Systems, D. M. Balston and R. C. V. Macario, editors

CDMA for Wireless Personal Communications, Ramjee Prasad

Client/Server Computing: Architecture, Applications, and Distributed Systems Management, Bruce Elbert and Bobby Martyna

Communication and Computing for Distributed Multimedia Systems, Guojun Lu

Communications Technology Guide for Business, Richard Downey, et al.

Community Networks: Lessons from Blacksburg, Virginia, Andrew Cohill and Andrea Kavanaugh, editors

Computer Networks: Architecture, Protocols, and Software, John Y. Hsu

Computer Mediated Communications: Multimedia Applications, Rob Walters

Computer Telephone Integration, Rob Walters

Convolutional Coding: Fundamentals and Applications, Charles Lee

Corporate Networks: The Strategic Use of Telecommunications, Thomas Valovic

The Definitive Guide to Business Resumption Planning, Leo A. Wrobel

Digital Beamforming in Wireless Communications, John Litva and Titus Kwok-Yeung Lo

Digital Cellular Radio, George Calhoun

Digital Hardware Testing: Transistor-Level Fault Modeling and Testing, Rochit Rajsuman, editor

Digital Switching Control Architectures, Giuseppe Fantauzzi

Digital Video Communications, Martyn J. Riley and Iain E. G. Richardson

Distributed Multimedia Through Broadband Communications Services, Daniel Minoli and Robert Keinath

Distance Learning Technology and Applications, Daniel Minoli

EDI Security, Control, and Audit, Albert J. Marcella and Sally Chen

Electronic Mail, Jacob Palme

Enterprise Networking: Fractional T1 to SONET, Frame Relay to BISDN, Daniel Minoli

Expert Systems Applications in Integrated Network Management, E. C. Ericson, L. T. Ericson, and D. Minoli, editors

FAX: Digital Facsimile Technology and Applications, Second Edition, Dennis Bodson, Kenneth McConnell, and Richard Schaphorst

FDDI and FDDI-II: Architecture, Protocols, and Performance, Bernhard Albert and Anura P. Jayasumana

Fiber Network Service Survivability, Tsong-Ho Wu

Future Codes: Essays in Advanced Computer Technology and the Law, Curtis E. A. Karnow

Guide to Telecommunications Transmission Systems, Anton A. Huurdeman

A Guide to the TCP/IP Protocol Suite, Floyd Wilder

Implementing EDI, Mike Hendry

Implementing X.400 and X.500: The PP and QUIPU Systems, Steve Kille

Inbound Call Centers: Design, Implementation, and Management, Robert A. Gable

Information Superhighways Revisited: The Economics of Multimedia, Bruce Egan

Integrated Broadband Networks, Amit Bhargava

International Telecommunications Management, Bruce R. Elbert

International Telecommunication Standards Organizations, Andrew Macpherson

Internetworking LANs: Operation, Design, and Management, Robert Davidson and Nathan Muller

Introduction to Document Image Processing Techniques, Ronald G. Matteson

Introduction to Error-Correcting Codes, Michael Purser

An Introduction to GSM, Siegmund Redl, Matthias K. Weber and Malcom W. Oliphant

Introduction to Radio Propagation for Fixed and Mobile Communications, John Doble

Introduction to Satellite Communication, Bruce R. Elbert

Introduction to T1/T3 Networking, Regis J. (Bud) Bates

Introduction to Telephones and Telephone Systems, Second Edition, A. Michael Noll

Introduction to X.400, Cemil Betanov

LAN, ATM, and LAN Emulation Technologies, Daniel Minoli and Anthony Alles

Land-Mobile Radio System Engineering, Garry C. Hess

LAN/WAN Optimization Techniques, Harrell Van Norman

LANs to WANs: Network Management in the 1990s, Nathan J. Muller and Robert P. Davidson

Minimum Risk Strategy for Acquiring Communications Equipment and Services, Nathan J. Muller

Mobile Antenna Systems Handbook, Kyohei Fujimoto and J. R. James, editors

Mobile Communications in the U.S. and Europe: Regulation, Technology, and Markets, Michael Paetsch

Mobile Data Communications Systems, Peter Wong and David Britland

Mobile Information Systems, John Walker

Networking Strategies for Information Technology, Bruce Elbert

Packet Switching Evolution from Narrowband to Broadband ISDN, M. Smouts

Packet Video: Modeling and Signal Processing, Naohisa Ohta

Personal Communication Networks: Practical Implementation, Alan Hadden

Personal Communication Systems and Technologies, John Gardiner and Barry West, editors

Practical Computer Network Security, Mike Hendry

Principles of Secure Communication Systems, Second Edition, Don J. Torrieri

Principles of Signaling for Cell Relay and Frame Relay, Daniel Minoli and George Dobrowski

Principles of Signals and Systems: Deterministic Signals, B. Picinbono

Private Telecommunication Networks, Bruce Elbert

Radio-Relay Systems, Anton A. Huurdeman

RF and Microwave Circuit Design for Wireless Communications, Lawrence E. Larson

The Satellite Communication Applications Handbook, Bruce R. Elbert

Secure Data Networking, Michael Purser

Selections from Cryptologia: History, People, and Technology, Cipher A. Deavours, et al.

Service Management in Computing and Telecommunications, Richard Hallows

Signaling in ATM Networks, Raif O. Onvural, Rao Cherukuri

Smart Cards, José Manuel Otón and José Luis Zoreda

Smart Card Security and Applications, Mike Hendry

Smart Highways, Smart Cars, Richard Whelan

Successful Business Strategies Using Telecommunications Services, Martin F. Bartholomew

Super-High-Definition Images: Beyond HDTV, Naohisa Ohta, et al.

Telecommunications Deregulation, James Shaw

Television Technology: Fundamentals and Future Prospects, A. Michael Noll

Telecommunications Technology Handbook, Daniel Minoli

Telecommuting, Osman Eldib and Daniel Minoli

Telemetry Systems Design, Frank Carden

Teletraffic Technologies in ATM Networks, Hiroshi Saito

Toll-Free Services: A Complete Guide to Design, Implementation, and Management, Robert A. Gable

Transmission Networking: SONET and the SDH, Mike Sexton and Andy Reid

Troposcatter Radio Links, G. Roda

Understanding Emerging Network Services, Pricing, and Regulation, Leo A. Wrobel and Eddie M. Pope

Understanding GPS: Principles and Applications, Elliot D. Kaplan, editor

Understanding Networking Technology: Concepts, Terms and Trends, Mark Norris

UNIX Internetworking, Second Edition, Uday O. Pabrai

Videoconferencing and Videotelephony: Technology and Standards, Richard Schaphorst

Voice Recognition, Richard L. Klevans and Robert D. Rodman

Wireless Access and the Local Telephone Network, George Calhoun

Wireless Communications in Developing Countries: Cellular and Satellite Systems, Rachael E. Schwartz

Wireless Communications for Intelligent Transportation Systems, Scott D. Elliot and Daniel J. Dailey

Wireless Data Networking, Nathan J. Muller

Wireless LAN Systems, A. Santamaría and F. J. López-Hernández

Wireless: The Revolution in Personal Telecommunications, Ira Brodsky

Writing Disaster Recovery Plans for Telecommunications Networks and LANs, Leo A. Wrobel

X Window System User's Guide, Uday O. Pabrai

For further information on these and other Artech House titles, including previously considered out-of-print books now available through our In-Print-Forever™ (IPF™) program, contact:

Artech House
685 Canton Street
Norwood, MA 02062
781-769-9750
Fax: 781-769-6334
Telex: 951-659
email: artech@artech-house.com

Artech House
Portland House, Stag Place
London SW1E 5XA England
+44 (0) 171-973-8077
Fax: +44 (0) 171-630-0166
Telex: 951-659
email: artech-uk@artech-house.com

Find us on the World Wide Web at:
www.artech-house.com